Table for flow conversion

Unit	m³/sec	m³/day	ℓ/sec	ft³/sec	ft³/day	ac-ft/day	gal/min	gal/day	mgd
1 cubic meter/ second	1	8.64×10^4	10^3	35.31	3.051×10^6	70.05	1.58×10^4	2.282×10^7	22.824
1 cubic meter/ day	1.157×10^{-5}	1	0.0116	4.09×10^{-4}	35.31	8.1×10^{-4}	0.1835	264.17	2.64×10^{-4}
1 liter/ second	0.001	86.4	1	0.0353	3051.2	0.070	15.85	2.28×10^4	2.28×10^{-2}
1 cu. foot/ second	0.0283	2446.6	28.32	1	8.64×10^4	1.984	448.8	6.46×10^5	0.646
1 cu. foot/ day	3.28×10^{-7}	0.02832	3.28×10^{-4}	1.16×10^{-5}	1	2.3×10^{-5}	5.19×10^{-3}	7.48	7.48×10^{-6}
1 acre-foot/ day	0.0143	1233.5	14.276	0.5042	43,560	1	226.28	3.259×10^5	0.3258
1 gallon/ minute	6.3×10^{-5}	5.451	0.0631	2.23×10^{-3}	192.5	4.42×10^{-3}	1	1440	1.44×10^{-3}
1 gallon/ day	4.3×10^{-8}	3.79×10^{-3}	4.382×10^{-5}	1.55×10^{-6}	0.11337	3.07×10^{-6}	6.94×10^{-4}	1	10^{-6}
1 million gallons/ day	4.38×10^{-2}	3785	43.82	1.55	1.337×10^5	3.07	694	10^6	1

Applied Hydrogeology

Third Edition

C. W. Fetter
University of Wisconsin—Oshkosh

Macmillan College Publishing Company
New York

Maxwell Macmillan Canada
Toronto

Maxwell Macmillan International
New York   Oxford   Singapore   Sydney

Cover Illustrator: T. S. Jobst
Cover art: Sue Birch
Editor: Robert A. McConnin
Production Editor: Sheryl Glicker Langner
Art Coordinator: Peter A. Robison
Text Design Coordinator: Jill E. Bonar
Cover Designer: Thomas Mack
Production Buyer: Patricia A. Tonneman

This book was set in Times Roman and Optima by The Clarinda Company and was printed and bound by Book Press, Inc., a Quebecor America Book Group Company. The cover was printed by Phoenix Color Corp.

Copyright © 1994 by Macmillan College Publishing Company, Inc.

Printed in the United States of America

Macmillan College Publishing Company
866 Third Avenue
New York, New York 10022

Macmillan College Publishing Company is part of the
Maxwell Communication Group of Companies.

Maxwell Macmillan Canada, Inc.
1200 Eglinton Avenue East, Suite 200
Don Mills, Ontario M3C 3N1

Library of Congress Cataloging-in-Publication Data
Fetter, C. W. (Charles Willard), 1942–
    Applied hydrogeology / C.W. Fetter. — 3rd ed.
        p.   cm.
    Includes bibliographical references and index.
    ISBN 0-02-336490-4
    1. Hydrogeology.   2. Water-supply.   I. Title.
GB1003.2.F47   1994
551.49—dc20

93-22893
CIP

Printing: 1   2   3   4   5   6   7   8   9   Year: 4   5   6   7

This book is dedicated to my wife, Nancy, and my children, Bill, Rob, and Elizabeth.

Preface

Hydrogeology continues to be the fastest growing branch of geology in the United States. There is ongoing demand for trained hydrogeologists by consulting firms, state regulatory agencies and industrial firms. Most of the employment in hydrogeology is in the environmental area. Most recent graduates who may be well trained in the science of hydrogeology have little, if any, knowledge about the profession of hydrogeology. I have added a new section in the introductory chapter on the type of work done by hydrogeologists, the business of hydrogeology, and the ethics of hydrogeology. In Chapter 12 I have included a section on the responsibilities of the field hydrogeologist. I hope that this material will better prepare hydrogeology graduates for their future careers.

Applied Hydrogeology is intended as an introductory textbook for courses in hydrogeology at either advanced undergraduate or dual graduate/undergraduate levels and as a reference book for the working professional. The reader is expected to have a working knowledge of college algebra; calculus is helpful, but not necessary for practical understanding of the material. A background in college chemistry is necessary to understand the chapter on water chemistry. The book stresses application of mathematics to problem solving rather than derivation of theory. To this end, you will find many example problems with step-by-step solutions. Case studies in many chapters enhance understanding of the occurrence and movement of ground water in a variety of geologic settings. The expanded appendices, which are tables of various functions and tables for unit conversions, provide additional data for solving problems in well hydraulics, water chemistry, and contaminant transport. A glossary of hydrogeological terms makes the book a valuable reference.

This edition has many more end-of-chapter problems, with problems given for all but two chapters. In most cases the problems are paired. An odd-numbered problem will have the answer given in a section at the back of this text, and a similar even-numbered problem will follow without an answer. Instructors who have adopted the text can request a solutions manual from Robert McConnin, Senior Editor, Macmillan College Publishing Company, 866 Third

Avenue, New York, New York 10022. Please send your request on university letterhead.

Another new feature of this edition is the inclusion of a computer disk. The use of this disk is not necessary to master the basic elements of hydrogeology. It is intended as an enrichment activity and as an introduction to the use of computers in hydrogeology. The three programs on this disk are working student versions of programs used by ground-water professionals. They have been furnished free of charge by the software publishers. No technical support is furnished for these programs, either by the author, Macmillan College Publishing Company, or the software publishers. I have found them easy to use and very helpful in teaching students some basic computer applications.

Still more new features in this edition include a list of the notation used in each chapter, a discussion of problem solving by dimensional analysis and significant digits, and the grouping of all of the references cited at the end of the text. Most equations have been expressed in a form that can be solved with consistent units in either the English or SI system. For those few equations not in consistent unit form, both English and metric versions are given.

I would like to thank the many people who have helped me with the preparation of this edition. Comments on the second edition and suggestions for the third edition were given by: Dr. Hund-Der Yeh, National Chiao Tung University, Taiwan; and E. Scott Bair, The Ohio State University. The following reviewers provided valuable input for the third edition: George M. Hornberger, University of Virginia; Laura L. Sanders, Northeastern Illinois University; William W. Simpkins, Iowa State University; Stephen E. Silliman, University of Notre Dame; H. Leonard Vacher, University of South Florida; Edward A. Keller, University of California—Santa Barbara; William H. Harris, Brooklyn College—CUNY; John C. Ridge, Tufts University; Regina M. Capuano, University of Houston; Thomas R. Wood, University of Idaho; James A. Saunders, Auburn University; Leonard R. Gardner, University of South Carolina; Arthur Sengupta, Florida International University; Allen H. Johnson, West Chester University; Mark L. Lord, Allegheny College; Stephan G. Custer, Montana State University; and Keenan Lee, Colorado School of Mines. The following individuals provided reviews and comments on individual sections of text: E. Scott Bair, The Ohio State University; Nicholas Valkenberg, Geraghty and Miller, Inc.; Richard Stoll, Wisconsin Department of Natural Resources; Dr. William Ganus, Kerr McGee Corporation; and James Rumbaugh, Geraghty and Miller, Inc. I am grateful to Dr. C. J. Hemker for allowing me to give you a copy of FLOWPATHD and to Geraghty and Miller Software Group for creating and furnishing student versions of QUICKFLOW and AQTESOLV. Dr. Richard Parizek furnished photographs for Chapter 13. Mary Dommer was a great help in assembling the manuscript. Sue Birch helped me with ideas for the cover art. I thank the many students that I have had in my classes for their suggestions for improvement of the second edition and for finding typographical errors.

Contents

CHAPTER FIVE
Principles of Ground-Water Flow 131

CHAPTER SIX
Soil Moisture and Ground-Water Recharge 175

CHAPTER SEVEN
Ground-Water Flow to Wells 197

CHAPTER EIGHT
Regional Ground-Water Flow   275

CHAPTER NINE
Geology of Ground-Water Occurrence   319

CHAPTER TWELVE
Ground-Water Development and Management   511

CHAPTER THIRTEEN
Field Methods   543

CHAPTER FOURTEEN
Ground-Water Models   593

Appendices   617

Glossary   635

Answers to Selected Problems   651

References   655

Index   681

1 Water

If anyone be too lazy to keep his dam in proper condition, and does not keep it so; if then the dam breaks and all the fields are flooded, then shall he in whose dam the break occurred be sold for money and the money shall replace the corn which he has caused to be ruined.

Code of Hammurabi, Section 53 (1760 B.C.)

1.1 WATER

Water is the elixir of life; without it life is not possible. Although many environmental factors determine the density and distribution of vegetation, one of the most important is the amount of precipitation. Agriculture can flourish in some deserts, but only with water either pumped from the ground or imported from other areas. Civilizations have flourished with the development of reliable water supplies—and then collapsed as the water supply failed. This is a book about the occurrence of water, both at the surface and in the ground.

A person requires about 3 quarts (qt) (3 liters (L)) of potable water per day to maintain the essential fluids of the body. Primitive people in arid lands exist with this amount as their total consumption. A single cycle of a flush toilet may use 6 gallons (gal) (23 L) of water. In New York City the per capita water usage exceeds 260 gal (1000 L) daily. Much of this use is for industrial, municipal, and commercial purposes; for personal purposes, the typical American uses 50 to 80 gal (200 to 300 L) per day. Even greater quantities of water are required for energy and food production.

In 1990 the total offstream water use in the United States has been estimated to be 408 billion gal (1530 billion L) per day of fresh and saline water. This does not include water used for hydroelectric power generation and other instream uses, but does include water used for thermoelectric power plant cooling. Fresh-water use in 1990 included 79.4 billion gal (300 billion L) per day of ground water and 259 billion gal (980 billion L) per day of surface water. Per capita fresh-water use was 1440 gal (5450 L) per day. Consumptive use of water—that is, water evaporated during use—was about 94 billion gal (355 billion L) per day (Solley & Pierce 1992).

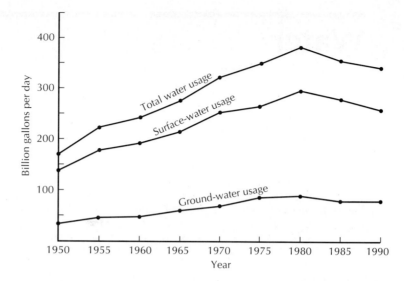

FIGURE 1.1   Withdrawal of fresh water in the United States for all uses except hydroelectric-power generation. Data from Solley & Pierce 1992; Solley, Merk, & Pierce 1988; Solley, Chase, & Mann 1983; Murray & Reeves 1977; Murray 1968; MacKichan & Kammerer 1961.

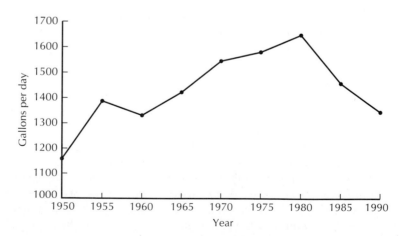

FIGURE 1.2   Per capita fresh-water usage in the United States for all uses except hydroelectric-power generation. Data from: Solley & Pierce 1992; Solley, Merk, & Pierce 1988; Solley, Chase, & Mann 1983; Murray & Reeves 1977; Murray 1968; MacKichan & Kammerer 1961.

A common goal of all countries is to increase economic production. It is generally thought this will result in a better lifestyle for the citizens. Although the validity of this assumption has been questioned by some individuals in the heavily industrialized countries, the goal of governments at all levels still remains to promote economic expansion. This will naturally increase per capita water usage, although water and energy conservation may help. Inasmuch as the populations of most countries are growing, it is likely the total use of water will increase, even with conservation measures.

The United States has had a history of increasing water use. Figure 1.1 illustrates the withdrawal of fresh water in the United States for offstream uses such as municipal supply, industrial, domestic, irrigation, and electrical power plant cooling. Between 1950 and 1980 total fresh-water use increased by 117%, whereas per capita fresh-water use increased by 42% (Figure 1.2). Between 1980 and 1985 fresh-water use declined for the first time, from 373 billion gal (1412 billion L) per day to 345 billion gal (1306 billion L) per day. The 1985 fresh ground-water use of 75 billion gal (284 billion L) was less than in either 1980 (83 billion gal per day) or 1975 (82 billion per day) (Solley & Pierce 1992; Murray 1968).

This decline in water usage continued between 1985 and 1990. Total fresh-water use in 1990 was 338 billion gal (1280 billion L) per day. The use in 1990 of 79.4 billion gal (300 billion L) of fresh ground water per day was up slightly from 1985 but was still lower than either 1975 or 1980 pumpage.

1.2 HYDROLOGY AND HYDROGEOLOGY

As viewed from a spacecraft, the earth appears to have a blue-green cast owing to the vast quantities of water covering the globe. The oceans may be obscured by billowing swirls of clouds. These vast quantities of water distinguish Earth from the other planets in the solar system.

Hydrology is the study of water. In the broadest sense, hydrology addresses the occurrence, distribution, movement, and chemistry of all waters of the earth. **Hydrogeology** encompasses the interrelationships of geologic materials and processes with water. (A similar term, **geohydrology,** is sometimes used as a synonym for hydrogeology, although it more properly describes an engineering field dealing with subsurface fluid hydrology.)

The physiography, surficial geology, and topography of a drainage basin, together with the vegetation, influence the relationship between precipitation over the basin and water draining from it. The creation and distribution of precipitation is heavily influenced by the presence of mountain ranges and other topographic features. Running water and ground water are geologic agents that help shape the land. The movement and chemistry of ground water is heavily dependent upon geology.

Hydrogeology is both a descriptive and an analytic science. The development and management of water resources are important parts of hydrogeology as well.

1.3   THE HYDROLOGIC CYCLE

An account of the water supply of the world would reveal that saline water in the oceans accounts for 97.2% of the total. Land areas hold 2.8% of the total. Ice caps and glaciers hold 2.14%; ground water to a depth of 13,000 feet (ft) (4000 meters (m)) accounts for 0.61% of the total; soil moisture, 0.005%; fresh-water lakes, 0.009%; rivers, 0.0001%; and saline lakes, 0.008% (Feth 1973). More than 75% of the water in land areas is locked in glacial ice or is saline (Figure 1.3).

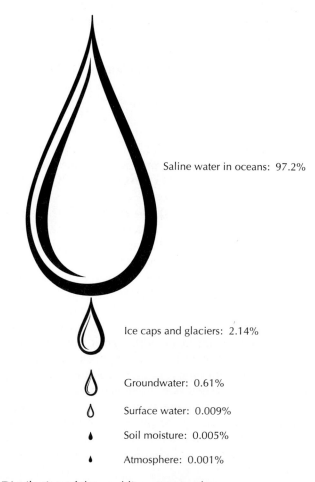

FIGURE 1.3   Distribution of the world's water supply.

Only a small percentage of the world's total water supply is available to humans as fresh water. More than 98% of the available fresh water is ground water, which far exceeds the volume of surface water. At any given time, only 0.001% of the total supply of water is in the atmosphere. However, atmospheric water circulates very rapidly, so that each year enough water falls to cover the conterminous United States to a depth of 30 inches (in.) (75 centimeters (cm)). Of this amount, 22 in. (55 cm) are returned to the atmosphere through evaporation and transpiration by growing plants, whereas 8 in. (20 cm) flow into the oceans as rivers (Federal Council for Science and Technology 1962). Although the previous sentence implies that the **hydrologic cycle** begins with water from the oceans, the cycle actually has no beginning and no end. As most of the water is in the oceans, it is convenient to describe the hydrologic cycle as starting with the oceans.

Water evaporates from the surface of the oceans. The amount of evaporated water varies, being greatest near the equator, where solar radiation is more intense. Evaporated water is pure, since when it is carried into the atmosphere the salts of the sea are left behind. Water vapor moves through the atmosphere as an integral part of the phenomena we term "the weather." When atmospheric conditions are suitable, water vapor condenses and forms droplets. These droplets may fall to the sea or onto land or may revaporize while still aloft.

Precipitation that falls on the land surface enters into a number of different pathways of the hydrologic cycle. Some water may be temporarily stored on the land surface as ice and snow or water in puddles, which is known as **depression storage.** Some of the rain or melting snow will drain across the land to a stream channel. This is termed **overland flow.** If the surface soil is porous, some rain or melting snow will seep into the ground by a process called **infiltration.**

Below the land surface the soil pores contain both air and water. The region is known as the **vadose zone,** or **zone of aeration.** Water stored in the vadose zone is called **soil moisture,** or **vadose water.** Soil moisture is drawn into the rootlets of growing plants. As the plant uses the water, it is transpired as vapor to the atmosphere. Under some conditions water can flow laterally in the vadose zone, a process know as **interflow.** Water vapor in the vadose zone can also migrate back to the land surface to evaporate.

Excess soil moisture is pulled downward by gravity, a process known as **gravity drainage.** At some depth, the pores of the soil or rock are saturated with water. The top of the **zone of saturation** is called the **water table.** Water stored in the zone of saturation is known as **ground water.** It then moves as **ground-water flow** through the rock and soil layers of the earth until it discharges as a spring, or as seepage into a pond, lake, stream, river, or ocean (Figure 1.4).

Water flowing in a stream can come from overland flow or from ground water that has seeped into the stream bed. The ground-water contribution to a stream is termed **baseflow,** while the total flow in a stream is **runoff.** Water stored in ponds, lakes, rivers, and streams is called **surface water.**

Evaporation is not restricted to open water bodies, such as the ocean, lakes, streams, and reservoirs. Precipitation intercepted by leaves and other vegetative surfaces can also evaporate, as can water detained in land-surface

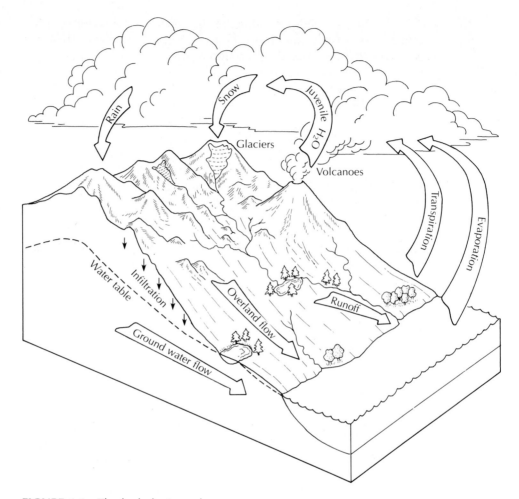

FIGURE 1.4   The hydrologic cycle.

depressions or soil moisture in the upper layers of the soil. Direct evaporation of ground water can take place when the saturated zone is at or near the land surface.

Magmatic water is contained within magmas deep in the crust. If the magma reaches the surface of the earth or the ocean floor, the magmatic water is added to the water in the hydrologic cycle. Much of the steam seen in some volcanic eruptions is ground water that comes into contact with the rising magma and is not magmatic water. Some of the water in the ocean sediments is subducted along with the sediments and is withdrawn from the hydrologic cycle. This water may eventually become part of a magma.

Figure 1.5 is a schematic drawing of the hydrologic cycle showing the major reservoirs where water is stored and the pathways by which water can move from one reservoir to others.

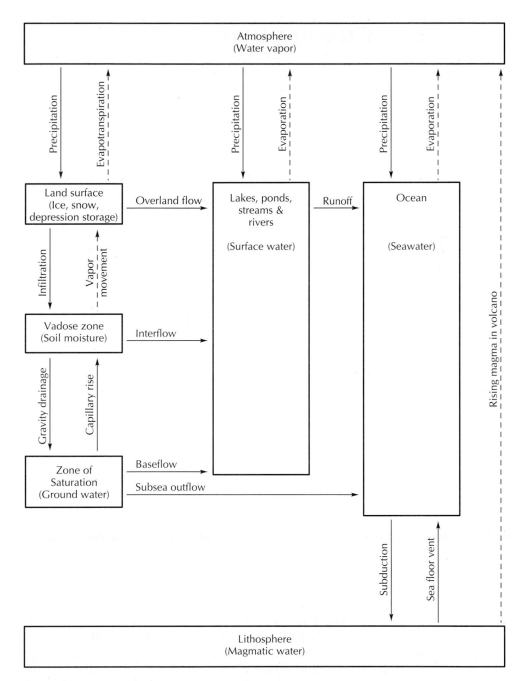

FIGURE 1.5 Schematic drawing of the hydrologic cycle. Movement of liquid water is shown with a solid line and movement of water vapor is shown with a dashed line.

1.4   ENERGY TRANSFORMATIONS

The hydrologic cycle is an open system in which solar radiation serves as a source of constant energy. This is most evident in the evaporation and atmospheric circulation of water. The energy of a flowing river is due to the work done by solar energy evaporating water from the ocean surface and lifting it to higher elevations, where it falls to earth.

When water changes from one state to another (liquid, vapor, or solid), there is an accompanying change in the heat energy of the water. The *heat energy* is the amount of thermal energy contained by a substance. A *calorie* (cal) of heat is defined as the energy necessary to raise the temperature of one gram of pure water from 14.5°C to 15.5°C. At other temperatures it takes approximately 1 cal to change the temperature of 1 gram (g) of water 1°C.

The evaporation of water requires an input of energy, the *latent heat of vaporization*. At environmental temperatures (0 to 40°C) the latent heat of vaporization H_v, in calories per gram of water, can be found by

$$H_v = 597.3 - 0.564T \tag{1-1}$$

where T is the temperature in degrees Celsius.

When water vapor condenses to a liquid form, an equivalent heat amount, the *latent heat of condensation,* is released. This factor can also be obtained from Equation 1–1.

In order to melt 1 g of ice at 0°C, 79.7 cal of heat must be added, the *latent heat of fusion*. The resulting water also has a temperature of 0°C, although the gram of water holds more heat energy than the gram of ice.

Water can also pass directly from a solid state to a vapor state by a process called *sublimation*. The energy necessary to accomplish this is the sum of the latent heat of vaporization and the latent heat of fusion. At 0°C this is 677 cal/g. Freezing of water releases 79.7 cal/g, and the formation of frost at 0°C releases 677 cal/g.

The transportation of water through the hydrologic cycle and the accompanying heat transfers are vital to the heat balance of the earth. At the equator, the amount of solar radiation is fairly constant through the year, whereas at the poles it varies from zero during the polar winter to amounts greater than those at the equator during the polar summer. During polar winters the land is in shadow, as the sun does not strike the ground. On the other hand, during the summers, the sun shines continuously. Over the year, the Northern Hemisphere northward of 38° latitude has a net heat loss, as the outgoing terrestrial radiation to space exceeds the incoming solar radiation that is absorbed. Between the equator and 38° N, there is more solar radiation absorbed than terrestrial radiation lost to space. In order to balance these anomalies, heat is transferred by currents in the oceans and through the atmosphere as movement of air masses and water vapor. This creates climatic conditions and changing weather patterns that profoundly affect the hydrologic cycle.

1.5   THE HYDROLOGIC EQUATION

The hydrologic cycle is a useful concept but is quantitatively rather vague. The **hydrologic equation** provides a quantitative means of evaluating the hydrologic cycle. This fundamental equation is a simple statement of the *law of mass conservation*. It may be expressed as

$$\text{Inflow} = \text{Outflow} \pm \text{Changes in Storage}$$

If we consider any hydrologic system—for instance, a lake—it has a certain volume of water at a given time. There are a number of inflows that add water: precipitation that falls on the lake surface, streams that flow into the lake, ground water that seeps into the lake, and overland flow from nearby land surfaces. Water also leaves the lake through evaporation, transpiration by emergent aquatic vegetation, outlet streams, and ground-water seepage from the lake bottom. If, over a given period of time, the total inflows are greater than the total outflows, the lake level will rise as more water accumulates. If the outflows exceed the inflows over a time period, the volume of water in the lake will decrease. Any differences between rates of inflow and outflow in a hydrologic system will result in a change in the volume of water stored in the system.

The hydrologic equation can be applied to systems of any size. It is as useful for a small reservoir as it is for an entire continent. The equation is time-dependent. The elements of inflow must be measured over the same time periods as the outflows.

The basic unit of surface-water hydrology is the **drainage basin,** which consists of all the land area sloping toward a particular discharge point. It is outlined by surface water or **topographic divides.** In ground-water hydrology, we utilize the concept of a **ground-water basin,** which is the subsurface volume through which ground water flows toward a specific discharge zone. It is surrounded by **ground-water divides.** The boundaries of a surface-water basin and the underlying ground-water basin do not necessarily coincide, although the water budget of the area must account for both ground and surface water. Many times hydrologic budgets are made for areas surrounded by political boundaries and not hydrologic boundaries. However, one still must know the location of the hydrologic boundaries, both surface and subsurface, in order to perform a water-budget analysis. Water will flow from the hydrologic boundary toward the point of discharge and hence may flow into the study area if the boundary of the study area does not coincide with the hydrologic boundary.

The hydrologic inputs to an area may include (1) precipitation; (2) surface-water inflow into the area, including streamflow and overland flow; (3) ground-water inflow from outside the area; and (4) artificial import of water into the area through pipes and canals.

The hydrologic outputs from an area may include (1) evapotranspiration from land areas; (2) evaporation of surface water; (3) runoff of surface water;

(4) ground-water outflow; and (5) artificial export of water through pipes and canals.

The changes in storage necessary to balance the hydrologic equation include changes in the volume of (1) surface water in streams, rivers, lakes, and ponds; (2) soil moisture in the vadose zone; (3) ice and snow at the surface; (4) temporary depression storage; (5) intercepted water on plant surfaces; and (6) ground water below the water table.

The application of the hydrologic equation to a watershed is illustrated in the following case study.

CASE STUDY: MONO LAKE

Half a dozen little mountain brooks flow into Mono Lake, but not a stream of any kind flows out of it. What it does with its surplus water is a dark and bloody mystery.

Mark Twain

Mono Lake lies on the eastern slope of the Sierra Nevada near the east entrance to Yosemite National Park. Mono Lake is a terminal lake, which means that although water enters the lake by precipitation and by streams and ground water flowing into the lake, it can leave only by evaporation. The lake level fluctuates with climatic changes. The volume of water that leaves the lake by evaporation is the product of the surface area times the depth of evaporation. If the volume that leaves by evaporation is exactly balanced by the inflow, the lake level will not change. If the inflow exceeds evaporation, the water level will rise. If the inflow is less than evaporation, the lake level will fall.

The Mono Lake basin has an area of 695 square miles (mi^2) (180,000 hectares (ha)). Inputs to the lake under natural conditions are direct precipitation, with an estimated annual average of 8 in. (0.2 m); runoff from the land areas via gaged streams, which is estimated to average 150,000 acre-feet (ac-ft)* per year (y) (1.85×10^8 cubic meters (m^3)); and ungaged runoff and ground-water inflow, which is estimated to average 37,000 ac-ft per year (4.56×10^7 m^3). The average annual rate of lake evaporation is about 45 in. (1.1 m) (Vorster 1985).

When it was first surveyed in 1856, the elevation of Mono Lake was 6407 ft (1953 m) above sea level. Climatic effects of moister and drier periods caused the lake level to rise to as much as 6428 ft (1959 m) in 1919 and then to fall to 6410 ft (1954 m) by 1941. In that year water was first diverted from four of the five major streams feeding Mono Lake into the Los Angeles Aqueduct and thence to southern California.

Since the beginning of diversions in 1941, the surface elevation of Mono Lake has declined substantially (Figure 1.6). Diversions amounted to as much as 100,000 ac-ft (1.23 $\times 10^8$ m^3) per year. The historic low was reached in December 1981, when the lake elevation was 6372.0 ft (1942.2 m). The decline was arrested and the level rose to 6381 ft (1945 m) during a very wet period from 1982 through 1984. A return to more normal precipitation conditions meant that the lake level began to fall again. In 1989 the diversions were halted under a temporary court restraining order that prohibited any diversions that

---

*An acre-foot is a measure of the volume of water that is commonly used in the western United States. It is the amount of water that will cover an acre of land to a depth of 1 ft (43,560 ft^3).

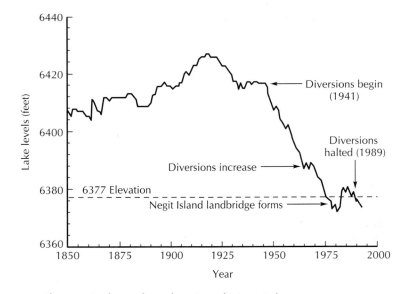

FIGURE 1.6 Changes in the surface elevation of Mono Lake, 1850 to 1992.

would result in a lake level of less than 6377 ft (1944 m). However, even without any diversions the level of Mono Lake still declined due to very dry conditions in the eastern Sierra Nevada, so that by the end of 1992 it was 6373.5 ft (1942.6 m).

In 1941, the year that diversions began, the surface area of Mono Lake was 53,500 ac (21,670 ha). When the lake elevation declined by 38 ft (12 m) from 1941 to 1981, the surface area shrank to about 40,000 ac (16,200 ha). The annual diversion of 100,000 ac-ft (1.23 × 10⁸ m³) would cover the 40,000-ac (16,200-ha) lake to a depth of 2.5 ft (0.76 m). The water level fell because the amount of the diversion plus the natural evaporation from the lake is far in excess of the amount of precipitation onto the lake surface plus the remaining surface inflow and the ground-water inflow. If the courts were again to permit the unrestrained diversion of 100,000 ac-ft (1.23 × 10⁸ m³) of water from the Mono Lake basin, the lake level could fall to as low as 6330 ft (1930 m), with a range of 6330 ft (1930 m) to 6346 ft (1934 m), depending upon climatic conditions. At these elevations the lake would only be 23,500 to 27,600 ac (9,520 to 11,200 ha) in size, and lake evaporation would be reduced enough to balance inflows with outflows (Vorster 1985).

One consequence of the reduction of the volume of Mono Lake has been an increase in the salinity of the lake. In its original, natural condition with a surface elevation of 6410 ft (1954 m), the salinity of Mono Lake was 5.4%. At an elevation of 6377 ft (1944 m), the salinity was 9.3%. The increase in salt concentration has resulted in a reduction of the brine shrimp population of the lake. There is a commercial fishery in Mono Lake for brine shrimp, and the shrimp also serve as an important food source for nesting and migratory birds. Brine flies also inhabit the shallow waters of the lake edge and provide a second food source for the many species of birds that migrate through the area and the nesting colonies of California gulls and snowy plovers.

If the lake levels were allowed to fall to 6335 ft (1931 m), the salt content could reach 22%, which would eliminate the brine shrimp and flies. This would be catastrophic for the hundreds of thousands of migratory birds that stop every year at Mono Lake to fill up on brine shrimp and brine flies. As many as 750,000 eared grebes have been estimated to stop at Mono Lake during their annual migration to Central America.

California gulls have nested on two islands in Mono Lake. When the lake level dropped to 6375 ft (1943 m), one of the islands, Negit Island, became connected to the mainland by a land bridge. In 1979 coyotes crossed from the mainland to the island and disrupted 34,000 pairs of nesting California gulls. Unless lake elevations are maintained at a level that submerges this land bridge, the nesting gulls are not safe from predation.

1.6   HYDROGEOLOGISTS

The professional hydrogeologist has a wide variety of occupations from which to choose. Employment may be found with federal agencies, United Nations groups, state agencies, and local government. Energy and mining companies may call upon the services of hydrogeologists to help provide water where it is needed or perhaps remove it where it is unwanted. Private consulting organizations also employ many individuals trained in hydrogeology. Water resource management districts and planning agencies often include hydrogeologists on their staffs.

Hydrogeology is an interdisciplinary field. The hydrogeologist usually has training in geology, hydrology, chemistry, mathematics, and physics. Hydrogeologists are also being trained in such areas of engineering as fluid mechanics and flow through porous media, as well as in computer science. Such training is necessary, as hydrogeologists must be able to communicate effectively with engineers, planners, ecologists, resource managers, and other professionals. By the same token, an understanding of the basic principles of hydrogeology is useful to soil scientists, engineers, planners, foresters, and others in similar fields. Modeling of hydrologic systems is another area requiring knowledge of a number of disciplines.

1.7   APPLIED HYDROGEOLOGY

The argument has been made that "hydrogeology is more than a classical science" (Stephenson 1974). Traditional studies in hydrogeology have focused on either the mathematical treatment of flow through porous media or on a general geologic description of the distribution of rock formations in which ground water occurs. One occasionally even finds a paper describing the theoretical flow of fluids through an idealized porous medium that probably does not occur in nature. Likewise, many reports on the ground-water geology of an area make no attempt to evaluate how much water is available for use. Neither type of study has much practical value in and of itself.

Hydrogeologists are being employed as problem solvers and decision makers (Sommers 1970). They need to identify a problem, define the data needs, design a field program for collection of data, propose alternative solutions to the problem, and implement the preferred solution.

Hydrogeology is also recognized by both hydrogeologists (Sommers 1970) and planners (McHarg 1969) as an important part of environmental planning. As population and economic growth expand, human use of the natural environment is becoming more intense. Incidents of environmental degradation are less likely to occur when the planner and the hydrogeologist cooperate.

The need for hydrogeological services has never been greater (Stephenson, Cutright, & Woessner 1991). At the same time, geology departments have not focused on the training of hydrogeologists as a priority need (Farvolden & Cherry 1991). As a result, there has been a persistent shortage of trained hydrogeologists. The Association of Ground Water Scientists and Engineers, a division of the National Ground Water Association, is the largest organization of ground-water professionals in the United States and Canada. In 1982 its membership was estimated to be 3000 and by 1992 had grown to 18,000 (National Ground Water Association 1992). This 500 percent growth in 10 y illustrates the growing demand for trained hydrogeologists and ground-water engineers.

A glance toward the future suggests there will be fewer natural resources, more intensive use of the land, and a greater desire to promote environmental conservation along with economic growth. The challenges to the hydrogeologist will grow, as will the opportunities.

1.8 THE BUSINESS OF HYDROGEOLOGY (WHAT DO HYDROGEOLOGISTS DO ALL DAY?)

1.8.1 Application of Hydrogeology to Human Concerns

The work done by applied hydrogeologists can be divided into three realms: research, solving problems involving ground-water supply and control, and solving problems involving ground-water contamination. In the first realm, research, the basic principles of hydrogeology, such as the equations governing fluid flow and movement of contaminants through porous media, have been determined. Basic research is conducted by professors and graduate students at universities and by hydrogeologists employed by nonprofit organizations such as the United States Geological Survey. Basic research involves the search for first principles, which may or may not have either an immediate or an apparent practical application. Applied research is performed to solve a specific problem. The applied researcher utilizes the same research techniques as are appropriate for basic research. However, the planned result of the research program is the development of knowledge that can be applied immediately to the solution of known problems. Applied research may be conducted at universities and by nonprofit organizations, but applied researchers can also be found working for profit-making organizations. The profit-making organization may be trying to develop a product, such as a computer program or a new device to withdraw a water sample from a well, for which there may be a market. These organizations may also have been contracted to solve some hydrogeological problem that a second party is experiencing. An example of this might be the development of a new method of detecting soil contamination under particular geologic conditions. The results of both basic and applied research may be published in journals and formal reports of nonprofit agencies. Presentations of research results may also be given at professional meetings.

The second realm is the application of principles of hydrogeology to realize some economic benefit. These are sometimes called "clean-water" projects; they deal with ground-water supply and control. As an example, a community or an industry may hire a hydrogeologist to locate and develop a source of ground water. The project is driven by economics. The hydrogeologist will identify a source of ground water, determine if there is enough water of acceptable quality available, and work with an engineer to develop cost estimates. If there are several potential sources of water, the hydrogeologist and engineer will try to determine the best alternative from a standpoint of availability, quality and cost.

Another project of this type is ground-water control. For some construction projects and most mining projects, excavations must take place below the water table. Dewatering wells can be placed around a project to lower the water level in order to provide a dry working environment. The cost of the dewatering project must be considered in determining if the project is economically viable. The hydrogeologist will determine how many wells are needed, where they are to be placed, how much water must be pumped, and if the dewatering well will have an undesirable effect on nearby users of ground water, such as their wells going dry.

Hydrogeologists are also involved with issues of aquifer protection and water conservation. It is clear that it is far better to prevent the contamination of an aquifer in the first place than to remediate a contaminated aquifer. Many communities are adopting zoning for the recharge areas of important aquifers in order to prohibit activities that pose a threat to ground-water quality.

If a municipality or private company has numerous projects requiring hydrogeological work, they may employ hydrogeologists on their permanent staff. However, this usually is not the case; instead, the municipality or private company owner will hire a firm of consulting hydrogeologists or engineers. The actual work is then performed by employees of the consulting company. The consulting company will do the hydrogeological work using its own employees, but it may subcontract with a well-drilling company to install both test wells and permanent wells. If a well is for a public water supply, a hydrogeologist working for a state agency will most likely review the preliminary siting studies and final design to determine if the project meets state standards.

In order to illustrate the type of work that is done by a hydrogeologist, we will look in some detail at what might be involved with the search for a new water-supply well. For such a project a hydrogeologist first gathers all available information of the area's hydrogeology, such as well logs and published geological and water supply reports from various sources. Geological mapping and examination of air photos might supplement the review of available data. The data are then evaluated to give an initial idea of where to locate a test well. The hydrogeologist works with the project owners to locate available land that might have a potential as a well site. He or she then prepares the plans and specifications that are used to obtain bids from drilling contractors for constructing a test boring and well. The hydrogeologist spends time in the field overseeing the drilling of a test boring in order to obtain geologic samples. The geologic samples are described by the hydrogeologist, and subsequent laboratory tests may also be performed to evaluate them in order to identify a promising aquifer. A test well is then designed based on the geologic samples and is installed under the field supervision of the hydrogeologist. An aquifer test is then conducted using the test well and (perhaps) nearby observation wells. The data from the aquifer test are analyzed and the hydraulic properties of the aquifer are determined. The hydrologist may then make a computer model of the aquifer in order to determine if a permanent well or well field could yield the desired amount of water on a sustained basis. He or she then prepares a written report for the project owners to inform them of the results of all the work to date and the recommendations for a permanent well. The report is probably also submitted to the proper state agency for approval. If the state approves, then the owners must decide whether or not to go ahead with a permanent well. If the decision is affirmative, the hydrogeologist prepares the plans and specifications for the new well. He or she also works with the engineer who is selecting the pump, determining the layout of the above-ground piping for the well, and designing the building to house the well head. Approval of the design by a hydrogeologist or engineer working for a state agency might be necessary. The final steps involve overseeing the construction of the well to ensure that the contractor follows the plans and specifications and to conduct a performance test of the completed well.

The third and newest realm of the hydrogeologist is the application of hydrogeology in order to satisfy some regulatory or legal requirement. These projects are usually "dirty-water" projects. There are a host of recent federal and state laws and regulations that require studies to be conducted by hydrogeologists and reports to be prepared for submittal to state agencies or the federal government. For example, the federal Resource Conservation and Recovery Act (RCRA) requires owners or operators of facilities that treat, store or dispose of hazardous waste to have a ground-water-monitoring plan in place. Ground-water samples must be collected quarterly, and an annual report must be filed with either the Environmental Protection Agency or a state agency that has been designated to act for the Environmental Protection Agency. The collection of the ground-water samples and the preparation of the reports are frequently done by hydrogeologists. If ground-water monitoring shows that a release to ground water of hazardous waste or hazardous waste constituents has occurred, the site owner or operator must submit a plan for a ground-water-quality assessment program that has been certified by a qualified geologist or geotechnical engineer. The plan must be approved by a hydrogeologist or engineer working for the regulatory agency and then carried out by hydrogeologists working for a consulting company that has been contracted by the site owner or operator.

RCRA is only one of many state and federal regulations that require work to be performed by hydrogeologists. This type of work requires that the hydrogeologist become thoroughly familiar with the specific requirements of existing regulations. Work must be scientifically correct, and it must also conform to the exact letter of the regulatory code. These regulations cover activities such as treatment, storage, and disposal of hazardous and radioactive waste and tailings from the mining and milling of radioactive minerals; the disposal of municipal waste; the injection of liquid waste into deep injection wells; the investigation and remediation of contaminated soil at abandoned hazardous waste sites; and the removal of underground storage tanks. Hydrogeologists working for owners or consulting firms conduct the field studies, analyze the field and laboratory data, and write the final reports that are required. Hydrogeologists working for regulatory agencies conduct field inspections of regulated facilities and review and approve the reports that are submitted. Work is often done under the cloak of a legal document, such as a consent decree. Such a document specifies the scope of the work, the procedures to be utilized, and the schedule that must be followed. Failure to comply with the document may result in the assessment of fines and penalties against the site owner. This creates an incentive for the hydrogeologist to work in an effective and efficient manner.

Litigation may arise from cases involving contamination of soil and ground water. The litigation may be initiated by the federal government or a state in order to compel the cleanup of a contaminated site. A lawsuit can also be filed by a private party in order to recover damages due to the release of a toxic or hazardous substance to soil or ground water. The damages may be to property or to the health of the plaintiff. In all such lawsuits, both sides will generally rely upon the expert testimony of hydrogeologists.

1.8.2   Business Aspects of Hydrogeology

Hydrogeologists who perform all the interesting work just described also expect to be compensated for their time. There must be a source of funds to pay their wages.

The salary of a hydrogeologist working for a project owner is paid by the owner out of the revenue that is generated by the business. A hydrogeologist working for a consulting company is paid by the company with funds that come from the fees that are charged to the project owner for the work that is contracted. The salary of a hydrogeologist working for the state review agency is paid with either tax revenue or a combination of tax revenue and fees charged to project owners who submit plans to the agency for review and costs recovered from polluters by means of lawsuits filed by the state or federal government.

In all cases a hydrogeologist working for a consulting firm is required to keep track of the amount of time spent working directly on each project with which he or she is associated. The time spent is recorded daily on a time sheet that is turned in once every week or two. Consulting work can be contracted in several ways. One method is a lump-sum cost for an entire project. Alternatively, the fee may be based on total hours spent plus direct expenses, such a travel costs and drilling subcontracts. For lump-sum projects, the consulting company management needs to know how many hours employees spend on a project in order to determine if the project made or lost money. For hourly rate projects, the management needs to know how many hours were spent on a project in order to know how much to charge the client. The hourly rate charged is usually several times the actual hourly wage paid to the employee, because it must also cover employee benefits, such as insurance, social security, paid holidays and vacations; fixed overhead, such as rent, insurance, and utilities; office overhead, such as office furniture, computers, field equipment, copiers, and secretarial help; administrative overhead for the salaries of management; sales overhead for the time the staff spends preparing proposals for new work and calling on prospective clients; training costs, including both salaries and expenses for employees to receive both safety training and advanced education; and, finally, some profit for the firm's owners. Many of the hours that a firm's employees spend cannot be billed to a client. As just mentioned, these hours include general company management as opposed to project management, new business development, and training. However, if a consulting firm is to succeed, the employees must be able to bill a certain number of hours a year.

In order to remain profitable, a firm must bring in enough business at a fair price, and the employees must work hard and efficiently in order to complete the work within the budget. If the firm doesn't bring enough new business, there might not be enough billable work available. This could lead to losses for the firm's owners and layoffs of employees. If the employees do not perform up to expectations, too many projects might result in financial loss, and the firm might not be able to remain in business.

Employees must also be thorough and careful in their work. If sloppy work is performed, a firm can be sued for malpractice. The legal costs of defending a firm against a malpractice suit can be great, and the damage awards if a firm loses a suit can jeopardize the economic viability of the business. Many firms carry what is known as *errors and omissions insurance* to protect themselves from malpractice lawsuits. This type of insurance is expensive and difficult to obtain. The policy has a deductible amount (the minimum sum the firm must pay if the lawsuit is lost) and a maximum amount (the greatest amount the insurance company must pay, no matter how high the award to the plaintiff). If a firm has a history of being sued for malpractice and losing, then such insurance might be impossible to obtain.

1.8.3 Ethical Aspects of Hydrogeology

Hydrogeologists have ethical responsibilities to their employers, to clients, and to the general public. As an employee, a hydrogeologist has a responsibility to do his or her best work at all times, to be diligent in his or her work habits, and to be honest in financial matters with the employer. Hydrogeologists who work for consulting firms must treat the firm's clients fairly in financial matters and do work that is as precise and correct as is possible. Only that work necessary to fulfill the contractual obligations should be conducted, even if the project still has money available in the budget. If the contracted work cannot be completed within the budget because of unforeseen or changed circumstances, the client can be approached for a modification of the terms of the contract.

Ethical problems can potentially arise for employees if a firm has obtained a fixed-price contract or a contract that calls for an hourly rate plus expenses with a maximum amount that can be billed to the client. If all the project funds are expended before the work is completed and the employees working on the project keep charging hours to that project, it will become a money loser. This does not look good for the employees, since the objective of a consulting business is to make money, not to lose it. There might be a temptation to keep working on a losing project but to charge one's time to a project that still has some funds available. However, this not only is unethical, but it could also be construed as fraud. The manager of one consulting project where the client was the federal government was convicted of falsifying time cards in such a case and was sentenced to a federal penitentiary.

If a hydrogeologist determines that there is a situation that may adversely affect the client, such as contamination at a site that was thought to be clean, the client should be advised promptly. This information is confidential and should not be disclosed to others unless it is required by law to do so. Once advised by the hydrogeologist, the property owner may have the legal requirement to report the contamination to the appropriate regulatory agency.

The hydrogeologist must also be aware of and avoid any conflicts of interest. For example, an employee of a state regulatory agency should not accept any private consulting work on projects within the state, nor should he or she accept any consulting work from firms that are regulated within the state, even if the work is on an out-of-state project. A consulting firm that works for a client who has a contamination plume extending off site to an adjacent property should not work for the owner of the adjacent property unless both parties agree to be bound by the findings of a single study covering both properties.

If a hydrogeologist becomes aware of any decision or action of a client or of his or her employer that violates any law or regulation, then the hydrogeologist should advise against that action or decision. If the violation continues and appears materially to affect the public health, safety, or welfare, the hydrogeologist should promptly advise the proper authorities. (As a practical matter, although it is illegal, such action might cause the hydrogeologist to lose his or her job. Being ethical sometimes comes with a price.)

If a report for submission to a regulatory agency is being prepared, it should honestly report all the findings of a study, even if parts may be adverse to the client's position. Many times a client is given a draft of a report to review prior to submittal. If this is the case, the hydrogeologist must resist suggestions to delete or change data or conclusions that he or she feels should be contained in an honest report, because the hydrogeologist has an ethical responsibility to the general public to protect human health, safety, and welfare as well as the environment.

Hydrogeologists who work for review agencies have an ethical responsibility to base their reviews only on scientific considerations. Such agencies can be politicized, and pressure may be put on a reviewer to base a decision on political rather than scientific grounds. Just as the hydrogeologist working for the client must be honest in preparing a report, the reviewer who reads it has an ethical responsibility to be equally honest.

Hydrogeologists working for review agencies also usually keep time sheets. Certain enforcement actions allow the regulatory agencies to recover their costs from the polluters, including the wages and benefits of employees. The regulatory hydrogeologist has an ethical responsibility to be as accurate as possible in assigning time to specific enforcement actions for which cost recovery is anticipated.

1.9   SOURCES OF HYDROGEOLOGIC INFORMATION

Hydrogeologic information is available from a wide range of sources. In terms of sheer volume, the Water Resources Division of the U.S. Geological Survey (USGS) is the leading source in the United States. This agency collects basic data on streamflow, surface-water quality, ground-water levels, and ground-water quality. The USGS also conducts water resources investigations and basic research. USGS publications are available in libraries that are designated depositories of federal documents; these publications are also available from the U.S. Government Printing Office.

The U.S. Geological Survey maintains a computerized central storage facility for water resources data called the National Water Data Storage and Retrieval System, generally known by the acronym WATSTORE (Baker & Foulk 1980). Ground-water data in WATSTORE can be accessed by authorized users by means of an on-line computer retrieval system known as the Ground-Water Site Inventory (GWSI) file (Mercer & Morgan 1982). The ability to access the tens of thousands of data files on both ground and surface water held by the USGS is a powerful tool. The GWSI system has the ability to reduce the data to make x-y plots, graphs, and tables.

The National Ground Water Association maintains Ground Water On-Line, a computerized data base of bibliographic information (National Ground Water Association 1986). As of 1992 there were more than 67,000 documents indexed by more than 700 hydrogeological descriptors. A bibliographic search can be conducted by computer that will seek out all the indexed documents that correspond to the selected descriptors. The search can be conducted on-line by using a personal computer and a modem. An interlibrary loan and photocopying service is also available to obtain copies of the articles the data base search found. There are many other bibliographic data base search services as well that might be helpful to the hydrogeologist.

The National Oceanic and Atmospheric Administration is the parent organization of the National Weather Service. *The Climatic Record of the United States* is published for each state and contains precipitation, temperature, evaporation, and other climatic data. Other U.S. federal agencies that may conduct studies related to hydrogeology include the Corps of Engineers, Bureau of

Land Management, Bureau of Reclamation, Soil Conservation Service, Environmental Protection Agency, Nuclear Regulatory Agency, and Department of Energy.

In most states, there are one or more agencies responsible for water-oriented research and other activities. The functions, responsibilities, and organizational formats of state agencies in water resources activities vary from state to state. Typical agency designations include State Department of Water Resources or Water Survey, State Geological Survey, Department of Conservation or Natural Resources, and State Department or Board of Health. In many states, various responsibilities are allocated among several agencies. In addition, Congress has established provisions for a water resources research center or institute in each state and Puerto Rico. These are associated with a major university in each state.

Reports of current research and recent developments in hydrogeology and ground water are included in the following journals:

☐ *Bulletin, International Association of Scientific Hydrology*
☐ *Ground Water*
☐ *Ground Water Monitoring and Remediation*
☐ *Journal American Water Works Association*
☐ *Journal of Applied Hydrogeology*
☐ *Journal of Contaminant Hydrology*
☐ *Journal of Hydrology*
☐ *Memoirs, International Association of Hydrogeologists*
☐ *Transactions, American Society of Civil Engineers*
☐ *Water Resources Bulletin*
☐ *Water Resources Research*

A number of professional organizations sponsor symposia and meetings where technical sessions on hydrogeology or ground water are held. These include the following:

☐ American Geophysical Union
☐ American Institute of Hydrology
☐ American Society of Civil Engineers
☐ American Water Resources Association
☐ Geological Society of America
☐ Geological Society of Canada
☐ International Association of Hydrogeologists
☐ International Association of Scientific Hydrology
☐ International Water Resources Association
☐ National Ground Water Association

COMPUTER NOTES

There is a vast amount of environmental data available on compact-disc read-only memory (CD-ROM) technology. A CD-ROM drive can be added to most personal computers. One CD-ROM disc can store up to 660 megabytes (MB) of data. The following data sets are available:

- [ ] USGS WATSTORE data base. Daily discharge records from all USGS gaging stations going back to 1850; annual and partial flood peaks at more than 25,000 USGS stream-gaging stations; and water quality data.

- [ ] Environment Canada HYDAT data base. Daily discharge, instantaneous peak flow, sediment data, and lake levels from Canadian stations going back to 1900.

- [ ] National Climatic Data Center TD-3200 data base. Daily observations of precipitation, snowfall, maximum and minimum temperature, and evaporation from more than 25,000 present and historical cooperative stations in the United States.

- [ ] National Climatic Data Center TD-3240 and TD-3260 data bases. Hourly precipitation data from 5000 stations and 15-minute (min) precipitation data from 2,700 stations.

- [ ] National Climatic Data Center TD-3280 data base. Hourly observations of 15 different climatological parameters, including wind, sky cover, wet-bulb/dry-bulb temperature, relative humidity and global radiation from 300 first-order stations.

Vendors of these databases sell discs with the various databases for different regions of the United States and, if available, for Canada. They also provide an annual subscription service to update the databases. The vendors have software available to analyze the data or export it in spreadsheet format.

These databases could be very useful to hydrogeologists working on a routine basis with streamflow and/or precipitation data. If a large amount of data must be analyzed, the cost of the database would be quickly recovered, compared with the cost of obtaining the same data from government reports in the library. Vendors of these databases include:

- [ ] EarthInfo Inc., 5541 Central Avenue, Boulder, CO 80301. (303)938-1788

- [ ] Hydrosphere Data Products Inc., 1002 Walnut, Suite 200, Boulder, CO 80302. (800)949-4937

1.10   WORKING THE PROBLEMS

Most of the chapters have end-of-chapter problems for students to work. The answers to most of the odd-numbered problems are given. These problems are designed so that students can work them using calculators, graph paper, and tables found in the appendices. Many solved example problems will be found throughout the book. By working the problems, students will gain a much deeper understanding of the material.

In working the problems, students should pay attention to the number of *significant digits* that are used. Significant digits arise when using measured values. Although we can count objects exactly, a measurement is always an approximation. The last digit in the measurement shows the degree of approximation. For example, a measurement of 17.63 cm is only an estimate. All we know for sure is that the object measured is actually somewhere between 17.625 and 17.635 cm long. The measurement 17.63 cm has four significant digits. If someone else measures the same object and says that it is 18 cm long, the actual length is between 17.5 and 18.5 cm, and the measurement has been made to two significant digits.

When two or more numbers are multiplied (or divided), their product (or quotient) should have the same number of significant digits as the multiplier (or divisor) with the least number. For example, if we measure the sides of a rectangle as 17.63 cm and 14.2356 cm, the area of the rectangle is 251.0 cm^2, not 250.97363 cm^2. We use the number of significant digits of the least precise measurement, in this case four. We report the number as 251.0, not 251, to show the number of significant digits.

When measurements are added (or subtracted), the sum (or difference) should not have any significant digits to the right of the last significant digit of any of the addends (or subtrahends). For example, if we add 17 + 2.35 + 1.346 + 0.072, the sum is 21, not 20.768. Since 17 has only two significant digits, the sum can only have two significant digits. Notice that we have rounded the number that is obtained from the calculator to the appropriate number of significant digits. However, if the measurements were 17.0, 2.35, 1.346, and 0.072, the sum would be 20.8, not 21, because 17.0 has three significant digits. Be aware that the numbers 17, 17.0, 17.00, and 17.000 differ in the number of significant digits. When zeros occur to the left of the decimal, it is harder to determine the number of significant digits. For example, 100.0 has four significant digits, but does 100 have one, two, or three significant digits? Unless an uncertainty range is specified, this question is unanswered. For example, 100 ± 1 has three significant digits and 100 ± 10 has two significant digits. For purposes of working problems in the text, for a number such as 100 (or 2500 or 10,000) assume that it is exact and determine the number of significant digits from other numbers in the problem.

In solving the problems in the text we frequently employ the concept of *dimensional analysis*. In dimensional analysis the units of measurement are used as a guide in the calculations to obtain the desired units for the answer.

A simple example of dimensional analysis is in calculating the number of inches in a measured distance of 1.7 mi. We know that there are exactly 12 in. in 1 ft and exactly 5280 ft in 1 mi. The problem can be set up as follows:

$$1.7 \text{ mi} \times \frac{5280 \text{ ft}}{1 \text{ mi}} \times \frac{12 \text{ in.}}{1 \text{ ft}} = 107{,}712 \text{ in.}$$

Some of the units cancel each other, since they appear in both the numerator and the denominator. In this example miles and feet cancel, leaving inches as the unit.

The answer of 107,712 in. then must be adjusted to the proper number of significant digits.

EXAMPLE PROBLEM

How many significant digits are in 107,712 in.?

The mileage measurement of 1.7 has only two significant digits, so we round the answer to 110,000 in., a number with two significant digits.

In working the problems, you will also need to use *conversion factors*. These are shortcuts to dimensional analysis when converting units from one system of measurement to another. Conversion factors for length, area, volume, and time are found in the appendices; those of flow are inside the front cover.

EXAMPLE PROBLEM

Use a conversion factor to find the number of inches in 1.7 miles.

In Appendix 7 we see that 1 mi is equal to 63,360 in.

$$1.7 \text{ mi} \times \frac{63,360 \text{ in.}}{1 \text{ mi}} = 107,710 \text{ in.}$$

$$\approx 110,000 \quad \text{(rounded to the correct number of significant digits)}$$

The units with which you will work in this book have all been derived from three basic factors, mass *(M)*, time *(T)*, and length *(L)*. Areas are in units of length times length (L^2) and volumes are in units of length cubed (L^3). For example, velocity is length divided by time *(L/T)* and is expressed in such units as feet per day. Concentration is mass of solute per unit volume of solution (M/L^3) and is expressed in units such as milligrams per liter. Density is the mass of an object per unit volume (M/L^3) and is expressed in units such as kilograms per cubic meter. As a check on dimensional analysis using units, it is often helpful to conduct dimensional analysis using *M*, *L*, and *T*.

NOTATION

H_v Heat of vaporization   **T** Temperature

PROBLEMS

Answers to odd-numbered problems will appear at the end of the book.

1. A farmer has a reservoir with vertical sides and a surface area of 2.5 ac. Following the rainy season, the reservoir is filled to a depth of 3.0 m. During the dry season the reservoir loses 2.5 in. of water per week (wk) to evaporation. If the average irrigation demand during the dry season is 0.23 ac-ft per day, for how many weeks can the farmer irrigate from the reservoir?

2. How long must a pump with a capacity of 12 gal/min pump to fill a tank with a capacity of 37 m^3?

3. A circular water transmission pipe has a diameter of 1.0 ft and is 8.3 mi long. How much water does it take to fill the pipe?

4. If the water is flowing into the pipe of Problem 3 at a velocity of 1.3 feet per second (ft/s), what is the rate at which the pipe is transmitting water?

5. A small urban watershed has an area of 16.34 mi^2. A summer storm drops an average of 1.50 in. of rain over the entire watershed. If 50% of the rainfall runs off the watershed into surface-water bodies, what is the volume of runoff:
   A. In cubic inches?
   B. In cubic feet?
   C. In cubic meters?

6. A ground-water basin in a coastal area has an area of 200 mi^2. The land area is 195 mi^2 and the area of the rivers is 5 mi^2. A water budget for the basin has the following long-term average annual values: precipitation, 35 in./y; evapotranspiration, 23 in./y; direct runoff, 3 in./y; baseflow, 6 in./y; streamflow, 9 in./y; subsea outflow, 3 in./y.
   A. Prepare an annual water budget for the basin as a whole, listing inputs in one column and outputs in a second. Make sure the two balance, because these are long-term values and we assume that no changes in storage take place.
   B. Prepare an annual water budget for the streams.
   C. Prepare an annual water budget for the ground-water reservoir.
   D. What is the annual streamflow from the basin expressed as an average rate in cubic feet per second?
   E. What is the average rate of ground water recharge expressed in units of millions of gallons per day per square mile of surface area?

7. What mass of water at 15°C can be cooled 1°C by the heat needed to melt 43 g of ice at 0°C?

8. What mass of water at 15°C can be cooled 1°C by the heat needed to sublime (go from a solid to a vapor phase) 153 g of ice at 0°C?

9. By how much will 153 g of water warm as 1.5 g of water vapor condenses?

10. A. How long will it take to form 5 g of steam at 100°C from 5 g of ice at
         −15°C if heat is added to the system at a rate of 10 cal per minute?

     B. Make a plot of temperature (y-axis) as a function of time (x-axis) for part
         (a). Indicate when phase changes occur.

11. A 750-milliliter (mL) bottle of mineral water, which is at a temperature of
     25°C, is poured over 45 g of ice, which is at −10°C. What is the temperature
     of the water when the ice melts, assuming that the water is in an insulated
     container that does not change temperature?

2 Evaporation and Precipitation

Rivers depend for their existence on the rains and on the waters within the earth, as the earth is hollow, and has water in its cavities.

Anaxagoras of Clazomenae (500–428 B.C.)

2.1   EVAPORATION

Water molecules are continually being exchanged between a liquid and atmospheric water vapor. If the number passing to the vapor state exceeds the number joining the liquid, the result is **evaporation.** When water passes from the liquid to the vapor state, it will absorb 590 cal of heat from the evaporative surface for every gram of water evaporated. The vapor pressure of the liquid is directly proportional to the temperature. Evaporation will proceed until the air becomes saturated with moisture. The **absolute humidity** of a given air mass is the number of grams of water per cubic meter of air.

At any given temperature, air can hold a maximum amount of moisture: the **saturation humidity.** This is directly proportional to the temperature of the air. Table 2.1 gives the saturation humidity for several environmental temperatures. The **relative humidity** for an air mass is the percent ratio of the absolute humidity to the saturation humidity for the temperature of the air mass. As the relative humidity approaches 100%, evaporation ceases.

Condensation occurs when the air mass can no longer hold all of its humidity. This happens when an air mass is cooled and the saturation humidity value drops. If the absolute humidity remains constant, the relative humidity will rise. When it reaches 100%, any further cooling will result in condensation. The **dew point** for an air mass is the temperature at which condensation will begin. As condensation is the reverse of evaporation, the process of condensation releases 590 calories of heat to the surroundings per gram of water: the latent heat of condensation.

Evaporation of water takes place from free-water surfaces—lakes, reservoirs, puddles, dew droplets, etc. The rate is dependent upon factors such as the water temperature and the temperature and absolute humidity of the layer of air just above the free-water surface. Solar radiation is the driving energy force behind evaporation, as it warms both the water and the air. The rate of evapora-

TABLE 2.1   Saturation humidity of air (grams per
cubic meter)

Temperature (°C)	Humidity
−25	0.705
−20	1.074
−15	1.605
−10	2.358
−5	3.407
0	4.874
5	6.797
10	9.399
15	12.83
20	17.30
25	23.05
30	30.38

Source: *Handbook of Chemistry and Physics* (Cleveland, Ohio: CRC
Publishing Company, 1976).

tion is also related to the wind—especially over land. The wind carries vapor
away from the free-water surface and keeps absolute humidity low. By disturbing
the water surface, the wind may also increase the rate of molecular diffusion
from it.

Evaporation from lakes and reservoirs is an important consideration in
water-budget studies. It can be computed for a lake or reservoir if all of the inflows
(precipitation over the surface, surface-water inflow, and ground-water inflow)
and the outflows (ground-water outseepage, spillway discharge, and pumpage)
and change in storage are known. The hydrologic equation (inflow = outflow ±
changes in storage) is used. All these factors, with the exception of the ground-
water flux, can be measured with an error of perhaps ± 10 percent. In a carefully
prepared water budget study for Lake Hefner, Oklahoma, daily evaporation was
computed to an accuracy of 5% to 10% (Harbeck & Kennon 1954). For many
reservoirs, monthly or annual evaporation can be computed fairly easily. The
most difficult factor to measure is the ground-water flux.

Free-water evaporation is measured quite simply by using shallow pans.
The most commonly used is the **land pan.** The U.S. National Weather Service
maintains about 450 evaporation stations using Class A land pans. Similar pans
are used in Canada. They are 4 ft (122 cm) in diameter and 10 in. (25.4 cm) deep,
made of unpainted galvanized metal. Land pans are placed on supports so that air
can circulate all around. Water depths from 7 to 8 in. (18 to 20 cm) are maintained.
Records are kept of the daily depth of water, the volume of water added to replace
evaporated water, and the daily precipitation into the pan. Using the hydrologic
budget, the daily evaporation can be computed. Errors may result from splash
caused by heavy rainfall and drinking by birds. The wind movement is also
measured and expressed in units of miles per day (mi/day) (A steady wind blowing
at a velocity of 10 miles per hour (mi/h) would have a 24-h wind movement of 240
mi/day).

TABLE 2.2 Class A land pan coefficients for midwestern United States

January	0.62	July	0.76
February	0.72	August	0.75
March	0.77	September	0.73
April	0.77	October	0.69
May	0.78	November	0.63
June	0.77	December	0.58
	Annual 0.75		

Source: W. J. Roberts & J. B. Stall, Illinois State Water Survey Report of Investigation 57, 1967.

Research has shown that the manner in which precipitation is measured can affect the amount of evaporation that is calculated at an evaporation station. Precipitation can be determined by a standard rain gage placed next to the evaporation pan. Daily measurements of water level in the pan and precipitation as measured by the rain gage are made. An alternative method is to use a sensitive water-level recorder and continuously record the water level in the evaporation pan. Net evaporation is determined by summing all the measured declines in the water level. In this instance, the evaporation pan is acting as the rain gage. In a study of evaporation in the Florida Everglades, during rain-free periods both types of evaporation stations yielded similar results. However, during rainy periods, evaporation at the stations that utilized rain gages was significantly greater than the station with a level recorder. The conclusion of the study was that the rain gage caught more rain than the pan. As a result, the calculated evaporation was higher as the input to the hydrologic equation was greater (Gunderson 1989).

The water in a Class A land pan will be warmed much more readily by solar radiation than the surface waters of a lake or reservoir. The chief reason is the difference between the water depth in the pan and the depth of the surface layer of reservoir water. The pan may also gain or lose heat through the sides and bottom, a process that does not occur in reservoirs. For these reasons, observed pan evaporation is multiplied by a factor with a value less than 1.0, the pan coefficient, to estimate reservoir evaporation during the period of observation. Detailed studies in the United States Midwest have yielded monthly pan coefficients ranging from 0.58 in December to 0.78 in May, with an annual value of 0.75 (Roberts & Stall 1967) (see Table 2.2).

The National Weather Service has developed a lake evaporation nomograph (Kohler, Nordenson, & Fox 1955). From this diagram, daily lake evaporation can be determined using mean daily temperature, solar radiation in langleys* per day, mean daily dew point temperature, and wind movement in miles per day.

---

*A langley is a measure of solar radiation equal to one calorie per square centimeter of surface. In the SI system (International System of Units, based on the meter, kilogram, second, and ampere), the unit is the joule per square meter, which is equal to 4.184×10^4 langleys.

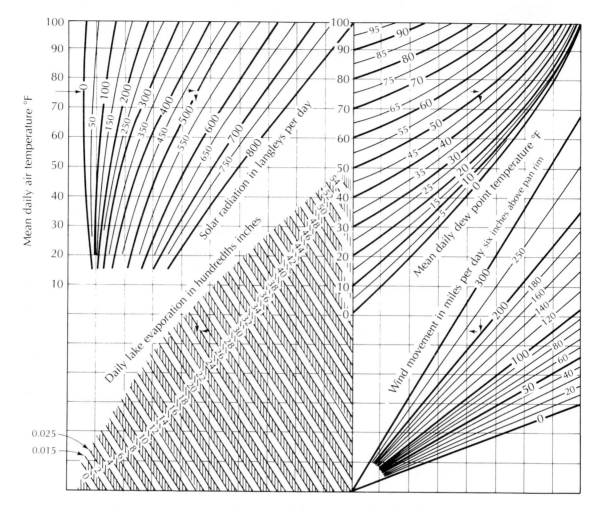

FIGURE 2.1   Nomograph used to determine the value of daily lake evaporation for shallow lakes if solar radiation, mean daily air temperature, mean daily dew point temperature, and wind movement are known. Source: Roberts & Stall 1967.

The graph in Figure 2.1 is entered from the left side at the mean daily air temperature. As an example, this is 75°F. A horizontal line is drawn across the chart along the 75°F axis. Perpendicular lines are dropped at the intersections of the values of solar radiation and mean daily dew point temperature. In the example, these are 500 langleys per day and 50°F. The right-hand perpendicular extends from the mean daily dew-point temperature to the total daily wind movement. The example value is 200 mi/day. From this intersection, a horizontal line is drawn toward the left. This horizontal line and the left-hand perpendicular will intersect in a field indicating the mean daily lake evaporation. For the example in Figure 2.1, this is 0.25 in./day.

In some instances, it may be necessary to estimate evaporation without the availability of evaporation pan data. Such estimates are possible via methods based on heat budgets (Penman 1948). The energy budget for a reservoir may be used to find the amount of energy used for evaporation, which in turn can yield the amount of evaporation. Other methods use aerodynamic data and vapor pressures of the water and air in empirical formulas (Harbeck 1962). Some investigators have been able to combine these two approaches (Kohler & Parmele 1967).

2.2   TRANSPIRATION

Free-water evaporation is only part of the mechanism for mass transfer of water to the atmosphere. Growing plants are continuously pumping water from the ground into the atmosphere through a process called **transpiration** (Hendricks & Hansen, 1962). Water is drawn into a plant rootlet from the soil moisture owing to osmotic pressure, whereupon it moves through the plant to the leaves. The turgidity of nonwoody vascular plants is caused by the cellular pressures of the contained water. The water is passed as vapor through openings in the surface of the leaves known as *stomata*. Air also passes through these openings. A small portion (less than 1%) of the water is used to manufacture plant tissue, but most is transpired to the atmosphere. The process of transpiration accounts for most of the vapor losses from a land-dominated drainage basin.

The amount of transpiration is a function of the density and size of the vegetation. As an example, transpiration from a cornfield in May, when the plants are a few centimeters high, is much less than in August, when they may exceed 7 ft (2 m) in height. Transpiration is obviously important only during the growing season; about 95% takes place during the daylight hours, when photosynthesis is occurring. Transpiration is also limited by available soil moisture. When the soil-moisture content becomes so low that the surface tension of the soil–water interface exceeds the osmotic pressure of the roots, water will no longer enter the roots. This is termed the **wilting point** of the soil.

When available water becomes limited, deep-rooted plants are more resistant to drought wilting than shallow-rooted plants, as the former can draw moisture from deeper layers. Also, some plants have fewer stomata and can close them through the use of special cells to reduce water loss during drought periods. Such drought-resistant species can transpire less water during periods of stress. **Phreatophytes** are plants with a tap root system extending to the water table. They can transpire at a high rate even in the desert, so long as the water table does not drop below the tap root. In areas of low precipitation, the native vegetation is adapted to existing with minimal water. These desert plants are called **xerophytes.** They have a shallow root system that spreads out away from the plant.

Aquatic plants, or **hydrophytes,** are a special case. They exist with their root systems submerged, and the special cells some plants have to close the stomata are lacking. As long as adequate water is available, transpiration proceeds at a high rate. The rate of transpiration is controlled by the amount of solar energy and the heat content of the water. The water loss from a pond is about the same, whether or not emergent aquatic vegetation is present.

Measurement of transpiration can be performed under carefully controlled laboratory conditions. A *phytometer* is a sealed container partially filled with soil. Transpiration by plants rooted in the soil causes an increase in the humidity, which can be measured in the air space around the plant. However, such laboratory studies reveal little about the behavior of plants in natural or agricultural conditions.

2.3   EVAPOTRANSPIRATION

Under field conditions it is not possible to separate evaporation from transpiration totally. Indeed, we are generally concerned with the total water loss, or **evapotranspiration,** from a basin. Whether the loss is due to free-water evaporation, plant transpiration, or soil-moisture evaporation is of little importance.

The term **potential evapotranspiration** was introduced by Thornthwaite (1944) as equal to "the water loss which will occur if at no time there is a deficiency of water in the soil for the use of vegetation." Thornthwaite recognized that there is an upper limit to the amount of water an ecosystem will lose by evapotranspiration. The majority of the water loss due to evapotranspiration takes place during the summer months, with little or no loss during the winter. Because there is often not sufficient water available from soil moisture, the term **actual evapotranspiration** is used to describe the amount of evapotranspiration that occurs under field conditions. Figure 2.2 shows potential evapotranspiration and actual evapotranspiration for a region with a warm, dry summer and a cool, moist fall, winter, and spring. Under these conditions the actual evapotranspiration is much less than the potential, especially if the soil-moisture storage capacity is limited. In months when the potential evapotranspiration is less than the rainfall, some of the demand will be met by drawing upon moisture stored in the soil. When available soil moisture is depleted, the actual evapotranspiration will be limited to the monthly precipitation. Figure 2.3 shows potential and actual evapotranspiration in an area where the precipitation is more or less evenly distributed through the year. This circumstance results in the actual evapotranspiration being closer to the potential value.

Thornthwaite's method is based upon the assumption that potential evapotranspiration was dependent only upon meteorological conditions and ignored the effect of vegetative density and maturity. While this assumption is not correct, the method devised by Thornthwaite to compute potential evapotranspiration is still useful. The only necessary factors to input are mean monthly air temperature, latitude, and month (Thornthwaite & Mather 1955; Thornthwaite & Mather 1957). The last two factors yield average monthly sunlight. The Thornthwaite method is reasonably accurate in determining annual values, especially in humid areas. As no factor for vegetative growth is included, values computed for spring and early summer are too high, as the crop is just emerging; midsummer values may be too low.

Another method of estimating potential evapotranspiration was developed by Blaney (1959) and Criddle (1958). This method introduces a crop factor, which varies as the growing season progresses. Thus, some of the objections to the Thornthwaite method are overcome, but the effects of wind and relative humidity on evapotranspiration remain unaccounted for.

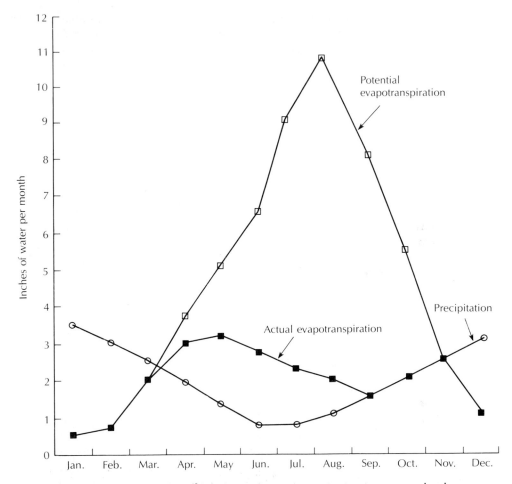

FIGURE 2.2 Diagram of potential and actual evapotranspiration in an area that has coarse soils with limited soil-moisture storage, warm, dry summers, and cool, moist winters.

Penman (1956; Blaney 1956) derived a theoretical equation for the estimation of free water evaporation and evapotranspiration. This equation uses climatic data, including vapor pressures, sunshine duration, net radiation, wind speed, and mean temperature. Many of these factors are not regularly collected at most sites and must be estimated if the Penman method is to be used.

Evapotranspiration can be measured directly using a **lysimeter**—a large container holding soil and plants. The lysimeter is set outdoors, and the initial soil-moisture content is determined. Precipitation into the lysimeter and any irrigation water added are measured. Changes in soil-moisture storage reveal how much of the added water is lost to evapotranspiration. It is necessary to design the lysimeter so that any moisture in excess of that specifically retained by the soil is collected. The following equation can be used with the lysimeter:

$$E_T \; = \; S_i + \; P + \; I \; - \; S_f \; - \; D \tag{2-1}$$

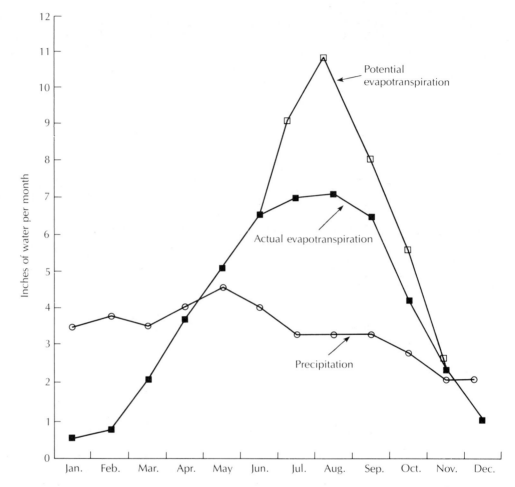

FIGURE 2.3   Diagram of potential and actual evapotranspiration in an area that has fine soils with ample soil-moisture storage, warm summers, cool winters, and little change in precipitation throughout the year.

where

E_T is the evapotranspiration for a period

S_i is the volume of initial soil moisture

S_f is the volume of final soil moisture

P is the precipitation into the lysimeter

I is the irrigation water added to the lysimeter

D is the excess moisture drained from the soil

Lysimeters should be designed so that they accurately reproduce the soil type and profile, moisture content, and type and size of vegetation of the surrounding area. They should be buried so that the soil surface is at the same level inside and outside the container. Soil-moisture changes can be determined by sampling the soil, by means of moisture meters, or by weighing the entire mass of soil, water, and plants. Whatever method is employed, operation of a lysimeter is both time-consuming and expensive. If water is applied to the lysimeter at a rate sufficient to keep the soil at, or nearly at, field capacity (see Section 6.5), the lysimeter will measure potential evapotranspiration.

When the soil moisture drops below the amount that the soil can hold against gravity by surface tension (the field capacity), available water may limit evapotranspiration to some value less than the potential evapotranspiration. The plants are required to draw upon soil moisture and, as this diminishes and less water is extracted, actual evapotranspiration falls below the potential. If the soil moisture drops too low, the plants may wither and die. The soil-moisture content below which plants can no longer obtain moisture is the wilting point.

There is some uncertainty about the rate of evapotranspiration when the soil moisture is between the wilting point and the field capacity. Some have suggested that it proceeds at a rate equal to the potential evapotranspiration until the wilting point is reached (Veihmeyer & Hendirckson 1955), but others have suggested that the evapotranspiration rate is linearly proportional to the ratio of the remaining available soil moisture to the initial available soil moisture (Thornthwaite & Mather 1955). Soil texture and unsaturated soil permeability play major roles in determining the rate of actual evapotranspiration (Molz et al. 1968).

Evapotranspiration is the major use of water in all but extremely humid, cool climates. If evapotranspiration were reduced, then runoff or ground-water infiltration or both could increase. This would increase the available water supply. Studies have shown that basin runoff from a forested watershed has increased following the timbering of the forest (Hibbert 1967). The increase is greatest during the first year, when there is little reforestation. As the forest regrows, the runoff again decreases. Cutting of forests to increase runoff may also result in increased erosion from the uplands and concurrent sedimentation in the lowlands. Conversion of one plant cover to another can also affect the evapotranspiration rate. In arid Arizona, the conversion of a plot of land formerly covered with chaparral to grasses resulted in streamflow increases of several hundred percent. This was due in part to lower evapotranspiration, as the grass was not as deep-rooted as the chaparral (Hibbert 1971). However, in Colorado, the conversion of sagebrush to bunchgrass had no appreciable effect on the amount of watershed runoff, although an increase in cattle forage did result (Shown, Lusby, & Branson 1972).

In some areas of the humid eastern United States, which were originally wooded, marginal farms are being abandoned. The old fields are gradually reverting to forest. There has been a concomitant decrease in streamflow from these watersheds. The replacement of deciduous forests with conifers results in an increase in evapotranspiration (Urie 1967).

In an urbanized watershed one would naturally expect that the flood flows would increase as pervious soil is replaced by impervious pavement. A surprising effect of urbanization is that in dry periods total runoff appears to be reduced (Ferguson & Suckling 1990). This means that urbanization has actually increased evapotranspiration, even as the vegetative cover has decreased. This may be due to the pattern of vegetation surrounded by pavement. Heat from the pavement areas causes overlying air to warm and rise, which can increase the evapotranspiration from the vegetated areas.

Experiments have shown that evaporation from small lakes and reservoirs can be reduced by applying a monolayer of a fatty alcohol to the water surface (Bartholic, Runkels, & Stenmark, 1967). This has not proven to be practical, however, owing to the cost of a treatment and the rapid rate at which the fatty alcohol dissipates. Likewise, fatty alcohols have been used as antitranspirants in treating plants and soils. However, concentrations high enough to reduce transpiration also reduce crop growth (Gale, Roberts, & Hagen 1967). Chemical antievaporants and antitranspirants have not yielded the hoped-for success.

2.4   CONDENSATION

When an air mass with a relative humidity lower than 100 percent is cooled without losing moisture, the relative humidity will approach 100 percent as the dew point temperature is approached. When the air mass is saturated, **condensation** may start to occur. Condensation generally requires a surface or nucleus on which to form. The morning dew or frost is the result of condensation taking place on plants or other surfaces. Rain or ice needs nuclei in the range of 0.1 to 10 μm. Particles serving as nuclei include clay minerals, salt, and combustion products.

In the absence of sufficient nuclei, the air mass may become supersaturated without the formation of raindrops or ice crystals. This is the theory behind artificial precipitation augmentation. "Cloud-seeding" procedures involve the addition of artificial nuclei, including silver iodide and dry ice, to the atmosphere. Research has shown that even in severe droughts, atmospheric conditions conducive to successful seeding may sometimes occur during the summer (Huff & Semonin 1975).

Once droplets or ice crystals have formed, they initially grow by attraction (diffusion) of water vapor as well as additional condensation. Rising air masses or upward movements of clouds tend to keep newly formed fog and cloud elements aloft. These elements are in the size range of 10 to 50 micrometers (μm). As cloud elements collide and coalesce, raindrops begin to form. When the raindrops start to fall, further collisions occur, so that some raindrops may grow as large as 0.2 in. (6 millimeters (mm)) in diameter. Rain that falls through an unsaturated air mass may evaporate before it reaches the ground. Falling ice crystals grow by diffusion and collision to form snowflakes. The largest snowflakes form when temperatures are close to freezing.

2.5   FORMATION OF PRECIPITATION

In order for precipitation to occur, several conditions must be met: (1) a humid air mass must be cooled to the dew point temperature, (2) condensation or freezing nuclei must be present, (3) droplets must coalesce to form raindrops, and (4) the raindrops must be of sufficient size when they leave the clouds to ensure that they will not totally evaporate before they reach the ground.

Air masses are cooled by a process known as **adiabatic expansion,** which occurs when the air mass rises in the atmosphere. Since the atmosphere becomes less dense with altitude, a rising air mass must expand owing to the lower pressure. If there is no exchange of heat between the air mass and its surroundings, the laws of thermodynamics dictate that the temperature will fall.

When the rising air mass is dry—that is, the relative humidity is lower than 100%—the rate of cooling is 1°C for every 328-ft (100-m) rise in height. This is the **dry adiabatic lapse rate.** When the air mass reaches the dew point temperature, further lifting and cooling will cause condensation. The latent heat of vaporization is released; hence, the **wet adiabatic lapse rate** is lower than the dry rate. The exact value depends upon the amount of condensation occurring.

Under normal conditions, air temperature decreases with increasing altitude at a mean rate of 0.7°C for every 328 ft. Owing to uneven or unsteady heating or cooling, the temperature gradient or lapse rate may be more or less than 0.7°C per 328 ft. **Temperature inversions,** or layers of warm air overlying cooler air, exist and are typically caused by warm air masses overriding cold fronts or by conductive cooling of the earth's surface. Solar radiation during the day causes high temperature gradients.

Most rising air masses can be attributed to one of three causal factors: movement of weather fronts, convective processes, and orographic effects. *Frontal precipitation* is caused by the lifting of an air mass by a moving weather front. If a warm front is moving upward over a colder, more dense air mass, precipitation and cloudiness will extend for several hundred miles ahead of the surface front (Figure 2.4A). The slope of a front of this type is small and the rate of ascent of the warm air mass is slow; hence, precipitation is generally light. Should a cold front be moving, it will typically be faster than a warm front. The cold front is steeper, warm air is forced upward more rapidly, and heavier rain amounts may be recorded—especially near the surface front (Figure 2.4B).

Uneven heating of an air mass at the surface, or cooling at the top of an air mass, may cause it to be warmer than the surrounding air. The denser air will flow beneath it, causing the air mass to rise. It will continue to rise, and cool adiabatically, until the temperature is equal to its surroundings. This can cause *convectional* rising. Summer thunderstorms and associated cumulus clouds are a result of this process (Figure 2.4C).

If a moving air mass is forced upward over a mountain range, it must gain altitude. *Orographic* cooling and precipitation may result. The vegetation of the Black Hills of South Dakota is different from the surrounding grassland prairie. There is sufficient precipitation for forest to grow and, from a distance, clouds can often be seen hovering over the hills. This is orographically caused condensation, as air masses are forced upward as they move from west to east over the central Black Hills (Figure 2.4D).

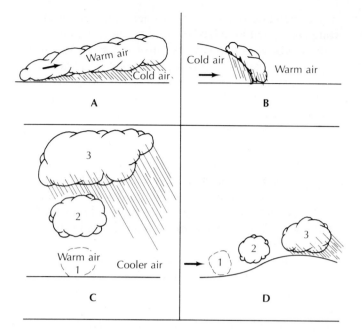

FIGURE 2.4   Precipitation caused by adiabatic lifting of an air mass may be the result of the following activity: **A.** A warm front pushing over a cold front. **B.** A cold front colliding with a warm front. **C.** Uneven heating near the surface causing a warm air mass to rise convectionally. **D.** Orographic lifting caused by prevailing winds blowing over a topographic high.

2.6   MEASUREMENT OF PRECIPITATION

Any open container can be used to catch and measure rainfall. Experiments have shown that the size of the opening has little effect on the catch, except for very small (less than 1-inch-diameter) gages (Huff 1955). The United States Standard Rain Gage has an opening 8 in. (20.3 cm) in diameter, whereas the Canadian standard gage is 9 cm (3.57 in.) in diameter. These are manually read gages; the water is emptied and the gages are read once a day. The catch of precipitation gages is affected by high winds. Such gages generally catch less than the true amount of rainfall because of updrafts around the gage opening. The location of the gage is also critical. In one study, two identical 8-in. gages were placed 10 ft (3 m) apart on a ridge. One gage consistently caught 50% more rainfall than the other (Court 1960). Gages should be placed as close to the ground as possible in order to avoid wind. They should be in the open, away from trees and buildings. Low bushes and shrubs can provide a windbreak. Level ground is best, with the top of the gage horizontal. On steep slopes, it may be desirable to have the orifice opening parallel to the slope.

   The effect of wind is greatest for light rain or snow. Some rain gages are equipped with a shield, or wind deflector, around the opening in order to overcome wind problems. This will improve the catch of snow, but it will still be less than 100% effective in substantial winds.

There are a number of different types of recording rain gages available that can automatically measure or weigh the precipitation. The temporal distribution of precipitation through a day can thus be obtained. Such data are necessary for any studies of precipitation intensity. For remote areas, recording rain gages can be used to record daily precipitation for long time periods. In such circumstances, manual gages could provide only a total rainfall for the period between readings.

In the United States there are some 13,500 precipitation stations, for the most part operated by trained volunteers. Daily records from these weather stations are published monthly on a state-by-state basis in *Climatological Data;* data from recording stations are published in *Hourly Precipitation Data.* Both of these are publications of the U.S. Environmental Data Service. Canada has about 2000 precipitation stations, the data from which are published by the Canadian Atmospheric Environment Service in the *Monthly Record of Observations.*

As every viewer of local television news and weather programs knows, radar can be used to detect areas of precipitation. Rain droplets or snow particles reflect part of the directed radar beam back to the originating station. The amount of reflected energy is directly proportional to the intensity of the precipitation. The radar apparatus measures precipitation in the atmosphere. As the beam is at an oblique angle to the ground, the farther the distance from the station, the greater the altitude of the precipitation being measured. Radar measurements of precipitation may not accurately indicate ground precipitation. Evaporation may occur between the point of measurement and the ground, or wind may cause the precipitation to drift so that it falls to earth at some place other than that indicated by the radar (Stout & Mueller 1968).

Some special radar equipment can convert the intensity of the radar reflection into precipitation rates. The rates are integrated over time to yield a depth of total precipitation over an area. This yields data about precipitation rates between ground stations. The use of radar in combination with conventional ground-station rain gages can give improved areal measurement of precipitation (Wilson 1970; Mylne 1989).

2.7   SNOW MEASUREMENTS

The measurement of snowfall in standard rain gages is subject to error due to turbulence around the gage. The snow that is caught is melted and the **water equivalent** reported. If only an approximation is required, a water content of 10 percent of the snow depth can be assumed. However, as anyone who regularly shovels snow knows, the density of newly fallen snow can vary considerably.

In northern and mountainous climates, the accumulation of snow on the ground is an important hydrologic parameter. In some areas, the runoff of melting snow in the spring is a predominant source of water for reservoirs used for water supply, irrigation, and power generation. A thick accumulation of snow can also mean a high flood potential when snowmelt occurs in the spring. Melting snow also recharges soil moisture and the water table.

Snow surveys are made periodically through the winter to measure the thickness and water content of accumulated snow. A thin-walled tube with a sharp leading edge is driven through the snow to the ground. The tube and the snow contained within are weighed, and the weight of the empty tube subtracted to determine the weight of the snow. A snow survey requires that someone make traverses, stopping at predetermined stations to make measurements. The snow courses should sample representative terrain, vegetative cover, and altitude of the catchment area.

The extent of snow cover can be mapped using satellite photography (Barnes & Bowley 1968). The resulting data, combined with data from snow-course surveys, can be used to determine the total volume of water in the snowpack. Melting of the snowpack can begin only when the temperature of the snow has risen to 0°C. Initial meltwater clings to snow granules by surface tension, so that at least 2 to 8 percent of the snowpack must melt before runoff begins. Energy-balance methods can be used to predict daily snowmelt (Price & Dunn 1976).

In a study of the hydroclimatic elements of the northern Rocky Mountains, it was found that the amount of winter snowpack was the best predictor of annual streamflow (Chagnon, McKee, & Doesken 1991). Hence winter snowpack is a key factor in water basin planning.

2.8   EFFECTIVE DEPTH OF PRECIPITATION

In water budget studies, it is necessary to know the average depth of precipitation over a drainage basin. This may be determined for time periods ranging from the duration of part of a single storm to a year. The data are generally measurements of precipitation and/or equivalent snowfall at a number of points throughout the drainage basin.

A problem is created if data are missing at one or more stations. This can occur as a result of equipment malfunction or operator absence. To solve the problem, three close precipitation stations with full records that are evenly spaced around the station with a missing record are used. The following equation yields an estimate of the missing data at Station Z. The mean annual precipitation (N) at Station Z and the three index stations, A, B, and C, as well as the actual precipitation (P) at the index stations for the time period over which data are missing are needed:

$$P_Z = \frac{1}{3}\left[\frac{N_Z}{N_A}P_A + \frac{N_Z}{N_B}P_B + \frac{N_Z}{N_C}P_C\right] \qquad (2\text{--}2)$$

If the rain-gage network is of uniform density, then a simple arithmetic average of the point-rainfall data for each station is sufficient to determine the **effective uniform depth (EUD)** of precipitation over the drainage basin (Figure 2.5).

If the rain-gage network is not uniform, then some adjustment is necessary. The most accurate method, excluding use of radar data, is to draw a

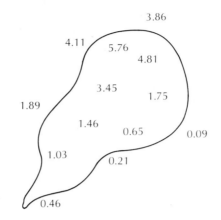

FIGURE 2.5 Precipitation-gage network over a drainage basin. Precipitation amounts are given in inches. Station locations are at decimal points.

precipitation contour map with lines of equal rainfall **(isohyets)**. In drawing the isohyets, such factors as known influence of topography on precipitation can be taken into account. Simple linear interpolation between precipitation stations can also be used. The area bounded by adjacent isohyets is measured with a planimeter, and the average depth of precipitation over the area is the mean of the bounding isohyets. The effective uniform depth of precipitation is the weighted average based on the relative size of each isohyetal area (Figure 2.6). The drawback of the isohyetal method is that the isohyets must be redrawn and the areas remeasured for each analysis.

The **Thiessen method** to adjust for nonuniform gage distribution uses a weighing factor for each rain gage. The factor is based on the size of the area

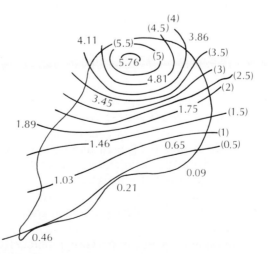

FIGURE 2.6 Isohyetal lines for the precipitation-gage network of Figure 2.5. The isohyets show contours of equal rainfall depth with a contour interval of 0.5 in. The contours are based on simple linear interpolation.

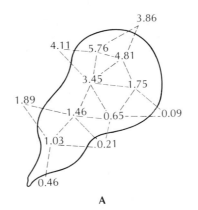

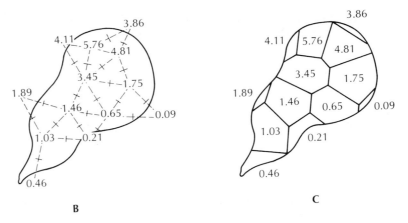

FIGURE 2.7   Thiessen polygons based on the rain-gage network of Figure 2.5. **A.** The stations are connected with lines. **B.** The perpendicular bisector of each line is found. **C.** The bisectors are extended to form the polygons around each station.

within the drainage basin that is closest to a given rain gage. These areas are irregular polygons. The method of constructing them can be described rather easily; however, it takes a bit of practice to master the technique. The rain-gage network is drawn on a map of the drainage basin. Adjacent stations are connected by a network of lines (Figure 2.7A). Should there be doubt as to which stations to connect, lines should be between the closest stations. A perpendicular line is then drawn at the midpoint of each line connecting two stations (Figure 2.7B), and extensions of the perpendicular bisectors are used to draw polygons around each station (Figure 2.7C). It is best to start with a centrally located station and then expand the polygonal network outward. The area of each polygon is measured, and a weighted average for each station's precipitation is used to find the EUD.

In mountainous areas, orographic effects can create vastly different microclimates over small distances. Significant precipitation can fall on one side of a ridge but little on the other. In such regions the Thiessen method and contouring by linear interpolation can yield erroneous results. Detailed study of the vegetation can identify wet and dry slopes. This information, in conjunction with topographic maps, can be used to make interpreted contour maps with isohyetal lines reflecting the presence of wet and dry slopes.

EXAMPLE PROBLEM Determine the effective uniform depth of precipitation using the arithmetic mean, isohyetal, and Thiessen methods.

Arithmetic Mean Method

Figure 2.5 shows a drainage basin with seven stations in its boundaries. An additional six stations are located outside the drainage divide. In the arithmetic mean method, only the gages inside the drainage basin boundary are considered.

$$\text{Arithmetic mean} = \frac{1.03 + 0.65 + 1.46 + 1.75 + 4.81 + 3.45 + 5.76}{7}$$
$$= 2.70 \text{ in.}$$

Isohyetal Method

The first step is to draw lines of equal precipitation (isohyets) on the drainage basin map. Isohyets are usually whole numbers or decimals (every 0.1 in., every 0.5 in., every 1 mm, etc.). The following rules apply:

1. Isohyets never cross.
2. Isohyets never split.
3. Isohyets never meet.
4. A station that does not fall on an isohyet will be between two isohyets. The isohyets will both be equal (either larger or smaller than the station value) or one will be larger and one smaller.
5. Adjacent isohyets must be equal or only one contour interval different in value.
6. Isohyets should be scaled between stations using linear interpolation.

Figure 2.6 shows the isohyetal map of the problem area.

The area between adjacent isohyets is determined by use of a planimeter. The equivalent uniform depth of precipitation between isohyets is usually assumed to be equal to the median value of the two isohyets. For example, the EUD between a 1-in. isohyet and a 2-in. isohyet is 1.5 in. For areas enclosed by a single isohyet, judgment should be used to estimate the equivalent uniform depth. The weighted average precipitation is based on the equivalent uniform depth of precipitation between adjacent isohyets and their areas.

A	B	C Net Area (sq mi)	D Percent of Total Area	E Weighted Precipitation (in.) (B × D)
Isohyet (in.)	Estimated EUD			
5.5+	5.6	1.1	0.8	0.045
5.0–5.5	5.25	7.6	5.3	0.278
4.5–5.0	4.75	10.6	7.4	0.352
4.0–4.5	4.25	9.5	6.7	0.285
3.5–4.0	3.75	8.6	6.0	0.225
3.0–3.5	3.25	8.3	5.8	0.189
2.5–3.0	2.75	10.7	7.5	0.206
2.0–2.5	2.25	12.3	8.6	0.194
1.5–2.0	1.75	15.1	10.6	0.186
1.0–1.5	1.25	23.8	16.7	0.209
0.5–1.0	0.75	31.2	21.8	0.164
<0.5	0.3	4.0	2.8	0.008
TOTAL		142.8 sq mi		2.34 in. NET EUD

Thiessen Method

The Thiessen method provides for the nonuniform distribution of gages by determining a weighting factor for each gage. A weighted mean of the precipitation values can then be computed. Thiessen polygons for the example problem are shown in Figure 2.7C. The area of each polygon is determined by a planimeter.

A Station Precipitation (in.)	B Net Area (sq mi)	C Percent of Total Area	D Weighted Precipitation (in.) (A × C)
5.76	16.9	11.9	0.686
4.81	16.1	11.4	0.546
4.11	3.4	2.4	0.099
3.86	1.6	1.1	0.044
3.45	19.3	13.6	0.470
1.89	2.5	1.8	0.033
1.75	12.0	8.5	0.148
1.46	19.8	14.0	0.204
1.03	18.0	12.7	0.131
0.65	17.0	12.0	0.078
0.46	6.0	4.2	0.019
0.21	7.2	5.1	0.011
0.09	2.0	1.4	0.001
TOTAL	141.8 sq mi		2.47 in. NET EUD

A weighted mean of the EUD is found, based on the depth of precipitation and the area of the polygon within the basin boundary.

NOTATION

E_T Evapotranspiration

D Excess moisture

I Irrigation water

P Precipitation

S_f Volume of final soil moisture

S_i Volume of initial soil moisture

PROBLEMS

Answers to odd-numbered problems will appear at the end of the book.

1. A swimming pool has a length of 50 m and a width of 25 m. During July the Class A land pan evaporation is 17.0 cm. If the pan coefficient is 0.80, what is the monthly water loss from the pool due to evaporation?

2. A reservoir has a surface area of 690 ac. The table on page 46 shows the monthly inflow of surface water, outflow as releases from the reservoir via the spillway, direct precipitation into the reservoir, and evaporation from the reservoir. The reservoir elevation was 701.0 ft on January 1. Compute the reservoir elevation at the end of each month.

3. Figure 2.8 is a map of a drainage basin and the rainfall amounts during a storm at a number of precipitation stations both within and outside the drainage basin. Make a Thiessen network drawing for the drainage basin. The exact station location is the decimal point in the rainfall amount. The relative size of the area associated with each Thiessen polygon can be measured with a planimeter or estimated by tracing the Thiessen network on cross-section paper and counting the number of squares in each polygon. Estimate the effective uniform depth of precipitation over the drainage basin.

4. Make a copy of the drainage basin in Figure 2.8. Contour the precipitation data to create isohyetal lines and determine the effective uniform depth of precipitation.

5. A pond has a surface area of 22 ac. If the mean daily air temperature is 70°F, the mean daily dew point temperature is 55°F, the solar radiation is 450 langleys, and the daily wind movement is 70 mi, what is the daily lake evaporation?

6. Make a cross-sectional plot of saturation humidity as a function of temperature using the data in Table 2.1. Label the area of the graph that is undersaturated and the area that is supersaturated.

7. Consider an air mass that has an absolute humidity of 8 g/m^3 at a temperature of 20°C. Using the graph that you created for Problem 2.6 find:

    A. The dew point.

    B. The relative humidity.

8. A reservoir with a surface area of 43 mi^2 lies in a desert valley with a mean daily air temperature of 80°F, mean daily dew point temperature of 45°F, solar radiation of 650 langleys/day, and a mean wind movement of 180 mi/day.

    A. Compute the mean daily lake evaporation in inches.

B. What is the volume of water, in cubic feet, lost per day from the lake due to evaporation? (Be sure to convert the area in square miles to square feet.)

Month	Inflow (acre-feet)	Outflow (acre-feet)	Precipitation		Evaporation		Net Change		Elevation (feet)
			(inches)	(acre-feet)	(inches)	(acre-feet)	(acre-feet)	(feet)	
Dec									701.0
Jan	1732	175	2.75	158	1.05	60	+1655	+2.4	703.4
Feb	1755	190	3.05		1.55				
Mar	872	232	3.76		2.05				
Apr	955	375	4.11		2.80				
May	708	525	2.70		3.75				
Jun	312	955	1.05		4.25				
Jul	102	1720	.75		5.15				
Aug	37	2250	1.25		5.76				
Sep	175	1575	1.55		4.92				
Oct	575	550	3.79		3.02				
Nov	1250	175	4.53		1.75				
Dec	1875	125	5.01		0.60				

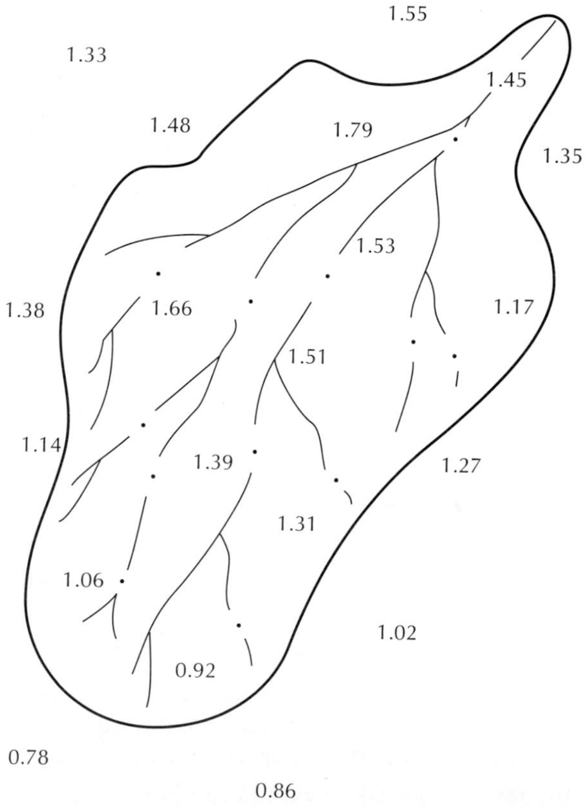

FIGURE 2.8 Base map for Problems 2.3 and 2.4.

3 Runoff and Streamflow

The land had great depth of soil and gathered the water into itself and stored it up into the soil . . . as though it were a sort of natural water jar; it drew down into the natural hollow the water which it had absorbed from the high ground and so afforded in all districts of the country liberal sources of springs and rivers. . . .

Critias, Plato (427–346 B.C.)

3.1   EVENTS DURING PRECIPITATION

During a precipitation event, some of the rainfall is intercepted by vegetation before it reaches the ground (**interception**). This may later fall to the ground or evaporate. In a heavily forested area, most of the precipitation is caught by leaves and twigs. For a period at the start of a summer thunderstorm, no raindrops reach the forest floor, although drops can be heard striking the leaves overhead. When the storage capacity of the leaf surfaces is exhausted, water will run down tree trunks and drip downward (**stem flow**) (Brown & Barker 1970; Rogerson & Byrnes 1968; Helvey 1967). The amount of water intercepted by dense forests ranges from 8% to 35% of total annual precipitation (Dunne & Leopold 1978). In a mixed hardwood forest in the northeastern United States, it averaged 20 percent in the summer and winter seasons (Trimble & Weitzman 1954). Although evaporation of intercepted water reduces the net transpiration by the plants, in some cases most of the evaporated water is simply lost. One study concluded that only about 10% of the intercepted water actually reduced evapotranspiration (Thorud 1967).

The rate of interception is greatest at the beginning of a precipitation event and declines exponentially with time. If the rain is short-lived and light, a large percentage of the precipitation may be intercepted. If it is heavy and long-lived, only a small percentage may be intercepted.

Rainfall reaching the land surface can *infiltrate* into pervious soil. Soil has a finite capacity to absorb water. The **infiltration capacity** varies not only from soil to soil but is also different for dry versus moist conditions in the same soil.

If a soil is initially dry, the infiltration capacity is high. Surface effects between the soil particles and the water exert a tension that draws the moisture downward into the soil through labyrinthine capillary passages. As the capillary

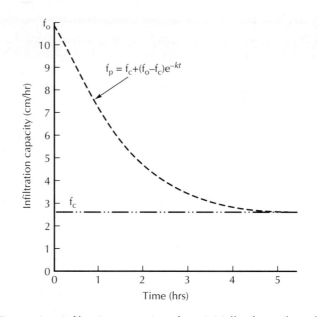

FIGURE 3.1 Decreasing infiltration capacity of an initially dry soil as the soil-moisture content of the surface layer increases.

forces diminish with increased soil-moisture content, the infiltration capacity drops (Figure 3.1). In addition, colloidal particles in the soil swell as the moisture content increases. Eventually, the infiltration capacity reaches a more or less constant, or equilibrium, value.

The infiltration capacity curve can be described by Equation 3–1 (Horton 1933, 1940):

$$f_p = f_c + (f_o - f_c)e^{-kt} \tag{3–1}$$

where

f_p is the infiltration capacity (L/T; ft/s or m/s) at time t (T; s)

f_c is the equilibrium infiltration capacity (L/T; ft/s or m/s)

f_o is the initial infiltration capacity (L/T; ft/s or m/s)

k is a constant representing the rate of decreased infiltration capacity ($1/T$; 1/s)

If the precipitation rate is lower than the equilibrium infiltration capacity, then all the precipitation reaching the land surface will infiltrate (Figure 3.2A). If the precipitation rate is greater than the equilibrium infiltration capacity but less than the initial infiltration capacity, at the beginning all the precipitation will infiltrate, but when the infiltration rate drops below the precipitation rate, some of the precipitation will remain on the ground surface (Figure 3.2B). Finally, if the precipitation rate is greater than the initial infiltration capacity, some water will immediately remain on the land surface (Figure 3.2C).

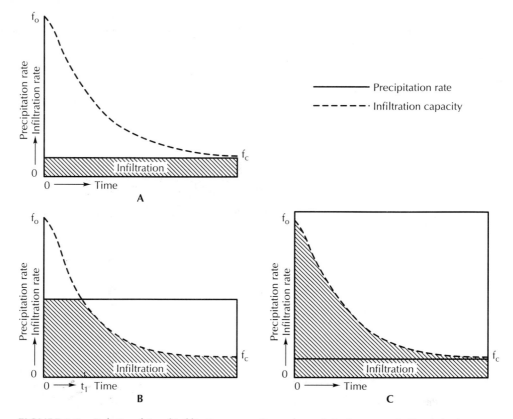

FIGURE 3.2 Relationship of infiltration capacity and precipitation rate. **A.** Precipitation rate less than equilibrium infiltration capacity. **B.** Precipitation rate greater than equilibrium infiltration capacity but less than initial infiltration capacity. **C.** Precipitation rate greater than initial infiltration capacity.

Conditions that encourage a high infiltration rate include coarse soils, well-vegetated land, low soil moisture, and a topsoil layer made porous by insects and other burrowing animals, in addition to land-use practices that avoid soil compaction. Once the final infiltration rate is reached, the depth of ponded water also promotes high infiltration.

The water reaching the ground can infiltrate into the soil, form puddles, or flow as a thin sheet of water across the land surface. Hydrologists refer to the water trapped in puddles as *depression storage*. It ultimately evaporates or infiltrates.

The overland flow process, sometimes called *Horton overland flow* after Robert Horton (Horton 1933, 1940), occurs only when the precipitation rate exceeds the infiltration capacity. In areas in which soils have a high infiltration capacity, this process may occur only during very intense storms or when the soil is saturated or frozen. In order for overland flow to occur, the infiltration capacity of the soil must first be exceeded; then the depression storage must be filled (Figure 3.3).

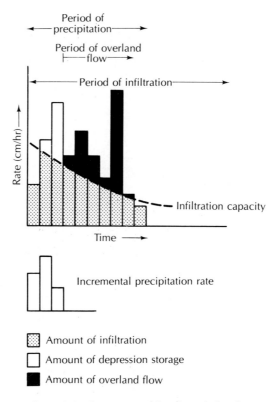

FIGURE 3.3 Incremental precipitation rate and its dissociation into amounts of infiltration, depression storage, and overland flow. Infiltration begins when the precipitation does. Overland flow does not begin until the depression storage is exhausted. Overland flow continues past the termination of precipitation. Infiltration will continue as long as there is any water in depression storage—usually past the period of overland flow.

If the unsaturated zone is uniformly permeable, most of the infiltrated water percolates vertically. Should layers of soil with a lower vertical hydraulic conductivity occur beneath the surface, then infiltrated water may move horizontally in the unsaturated zone. This *interflow* may be substantial in some drainage basins and contribute significantly to total streamflow. Thin permeable soil overlying fractured bedrock of low permeability would provide a geologic condition contributing to significant interflow (Figure 3.4).

Water will fall directly onto the surfaces of lakes and reservoirs during the period of precipitation. This amount might not be considerable for streams, but for lakes and reservoirs it could be. Lake Michigan and its associated water bodies have a surface area of 22,300 mi^2. The land area of the surrounding drainage basin is 45,000 mi^2 (International Great Lakes Levels Board 1973). Assuming equal distribution of precipitation over the entire Lake Michigan basin, about one-third falls as **direct precipitation** on a water body.

Infiltrated water that reaches the water table becomes stored in the ground-water reservoir. This is not static storage, as ground water is in constant

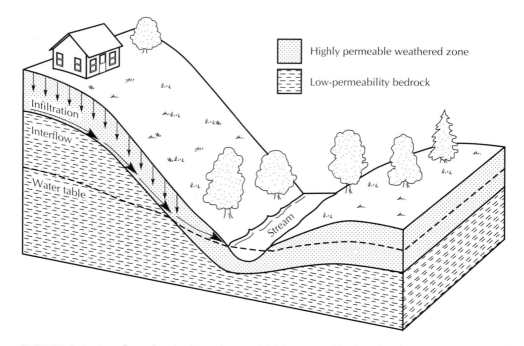

Highly permeable weathered zone

Low-permeability bedrock

FIGURE 3.4 Interflow developing where a highly permeable but thin layer of weathered rock overlies a bedrock unit of lower permeability.

movement. While freshly infiltrated precipitation is entering the ground-water reservoir, other ground water, known as *baseflow,* is discharging into a stream. If infiltration causes the water table to rise, ground-water discharge into nearby streams will also increase. For baseflow streams, the amount of ground-water discharge is directly proportional to the hydraulic gradient toward the stream (Figure 3.5).

The runoff cycle in which so much emphasis is placed on Horton overland flow has been criticized on several fronts (Chorley 1978). Horton overland flow is rarely observed in the field, except after very heavy precipitation events. This is especially true if the ground is covered with vegetation or humus, such as leaf litter (Kirkby & Chorley 1967). Horton overland flow appears to be more common in arid regions or areas in which the soil has been compacted by vehicles, animals, etc. (Dunne 1978). Overland runoff can also occur when precipitation falls on soils that are saturated.

Water that infiltrates into the soil on a slope can move downslope as lateral unsaturated flow in the soil zone. This has been called **throughflow** (Kirkby & Chorley 1967). The difference between throughflow and interflow is that throughflow emerges as seepage at the foot of the slope rather than entering a stream, as does interflow. Thus, the throughflow appears as overland flow before entering a stream channel. This overland flow is called **return flow** (Dunne & Black 1970) to distinguish it from Horton overland flow.

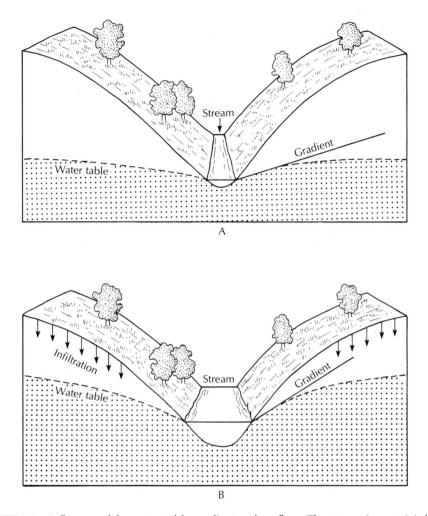

FIGURE 3.5 Influence of the water-table gradient on baseflow. The stream in part A is being fed by ground water with a low hydraulic gradient. A gentle rain does not produce overland flow, but infiltration raises the water table. The increased hydraulic gradient of part B causes more baseflow to the stream, which is now deeper and has a greater discharge.

A more comprehensive concept of the hydrologic cycle has been proposed by Dunne (1978). In arid to subhumid climates where there is thin vegetation, Horton overland flow is the main contributor to the storm peak and comprises most of the streamflow. In humid climates, Horton overland flow is not significant, but interflow, return flow, and direct precipitation on the channel are important. Where there are thin soils and gentle, concave slopes, direct precipitation and return flow are more important than interflow. On steep, straight slopes, interflow becomes much more important, although return flow and direct precipitation still cause the peaks.

3.2   HYDROGRAPH SEPARATION

A stream **hydrograph** shows the discharge of a river at a single location as a function of time. While the total streamflow shown on the hydrograph gives no indication of its origin, it is possible to break down the hydrograph into components such as overland flow, baseflow, interflow, and direct precipitation. The model presented in this section is based on the Horton runoff cycle; it would be most useful for arid-zone hydrology.

3.2.1   Baseflow Recessions

The hydrograph of a stream during a period with no excess precipitation will decay, following an exponential curve. The discharge is composed entirely of ground-water contributions. As the stream drains water from the ground-water reservoir, the water table falls, leaving less and less ground water to feed the stream. If there were no replenishment of the ground-water reservoir, baseflow to the stream would become zero. Figure 3.6 shows a **baseflow recession hydrograph** for a stream in a climate with a dry summer season.

The **baseflow recession** for a drainage basin is a hydromorphic characteristic. It is a function of the overall topography, drainage pattern, soils, and geology of the watershed. Figure 3.7 illustrates this by showing the annual summer recession of a river for six consecutive years. The start of the baseflow

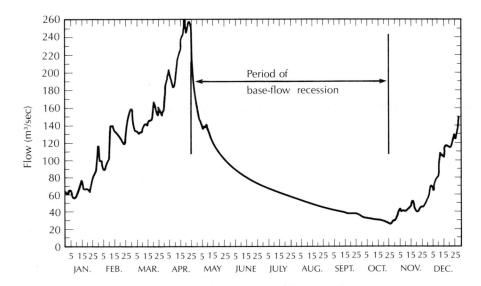

FIGURE 3.6 Typical annual hydrograph for a river with a long dry summer season: Lualaba River, Central Africa. Source: C. O. Wisler & E. F. Brater, eds., *Hydrology,* 2nd ed. (New York: John Wiley, 1959). Used with permission.

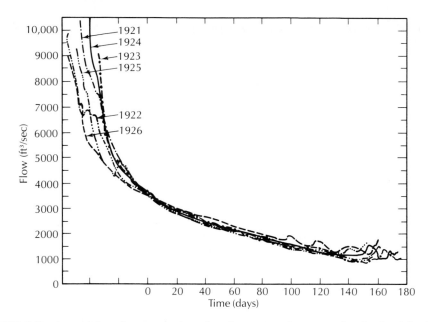

FIGURE 3.7 Annual baseflow recessions for six consecutive years for the Lualaba River, Central Africa. Source: C. O. Wisler & E. F. Brater, eds., *Hydrology*, 2nd ed. (New York: John Wiley, 1959). Used with permission.

recession was considered to be the day when the annual discharge dropped below 3500 ft³/s. The recession is similar from year to year. The baseflow of the stream decreases during a dry period because as ground water drains into the stream, the water table falls. A lower water table means that the rate at which ground water seeps into the stream declines. Picture in your mind a bucket with a hole near the bottom. As the water drains from the bucket, the water level (water table) falls and the stream of water draining from the bucket (baseflow to streams) declines in volume. The stream of water draining (baseflow) will not increase until the water in the bucket is replenished (recharge) and the water level (water table) rises.

The baseflow recession equation is

$$Q = Q_0 e^{-at} \tag{3-2}$$

where

Q is the flow at some time t after the recession started (L^3/T; ft³/s or m³/s)

Q_0 is the flow at the start of the recession (L^3/T; ft³/s or m³/s)

a is a recession constant for the basin ($1/T$; d^{-1})

t is the time since the recession began (T; d)

EXAMPLE **PROBLEM**	**Part A:** Find the recession constant for the basin of Figure 3.7.

If

$$Q = Q_0 e^{-at}$$

then

$$e^{-at} = Q/Q_0$$
$$-at = \ln Q/Q_0$$
$$a = -(1/t) \ln Q/Q_0$$

From Figure 3.7, $Q_0 = 3500$ ft^3/s. After 100 d, $Q = 1500$ ft^3/s.

$$a = -\frac{1}{t} \ln \frac{Q}{Q_0}$$

$$= -\frac{1}{100 \text{ d}} \ln \frac{1500 \text{ ft}^3/\text{s}}{3500 \text{ ft}^3/\text{s}}$$

$$= -0.01 \text{ d}^{-1} \ln 0.4286$$

$$= -0.01 \text{ d}^{-1} \times (-0.847)$$

$$= 8.47 \times 10^{-3} \text{ d}^{-1}$$

Part B: What would the baseflow be after 40 d of recession?

$$Q = Q_0 e^{-at}$$

$$= 3500 \text{ ft}^3/\text{s} \exp (-8.47 \times 10^{-3} \text{ d}^{-1} \times 40 \text{ d})$$

$$= 3500 \text{ ft}^3/\text{s} \times 0.713$$

$$= 2500 \text{ ft}^3/\text{s}$$

3.2.2 Storm Hydrograph

Although the baseflow component of a stream is somewhat constant, the total discharge of the stream may fluctuate greatly through the year. The difference is due to the episodic nature of precipitation events that contribute overland flow, interflow, and direct precipitation. For most drainage basins, direct precipitation adds only a modest amount of water to the stream. Interflow is a factor that can be highly variable, depending upon the geology of the drainage basin. A deep, sandy soil might not induce any interflow; on the other hand, a lava landscape covered by loose rubble might have no overland flow but great amounts of interflow at the base of the rubble where it overlies a hard, low-permeability lava flow. Steeply sloping land also promotes interflow. The most consistent factor in the storm hydrograph is overland flow. Figure 3.8 shows a hypothetical storm hydrograph broken down into overland flow, interflow, direct precipitation, and

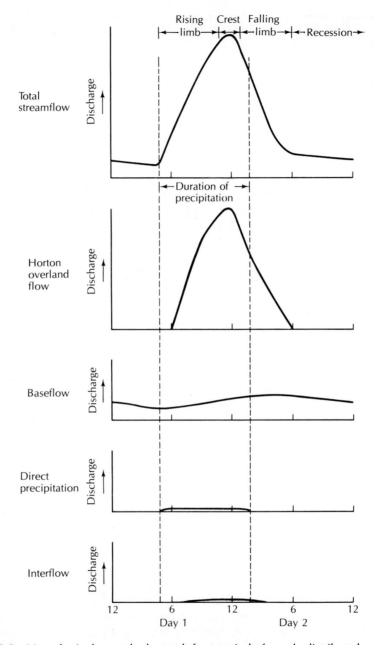

FIGURE 3.8 Hypothetical storm hydrograph for a period of evenly distributed precipitation, separated into Horton overland flow, direct precipitation, and interflow.

baseflow recession. The baseflow component is given for a stream that continues to receive ground-water discharge through the duration of the overland-flow peak.

One of the tasks in analyzing a storm hydrograph is to separate the overland-flow component from the baseflow. Generally, it is first assumed that both the direct precipitation and the interflow components are inconsequential; however, the hydrogeologist should be aware of the general geology and surface slope of the drainage area before assuming the inconsequence of the latter component. The overland flow is assumed to end some fixed time after the storm peak. As a general rule of thumb, this can be approximated by the formula (Linsley, Kohler, & Paulhus 1975)

$$D = A^{0.2} \qquad \textbf{(3-3A)}$$

where

> D is the number of days between the storm peak and the end of overland flow
>
> A is the drainage basin area in square miles

or

$$D = 0.827A^{0.2} \qquad \textbf{(3-3B)}$$

where A is the drainage basin area in square kilometers. Note that Equations 3-3A and 3-3B are empirical relationships and are not dimensionally correct.

The exponential constant of 0.2 is somewhat arbitrary; thus, blind use of the preceding formula could result in error. The value will depend upon many drainage basin characteristics, such as mean slope, vegetation, drainage density, roughness, etc.

The baseflow recession that existed prior to the storm peak is extended until it is approximately under the storm peak. It is then drawn so as to rise to meet the stream hydrograph at a point D days after the peak. In Figure 3.9, a

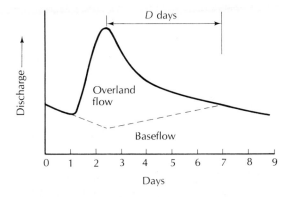

FIGURE 3.9 Hydrograph separation into overland-flow component and baseflow component for a stream receiving Horton overland flow.

storm hydrograph for a drainage basin of 2100 mi² has been separated. For the given basin, D is equal to 4.6 d.

3.2.3 Gaining and Losing Streams

The typical stream of a humid region receives ground-water discharge; therefore, as one goes downstream the baseflow increases, even if no tributaries enter. This is a **gaining,** or **effluent,** stream. The water table slopes toward the stream, so that the hydraulic gradient of the aquifer is toward the stream. Figure 3.10A shows a cross section through a gaining reach of a stream.

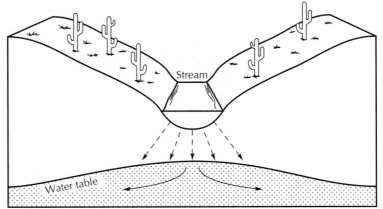

A

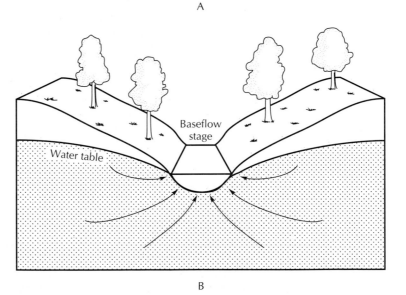

B

FIGURE 3.10   A. Cross section of a gaining stream, which is typical of humid regions, where ground water recharges streams. **B.** Cross section of a losing stream, which is typical of arid regions, where streams can recharge ground water.

In arid regions, many rivers are fed by overland flow, interflow, and baseflow at high altitudes. As they wind their way to a lower elevation, the local precipitation amounts decrease; consequently, there is less infiltration and a lower water table. There may also be a dramatic change in the depth to ground water when a stream draining a high-altitude basin of lower-permeability material flows out onto coarse alluvial materials. For whatever reason, if the bottom of the stream channel is higher than the local water table, water may drain from the stream into the ground (Figure 3.10B). As one goes downstream, less and less water will be found in the channel. The stream is **losing,** or *influent*. The rate of water loss is a function of the depth of water and the hydraulic conductivity of the underlying alluvium. Fine-grained deposits on the channel bottom will retard the rate of loss to the ground water.

A stream that is normally a gaining stream during baseflow recessions may temporarily become a losing stream during floods. If the flood-crest depth in the channel is greater than the local water-table elevation, the hydraulic gradient in the aquifer next to the stream is reversed. Water flows from the stream into the ground (Figure 3.11). The result is a temporary storage of flood water in the aquifer next to the stream. When the flood crest passes, the hydraulic gradient again reverses, and the stream is once again gaining (Figure 3.12).

Tabidian, Pederson, and Tabidian (1992) studied the impact on ground-water levels of a 75- to 100-y flood event of the Big Blue River in Nebraska. The river is connected to an aquifer consisting of alluvium and glacial deposits, which vary from gravel to clay in nature. The river is normally gaining, but during the flood the gradient became reversed and water drained from the river into the aquifer. In some areas ground-water levels rose more than 10 ft in less than a week, and rises in ground-water level were recorded as much as 2 mi from the river. Due to the heterogeneous nature of the aquifer, water-level rises were not uniform along the river, nor were they symmetrical across the river.

Heavy ground-water pumping near a stream can lower the water table to an elevation below the level of the stream bottom. The reach of the stream affected by the lowered water table will become a losing stream, while upstream

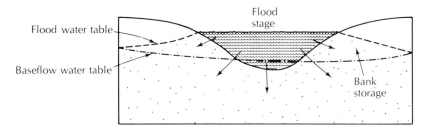

FIGURE 3.11 A stream that is gaining during low-flow periods can temporarily become a losing stream during flood stage.

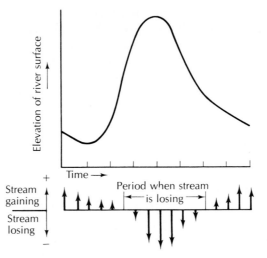

FIGURE 3.12 Effect of flood state on the ground-water regime adjacent to the river. As the flood peak passes, the normal direction of ground-water flow into the stream is reversed.

and downstream reaches can still be gaining. Figure 3.13 is an idealization of this phenomenon, based on the behavior of the well field along the Fenton River of the University of Connecticut at Storrs.

Conjunctive use of ground and surface water is discussed in Section 12.10. Legal issues have arisen when ground-water pumping has depleted stream-

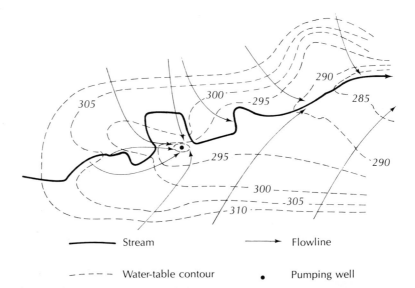

FIGURE 3.13 Induced stream-bed infiltration caused by a pumping well.
Source: P. Rahn, *Ground Water* 6, no. 3 (1968):21–32.

flow in states where the water in the stream has been appropriated to other uses (Section 12.6).

3.3   RAINFALL-RUNOFF RELATIONSHIPS

One of the basic problems of hydrology is predicting the amount of runoff that will occur from a given storm. Structures that carry runoff water are designed on the basis of the peak runoff rate. A maximum expected rainfall rate is determined based on climatic records. Structures such as storm sewers and highway culverts are then designed to carry this expected runoff. There are a number of equations that have been developed to make this prediction. The most simple is the *rational equation*.

The rational equation states that if it rains long enough, the peak discharge from the drainage basin will be the average rate of rainfall times the drainage basin area, reduced by a factor to account for infiltration. The **time of concentration** is the length of time necessary for water to flow from the most distant part of the watershed to the point of discharge. If the period of precipitation exceeds the time of concentration, then the rational equation will apply. The velocity of the water in the stream channel can be calculated with the Manning equation (Section 3.7). The time of concentration is then the length of the stream channel divided by the water velocity, plus the estimated time for overland flow to reach the channel.

The rational equation assumes that the rainfall rate is constant and that the rate of infiltration is constant. The rational method is of greatest validity when used in analysis of small drainage basins of 200 ac (100 ha) or less. The rational equation is

$$Q = CIA \tag{3-4}$$

where

Q is the peak runoff rate (L^3/T; ft^3/s or m^3/s)

I is the average rainfall intensity (L/T; ft/s or m/s)

A is the drainage area (L^2; ft^2 or m^2)

C is a runoff coefficient from Table 3.1 (dimensionless)

In Table 3.1 values of C are given for many different land uses to account for differing rates of infiltration. The more urbanized the land use, the greater the percentage of the impervious surface and the greater the percentage of runoff. For each land use, a range of the value of C is given. The lower number is used for storms of low intensity; storms of greater intensity will have proportionally more runoff, justifying the use of a higher C factor.

TABLE 3.1   Runoff Factor for Rational Equation

Description of Area	C
Business	
Downtown	0.70–0.95
Neighborhood	0.50–0.70
Residential	
Single-family	0.30–0.50
Multiunits, detached	0.40–0.60
Multiunits, attached	0.60–0.75
Residential suburban	0.25–0.40
Apartment	0.50–0.70
Industrial	
Light	0.50–0.80
Heavy	0.60–0.90
Parks, cemeteries	0.10–0.25
Playgrounds	0.20–0.35
Railroad yard	0.20–0.35
Unimproved	0.10–0.30
Character of surface	
Pavement	
Asphalt and concrete	0.70–0.95
Brick	0.70–0.85
Roofs	0.75–0.95
Lawns, sandy soil	
Flat, up to 2% grade	0.05–0.10
Average, 2%–7% grade	0.10–0.15
Steep, over 7%	0.15–0.20
Lawns, heavy soil	
Flat, up to 2% grade	0.13–0.17
Average, 2%–7% grade	0.18–0.22
Steep, over 7%	0.25–0.35

Source: American Society of Civil Engineers, "Design and Construction of Sanitary and Storm Sewers," *Manuals and Reports of Engineering Practice No. 37*, 1970.

3.4   DURATION CURVES

For design or regulatory purposes, it may be necessary to know how often the discharge of a stream may be less than or greater than a given value. As an example, if a river is considered for a water-supply source, it is necessary to know how much water can be obtained. The average flow is not a particularly useful value, in that possibly 50% of the time the river would carry less than the average discharge. Depending upon the available storage and other sources of supply, some flow duration is selected as the reliable flow. For example, the 90% duration is the flow that will be equaled or exceeded 90% of the time.

Duration curves are generally constructed for either daily flow or annual flow, although other time periods could also be considered. Data are ranked from greatest to least flow values. They are then assigned a serial rank, *m*, starting with 1 for the greatest flow and going to *n*, the number of data values. If two or more data values are equal, each should receive a different serial rank. The probability, *P*, as a percentage, that a given flow will be equaled or exceeded may be found by the equation

$$P = 100 \frac{m}{n + 1} \qquad (3\text{--}5)$$

A plot of *P* as a function of flow will yield a duration curve showing the percentage of time a given flow is equaled or exceeded. The curve can be plotted on a type of graph paper known as probability paper. This paper is constructed with a special abscissa and ordinates that may be either arithmetic or logarithmic.

Figure 3.14 shows duration curves of daily flow for three rivers in Wisconsin. In order to compare the three directly, the discharge has been

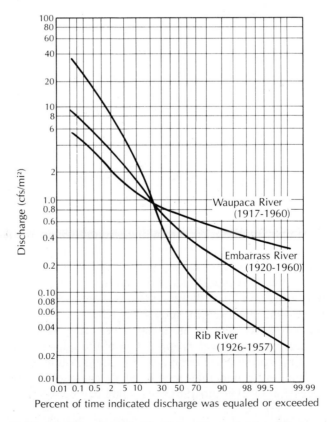

FIGURE 3.14 Daily duration curves for the three streams having different runoff characteristics owing to the differing geology of the drainage basins.
Source: U.S. Geological Survey.

computed as cubic feet per second per square mile ($ft^3/s/mi^2$) of drainage basin. This makes the flow independent of the size of the drainage basin. The three streams are all in central Wisconsin and the annual runoff (precipitation less evaporation) is about 11 in. (28 cm) per year for all three. Examination of Figure 3.14 reveals a great variability in the distribution of this annual runoff.

The Rib River has high flood values; 1.0% of the time flow equals or exceeds 12 $ft^3/s/mi^2$. On the other hand, the 1% value for the Waupaca River is 2.7 $ft^3/s/mi^2$, with the Embarrass River intermediate. All three rivers have the same 20% flow value: 0.9 $ft^3/s/mi^2$. Whereas the Rib River had the greatest flood flows, it has the smallest low flows. One percent of the time the Rib River discharge is less than 0.04 $ft^3/s/mi^2$. (The graph shows this as 99% of the time the flow equals or exceeds 0.04 $ft^3/s/mi^2$.) The Waupaca River has low flows an entire magnitude greater; the flow is less than 0.39 $ft^3/s/mi^2$ for 1% of the time. Again, the Embarrass River falls about evenly between the two.

This distribution of runoff is caused by the geology of the drainage basins. The Rib River is located in an area of crystalline bedrock, which has a very low hydraulic conductivity. Part of the drainage basin is in the driftless area, where superficial glacial deposits are lacking, overland flow and return flow are high, and baseflow is scant. The soils are thin, with little water-retaining capacity. The drainage basin of the Waupaca River has thick deposits of unconsolidated sand. Most of the potential overland flow is absorbed by the sand; hence, there are small flood peaks. This water can drain slowly and provide high baseflows. The Embarrass River has thick deposits of glacial drift—but it is till and lake clay—so the hydraulic response of the watershed is intermediate.

3.5   DETERMINING GROUND-WATER RECHARGE FROM BASEFLOW

A simple method of estimating ground-water recharge in a basin has been developed. It utilizes stream hydrographs from two or more consecutive years. The baseflow-recession equation (3–2) indicates that Q_0 varies logarithmically with time, t. A plot of a stream hydrograph with time on an arithmetic scale and discharge on a logarithmic scale will therefore yield a straight line for the baseflow recession. Figure 3.15 shows hypothetical stream hydrographs. The baseflow recessions are shown as dashed lines; they were considered to start when the summer stream level dropped below the adjacent water table and to end when the first spring flood occurred. The total potential ground-water discharge is the volume of water that would be discharged during a complete ground-water recession (Meyboom 1961). Its value can be found from:

$$V_{tp} = \frac{Q_0 t_1}{2.3} \qquad \textbf{(3–6)}$$

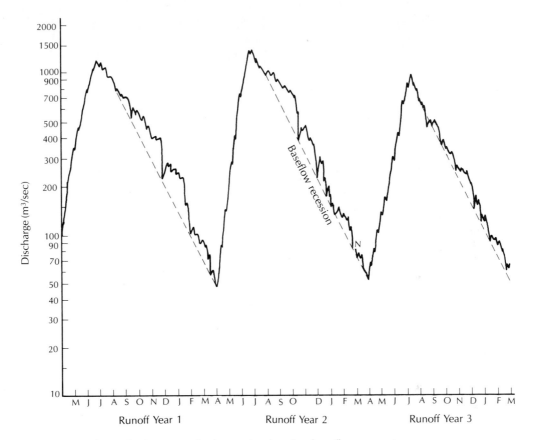

FIGURE 3.15 Semilogarithmic stream hydrographs showing baseflow recessions.

where

V_{tp} is volume of the total potential ground-water discharge (L^3; ft^3 or m^3)

Q_0 is the baseflow at the start of the recession (L^3/T; ft^3/s or m^3/s)

t_1 is the time that it takes the baseflow to go from Q_0 to $0.1Q_0$ (T; s)

If one determines the remaining potential ground-water discharge at the end of a recession and then the total potential ground-water discharge at the beginning of the next recession, the difference between the two is the ground-water recharge that has taken place between recessions. The amount of potential baseflow, V_t (L^3; ft^3 or m^3), remaining some time, t (T; s) after the start of a baseflow recession is given by

$$V_t = \frac{V_{tp}}{10^{(t/t_1)}} \qquad\qquad \textbf{(3–7)}$$

or

$$V_t = \frac{(Q_0 t_1)/2.3}{10^{(t/t_1)}} \qquad\qquad (3\text{–}8)$$

This analysis assumes that there are no consumptive uses of ground water in the basin so that all ground-water discharge is by means of baseflow to streams. If there are such uses as pumpage or evapotranspiration of ground water by phreatophytes, this use must be added to the amount determined by the baseflow recession method to get total recharge to the ground-water reservoir.

EXAMPLE PROBLEM

Refer to Figure 3.15. Determine the amount of ground-water recharge that takes place from the end of the baseflow recession of runoff year 1 to the start of the baseflow recession of runoff year 2.

The value of Q_0 for the first recession is 760 m^3/s and it takes 6.3 months (mo) for the discharge to reach $0.1Q_0$:

$$V_{tp} = \frac{Q_0 t_1}{2.3}$$

$$V_{tp} = \frac{760 \text{ m}^3 \text{ l/s} \times 6.3 \text{ mo} \times 30 \text{ d/mo} \times 1440 \text{ min/d} \times 60 \text{ s/min}}{2.3}$$

$$V_{tp} = 5.4 \times 10^9 \text{ m}^3$$

The value of V_t at the end of the recession, which lasts 7.5 mo, is

$$V_t = \frac{V_{tp}}{10^{(t/t_1)}} = \frac{5.4 \times 10^9 \text{ m}^3}{10^{(7.5/6.3)}} = \frac{5.4 \times 10^9 \text{ m}^3}{15.5} - 3.5 \times 10^8 \text{ m}^3$$

For the next year's recession, the value of Q_0 is 1000 m^3/s and t_1 is again 6.3 mo. Therefore,

$$V_{tp} = \frac{1000 \text{ m}^3/\text{s} \times 6.3 \text{ mo} \times 30 \text{ d/mo} \times 1440 \text{ min/d} \times 60 \text{ s/min}}{2.3}$$

$$= 7.1 \times 10^9 \text{ m}^3$$

The amount of recharge is equal to the total potential baseflow remaining at the end of the first baseflow recession subtracted from V_{tp} for the beginning of the next recession:

$$\text{Recharge} = 7.1 \times 10^9 \text{ m}^3 - 3.5 \times 10^8 \text{ m}^3$$
$$= 6.8 \times 10^9 \text{ m}^3$$

3.6   **MEASUREMENT OF STREAMFLOW**

3.6.1   Stream Gaging

Water flowing in an open channel is subject to friction as it comes in contact with the channel bottom and sides. As a result, the fastest current is at the surface in the center of the channel. If a series of careful measurements of flow velocity from the surface downward are made, a parabolic profile will emerge (Figure 3.16). Field studies have shown that the velocity at a depth equal to 0.6 times the total depth is very close to the average velocity for the entire section. The average of measurements made at 0.2 times depth and 0.8 times depth is also used to represent the average velocity of the entire profile.

The flow, Q (L^3/T; ft^3/s or m^3/s) in an open channel with a cross-sectional area, A (L^2; ft^2 or m^2) and average velocity, V (L/T; ft/s or m/s) can be found from the equation

$$Q = VA \qquad\qquad (3-9)$$

The velocity of flow can be measured by using a *current meter*. The U.S. Geological Survey has standard specifications for two types of meters. Each has a horizontal wheel with sets of small, cone-shaped cups attached. The wheel turns

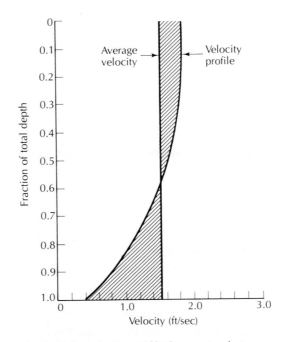

FIGURE 3.16   Typical parabolic velocity profile for a natural stream.

in the current and a cam attached to the spindle of the wheel makes an electrical contact once every revolution. The meter wheel is wired either to a set of headphones and a battery or to a direct readout meter. The operator with a headphone meter counts the clicks over a measured time period, usually thirty to sixty seconds, and uses a calibration curve furnished by the manufacturer to find the velocity.

The *Price-type meter* has a wheel about 5 in. (13 cm) in diameter; it is equipped with a vane to orient the meter perpendicular to the flow. The Price-type meter is usually suspended on a cable and lowered into a river, with a streamlined weight to pull it down. A *pygmy-Price meter* is smaller and is usually mounted on a graduated wading rod. The operator takes to the river with rubber boots and places the rod on the bottom of a stream in order to make the measurements.

In a typical stream, velocity will vary from bank to bank, necessitating a number of measurements. A straight reach of stream with a smooth shoreline, no brush hanging in the water, and no weeds or large rocks should be chosen. Places with back-eddies should be avoided; they will overestimate the total discharge, as the current meter will not distinguish the direction of flow. If a wading rod is to be used, the water must not be too deep or too swift for a person to wade. This is especially important to check when measuring peak flows.

A tape is stretched perpendicularly across the stream or along the bridge. The channel is subdivided into 15 to 30 segments. At the midpoint of each segment, the depth, d_i, is measured and recorded. The meter is then raised to 0.6 times the depth, and the average velocity, v_i, for that segment is measured. If the water is deep, the average velocity at 0.8 depth and 0.2 depth should be used. The discharge, q_i, for a segment of width, w_i, is given by

$$q_i = v_i d_i w_i \qquad\qquad \textbf{(3–10)}$$

The process is repeated for each segment of the cross section. The total discharge, Q, for the river is the sum of the discharge for each segment. For a measurement with m segments,

$$Q = \sum_{i=1}^{m} q_i \qquad\qquad \textbf{(3–11)}$$

If measurements are made from a bridge using a cable-suspended meter, a swift current may draw the meter downstream. The amount of line let out to measure the depth is thus too great. A correction must be made based on the angle of the cable from the vertical (Corbett et al. 1945).

Current measurements may be made through ice. A series of holes are cut in the ice across the river and the current measured by the preceding method. Because of friction between the ice and the underlying water, the velocity should be measured at 0.2 depth, 0.6 depth, and 0.8 depth and the results averaged.

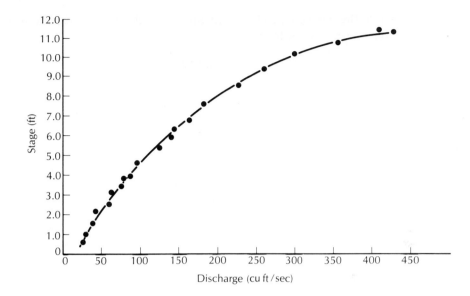

FIGURE 3.17 Typical stage-discharge rating curve.

It is possible to develop an empirical relationship between stream stages (elevation of the water surface above a datum) and discharge. As discharge measurements are slow and costly to make, knowledge of the preceding relationship is useful. A **rating curve** for a stream is made by simultaneously measuring the discharge of a stream and its stage and then repeating the measurements for a number of different stage heights. Stage versus discharge is then plotted as coordinates on graph paper to produce the rating curve (Figure 3.17). If the stream channel does not scour during flooding, and if the stage is not affected by such factors as tributary flow, a simple rating curve is sufficient. Otherwise, a rating curve must also include a factor for water-surface slope (Mitchell 1954). Automated stream-gaging stations employ a float device that measures the stage of a river by means of a stilling well connected to the stream. The stage data are transformed into discharge data by using either the rating curve or a rating table based on the curve. In the United States, most stream-stage measurements are recorded in digital form for automatic data processing.

3.6.2 Weirs

The discharge of small streams can be conveniently measured by use of a **weir.** This is a small dam with a spillway opening of specified shape. There are a number of standard shapes for sharp-crested weirs, the most common being a 90° V-notch or a rectangular cutout. A small earthen or concrete dam is built and the weir set into it. The dam will impound a small amount of water that should free fall over the weir crest, or lowest point of the spillway. The elevation of the backwater above the weir crest, H, is measured. The discharge over the weir can be found from the following formulas:

Rectangular weir
$$Q = 3\tfrac{1}{3}(L - 0.2H)H^{3/2} \tag{3-12A}$$

90° V-notch weir
$$Q = 2.5H^{5/2} \tag{3-13A}$$

where

Q is the discharge (ft^3/s)

L is the length of the weir crest (ft)

H is the head of the backwater above the weir crest (ft)

or

Rectangular weir with end contractions
$$Q = 1.84(L - 0.2H)H^{3/2} \tag{3-12B}$$

90° V-notch weir
$$Q = 1.379H^{5/2} \tag{3-13B}$$

where

Q is discharge (m^3/s)

L is length of weir crest (m)

H is the height of the backwater above weir crust (m)

Note that Equations 3–12A, 3–12B, 3–13A, and 3–13B are empirical and are not subject to dimensional analysis.

3.7   MANNING EQUATION

In open-channel hydraulics, the average velocity of flow of water may be found from the **Manning equation:**

$$V = \frac{1.49R^{2/3}S^{1/2}}{n} \tag{3-14A}$$

where

V is the average velocity (ft/s)

R is the hydraulic radius, or the ratio of the cross-sectional area of flow in square feet to the wetted perimeter (ft) (see Figure 3.18)

S is the energy gradient, which is the slope of the water surface

n is the Manning roughness coefficient

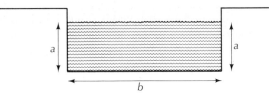

FIGURE 3.18 The cross-sectional area of flow is $a \times b$. The wetted perimeter is $a + b + a$.

or

$$V = \frac{1}{n} R^{2/3} S^{1/2}$$ **(3–14B)**

where

    V is velocity (m/s)

    R is hydraulic radius, the ratio of the cross-sectional area
        (m^2) divided by the wetted perimeter (m)

Note that Equations 3–14A and 3–14B are empirical and are not subject to dimensional analysis.

The velocity of flow is dependent upon the amount of friction between the water and the stream channel. Smoother channels will have less friction and, hence, faster flow. Channel roughness contributes to turbulence, which dissipates energy and reduces flow velocity. The following values for n are typical:

Mountain streams with rocky beds:	0.04–0.05
Winding natural streams with weeds:	0.035
Natural streams with little vegetation:	0.025
Straight, unlined earth canals:	0.020
Smoothed concrete:	0.012

The U.S. Geological Survey has published a series of photographs of rivers for which the value of the Manning roughness coefficient has been computed (Barns 1967). Field measurements of velocity, slope, area, and wetted perimeter were made and the value of n computed from Equation 3–14A. Careful study of the photographs can be used to obtain an estimate for the value of n for a given river under study.

The Manning equation can be used to determine flow in situations that preclude direct measurements. For example, if the current is changing rapidly (during a rising or falling flood peak, for example), conventional streamflow measurements would take too long. It might take the better part of an hour to make a discharge measurement, and flow velocity and discharge could change substantially during the period of measurement. Under these conditions, an instant computation can be made using river cross sections and a measured slope.

A drainage channel for storm water is lined with smooth concrete and has vertical sides with a flat bottom that is 3.50 ft wide. The bottom of the channel drops 1.50 ft over a distance of 500 ft. If the channel has 1.50 ft of water in it, what is its discharge?

The wetted perimeter is the sum of the bottom width and the depth of water on either side:

$$\text{Wetted perimeter} = 3.50 \text{ ft} + 1.50 \text{ ft} + 1.50 \text{ ft} = 6.50 \text{ ft}$$

The cross-sectional area is the product of the bottom width and the depth:

$$\text{Cross-sectional area} = 3.50 \text{ ft} \times 1.50 \text{ ft} = 5.25 \text{ ft}^2$$

The hydraulic radius is the ratio of the cross-sectional area to the wetted perimeter:

$$R = \frac{5.25 \text{ ft}^2}{6.50 \text{ ft}} = 0.808$$

The slope is the drop in elevation over the length of measurement:

$$S = \frac{1.50 \text{ ft}}{500 \text{ ft}} = 0.003$$

For smooth concrete the roughness factor is 0.012. The velocity of flow from Equation 3–14A is

$$V = \frac{1.49 R^{2/3} S^{1/2}}{n} = \frac{1.49 \, (0.808)^{2/3}(0.003)^{1/2}}{0.012}$$
$$= 5.90 \text{ ft/s}$$

The discharge from Equation 3–9 is the velocity of flow times the cross-sectional area:

$$Q = V \times A = 5.90 \text{ ft/s} \times 5.25 \text{ ft}^2 = 31.0 \text{ ft}^3/\text{s}$$

NOTATION

a Baseflow recession constant

A Area

C Runoff coefficient

d Stream depth

D Number of days

f_c Equilibrium infiltration capacity

f_0 Initial infiltration capacity

f_p Infiltration capacity

H Head of backwater over a weir crest

I Rainfall intensity

K Infiltration constant

L Length of a weir crest

n Manning roughness coefficient

P Probability

q Open-channel discharge

Q Total discharge of a stream

Q_0 Stream discharge at the start of a recession

R Hydraulic radius of a stream

S Slope of a stream

t Time

t_1 Time that it takes for baseflow to decline by 90%

V Average velocity of flow in an open channel

V_r Volume of remaining potential ground-water discharge

V_{tp} Volume of total potential ground-water discharge

w Stream width

PROBLEMS

Answers to odd-numbered problems will appear at the end of the book.

1. Analysis of baseflow recession curves from a drainage basin has yielded a recession constant of 1.2×10^{-2} when discharge is in cubic feet per second and time is in days.

   A. If a recession begins with a discharge of 2975 ft^3/s and t is in days, what will be the flow after 35 days and 70 days?

   B. If the recession begins with a discharge of 1165 ft^3, what would the flow be in 40 days?

2. The flow of a river at the start of a baseflow recession was 233 m^3/s; after 60 days the flow declined to 89.0 m^3/s.

   A. What is the recession constant?

   B. What would the flow be after 112 days?

3. Assume that the hydrograph in Figure 3.9 has a drainage basin area of 225 mi^2. Use Equation 3–3 to compute how long overland flow continues after the flood peak passes.

4. The hydrograph in Figure 3.19 is for a drainage basin with an area of 175 mi^2. Use Equation 3–3 to separate the overland flow component from the baseflow.

5. A V-notch weir is placed in a road culvert to measure the flow of a stream passing through the culvert. The value of H is 1.12 ft. Use Equation 3–13A to compute the discharge of the stream.

6. A rectangular weir is placed in a small stream to measure flow. The value of L is 3.5 ft and H is 0.42 ft. Use Equation 3–12A to compute the discharge of the stream.

7. An industrial park with flat-roofed buildings, large parking lots, and very little open area has a drainage basin area of 90 ac. The 25-y rainfall event (the amount that would on an average occur once in 25 y) has a precipitation intensity of 2 in./h.

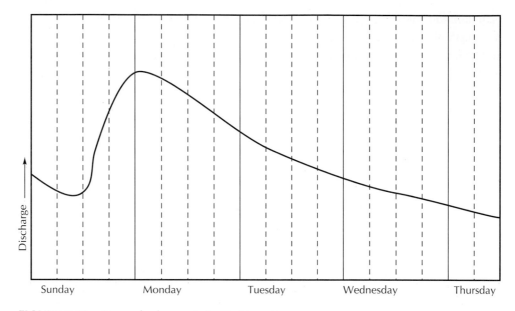

FIGURE 3.19 Stream hydrograph for Problem 4.

TABLE 3.2 Annual flow of the Colorado River at Lees Ferry (in millions of acre-feet)

Year	Flow	Year	Flow	Year	Flow
1896	10.089	1917	24.037	1938	17.545
1897	18.009	1918	15.364	1939	11.075
1898	13.815	1919	12.462	1940	8.601
1899	15.874	1920	21.951	1941	18.148
1900	13.228	1921	23.015	1942	19.125
1901	13.582	1922	18.305	1943	13.103
1902	9.393	1923	18.269	1944	15.154
1903	14.807	1924	14.201	1945	13.410
1904	15.645	1925	13.033	1946	10.426
1905	16.027	1926	15.853	1947	15.473
1906	19.124	1927	18.616	1948	15.613
1907	23.402	1928	17.279	1949	16.376
1908	12.856	1929	21.428	1950	12.894
1909	23.275	1930	14.888	1951	11.647
1910	14.248	1931	7.769	1952	20.290
1911	16.028	1932	17.243	1953	10.670
1912	20.520	1933	11.356	1954	7.900
1913	14.473	1934	5.640	1955	9.150
1914	21.222	1935	11.549	1956	10.720
1915	14.027	1936	13.800		
1916	19.201	1937	13.740		

    A. Use Equation 3–4 with a C factor of 0.75 to compute the runoff rate.

    B. The industrial park is drained by a drainage canal that is 12 ft wide, has vertical walls, and has a bottom slope of 0.005. Use Equation 3–14A to compute the depth of water that would flow in the canal.

8. The annual flow of the Colorado River at Lees Ferry for the period of 1896 to 1956 is given in Table 3.2 Use Equation 3–5 to construct a table of probability values. Use Figure 3.20 on page 76 (standard probability paper) to plot a duration curve showing the percent of the time an indicated discharge was equaled or exceeded.

9. An aqueduct has smooth earthen sides and bottom. The slope of the water surface is 5.0 ft/mi. The channel is trapezoidal in shape with a 45° angle to the sides of the trapezoid and a bottom segment that is 50 ft wide. The water in the aqueduct is 8.0 ft deep in the center. Use Equation 3–14A to compute the flow in the aqueduct.

10. A winding natural stream with weeds has an average depth of 1.5 m and is 23 m across. The stream channel drops 0.43 m/km. Use Equation 3–14B to compute the velocity of flow of the stream.

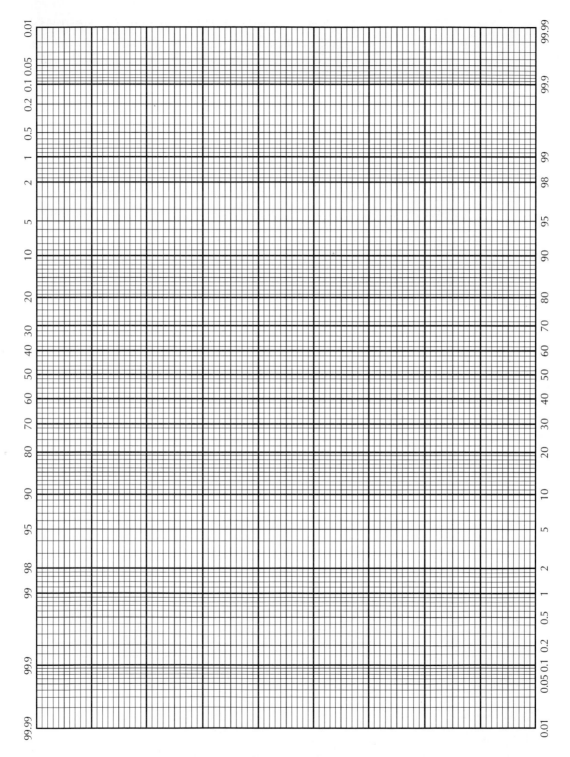

FIGURE 3.20 Standard probability paper for Problem 8.

4 Properties of Aquifers

Some of the vapour that is formed by day does not rise high because the ratio of the fire that is raising it to the water that is being raised is small. When this cools and descends at night, it is called dew and hoar-frost. When the vapour is frozen before it has condensed to water again it is hoar-frost . . . . It is dew when the vapour has condensed into water and the heat is not so great as to dry up the moisture that has been raised, nor the cold sufficient (owing to the warmth of the climate or season) for the vapour itself to freeze.

Meteorologica, Aristotle (384–322 B.C.)

4.1 MATTER AND ENERGY

In this discussion of some fundamental principles and definitions from physics, we will include both English and SI units. English units are in widespread use in the United States, but the remainder of the world uses the SI system.

Energy is the capacity to do work, which implies that some resistance to change in movement must be overcome. *Work* is done when a force is applied to a fluid while the fluid is moving. Work is equal to the product of the net force exerted and the distance through which the force moves:

$$W = FD \tag{4-1}$$

where

W is work (ML^2/T^2)

F is force (ML/T^2)

D is distance (L)

The *force* acting on a body is equal to the product of mass of the body and its acceleration (Newton's second law of motion):

$$F = ma \tag{4-2}$$

where

F is force (ML/T^2)

m is mass (M)

a is acceleration (L/T^2)

In the SI system the unit of *mass* is the kilogram (kg), the unit of *length* is the meter (m), and the unit of *time* is the second (s). In the English system the unit of mass is the slug, the unit of length is the foot (ft), and the unit of time is the second.

The unit of force in the SI system is the *newton* (N). Acceleration is expressed in meters per second per second (m/s^2) and a newton is 1 kg·m/s^2. In the English system the unit of force is a *pound* (lb), which is one slug-ft/s^2, since acceleration is expressed in feet per second per second.

The *weight* of a body is the gravitational force exerted on it by the earth. The gravitational acceleration, g, varies from place to place but is approximately 9.8 m/s^2 (32 ft/s^2). The weight of a body is given by

$$w = mg \qquad\qquad (4\text{--}3)$$

where

w is weight (ML/T^2)

g is acceleration of gravity (M/T^2)

m is mass (M)

Weight has the same units as force. The mass of a body, the weight of which is 1 N, at a place where $g = 9.80$ m/s^2, is

$$m = \frac{w}{g} = \frac{1 \text{ N}}{9.80 \text{ m/s}^2} = 0.102 \text{ kg}$$

In the English system the unit of weight is a pound. Confusion can arise because the pound—a unit of weight—is sometimes compared to the kilogram—a unit of mass. Balances and scales are even calibrated in both pounds and kilograms. If we say that a sample weighs 0.13 kg, we mean that the sample has a mass of 0.13 kg in the earth's gravitational field. The weight is 1.27 N (0.29 lb). A kilogram is equal to 1000 g.

The **density** of a fluid or a solid is its mass per unit volume. The units are N/m^3 or slugs/ft^3.

$$\rho = m/V \qquad\qquad (4\text{--}4)$$

where

ρ is density (M/L^3)

m is mass (M)

V is volume (L^3)

The specific weight of a substance is its weight per unit volume. The units are N/m^3 or lb/ft^3.

$$\gamma = w/V \qquad\qquad (4\text{--}5)$$

where

γ is specific weight (M/L^2T^2)

w is weight (ML/T^2)

By combining Equations 4–3, 4–4, and 4–5 we can arrive at an alternative definition of specific weight:

$$\gamma = \rho g \qquad\qquad \textbf{(4–6)}$$

Although it is not an official SI unit, density is frequently expressed in terms of grams per cubic centimeter (g/cm^3). If a substance has a density of 1 g/cm^3, the density is 1000 kg/m^3 in SI units.

EXAMPLE PROBLEM   A fluid has a density of 1.085 g/cm^3. If the acceleration of gravity is 9.81 m/s^2, what is the specific weight of the fluid?

$$\gamma = w/V \qquad w = mg \qquad \rho = m/V$$

$$\gamma = \rho g$$

$$\rho = 1.085 \text{ g/cm}^3 \times 1/1000 \text{ kg/g} \times 10^6 \text{ cm}^3/\text{m}^3$$

$$= 1.085 \times 10^3 \text{ kg/m}^3$$

$$\gamma = 1.085 \times 10^3 \text{ kg/m}^3 \times 9.81 \text{ m/s}^2$$

$$= 1.064 \times 10^4 \text{ N/m}^3$$

Pressure is the force applied to a unit area perpendicular to the direction of the force. In the English system the units are pounds per square foot. In the SI system, the units are newtons per square meter, which are also called *pascals* (Pa).

$$P = F/A \qquad\qquad \textbf{(4–7)}$$

where

P is pressure (M/LT^2)

F is force (ML/T^2)

A is cross-sectional area (L^2)

In hydrogeology, pressure is measured relative to atmospheric pressure, which varies with changing weather patterns. Standard atmospheric pressure is 1.013×10^5 Pa, or 2116 lb/ft^2.

Water is a Newtonian fluid, which means that its resistance to relative motion is proportional to a fluid property known as the **dynamic viscosity, μ.** Standard units of dynamic viscosity are $N\cdot s/m^2$ or $lb\text{-}s/ft^2$; the $g/(s\cdot cm)$ is also a convenient unit called the poise (P).

Water is a compressible fluid. If pressure is applied, the same mass of fluid will be contained within a smaller volume; in other words, the density

TABLE 4.1   English and SI Units

Parameter	English Unit	SI Unit	Conversion Factor	Dimensional Formula
Force	pound (lb)	newton (N)	1 lb = 4.448 N	ML/T^2
Mass	slug	kilogram (kg)	1 slug = 14.594 kg	M
Length	foot (ft)	meter (m)	1 ft = 0.3048 m	L
Time	second (s)	second	1 s = 1 s	T
Density	slug/ft^3	kg/m^3	1 slug/ft^3 = 515.4 kg/m^3	M/L^3
Specific weight	lb/ft^3	N/m^3	1 lb/ft^3 = 157.1 N/m^3	M/L^2T^2
Pressure	lb/ft^2	N/m^2	1 lb/ft^2 = 47.88 N/m^2	M/LT^2
Dynamic viscosity	lb-s/ft^2	N·s/m^2	1 lb-s/ft^2 = 47.88 N·s/m^2	M/LT
Bulk modulus	lb/ft^2	N/m^2	1 lb/ft^2 = 47.88 N/m^2	M/LT^2

increases. The change in density is proportional to the change in pressure, and the proportionality constant for compressibility is known as the *bulk modulus*. Water is elastic, so if the pressure is released, the volume will expand and the density will decrease proportional to the pressure change. Units of bulk modulus are N/m^2 or lb/ft^2.

Table 4.1 gives the conversion factors between English and SI units for these factors. Appendix 14 gives the density and dynamic viscosity of water at different temperatures.

4.2   POROSITY OF EARTH MATERIALS

At the time they are formed, some rocks contain void spaces while others are solid. Those rocks occurring near the surface of the earth are not totally solid. The physical and chemical weathering processes there continually decompose and disaggregate rock, thus creating voids. Slight movements of rock masses near the surface can cause rocks to crack or fracture. This also results in openings between rocks.

Sediments are assemblages of individual grains that were deposited by water, wind, ice, or gravity. There are openings called **pore spaces** between the sediment grains, so that sediments are not solid.

The cracks, voids, and pore spaces in earth materials are of great importance to hydrogeology. Ground water and soil moisture occur in the voids in otherwise solid earth materials.

4.2.1   Definition of Porosity

The **porosity** of earth materials is the percentage of the rock or soil that is void of material. It is defined mathematically by the equation

$$n = \frac{100V_v}{V}$$

(4–8)

where

n is the porosity (percentage)

V_v is the volume of void space in a unit volume of earth material (L^3; cm^3 or m^3)

V is the unit volume of earth material, including both voids and solids (L^3; cu^3 or m^3)

Laboratory porosity is determined by taking a sample of known volume (V). The sample is dried in an oven at 105°C until it reaches a constant weight. This expels moisture clinging to surfaces in the sample, but not water that is hydrated as a part of certain minerals. The dried sample is then submerged in a known volume of water and allowed to remain in a sealed chamber until it is saturated. The volume of the voids (V_v) is equal to the original water volume less the volume in the chamber after the saturated sample is removed.

This laboratory procedure yields a value of the effective porosity because it excludes pores that are not large enough to contain water molecules and those that are not interconnected. **Effective porosity,** n_e, is the porosity available for fluid flow. Peyton et al. (1986) studied the effective porosity of fine-grained sediments. One conclusion of that study was that the effective porosity of a sediment is a function of the size of the molecules that are being transported relative to the size of the passageways that connect the pores. These passageways, or pore throats, are typically smaller than the pores. If the molecule being transported has a greater diameter than some of the pore throats, this would limit the effective porosity with respect to that molecule. Peyton et al. (1986) found that even in a lacustrine clay, water molecules could pass through all the pore throats, so that the effective porosity was the same as the porosity. This suggests that at least in sediments all the pores are connected and we don't need to be concerned with effective porosity with respect to flow of water.

The total porosity can be computed from the relationship

$$n = 100 \left[1 - (\rho_b/\rho_d)\right] \tag{4-9}$$

where

n is the total porosity as a percentage

ρ_b is the bulk density of the aquifer material (M/L^3; g/cm^3 or kg/m^3)

ρ_d is the particle density of the aquifer material (M/L^3; g/cm^3 or kg/m^3)

The bulk density of the aquifer material is the mass of the sample after oven drying divided by the original sample volume (the sample can change volume upon oven drying). The particle density is the oven-dried mass divided by the volume of the mineral matter in the sample as determined by a water-displacement test. For most rock and soil the particle density is about 2.65 g/cm^3 (2650 kg/m^3).

4.2.2 Porosity and Classification of Sediments

The porosity of sediments consists of the void spaces between solid fragments. If the fragments are solid spheres of equal diameters, they can be put together in such a manner that each sphere sits directly on the crest of the underlying sphere (Figure 4.1). This is called *cubic packing,* with an associated porosity of 47.65% (Meinzer 1923a). If the spheres lie in the hollows formed by four adjacent spheres of the underlying layer, the result is *rhombohedral packing,* with a porosity of 25.95% (Meinzer 1923a).

These two configurations represent the extremes of porosity for arrangements of equidimensional spheres with each sphere touching all neighboring spheres. The diameter of the sphere does not influence the porosity. Thus, a room full of bowling balls in cubic packing would have the same porosity as a room full of 1-mm ball bearings. The volume of an individual pore would be much larger for the bowling balls. The porosity of well-rounded sediments, which have been sorted so that they are all about the same size, is independent of the particle size and falls in the range of about 26% to 48%, depending upon the packing.

If a sediment contains a mixture of grain sizes, the porosity will be lowered. The smaller particles can fill the void spaces between the larger ones. The wider the range of grain sizes, the lower the resulting porosity (Figure 4.2). Geologic agents can sort sediments into layers of similar sizes. Wind, running water, and wave action tend to create well-sorted sediments. Other processes, such as glacial action and landslides, result in sediments with a wide range of grain sizes. These poorly sorted sediments have low porosities.

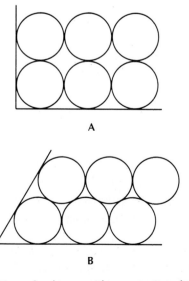

A

B

FIGURE 4.1 A. Cubic packing of spheres with a porosity of 47.65% **B.** Rhombohedral packing of spheres with a porosity of 25.95%.

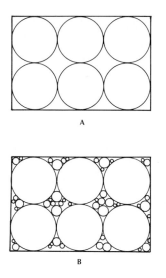

FIGURE 4.2 **A.** Cubic packing of spheres of equal diameter with a porosity of 47.65 percent. **B.** Cubic packing of spheres with void spaces occupied by grains of smaller diameter, resulting in a much lower overall porosity.

In addition to grain-size sorting, the porosity of sediments is affected by the shape of the grains. Well-rounded grains may be almost perfect spheres, but many grains are very irregular. They can be shaped like rods, disks, or books. Sphere-shaped grains will pack more tightly and have less porosity than particles of other shapes. The fabric or orientation of the particles, if they are not spheres, also influences porosity.

Sediments are classified on the basis of the size (diameter) of the individual grains. There are many classification systems in use. Figure 4.3 shows a common system frequently used by sedimentologists.

The engineering classification of sediments is somewhat different than the geological classification. The American Society of Testing Materials defines sediments on the basis of the grain-size distribution shown in Table 4.2.

The grain-size distribution of a sediment may be conveniently plotted on semilogarithmic paper. The cumulative percent finer by weight is plotted on the arithmetic scale and the grain size is plotted on the logarithmic scale. The grain size of the sand fraction is determined by shaking the sand through a series of sieves with decreasing mesh openings. The 200 mesh screen, with an opening of 0.075 mm, separates the sand fraction from the fines. The gradation of the fines is determined by a hydrometer test, which is based on the rate that the sediment settles in water. Figure 4.4 is a grain-size distribution curve for a silty fine to coarse sand. This sample is somewhat poorly sorted as there is a wide range of grain sizes present. Figure 4.5 (p. 86) is the grain-size distribution curve for a well-sorted fine sand. Less than 5% of the sample consisted of fines that pass the 200 mesh sieve. A hydrometer test was not performed on this sample because of the lack of fines.

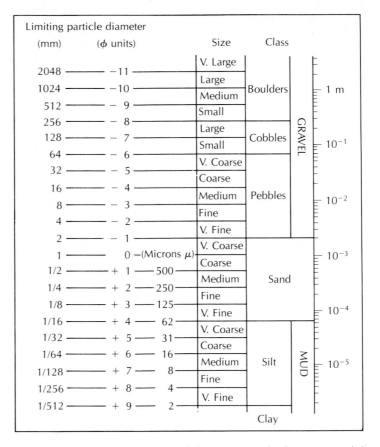

FIGURE 4.3 Standard sizes of sediments with limiting particle diameters and the ϕ scale of sediment size, in which ϕ is equal to $\log_2 s$ (the particle diameter). Source: G. M. Friedman and J. E. Sanders, *Principles of Sedimentology* (New York: John Wiley, 1978). Used with permission.

TABLE 4.2 Engineering grain-size classification

Name	Size Range (mm)	Example
Boulder	>305	Basketball
Cobbles	76–305	Grapefruit
Coarse gravel	19–76	Lemon
Fine gravel	4.75–19	Pea
Coarse sand	2–4.75	Water softener salt
Medium sand	0.42–2	Table salt
Fine sand	0.075–0.42	Powdered sugar
Fines	<0.075	Talcum powder

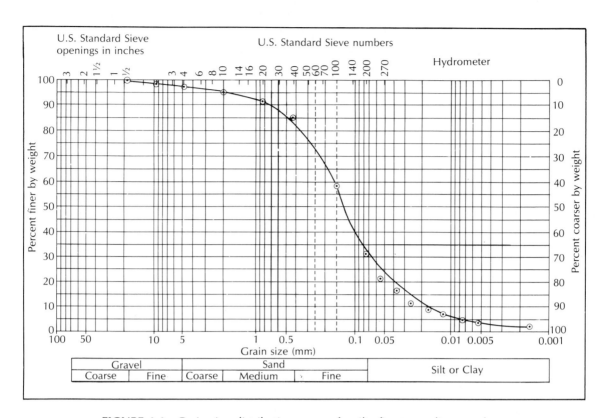

FIGURE 4.4 Grain-size distribution curve of a silty fine to medium sand.

The **uniformity coefficient** of a sediment is a measure of how well or poorly sorted it is. The uniformity coefficient, C_u, is the ratio of the grain size that is 60% finer by weight, d_{60}, to the grain size that is 10% finer by weight, d_{10}:

$$C_u = d_{60}/d_{10} \qquad\qquad (4\text{–}10)$$

A sample with a C_u less than 4 is well sorted; if the C_u is more than 6 it is poorly sorted. The poorly sorted silty sand in Figure 4.4 has a C_u of 8.3, whereas the well-sorted sand of Figure 4.5 has a C_u of 1.4.

The **effective grain size,** d_{10}, is the size corresponding to the 10% line on the grain-size curve.

Clays and some clay-rich or organic soils can have very high porosities. Organic materials do not pack very closely because of their irregular shapes. The dispersive effect of the electrostatic charge present on the surfaces of certain book-shaped clay minerals causes clay particles to be repelled by each other. The result is a relatively large proportion of void space.

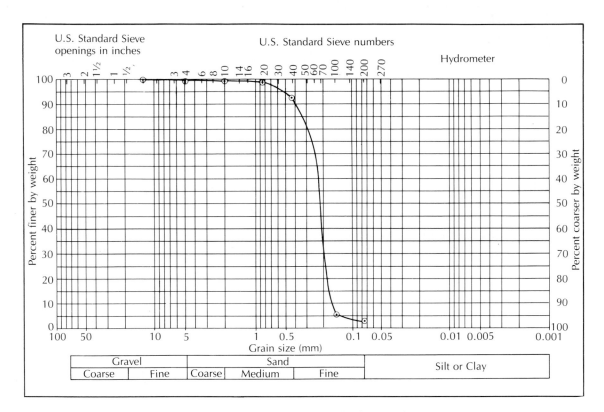

FIGURE 4.5 Grain-size distribution curve of a fine sand.

The general range of porosity that can be expected for some typical sediments is listed in Table 4.3.

4.2.3 Porosity of Sedimentary Rocks

Sedimentary rocks are formed from sediments through a process known as **diagenesis.** A sediment, which may be either a product of weathering or a chemically precipitated material, is buried. The weight of overlying materials and physicochemical reactions with fluids in the pore spaces induce changes in the

TABLE 4.3 Porosity ranges for sediments

Well-sorted sand or gravel	25–50%
Sand and gravel, mixed	20–35%
Glacial till	10–20%
Silt	35–50%
Clay	33–60%

Based on Meinzer (1923a); Davis (1969); Cohen (1965); and MacCary and Lambert (1962).

sediment. This includes compaction, removal of material, addition of material, and transformation of minerals by replacement or change in mineral phase. Compaction reduces pore volume by rearranging the grains and reshaping them. The deposition of cementing materials such as calcite, dolomite, or silica will reduce porosity, although the dissolution of material that is dissolved by the pore fluid will increase porosity. The primary structures of the sediment may be preserved in the sedimentary rock. The porosity of a sandstone, for instance, will be influenced by the grain size, size sorting, grain shape, and fabric of the original sediment. Diagenesis is a complex process, but in general the primary porosity of a sedimentary rock will be less than that of the original sediment. This is especially true of fine-grained sediments (silts and clays) (Figure 4.6).

Rocks at the earth's surface are usually fractured to some degree. The fracturing may be mild, resulting in widely spaced joints. At the other extreme, violent fracturing may completely shatter the rock, resulting in fault breccias. Fractures create secondary porosity in the rock. Ground water can be found in fractured sedimentary rocks in the pores between grains **(primary porosity)** as well as in fractures **(secondary porosity).** Ground water flowing through fractures may enlarge them by solution of material. Bedding planes in the sedimentary rocks may have primary porosity formed during deposition of the sediments and secondary porosity if the rock has moved along a bedding plane.

Some cohesive sediments (those rich in silt and/or clay) are also subject to fracturing. In some cases, this is merely from shrinkage cracks that develop when the sediment dries. However, slumping, loading, or tectonic activity can also cause fracturing in nonplastic cohesive sediments. This fracturing can be a significant source of secondary porosity in such deposits.

Limestones and dolomites are well-known and widespread examples of sedimentary rocks of chemical or biochemical origin. They are formed of calcium carbonate and calcium-magnesium carbonate, respectively. Gypsum, a calcium sulfate, and halite or rock salt (sodium chloride) are also widely distributed common examples of chemical precipitates.

The materials that formed these rocks were originally part of an aqueous solution. Inasmuch as the precipitation process is reversible, the rock can be redissolved. When these rock types are in a zone of circulating ground water, the

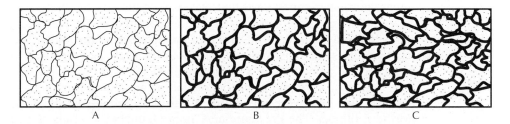

FIGURE 4.6   A. A clastic sediment with intergranular porosity. **B.** Reduction of porosity in the clastic sediment due to deposition of cementing material in the pore spaces. **C.** Further reduction in porosity due to compaction and cementation.

rock may be removed by solution. Ground water moves initially through pore spaces, as well as along fractures, joints, and bedding planes. As more water moves through the bedding planes, they are preferentially dissolved and enlarged, causing the rock to become very porous. Some limestone formations have openings large enough to permit thousand of tourists a day to pass through. The caverns at Carlsbad, New Mexico, and Ljubljana, Yugoslavia, exemplify such massive porosity. Gypsum and salt may also be cavernous (Meinzer 1923a).

The percent porosity of sedimentary rocks is highly variable. In clastic rocks, it can range from 3% to 30% (Davis 1969; Winsauer et al. 1952; Wyllie & Spangler 1952; Manger 1963). Reported values for limestones and dolomites range from less than 1% to 30% (Davis 1969; Manger 1963; Archie 1950; Murray 1960).

4.2.4 Porosity of Plutonic and Metamorphic Rocks

Plutonic rocks (those formed by intrusive igneous processes) and **metamorphic rocks** (those formed by applying heat and pressure to preexisting rocks) are typically thought to have a very low porosity as they are formed of interlocking crystals (Davis 1969). However, in a deep granite test hole in northern Illinois, porosity was measured at 1.42% at a depth of 5248 ft (1600 m). At this depth there were few fractures in the rock so some of the porosity was possibly primary. In the same borehole the porosity was as great as 2.15% where there were fractures present (Daniels, Olhoeft, & Scott 1983).

Two geologic processes, weathering and fracturing, increase overall rock porosity. Rock at depth is under pressure due to the weight of overlying materials. This rock may be fractured by expansion as the overlying weight is removed by erosion. Tectonic stresses in the earth can cause folding and faulting. Rock in a fault shear zone may be extensively fractured. Expansion cracks can form at the crest of a fold. Joint sets in crystalline rock are usually found in three mutually perpendicular directions (Krynine & Judd 1957). Fracturing increases porosity of crystalline rocks by about 2% to 5% (Davis 1969; Brace, Paulding, & Scholz 1966). Weathering due to chemical decomposition and physical disintegration operates with greater efficacy with increasing rock porosity. Weathered plutonic and metamorphic rocks can have porosities in the range of 30% to 60% (Stewart 1964). Owing to the sheetlike structure of some weathering minerals, such as the micas, porosities can exceed that of loosely packed spheres.

Porosity due to fracturing is concentrated in the rock along the sets of joints and is a function of the width of the openings in the joints. Weathered rock has the pore spaces distributed throughout the rock, although weathering may be more intense along joint or weathering planes.

Figure 4.7 contains histograms of the number of fractures as a function of depth in two deep core holes drilled into 1.47-billion-year-old biotite granite in northern Illinois. The Precambrian/Cambrian unconformity is at a depth of about 1970 ft (600 m), which is shown at the top of the histograms. Fractures can be seen to be concentrated near the top of the Precambrian basement rock, indicating that weathering occurred during the time the basement rock was at the surface.

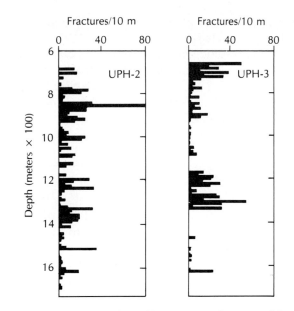

FIGURE 4.7 Histograms showing number of fractures as a function of depth at two deep core holes into Precambrian-age granite in northern Illinois. Source: B. C. Haimson & T. W. Doe, "State of Stress, Permeability, and Fractures in the Precambrian Granite of Northern Illinois," *Journal of Geophysical Research* 88, B9 (1983): 7355–71.

However, fracturing was also detected near the bottom of the test hole at a depth of 5248 ft (1600 m) (Haimson & Doe 1983). This indicates that unloading can create fractures to significant depths in some instances.

4.2.5 Porosity of Volcanic Rocks

Volcanic rocks (those formed by extrusive igneous activity) are similar in chemical composition to plutonic rocks because both are formed by the cooling of molten rock (magma). However, extrusive rocks cool and solidify quickly because they are formed in a surface environment; this gives them radically different porosity characteristics. Volcanic rocks include lava flows; deposits of ash and cinders, which can occur in loose, unconsolidated piles; and such rocks as welded tuff.

Lava cooling rapidly at the surface will trap degassing products, resulting in holes in the rock (vesicular texture). The holes create porosity, although they may not be interconnected. Shrinkage cracks that develop in the lava as it cools create joints. Flowing lava can form a crust, which then breaks apart to form a rubbly structure. The broken surface of buried lava flows, the remains of natural lava tubes and tunnels through which molten lava once poured, and stream gravels trapped between lava flows all produce a high porosity in some extrusive rocks. Porosity of basalt, a crystalline extrusive rock that is formed from magma with a low gas content, generally ranges from 1% to 12% (Schoeller 1962). Pumice, a glassy rock that is formed from a magma with a very high gas content, can have a porosity of as high as 87% (Davis 1969), although the vesicles are not well interconnected.

Pyroclastic deposits are formed by volcanic material thrown into the air when molten. They can have high porosities. Values of porosity of tuff ranging from 14% to 40% have been reported (Keller 1960). Recent volcanic ash may have a porosity of 50%. Weathering of volcanic deposits can increase the porosity to values in excess of 60% (Davis 1969).

4.3   SPECIFIC YIELD

Specific yield (S_y) is the ratio of the volume of water that drains from a saturated rock owing to the attraction of gravity to the total volume of the rock (Meinzer 1923b) (Figure 4.8).

Water molecules cling to surfaces because of surface tension of the water (Figure 4.9). If gravity exerts a stress on a film of water surrounding a mineral grain, some of the film will pull away and drip downward. The remaining film will be thinner, with a greater surface tension so that, eventually, the stress of gravity will be exactly balanced by the surface tension. **Pendular water** is the moisture clinging to the soil particles because of surface tension. At the moisture content of the specific yield, gravity drainage will cease.

If two samples are equivalent with regard to porosity, but the average grain size of one is much smaller than the other, the surface area of the finer sample will be larger. As a result, more water can be held as pendular moisture by the finer grains.

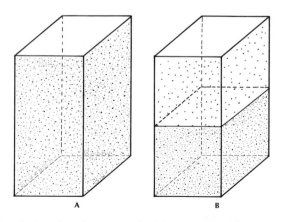

FIGURE 4.8   **A.** A volume of rock saturated with water. **B.** After gravity drainage, 1 unit volume of the rock has been dewatered with a corresponding lowering of the level of saturation. Specific yield is the ratio of the volume of water that drained from the rock, owing to gravity, to the total rock volume.

FIGURE 4.9 Pendular water clinging to spheres owing to surface tension. Gravity attraction is pulling the water downward.

The **specific retention** (S_r) of a rock or soil is the ratio of the volume of water a rock can retain against gravity drainage to the total volume of the rock (Meinzer 1923b). Since the specific yield represents the volume of water that a rock will yield by gravity drainage, with specific retention the remainder, the sum of the two is equal to porosity:

$$n = S_y + S_r \tag{4-11}$$

The specific retention increases with decreasing grain size, so that a clay may have a porosity of 50% with a specific retention of 48%.

Table 4.4 lists the specific yield, in percent, for a number of sediment textures. The data for this table were compiled from a large number of samples in various geographic locations. Maximum specific yield occurs in sediments in the

TABLE 4.4 Specific yields in percent

Material	Specific Yield Maximum	Minimum	Average
Clay	5	0	2
Sandy clay	12	3	7
Silt	19	3	18
Fine sand	28	10	21
Medium sand	32	15	26
Coarse sand	35	20	27
Gravelly sand	35	20	25
Fine gravel	35	21	25
Medium gravel	26	13	23
Coarse gravel	26	12	22

Source: Johnson (1967).

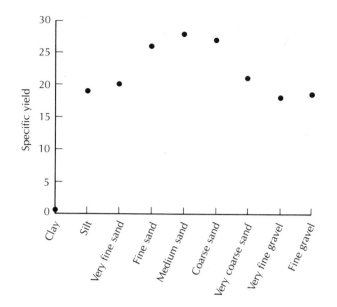

FIGURE 4.10   Specific yield of sediments from the Humboldt River Valley of Nevada as a function of the median grain size. Source: Data from P. Cohen, U.S. Geological Survey Water-Supply Paper 1975, 1965.

medium-to-coarse sand-size range (0.5 to 1.0 mm). This is shown graphically in Figure 4.10, which plots specific yield as a function of grain size for several hundred samples from the Humboldt River Valley of Nevada.

Both soil formed by weathering processes at the surface and sediments that are depositional generally contain a mixture of clay, silt, and sand. Figure 4.11 is a soil classification triangle showing lines of equal specific yield (Johnson 1967). It is apparent that the specific yield increases rapidly as the percentage of sand increases and as the percentages of silt, and especially clay, decrease.

Specific yield may be determined by laboratory methods. A sample of sediment of known volume is fully saturated. This is usually done in a soil column that is flooded slowly from the bottom, allowing air to escape upward. Water is then allowed to drain from the column (Johnson, Prill, & Morris 1963). Care must be taken to avoid evaporation losses; even for sand-sized grains, columns must be allowed to drain for very long time periods (months) before equilibrium is reached (Prill, Johnson, & Morris 1965). The ratio of the volume of water drained to the volume of the soil column is the specific yield (multiplied by 100 to express the value as a percentage).

The specific yield of sediment and rock can also be determined in the field. Water wells are pumped, and the rate at which the water level falls in nearby wells is measured (Wenzel 1942; Ferris et al. 1962; Prickett 1965). Chapter 7 includes a discussion of such pumping-test methods.

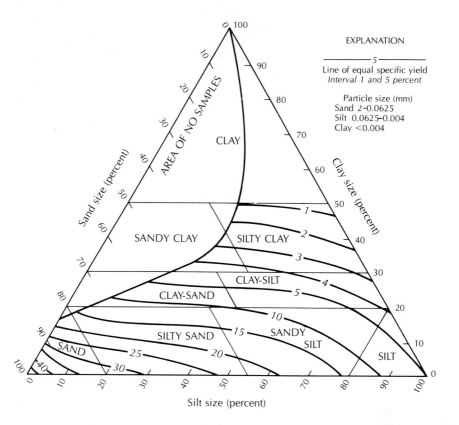

FIGURE 4.11 Textural classification triangle for unconsolidated materials showing the relation between particle size and specific yield. Source: A. I. Johnson, U.S. Geological Survey Water-Supply Paper 1662-D, 1967.

4.4 HYDRAULIC CONDUCTIVITY OF EARTH MATERIALS

We have seen that earth materials near the surface generally contain some void space and thus exhibit porosity. Moreover, in most cases, these voids are interconnected to some degree. Water contained in the voids is capable of moving from one void to another, thus circulating through the soil, sediment, and rock. It is the ability of a rock to transmit water that, together with its ability to hold water, constitute the most significant hydrologic properties. There are some rocks that exhibit porosity but lack interconnected voids, e.g., vesicular basalt. These rocks cannot convey water from one void to another. Some sediments and rocks have porosity, but the pores are so small that water flows through the rock with difficulty. Clay and shale are examples.

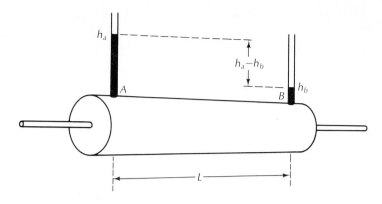

FIGURE 4.12   Horizontal pipe filled with sand to demonstrate Darcy's experiment. (Darcy's original equipment was actually vertically oriented.)

4.4.1   Darcy's Experiment

In the mid–nineteenth century, a French engineer, Henry Darcy, made the first systematic study of the movement of water through a porous medium (Darcy 1856). He studied the movement of water through beds of sand used for water filtration. Darcy found that the rate of water flow through a bed of a "given nature" is proportional to the difference in the height of the water between the two ends of the filter beds and inversely proportional to the length of the flow path. He determined also that the quantity of flow is proportional to a coefficient, K, which is dependent upon the nature of the porous medium.

Figure 4.12 illustrates a horizontal pipe filled with sand. Water is applied under pressure through end A. The pressure can be measured and observed by means of a thin vertical pipe open in the sand at point A. Water flows through the pipe and discharges at point B. Another vertical pipe or piezometer is present to measure the pressure at B.

Darcy found experimentally that the discharge, Q, is proportional to the difference in the height of the water, h (hydraulic head), between the ends and inversely proportional to the flow length, L:

$$Q \propto h_A - h_B \quad \text{and} \quad Q \propto 1/L$$

The flow is also obviously proportional to the cross-sectional area of the pipe, A. When combined with the proportionality constant, K, the result is the expression known as **Darcy's law:**

$$Q = -KA\left(\frac{h_A - h_B}{L}\right) \qquad \textbf{(4–12)}$$

This may be expressed in more general terms as

$$Q = -KA\left(\frac{dh}{dl}\right) \tag{4-13}$$

where dh/dl is known as the **hydraulic gradient.** The quantity dh represents the change in head between two points that are very close together, and dl is the small distance between these points. The negative sign indicates that flow is in the direction of decreasing hydraulic head. The use of the negative sign necessitates careful determination of the sign of the gradient. If the value of h_2 at point X_2 is greater than h_1 at point X_1, then flow is from point X_2 to X_1. If $h_1 > h_2$, then flow is from X_1 to X_2.

4.4.2 Hydraulic Conductivity

Equation 4–13 can be rearranged to show that the coefficient K has the dimensions of length/time (L/T), or velocity. This coefficient has been termed the **hydraulic conductivity.** In other works, it may be referred to as the coefficient of permeability:

$$K = \frac{-Q}{A(dh/dL)} \tag{4-14}$$

Discharge has the dimensions volume/time (L^3/T), area (L^2), and gradient (L/L). Substituting these dimensions into Equation 4–12, the dimensions of K are determined:

$$K = \frac{-(L^3/T)}{(L^2)(L/L)} = (L/T)$$

Hubbert (1956) pointed out that Darcy's proportionality constant, K, is a function of properties of both the porous medium and the fluid passing through it. It is intuitively obvious that a viscous fluid (one that is thick), such as crude oil, will move at a slower rate than water, which is thinner and has a lower viscosity. The discharge is directly proportional to the **specific weight,** γ, of the fluid. The specific weight is the force exerted by gravity on a unit volume of the fluid. This represents the driving force of the fluid. Discharge is also inversely proportional to the *dynamic viscosity* of the fluid, μ, which is a measure of the resistance of the fluid to the shearing that is necessary for fluid flow.

If experiments are performed with glass spheres of uniform diameter, the discharge is also proportional to the square of the diameter of the glass beads, d.

These proportionality relationships can be expressed as

$$Q \propto d^2$$
$$Q \propto \gamma$$
$$Q \propto \frac{1}{\mu}$$

Darcy's law can also be expressed as

$$Q = -\frac{Cd^2\gamma}{\mu}\frac{dh}{dl} \qquad (4-15)$$

The new proportionality constant, C, is called the *shape factor*. Both C and d^2 are properties of the porous media, whereas γ and μ are properties of the fluid. We can introduce a new constant, K_i, which is representative of the properties of the porous medium alone. It is termed the **intrinsic permeability.** This is basically a function of the size of the openings through which the fluid moves. The larger the square of the mean pore diameter, d, the lower the flow resistance. The cross-sectional area of a pore is also a function of the shape of the opening. A constant can be used to describe the overall effect of the shape of the pore spaces. Using this dimensionless constant, C, the intrinsic permeability is given by the expression

$$K_i = Cd^2 \qquad (4-16)$$

The dimensions of K_i are (L^2), or area. The relationship between hydraulic conductivity and intrinsic permeability is

$$K = K_i\left(\frac{\gamma}{\mu}\right) \qquad (4-17)$$

or

$$K = K_i\left(\frac{\rho g}{\mu}\right) \qquad (4-18)$$

where g is the acceleration of gravity and ρ is the density.

Units for K_i can be in square feet, square meters, or square centimeters. In the petroleum industry, the *darcy* is used as a unit of intrinsic permeability. (The petroleum engineer is similarly concerned with the occurrence and movement of fluids through porous media.) The darcy is defined as

$$1 \text{ darcy} = \frac{\dfrac{1 \text{ cP} \times 1 \text{ cm}^3/\text{s}}{1 \text{ cm}^2}}{1 \text{ atm}/1 \text{ cm}}$$

This expression can be converted to square centimeters, since

$$1 \text{ cP} = 0.01 \text{ dyn·s/cm}^2$$

and

$$1 \text{ atm} = 1.0132 \times 10^6 \text{ dyn/cm}^2$$

Substituting into the definition of the darcy, it may be seen that

$$1 \text{ darcy} = 9.87 \times 10^{-9} \text{ cm}^2$$

Both the viscosity and the density of a fluid are functions of its temperature. The colder the fluid, the more viscous it is. There is also a more complex relationship between temperature and density, as the density of water increases with temperature to 4°C, at which temperature it is at a maximum. The hydraulic conductivity of a rock or sediment will vary with the temperature of the water. As solutions become saline, this may also affect the values of specific gravity and viscosity, which will also cause the hydraulic conductivity to vary. The laboratory, or standard, value of hydraulic conductivity is defined for pure water at a temperature of 15.6°C. The units are defined in terms of length/time (Table 4.5). In the United States the unit feet per day is used in hydrogeological practice and centimeters per second is used in soils engineering practice. In the SI system of units hydraulic conductivity is in meters per day. A derived unit of gallons per day per square foot was used in the past. It can be converted to feet per day by dividing by 7.48.

TABLE 4.5 Conversion values for hydraulic conductivity

1 gal/day/ft²	= 0.0408 m/day
1 gal/day/ft²	= 0.134 ft/day
1 gal/day/ft²	= 4.72×10^{-5} cm/s
1 ft/day	= 0.305 m/day
1 ft/day	= 7.48 gal/day/ft²
1 ft/day	= 3.53×10^{-4} cm/s
1 cm/s	= 864 m/day
1 cm/s	= 2835 ft/day
1 cm/s	= 21,200 gal/day/ft²
1 m/day	= 24.5 gal/day/ft²
1 m/day	= 3.28 ft/day
1 m/day	= 0.00116 cm/s

TABLE 4.6 Ranges of intrinsic permeabilities and hydraulic conductivities for unconsolidated sediments

Material	Intrinsic Permeability (darcys)	Hydraulic Conductivity (cm/s)
Clay	$10^{-6}-10^{-3}$	$10^{-9}-10^{-6}$
Silt, sandy silts, clayey sands, till	$10^{-3}-10^{-1}$	$10^{-6}-10^{-4}$
Silty sands, fine sands	$10^{-2}-1$	$10^{-5}-10^{-3}$
Well-sorted sands, glacial outwash	$1-10^2$	$10^{-3}-10^{-1}$
Well-sorted gravel	$10-10^3$	$10^{-2}-1$

4.4.3 Permeability of Sediments

Unconsolidated coarse-grained sediments represent some of the most prolific producers of ground water. Likewise, clays are often used for engineering purposes, such as lining solid-waste disposal sites, because of their extremely low intrinsic permeability. There is obviously a wide-ranging continuum of permeability values for unconsolidated sediments (Table 4.6).

The intrinsic permeability is a function of the size of the pore opening. The smaller the size of the sediment grains, the larger the surface area the water contacts (Figure 4.13). This increases the frictional resistance to flow, which reduces the intrinsic permeability. For well-sorted sediments, the intrinsic permeability is proportional to the grain size of the sediment (Norris & Fidler 1965).

For sand-sized alluvial deposits, several factors relating intrinsic permeability to grain size have been noted (Masch & Denny 1966). These observations would hold true for all sedimentary deposits, regardless of origin of deposition.

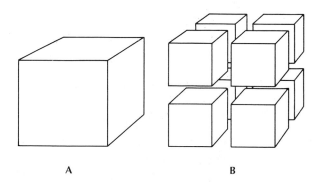

A   B

FIGURE 4.13 Relationship of sediment grain size to surface area of pore space. **A.** A cube of sediment with a surface area of 6 square units. **B.** The cube has been broken into 8 pieces, each with a diameter of one-half of the cube in part A. The surface area has increased to 12 square units—an increase of 100%.

1. As the median grain size increases, so does permeability. This is due to larger pore openings.

2. Permeability will decrease for a given median diameter as the standard deviation of particle size increases. The increase in standard deviation indicates a more poorly sorted sample, so that the finer material can fill the voids between larger fragments.

3. Coarser samples show a greater decrease in permeability with an increase in standard deviation than do fine samples.

4. Unimodal (one dominant size) samples have a greater permeability than bimodal (two dominant sizes) samples. This is again a result of poorer sorting of the sediment sizes, as the bimodal distribution indicates.

The hydraulic conductivity of sandy sediments can be estimated from the grain-size distribution curve by the **Hazen method** (Hazen 1911). The method is applicable to sands where the effective grain size (d_{10}) is between approximately 0.1 and 3.0 mm. The Hazen approximation is

$$K = C(d_{10})^2 \qquad (4\text{--}19)$$

where

K is hydraulic conductivity (cm/s)

d_{10} is the effective grain size (cm)

C is a coefficient based on the following table

Very fine sand, poorly sorted	40–80
Fine sand with appreciable fines	40–80
Medium sand, well sorted	80–120
Coarse sand, poorly sorted	80–120
Coarse sand, well sorted, clean	120–150

The work of Hazen (1911) demonstrated that hydraulic conductivity could be related to the square of a characteristic dimension of a sediment. Shepherd (1989) analyzed data from 18 published studies where hydraulic conductivity had been related to grain size. He found that all studies could be related to the general formula

$$K = Cd_{50}^{j} \qquad (4\text{--}20)$$

where

C is a shape factor

d_{50} is the mean grain size (mm)

j is an exponent

The shape factor, C, and the exponent, j, were greatest for sediments that were texturally mature, as evidenced by well-sorted samples with uniformly sized particles that had high roundness and sphericity. Both the shape factor and the exponent declined for sediments that were less texturally mature and were least for consolidated sediments.

Shepherd (1989) used the data sets to produce an idealized graph that relates hydraulic conductivity to the mean grain diameter for different sediment types (Figure 4.14). The *C* values on the graph are those that give the hydraulic

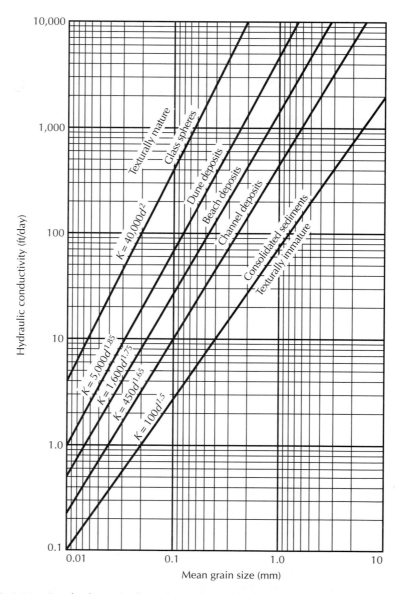

FIGURE 4.14 Graph showing the relationship of hydraulic conductivity to mean grain diameter for sediments of different textural maturity. Modified from R. G. Shepherd, *Ground Water* 27, no. 5 (1989): 633–638. Copyright © 1989 Ground Water Publishing Co.

conductivity in units of feet per day. An upper bound is presented for glass spheres, where the exponent is 2.0. Texturally mature sediments have an exponent of 1.75 or greater, whereas texturally immature sediments can have an exponent as low as 1.5.

CASE STUDY: HYDRAULIC CONDUCTIVITY ESTIMATES IN GLACIAL OUTWASH

A hazardous-waste-processing site was located on a level glacial outwash plain in southern Indiana. There were two aquifers present in the unconsolidated glacial deposits. The upper aquifer consisted of a well-sorted fine to medium sand. There were 27 ground-water monitoring wells installed in this aquifer. The lower aquifer was a poorly sorted fine to coarse sand. There were 9 monitoring wells installed in this aquifer. The grain-size analyses of the sand samples from the screen zones of the wells in each aquifer are summarized in the following table:

	Upper Aquifer		Lower Aquifer	
	Mean	Range	Mean	Range
d_{10}	0.14 mm	0.08–0.20 mm	0.16 mm	0.09–0.26 mm
d_{60}	0.31 mm	0.19–0.45 mm	2.04 mm	0.35–6.70 mm
C_u	2.29	1.50–3.89	11.01	3.89–33.50

The hydraulic conductivities of the sediments at each monitoring well were estimated by the Hazen method, using a coefficient of 100. The hydraulic conductivities of the sediments at each monitoring well were measured by means of a Hvorslev slug test performed on the well (see Section 7.5.3). The following table compares the results in centimeters per second.

	Geometric Mean (cm/s)	Range (cm/s)
	Upper Aquifer	
Hazen method	1.9×10^{-2}	$4.0 \times 10^{-2} - 6.4 \times 10^{-3}$
Hvorslev test	1.9×10^{-2}	$8.9 \times 10^{-2} - 4.2 \times 10^{-3}$
	Lower Aquifer	
Hazen method	1.2×10^{-2}	$2.6 \times 10^{-2} - 8.1 \times 10^{-3}$
Hvorslev test	1.4×10^{-2}	$1.7 \times 10^{-1} - 2.6 \times 10^{-3}$

The geometric means of the data sets were used to compare the Hvorslev test results with the Hazen method results in the above case study. Hydraulic conductivity values frequently vary by more than two orders of magnitude within the same hydrogeologic unit. An arithmetic mean of such a sample population tends to give more weight to the more permeable values. Some hydrogeologists believe that a more representative description of the average hydraulic conductivity of a hydrologic unit is the *geometric mean*. This is determined by taking the natural log of each value, finding the mean of the natural logs, and then obtaining the exponential (e^x) of that value to arrive at the geometric mean.

EXAMPLE PROBLEM

Find the geometric mean of the following set of hydraulic conductivity values and compare it to the arithmetic mean:

Hydraulic conductivity (K)	ln (K)
2.17×10^{-2} cm/s	-3.83
2.58×10^{-2} cm/s	-3.66
2.55×10^{-3} cm/s	-5.97
1.67×10^{-1} cm/s	-1.79
9.50×10^{-4} cm/s	-6.96
Sum: 2.18×10^{-1} cm/s	-22.21

Geometric mean: mean $\ln(K)$: $-22.21/5 = -4.44$

exp [mean $\ln(K)$]: $e^{-4.44} = 1.18 \times 10^{-2}$ cm/s

Arithmetic mean: $(2.18 \times 10^{-1})/5 = 4.36 \times 10^{-2}$ cm/s

---

Field studies of hydraulic permeability based on grain-size analysis of sediments from test borings and slug tests in monitoring wells yield information on the distribution of hydraulic conductivity across the site. Aquifer pumping tests, discussed in Chapter 7, are another way to determine the hydraulic conductivity of rock and soil in the field. Aquifer tests integrate the distributed permeability and give an average permeability over a large area.

4.4.4   Permeability of Rocks

The intrinsic permeability of rocks is due to primary openings formed with the rock and secondary openings created after the rock was formed. The size of openings, the degree of interconnection, and the amount of open space are all significant.

Clastic sedimentary rocks have primary permeability characteristics similar to those of unconsolidated sediments. However, diagenesis can reduce the size of the throats that connect adjacent pores through cementation and compaction. This could reduce permeability substantially without a large impact on primary porosity. Primary permeability may also be due to sedimentary structures, such as bedding planes.

Crystalline rocks, whether of igneous, metamorphic, or chemical origin, typically have a low primary permeability, in addition to a low porosity. The intergrown crystal structure contains very few openings, so fluids cannot pass through as readily. The exceptions to this are volcanic rocks, which can have a high primary porosity. If the openings are large and well connected, then high permeability may also be present.

Secondary permeability can develop in rocks through fracturing. The increase in permeability is initially due to the number and size of the fracture openings. As water moves through the fractures, minerals may be dissolved from the rock and the fracture enlarged. This increases the permeability. Chemically precipitated rocks (limestone, dolomite, gypsum, halite) are most susceptible to solution enlargement, although even igneous rocks may be so affected. Bedding-plane openings of sedimentary rocks may also be enlarged by solution.

Weathering is another process that can result in an increase in permeability. As the rock is decomposed or disintegrated, the number and size of pore spaces, cracks, and joints can increase.

EXAMPLE PROBLEM

The intrinsic permeability of a consolidated rock is 2.7×10^{-3} darcy. What is the hydraulic conductivity for water at 15°C?

At 15°C for water, from Appendix 14:

$$\rho = 0.999099 \text{ g/cm}^3$$

$$\mu = 0.011404 \frac{\text{g}}{\text{s·cm}}$$

The acceleration of gravity is given as

$$g = 980 \text{ cm/s}^2$$

As 1 darcy $= 9.87 \times 10^{-9}$ cm^2, the intrinsic permeability is 2.66×10^{-11} cm^2:

$$K = K_i\left(\frac{\rho g}{\mu}\right) = 2.66 \times 10^{-11} \text{ cm}^2 \times \frac{0.999099 \text{ g/cm}^3 \times 980 \text{ cm/s}^2}{0.011404 \text{ g/s·cm}}$$

$$K = 2.28 \times 10^{-6} \frac{\text{g/cm}^3 \times \text{cm/s}^2 \times \text{cm}^2}{\text{g/s·cm}}$$

$$= 2.28 \times 10^{-6} \text{ cm/s}$$

4.5   PERMEAMETERS

The value of the hydraulic conductivity of earth materials can be measured in the laboratory. Not surprisingly, the devices used to do this are called **permeameters.**

Permeameters all have some type of a chamber to hold a sample of rock or sediment. Rock permeameters hold a core of solid rock, usually cylindrical. Unconsolidated samples may be remolded into the permeameter chamber. It is also possible to make permeability analyses of "undisturbed" samples of unconsolidated materials if they are left in the field-sampling tubes, which become the permeameter sample chambers. If sediments are repacked into the permeameter, they will yield values of hydraulic conductivity that only approximate the value of K for undisturbed material. Recompacted hydraulic conductivities depend upon the density to which the sample is compacted.

The *constant-head permeameter* is used for noncohesive sediments, such as sand and rocks. A chamber with an overflow provides a supply of water at a constant head. Water moves through the sample at a steady rate. The hydraulic conductivity is determined from a variation of Darcy's law, which gives the flux of fluid per unit time, which we have called the discharge, Q. If we collect the fluid draining from a permeameter over some time, t, the total volume, V, is the product of the discharge and time. If we multiply both sides of Equation 4–12 by time, t, and rearrange, we obtain

$$Qt = -\frac{KAt(h_A - h_B)}{L} \tag{4–21}$$

If we substitute V for Qt and use h for $-(h_A - h_B)$, Equation 4–21 can be rearranged to form

$$\boxed{K = \frac{VL}{Ath}} \tag{4–22}$$

where

V is the volume of water discharging in time t (L^3; cm^2, and T; s)

L is the length of the sample (L; cm)

A is the cross-sectional area of the sample (L^2; cm^2)

h is the hydraulic head (L; cm)

K is the hydraulic conductivity (L/T; cm/s)

A constant-head permeameter is illustrated in Figure 4.15. It is critical in such permeameters to have hydraulic gradients approaching those in the field. The head should never be more than about 0.5 of the sample length. Some commercial permeameters permit heads of up to ten times the sample length. Under such conditions, the Reynolds number may become so high that Darcy's law is invalidated. If the permeameter is designed for upward flow, too great an upward-flow velocity may result in quicksand conditions in the permeameter.

For cohesive sediments with low conductivities, a *falling-head permeameter* is used (Figure 4.16). A much smaller volume of water moves through the sample. A falling-head tube is attached to the permeameter. The initial water level above the outlet in the falling-head tube, h_0, is noted. After some time period, t (generally several hours), the water level, h, is again measured. The

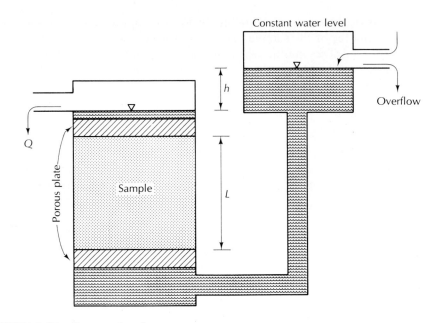

FIGURE 4.15   Constant-head permeameter apparatus.

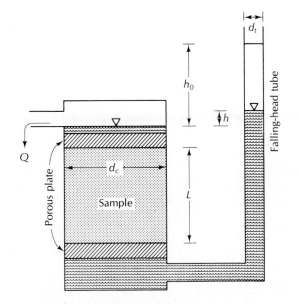

FIGURE 4.16   Falling-head permeameter apparatus.

inside diameter of the falling-head tube, d_t, the length of the sample, L, and the diameter of the sample, d_c, must also be known.

The rate at which water will drain from the falling-head tube into the sample chamber is the change in head with time multiplied by the cross-sectional area, A_t, of the falling-head tube.

$$q_{in} = -A_t \frac{dh}{dt} \tag{4-23}$$

If A_c is the cross-sectional area of the sample chamber, we can determine the volume of water draining from the sample chamber from Equation 4-22:

$$q_{out} = \frac{KA_c h}{L} \tag{4-24}$$

Under the principle of continuity, the volume of water entering the sample chamber must be equal to the volume draining from it, i.e., $q_{in} = q_{out}$.

$$-A_t \frac{dh}{dl} = \frac{KA_c H}{L} \tag{4-25}$$

Equation 4-25 can be rearranged to isolate K:

$$K = -\frac{A_t L}{A_c} \frac{dh}{dt} \tag{4-26}$$

If we integrate Equation 4-26 from $t = 0$ to $t = 1$ with the initial condition $h = h_0$ at $t = 0$, we obtain the formula by which hydraulic conductivity is determined by a falling-head permeameter:

$$K = \frac{A_t L}{A_c t} \ln \frac{h_0}{h} \tag{4-27}$$

It may be more convenient to use the diameters of the falling-head tube and the sample chamber rather than their cross-sectional areas. Since the area of a circle is proportional to the square of its diameter, the falling-head permeameter equation can also be expressed as

$$\boxed{K = \frac{d_t^2 L}{d_c^2 t} \ln\left(\frac{h_0}{h}\right)} \tag{4-28}$$

where

K is hydraulic conductivity (L/T; cm/s)

L is sample length (L; cm)

h_0 is initial head in the falling tube (*L;* cm)

h is final head in the falling tube (*L;* cm)

t is the time that it takes for the head to go from h_0 to h (*T;* s)

d_t is the inside diameter of the falling head tube (*L;* cm)

d_c is the inside diameter of the sample chamber (*L;* cm)

In using any permeameter, it is critical that the sample be completely saturated. Air bubbles in the sample will reduce the cross-sectional flow area, resulting in lowered measurements of conductivity. The sample must also be tightly pressed against the sidewall of the chamber. If it is not, water may move along the sidewall, avoiding the porous medium. In this case, measurements of conductivity may be too high.

EXAMPLE PROBLEM

A constant-head permeameter has a sample of medium-grained sand 15 cm in length and 25 cm^2 in cross-sectional area. With a head of 5.0 cm, a total of 100 mm of water is collected in 12 min. Find the hydraulic conductivity.

$$K = \frac{VL}{Ath}$$

$$K = \frac{100 \text{ cm}^3 \times 15 \text{ cm}}{25 \text{ cm}^2 \times 12 \text{ min} \times 60 \text{ s/min} \times 5 \text{ cm}}$$

$$= 1.7 \times 10^{-2} \text{ cm/s or 14 m/day}$$

EXAMPLE PROBLEM

A falling-head permeameter containing a silty, fine sand has a falling-head tube diameter of 2.0 cm, a sample diameter of 10 cm, and a flow length of 15 cm. The initial head is 5.0 cm. It falls to 0.50 cm over a period of 528 min. Find the hydraulic conductivity.

$$K = \frac{d_t^2 L}{d_c^2 t} \ln\left(\frac{h_0}{h}\right)$$

$$= \frac{2.0^2 \text{ cm}^2}{10^2 \text{ cm}^2} \times \frac{15 \text{ cm}}{528 \text{ min} \times 60 \text{ s/min}} \times \ln\frac{5.0 \text{ cm}}{0.50 \text{ cm}}$$

$$= 4.4 \times 10^{-5} \text{ cm/s or } 3.8 \times 10^{-2} \text{ m/day}$$

4.6   WATER TABLE

Water may be present beneath the earth's surface as a liquid, solid, or vapor. Other gases may also be present, either in vapor phase or dissolved in water. In the lower zone of porosity, generally all that is present is mineral matter and liquid water. The rock is saturated with water, and the water may also contain dissolved

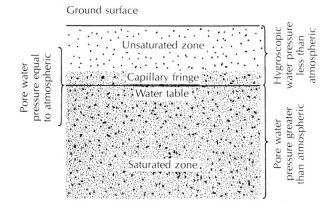

FIGURE 4.17   Distribution of fluid pressures in the ground with respect to the water table.

gas. The fluid pressure is greater than atmospheric pressure owing to the weight of overlying water. As the surface is approached, the fluid pressure decreases as the thickness of fluid above it decreases. At some depth, which varies from place to place, the pressure of the fluid in the pores is equal to atmospheric pressure. The undulating surface at which pore water pressure is equal to atmospheric pressure is called the *water table* (Figure 4.17).

Water in a shallow well (a few feet or so below the water table) will rise to the elevation of the water table at that location. The position of the water table often follows the general shape of the topography, although the water-table relief is not as great as the topographic relief. At all depths below the water table, the rock is generally saturated with water.*

A hypothetical experiment can serve to illustrate the formation of the water table. A box made of clear plastic is filled with sand. A notch is cut in one side of the plastic, and the surface of the sand is smoothed to model a valley draining toward the notch. A fine mist of water is then spread evenly over the surface of the sand, simulating rainfall. The precipitation rate is sufficiently low to preclude any overland flow. The water will move downward through the sand, so that, eventually, a zone of saturation will develop at the bottom. As shown in Figure 4.18A, this zone will have a level surface. As more rainfall is simulated, the water table will rise, continuing to be perfectly flat. It will follow this pattern until the water table reaches the lowest point in the valley.

Continuing rainfall will cause further increases in the height of the water table. In the valley, the water level will be above the surface, so that water will now flow through the notch. Elsewhere, the water table will be higher than the elevation of the notch, and ground water will begin to flow laterally because of the

---

*There are exceptions. The rocks may contain trapped liquid and gaseous hydrocarbons, for example. Or there may be isolated voids, which cannot fill with any fluid.

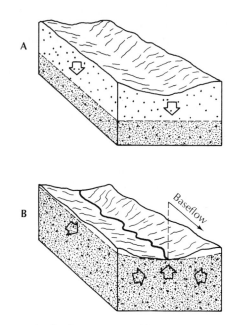

FIGURE 4.18 A. Diagram of a flat-lying water table in an aquifer where there is downward movement of water through the unsaturated zone but no lateral ground-water movement. **B.** Diagram of the water table in a region where water is moving downward through the unsaturated zone to the water table and moving as ground-water flow through the zone of saturation toward a discharge zone along the stream. Net discharge from the aquifer is occurring as baseflow from the stream.

hydraulic gradient. Now, water will flow through the saturated zone toward the point of discharge (Figure 4.18B).

We can make the following observations, of which items 4 and 5 pertain primarily to humid regions:

1. In the absence of ground-water flow, the water table will be flat.
2. A sloping water table indicates the ground water is flowing.
3. Ground-water discharge zones are in topographical low spots.
4. The water table has the same general shape as the surface topography.
5. Ground water generally flows away from topographical high spots and toward topographic lows.

4.7   AQUIFERS

Natural earth materials have a very wide range of hydraulic conductivities. Near the earth's surface there are very few, if any, geologic formations that are absolutely impermeable. Weathering, fracturing, and solution have affected most rocks to some degree. However, the rate of ground-water movement can be exceedingly slow in units of low hydraulic conductivity.

An **aquifer** is a geologic unit that can store and transmit water at rates fast enough to supply reasonable amounts to wells. The intrinsic permeability of aquifers would range from about 10^{-2} darcy upward. Unconsolidated sands and gravels, sandstones, limestones and dolomites, basalt flows, and fractured plutonic and metamorphic rocks are examples of rock units known to be aquifers.

A **confining layer** is a geologic unit having little or no intrinsic permeability—less than about 10^{-2} darcy. This is a somewhat arbitrary limit and depends upon local conditions. In areas of clay, with intrinsic permeabilities of 10^{-4} darcy, a silt of 10^{-2} darcy might be used to supply water to a small well. On the other hand, the same silt might be considered a confining layer if it were found in an area of coarse gravels with intrinsic permeabilities of 100 darcys. Ground water moves through most confining layers, although the rate of movement is very slow.

Confining layers are sometimes subdivided into aquitards, aquicludes, and aquifuges. An **aquifuge** is an absolutely impermeable unit that will not transmit any water. An **aquitard** is a layer of low permeability that can store ground water and also transmit it slowly from one aquifer to another. The term **leaky confining layer** is also applied to such a unit. Most authors now use the terms *confining layer* and *leaky confining layer*.

Confining layers can be important elements of regional flow systems, and leaky confining layers can transmit significant amounts of water if the cross-sectional area is large. On Long Island, New York, there is a deep aquifer, the Lloyd Sand Member of the Raritan Formation. The recharge to this aquifer passes downward across a confining layer 200 to 300 ft (60 to 100 m) thick, the Clay Member of the Raritan Formation. Over an areal extent measured in hundreds of square miles, a very low rate of vertical seepage provides recharge to the underlying aquifer. Some of the recharge to the principal artesian aquifer of the southeastern United States takes place by downward leakage through overlying confining layers, although much of the recharge is carried through sinkholes.

Aquifers can be close to the land surface, with continuous layers of materials of high intrinsic permeability extending from the land surface to the base of the aquifer. Such an aquifer is termed a *water-table aquifer* or **unconfined aquifer.** Recharge to the aquifer can be from downward seepage through the unsaturated zone (Figure 4.19). Recharge can also occur through lateral ground-water flow or upward seepage from underlying strata.

Some aquifers, called **confined,** or *artesian,* **aquifers,** are overlain by a confining layer. Recharge to them can occur either in a recharge area, where the aquifer crops out, or by slow downward leakage through a leaky confining layer (Figure 4.20). If a tightly cased well is placed through the confining layer, water from the aquifer may rise considerable distances above the top of the aquifer

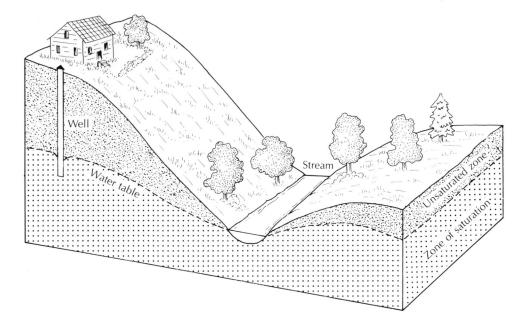

FIGURE 4.19   Unconfined, or water-table, aquifer.

(Figure 4.21). This indicates that the water in the aquifer is under pressure. The **potentiometric surface** for a confined aquifer is the surface representative of the level to which water will rise in a well cased to the aquifer. (The term *piezometric* was used in the past, but it has now been replaced by *potentiometric*.) If the potentiometric surface of an aquifer is above the land surface, a flowing artesian well may occur. Water will flow from the well casing without need for a pump. Of course, if a pump were installed, the amount of water obtained from the well could be increased.

In some cases, a layer of low-permeability material will be found as a lens in more permeable materials. Water moving downward through the unsaturated zone will be intercepted by this layer and will accumulate on top of the lens. A layer of saturated soil will form above the main water table. This is termed a **perched aquifer** (Figure 4.22). Water moves laterally above the low-permeability layer up to the edge and then seeps downward toward the main water table or forms a spring. Perched aquifers are common in glacial outwash, where lenses of clay formed in small glacial ponds are present. They are also often present in volcanic terranes, where weathered ash zones of low permeability can occur sandwiched between high-permeability basalt layers.

Perched aquifers are usually not very large; most would supply only enough water for household use. Some lakes are perched on low-permeability

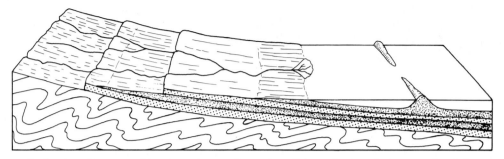

Confined aquifers created by alternating aquifers and confining units deposited on a regional dip.

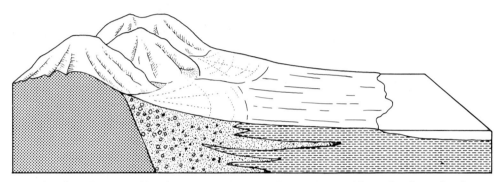

Confined aquifers created by deposition of alternating layers of permeable sand and gravel and impermeable silts and clays deposited in intermontane basins.

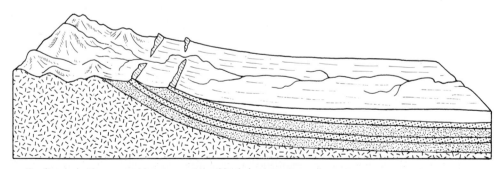

Confined aquifer created by upwarping of beds by intrusions.

FIGURE 4.20   Confined aquifers created when aquifers are overlain by confining beds.

sediments. Such ponds are especially vulnerable to widely fluctuating lake-stage levels with changes in the amount of rainfall.

      Confined ground water is found in a confined aquifer, **unconfined ground water** is found in a water table aquifer, and **perched ground water** is found in a perched aquifer.

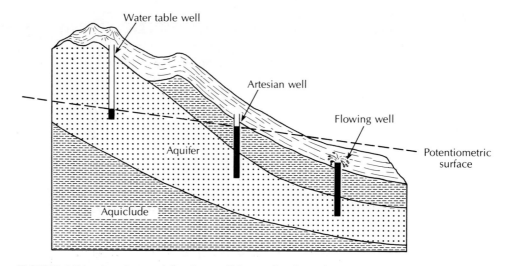

FIGURE 4.21 Artesian and flowing well in confined aquifer.

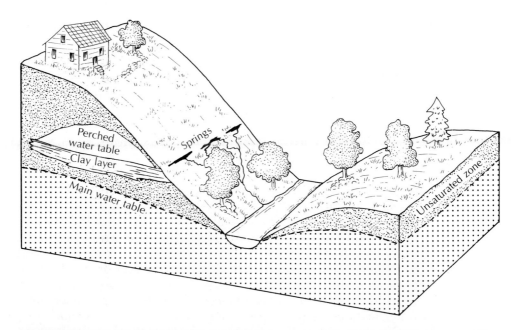

FIGURE 4.22 Perched aquifer formed above the main water table on a low-permeability layer in the unsaturated zone.

4.8 WATER TABLE AND POTENTIOMETRIC SURFACE MAPS

Maps of the water table for an unconfined aquifer and of the potentiometric surface of a confined aquifer are basic tools of hydrogeologic interpretation. These maps are two-dimensional representations of three-dimensional surfaces. Such maps can be shown as contour maps with lines of equal elevation. They can also be shown as perspective drawings representing a three-dimensional view of the surface.

The data used to construct water-table and potentiometric-surface maps are water-level elevations as measured in wells. However, not every well is useful for this purpose. Some water-supply wells are open borings in rocks that include both aquifers and confining beds. Other water-supply wells may have more than one well screen, each opposite a different aquifer. Since the water level in such wells is a reflection of the heads in several different units and not one specific unit, they are not useful in making water-level maps.

In order to make a ground-water level map, one needs water-level readings made in a number of wells, each of which is open only in the aquifer of interest. Since ground-water levels can change with time, all the readings should be made within a short period of time. Some measuring point on each well needs to be surveyed to a common datum so that the water levels can be referenced to a height above the datum; mean sea level is a common datum for this purpose.

If water levels are to be measured in a well that is normally used for water supply, one must make sure that the pump has been shut off long enough for the water level to recover to what is termed the static, or nonpumping, level. Depth to water measurements can be made every few minutes until the water level stops rising.

When making a water-table map, it is ideal if all the wells have an open borehole or a well screen at the depth of the water table. However, wells that are cased or screened below the water table can be used if they don't extend too far below the water table. For wells used to make a potentiometric-surface map, all aquifers above the aquifer of interest should be cased off.

Surface-water features such as springs, ponds, lakes, streams, and rivers can interact with the water table. In addition, the water table is often a subdued reflection of the surface topography. All this must be taken into account when preparing a water-table map. A base map showing the surface topography and the locations of surface-water features should be prepared. The elevations of lakes and ponds can be helpful information. The locations of the wells are then plotted on the base map, and the water-level elevations are noted. The datum for the water level in wells should be the same as the datum for the surface topography. Contours of equal ground-water elevations are then drawn, following the rules of contouring discussed for isohyets in Section 2.8. Interpolation of contours between data points is strongly influenced by the surface topography and surface-water features. For example, ground-water contours cannot be higher than the surface topography. The depth to ground water will typically be greater beneath hills than beneath valleys. If a lake is present, the lake surface is flat and the water table beneath it is also flat. Hence, ground-water contours must go around it. The

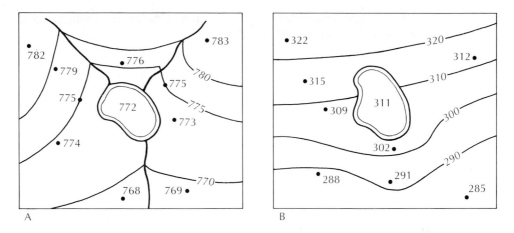

FIGURE 4.23 Maps showing construction of water-table maps in areas with surface-water bodies. **A.** A water-table lake with two gaining streams draining into it and one gaining stream draining from it. **B.** A perched lake that, through outseepage, is recharging the water table.

only exception to this rule is when the lake is perched on low-permeability sediments and has a surface elevation above the main water table. Ground-water contours form a V, pointing upstream when they cross a gaining stream. Ground-water contours bend downstream when they cross a losing stream.

Figure 4.23A is a water-table map where there is a gaining stream and a lake that is hydraulically connected with the water table. Figure 4.23B is a water-table map where there is a perched lake. Water is seeping from the perched lake, so the water-table contours bend down-gradient away from the lake.

In general, the potentiometric surface of a confined aquifer is not influenced by the surface topography and surface-water features. Because there is no hydraulic connection between a river and a confined aquifer beneath it, potentiometric-surface contours are not influenced by the presence of the river. Potentiometric-surface contours can even be above the land surface. This indicates that if a well were to be drilled at that location, it would flow.

In areas where the water table or potentiometric surface has a shallow gradient, the ground-water contours will be spaced well apart. If the gradient is steep, the ground-water contours will be closer together. Ground water will flow in the general direction that the water table or potentiometric surface is sloping.

4.9   AQUIFER CHARACTERISTICS

We have thus far considered the intrinsic permeability of earth materials and their hydraulic conductivity when transmitting water. A useful concept in many studies is aquifer **transmissivity,** which is a measure of the amount of water that can be transmitted horizontally through a unit width by the full saturated thickness of the aquifer under a hydraulic gradient of 1.

The transmissivity is the product of the hydraulic conductivity and the saturated thickness of the aquifer:

$$T = bK \tag{4-29}$$

where

T is transmissivity (L^2/T; ft^2/d or m^2/d)

b is saturated thickness of the aquifer (L; ft or m)

K is hydraulic conductivity (L/T; ft/d or m/d)

For a multilayer aquifer, the total transmissivity is the sum of the transmissivity of each of the layers:

$$T = \sum_{i=1}^{n} T_i \tag{4-30}$$

Aquifer transmissivity is a concept that assumes flow through the aquifer to be horizontal. In some cases, this is a valid assumption; in others, it is not.

When the head in a saturated aquifer or confining unit changes, water will be either stored or expelled. The *storage coefficient,* or **storativity** (*S*), is the volume of water that a permeable unit will absorb or expel from storage per unit surface area per unit change in head. It is a dimensionless quantity.

In the saturated zone, the head creates pressure, affecting the arrangement of mineral grains as well as the density of the water in the voids. If the pressure increases, the mineral skeleton will expand; if it drops, the mineral skeleton will contract. This is known as *elasticity.* Likewise, water will contract with an increase in pressure and expand if the pressure drops. When the head in an aquifer or confining bed declines, the aquifer skeleton compresses, which reduces the effective porosity and expels water. Additional water is released as the pore water expands due to lower pressure.

The **specific storage** (S_s) is the amount of water per unit volume of a saturated formation that is stored or expelled from storage owing to compressibility of the mineral skeleton and the pore water per unit change in head. This is also called the *elastic storage coefficient.* The concept can be applied to both aquifers and confining units.

The specific storage is given by the expression (Jacob 1940, 1950; Cooper 1966)

$$S_s = \rho_w g(\alpha + n\beta) \tag{4-31}$$

where

ρ_w is the density of the water (M/L^3; slug/ft^3 or kg/m^3)

g is the acceleration of gravity (L/T^2; ft/s^2 or m/s^2)

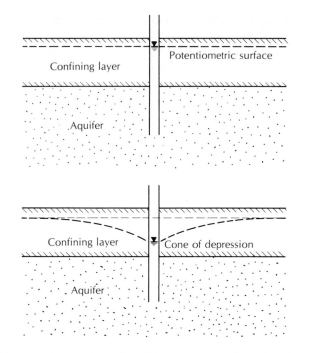

FIGURE 4.24 Diagram showing lowering of the potentiometric surface in a confined aquifer with the resultant water level still above the aquifer materials. In this circumstance, the aquifer remains saturated.

α   is the compressibility of the aquifer skeleton ($1/(M/LT^2)$; $1/(lb/ft^2)$ or $1/(N/m^2)$)

n is the porosity (L^3/L^3)

β   is the compressibility of the water* ($1/(M/LT^2)$; $1/(lb/ft^2)$ or $1/(N/m^2)$)

Specific storage has dimensions of $1/L$. The value of specific storage is very small, generally $0.0001 \ ft^{-1}$ or less.

In a confined aquifer, the head may decline—yet the potentiometric surface remains above the unit (Figure 4.24). Although water is released from storage, the aquifer remains saturated. The storativity (S) of a confined aquifer is the product of the specific storage (S_s) and the aquifer thickness (b):

$$S = bS_s \qquad\qquad (4-32)$$

Since specific storage has dimensions of $1/L$ and the aquifer thickness has dimensions of L, storativity is dimensionless.

All the water released is accounted for by the compressibility of the mineral skeleton and the pore water. The water comes from the entire thickness

---

*The compressibility of water at environmental temperatures is $4.6 \times 10^{-10} \ m^2/N$.

of the aquifer. The value of the storativity of confined aquifers is on the order of 0.005 or less.

In an unconfined unit, the level of saturation rises or falls with changes in the amount of water in storage. As the water level falls, water drains from the pore spaces. This storage or release is due to the *specific yield* (S_y) of the unit. Water is also stored or expelled depending on the specific storage of the unit. For an unconfined unit, the storativity is found by the formula

$$S = S_y + hS_s \qquad \qquad (4-33)$$

where h is the thickness of the saturated zone.

The value of S_y is several orders of magnitude greater than hS_s for an unconfined aquifer, and the storativity is usually taken to be equal to the specific yield. For a fine-grained unit, the specific yield may be very small, approaching the same order of magnitude as hS_s. Storativity of unconfined aquifers ranges from 0.02 to 0.30.

The volume of water drained from an aquifer as the head is lowered may be found from the formula

$$V_w = SA\,\Delta h \qquad \qquad (4-34)$$

where

V_w is the volume of water drained (L^3; ft^3 or m^3)

S is the storativity (dimensionless)

A is the surface area overlying the drained aquifer (L^2; ft^2 or m^2)

Δh is the average decline in head (L; ft or m)

EXAMPLE PROBLEM

An unconfined aquifer with a storativity of 0.13 has an area of 123 mi^2. The water table drops 5.23 ft during a drought. How much water was lost from storage?

$$V_w = SA\,\Delta h$$
$$= 0.13 \times 123\ \text{mi}^2 \times 2.7878 \times 10^7\ \text{ft}^2/\text{mi}^2 \times 5.23\ \text{ft}$$
$$= 2.3 \times 10^9\ \text{ft}^3$$

If the same aquifer had been confined with a storativity of 0.0005, what change in the amount of water in storage would have resulted?

$$V_w = 0.0005 \times 123\ \text{mi}^2 \times 2.7878 \times 10^7\ \text{ft}^2/\text{mi}^2 \times 5.23\ \text{ft}$$
$$= 9.0 \times 10^6\ \text{ft}^3$$

4.10   COMPRESSIBILITY AND EFFECTIVE STRESS

At a given plane in a saturated aquifer, a downward stress is placed on the aquifer skeleton by the weight of the overlying rock and water. This is called the *total stress*. There is an upward stress on the plane caused by the fluid pressure. The upward stress will, in part, counteract the total stress, so the resulting stress that is actually born by the aquifer skeleton, called the *effective stress,* is less than the total stress:

$$\sigma_T = \sigma_e + P \tag{4-35}$$

where

σ_T is total stress

P is pressure

σ_e is effective stress

If there is a change in total stress, the pressure and effective stress will also change.

$$d\sigma_T = d\sigma_e + dP \tag{4-36}$$

In confined aquifers, there can be significant changes in pressure with very little change in the actual thickness of the saturated water column. Under these conditions, the total stress remains essentially constant, and any change in pressure will result in a change in effective stress that is of equal magnitude but opposite in sign.

$$dP = -d\sigma_e \tag{4-37}$$

If pumping reduces the pressure head in a confined aquifer, the effective stress that acts on the aquifer skeleton will increase. The aquifer skeleton may consolidate or compact due to this increased stress. The consolidation occurs due to shifting of the mineral grains, which reduces the porosity.

Aquifer compressibility is defined as

$$\alpha = \frac{-db/b}{d\sigma_e} \tag{4-38}$$

where

α is aquifer compressibility ($1/(M/L^2)$; ft^2/lb or m^2/N)

db is change in aquifer thickness (L; ft or m)

b is original aquifer thickness ($L;$ ft or m)

$d\sigma_e$ is change in effective stress (M/LT^2; lb/ft^2 or N/m^2)

The negative sign indicates that the aquifer gets smaller with an increase in effective stress.

Since $dP = -d\sigma_e$, Equation 4–38 can also be written as

$$\alpha = \frac{db/b}{dP} \qquad (4-39)$$

EXAMPLE PROBLEM

A confined aquifer with an initial thickness of 45 m consolidates (compacts) 0.20 m when the head is lowered by 25 m.

Part A: What is the vertical compressibility of the aquifer?

The given parameter values are $dP = 25$ m, $b = 45$ m, and $db = 0.20$ m. A pressure head of 25 m of water can be converted to a fluid pressure by multiplying the pressure head by the density of water times the gravitational constant.

$$dP = 25 \text{ m} \times 1000 \text{ kg/m}^3 \times 9.8 \text{ m/s}^2 = 245{,}000 \text{ N/m}^2$$

From Equation 4–39,

$$\alpha = \frac{(0.20 \text{ m})/(45 \text{ m})}{245{,}000 \text{ N/m}^2}$$

$$\alpha = 1.8 \times 10^{-8} \text{ m}^2/\text{N}$$

Part B: If the porosity of the aquifer is 12 percent after compaction, calculate the storativity of the aquifer.

Aquifer storativity is found from Equations 4–31 and 4–32:

$$S = b[\rho_w g(\alpha + n\beta)]$$

The given parameter values are $b = 44.8$ m, $n = 0.12$, $\rho_w = 1000$ kg/m^3, $g = 9.8$ m/s^2, $\alpha = 1.8 \times 10^{-8}$ m^2/N, and $\beta = 4.6 \times 10^{-10}$ m^2/N.

$$S = (44.8 \text{ m})[1000 \text{ kg/m}^3 \times 9.8 \text{ m/s}^2(1.8 \times 10^{-8} \text{ m}^2/\text{N} + 0.13 \times 4.6 \times 10^{-10} \text{ m}^2/\text{N})]$$
$$= (44.8 \text{ m})(9800 \text{ N/m}^3)(1.806 \times 10^{-8} \text{ m}^2/\text{N})$$
$$= 7.9 \times 10^{-3}$$

4.11 HOMOGENEITY AND ISOTROPY

Hydrogeologists are interested in two key properties of geologic formations: hydraulic conductivity and specific storage or specific yield. A third property, the thickness, is also important, since the overall hydrogeologic response of a unit is a function of the product of the hydraulic parameters and the thickness.

A **homogeneous** unit is one that has the same properties at all locations. For a sandstone, this would indicate that the grain-size distribution, porosity, degree of cementation, and thickness are variable only within small limits. The values of the transmissivity and storativity of the unit would be about the same wherever present. A plutonic or metamorphic rock would have the same amount of fracturing everywhere, including the strike and dip of the joint sets. A limestone would have the same amount of jointing and solution openings at all locations.

In **heterogeneous** formations, hydraulic properties change spatially. One example would be a change in thickness. A sandstone that thickens as a wedge is nonhomogeneous, even if porosity, hydraulic conductivity, and specific storage remain constant. The change in thickness results in a change in the hydraulic properties of the unit. Layered units may also be nonhomogeneous. Most sedimentary units were deposited as successive layers of sediments with intervals of nondeposition. These layers are known to vary in thickness, from microscopic layers to those measured in meters. If the hydraulic properties of the layers are different, the entire unit is heterogeneous. The individual beds may be homogeneous. The third type of heterogeneity in a sedimentary unit occurs when a facies change in the unit involves a transformation of hydraulic characteristics as well as lithologic features. Figure 4.25 illustrates these examples of heterogeneity in clastic sedimentary units.

Carbonate units may be heterogeneous (1) if there is a change in thickness or (2) if the degree of solution openings of fractures varies. The formation of solution passageways by moving ground water is typically concentrated along

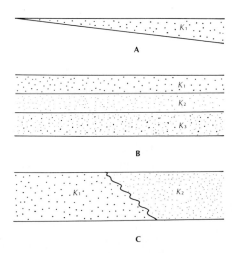

FIGURE 4.25 **A.** Heterogeneous formation consisting of a sediment that thickens in a wedge. **B.** Heterogeneous formation consisting of three layers of sediments of differing hydraulic conductivity. **C.** Heterogeneous formation consisting of sediments with different hydraulic conductivities lying next to each other.

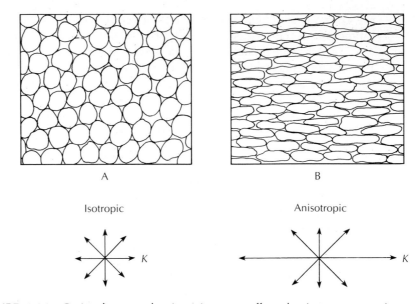

FIGURE 4.26 Grain shape and orientation can affect the isotropy or anisotropy of a sediment.

preferred fractures or bedding planes; thus, limestone formations are often heterogeneous. Plutonic rocks may have uneven fracturing or sporadic shear zones that render them heterogeneous. Basalt flows are virtually always heterogeneous by the very nature of the way they are formed. As might be expected, it is a very unusual geologic formation that is perfectly homogeneous. Geologic processes operate at varying rates and over uneven terrain, resulting in heterogeneity.

In a porous medium made of spheres of the same diameter packed uniformly, the geometry of the voids is the same in all directions. Thus, the intrinsic permeability of the unit is the same in all directions, and the unit is said to be **isotropic.** On the other hand, if the geometry of the voids is not uniform, there may be a direction in which the intrinsic permeability is greater. The medium is thus **anisotropic.** For example, a porous medium composed of book-shaped grains arranged in a subparallel manner would have a greater permeability parallel to the grains than crossing the grain orientation (Figure 4.26).

In fractured rock units, the direction of ground-water flow is completely constrained by the direction of the fractures. There may be zero intrinsic permeability in directions not parallel to a set of fractures (Figure 4.27). Basalt flows are highly anisotropic, as flow parallels the dip of the flow as it moves in the interflow zones. Shrinkage cracks in the basalt are vertical, yielding some vertical permeability.

In sedimentary units, there may be several layers, each of which is homogeneous. The equivalent vertical and horizontal hydraulic conductivity can be easily computed. Figure 4.28 shows a three-layered unit, each unit having a different horizontal and vertical hydraulic conductivity (K_h and K_v).

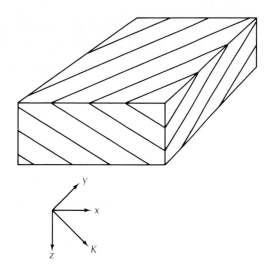

FIGURE 4.27 Anisotropy of fractured rock units due to directional nature of fracturing.

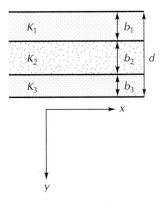

FIGURE 4.28 Heterogeneous formation consisting of three layers of differing hydraulic conductivity.

The average horizontal conductivity is found from the summation

$$K_h \text{ avg} = \sum_{m=1}^{n} \frac{K_{hm} b_m}{b} \qquad (4\text{–}40)$$

where

K_h avg is the average horizontal hydraulic conductivity (L/T; ft/d or m/d)

K_{hm} is the horizontal hydraulic conductivity of the mth layer (L/T; ft/d or m/d)

b_m is the thickness of the mth layer (L; ft or m)

b is the total aquifer thickness (L; ft or m)

The overall vertical hydraulic conductivity is given by

$$K_v \text{ avg} = \frac{b}{\sum\limits_{m=1}^{n} \dfrac{b_m}{K_{vm}}} \qquad \textbf{(4–41)}$$

where

K_v avg is the average vertical hydraulic conductivity (L/T; ft/d or m/d)

K_{vm} is the vertical hydraulic conductivity of the mth layer (L/T; ft/day or m/day)

b_m is the thickness of the mth layer (L; ft or m)

b is the total aquifer thickness (L; ft or m)

4.12   GRADIENT OF THE POTENTIOMETRIC SURFACE

On some occasions there may be too few wells in an area to make a full map of the water table or the potentiometric surface. For example, a waste-disposal site may have only three of four monitoring wells surrounding it. There are graphical methods available to determine the direction that the water table or potentiometric surface is sloping and the gradient within the area outlined by the wells. Vacher (1989) has presented an analytical solution to the three-point problem with a computer code for convenient solution.

Figure 4.29 illustrates the graphical method for three- and four-well situations. The first step is to make a sketch to scale showing the positions of the wells. This information can usually be traced from a base map of the site. Then follow these steps:

1. Draw a line that connects each well of the three-well setup (Figure 4.29A) or the corner wells for the four-well setup (Figure 4.29B).
2. Note the water elevation in each well.
3. Measure the map distance between a well pair.
4. Find the difference in elevation between a well pair.
5. Find map distance for each unit change in head for a well pair by dividing the head difference by the map distance between the well pairs.
6. Mark even increments along the line between the well pair. Select the increment length so that each increment is a convenient length, e.g., 0.5 ft, 1 ft, 5 ft, and 10 ft or 0.5 m, 1 m, 5 m, and 10 m.

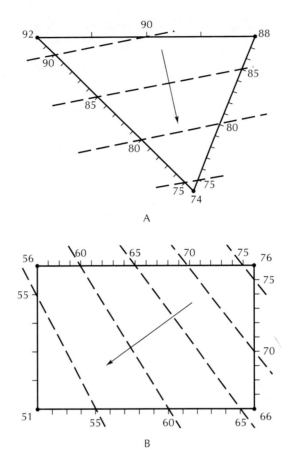

FIGURE 4.29 Graphical method for determining the slope of a potentiometric surface from **A.** three wells and **B.** four wells.

7. Repeat steps 3 to 6 for all well pairs.

8. Contour lines are created by joining all lines of equal head.

9. The gradient of the surface is in the direction of decreasing head and perpendicular to the contour lines.

If three wells form a right triangle, then the direction of the gradient can conveniently be determined mathematically (Fetter 1981b). Assume that the triangle formed by the wells has two legs, OX and OY, which form a 90° angle at well O.

Let dh/dx be the gradient measured from well O to well X. The gradient is the head difference in the two wells divided by the distance separating them. Let dh/dy be the gradient measured from well O to well Y. The gradient of head perpendicular to the equipotential lines is found from

$$\text{grad } h = \sqrt{(dh/dx)^2 + (dh/dy)^2} \qquad \textbf{(4–42)}$$

The angle θ that the direction of the gradient makes with line OX is

$$\theta = \arctan \frac{(dh/dy)}{(dh/dx)} \qquad\qquad (4\text{--}43)$$

NOTATION

a Acceleration

A Area

A_t Cross-sectional area of a falling-head tube

A_c Cross-sectional area of a permeameter sample chamber

b Aquifer thickness

c Shape factor

c_u Uniformity coefficient

d Grain size

D Distance

d_t Inside diameter of falling-head tube

d_c Inside diameter of a permeameter sample chamber

F Force

g Gravitational constant

h Head

j An exponent

K Hydraulic conductivity

K_h Horizontal hydraulic conductivity

K_i Intrinsic permeability

K_v Vertical hydraulic conductivity

L Length

m Mass

n Porosity

P Pressure

q Flux

Q Discharge (rate)

S Storativity

S_s Specific storage

S_r Specific retention

S_y Specific yield

T Transmissivity

w Weight

V Volume

V_v Volume of voids

V_w Volume of water

W Work

α Compressibility of aquifer skeleton

β Compressibility of water

γ Specific weight

Δh Decline in head

ρ Density

ρ_b Bulk density

ρ_d Mineral particle density

ρ_w Density of water

PROBLEMS

Answers to odd-numbered problems will appear at the end of the book.

1. What is the weight in newtons of an object with a mass of 32.1 kg?

2. What is the weight in pounds of an object with a mass of 12.4 slugs?

3. An object has a mass of 55.3 kg and a volume of 0.33 m^3.
   A. What is its density?
   B. What is its specific weight?
   C. Is the object more or less dense than water?

4. An object has a mass of 723 kg and a volume of 0.56 m^3.
   A. What is its density?
   B. What is its specific weight?
   C. Is it more or less dense than water?

5. The hydraulic conductivity of a silty sand was measured in a laboratory permeameter and found to be 1.36×10^{-5} cm/s at 25°C. What is the intrinsic permeability in cm^2? Refer to Appendix 14 for values of density and viscosity.

6. The hydraulic conductivity of a coarse sand was measured in a laboratory permeameter and found to be 7.92×10^{-3} cm/s at 25°C. What is the intrinsic permeability? Refer to Appendix 14 to obtain values for density and viscosity.

7. A constant-head permeameter has a cross-sectional area of 156 cm^2. The sample is 18 cm long. At a head of 5 cm, the permeameter discharges 50 cm^3 in 193 s.
   A. What is the hydraulic conductivity in centimeters per second and feet per day?
   B. What is the intrinsic permeability if the hydraulic conductivity was measured at 15°C?

8. A constant-head permeameter has a cross-sectional area of 225 cm^2. The sample is 25 cm long. At a head of 15 cm, the permeameter discharges 50 cm^3 in 456 s.
   A. What is the hydraulic conductivity in centimeters per second and feet per day?
   B. What is the intrinsic permeability if the hydraulic conductivity was measured at 20°C?

9. An aquifer has a specific yield of 0.19. During a drought period the following declines in the water table were noted:

Area	Size	Decline
A	14 mi^2	2.75 ft
B	7 mi^2	3.56 ft
C	28 mi^2	5.42 ft
D	33 mi^2	7.78 ft

What was the total volume of water represented by the decline in the water table?

10. A confined aquifer has a specific storage of 3.5×10^{-6} ft^{-1} and a thickness of 200 ft. How much water would it yield if the water declined an average of 2.5 ft over a circular area with a radius of 375 ft?

11. A confined aquifer has a specific storage of 7.5×10^{-6} m^{-1} and a porosity of 0.3. The compressibility of water is 4.6×10^{-10} m^2/N. What is the compressibility of the aquifer skeleton?

12. A confined aquifer has a specific storage of 8.8×10^{-6} m^{-1} and a porosity of 0.25. The compressibility of water is 4.6×10^{-10} m^2/N. What is the compressibility of the aquifer skeleton?

13. An aquifer has three different formations. Formation A has a thickness of 30 ft and a hydraulic conductivity of 7.0 ft/day. Formation B has a thickness of 15 ft and a conductivity of 78 ft/day. Formation C has a thickness of 22 ft and a conductivity of 17 ft/day. Assume that each individual formation is isotropic and homogeneous. Compute both the overall horizontal and vertical conductivity.

14. Use the Hazen method to estimate the hydraulic conductivity of the sediments graphed in Figures 4.4, 4.5, and 4.30.

15. Determine the effective grain size and uniformity coefficient of the sediments graphed in Figure 4.30.

16. Figure 4.31 is a map showing the ground-water elevations in wells screened in an unconfined aquifer at Milwaukee, Wisconsin. The aquifer is in good hydraulic connection with Lake Michigan, which has a surface elevation of 580 ft above sea level. Lakes and streams are also shown on the map.

    A. Make a water-table map with a contour interval of 50 ft, starting at 550 ft.

    B. Why do you suppose that ground-water levels are below the Lake Michigan surface elevation in part of the area?

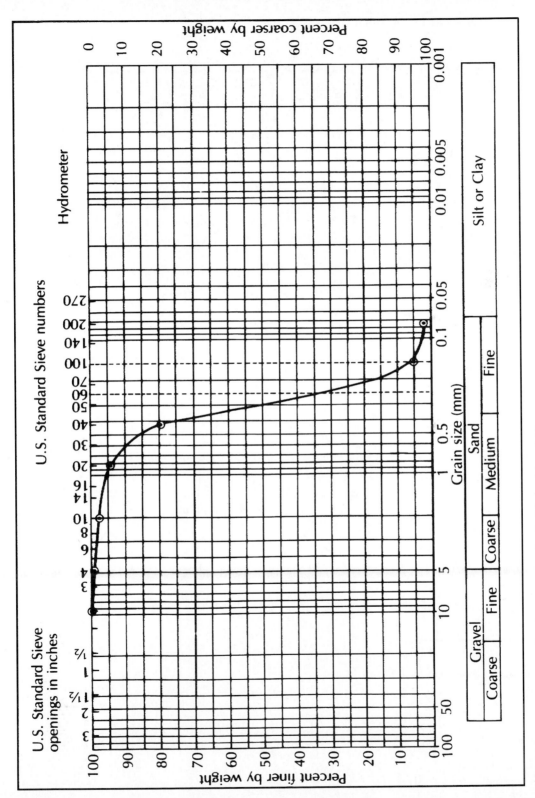

FIGURE 4.30 Grain-size distribution curve for Problems 14 and 15.

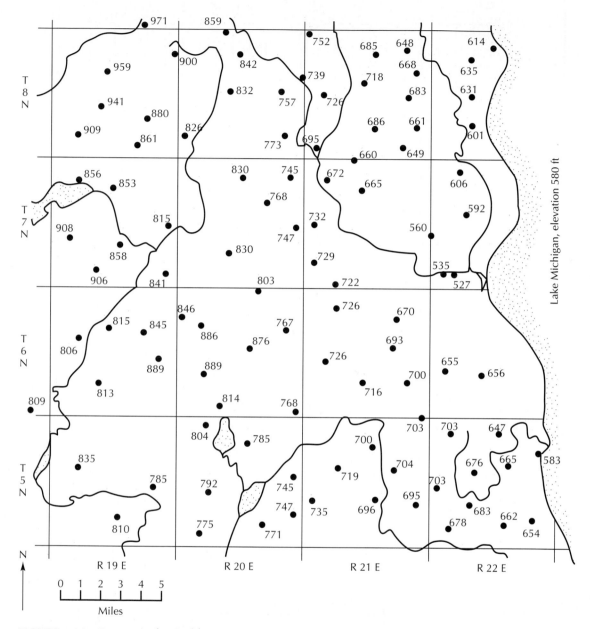

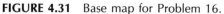

FIGURE 4.31 Base map for Problem 16.

5 Principles of Ground-Water Flow

Just as above the earth, small drops form and these join others, till finally water descends in a body as rain, so too we must suppose that in the earth the water at first trickles together little by little and that the sources of rivers drip, as it were, out of the earth and then unite.

Meteorologica, Aristotle (384–322 B.C.)

5.1 INTRODUCTION

Ground water possesses energy in mechanical, thermal, and chemical forms. Because the amounts of energy vary spatially, ground water is forced to move from one region to another in nature's attempt to eliminate these energy differentials. The flow of ground water is thus controlled by the laws of physics and thermodynamics. To enable a separate examination of mechanical energy, we will make the simplifying assumption that the water is of nearly constant temperature. Thermal energy must be considered, however, in such applications as geothermal flow systems and burial of radioactive heat sources.

There are three outside forces acting on the water contained in the ground. The most obvious of these is *gravity,* which pulls water downward. The second force is *external pressure*. Above the zone of saturation, atmospheric pressure is acting. The combination of atmospheric pressure and the weight of overlying water creates pressures in the zone of saturation. The third force is *molecular attraction,* which causes water to adhere to solid surfaces. It also creates surface tension in water when the water is exposed to air. The combination of these two processes is responsible for the phenomenon of capillarity.

When water in the ground is flowing through a porous medium, there are forces resisting the fluid movement. These consist of the *shear stresses* acting tangentially to the surface of the solid and *normal stresses* acting perpendicularly to the surface (Rumer 1969). We can think of these forces collectively as "friction." The internal molecular attraction of the fluid itself resists the movement of fluid molecules past each other. This shearing resistance is known as the viscosity of the fluid.

5.2   **MECHANICAL ENERGY**

There are a number of different types of mechanical energy recognized in classical physics. Of these, we will consider kinetic energy, gravitational potential energy, and energy of fluid pressures.

A moving body or fluid tends to remain in motion, according to Newtonian physics. This is because it possesses energy due to its motion—*kinetic energy*. This energy is equal to one-half the product of its mass and the square of the magnitude of the velocity:

$$E_k = \tfrac{1}{2}mv^2 \tag{5-1}$$

where

E_k is the kinetic energy (ML^2/T^2; slug-ft^2/s^2 or kg·m^2/s^2)

v is the velocity (L/T; ft/s or m/s)

If m is in kilograms and v in meters per second, then E_k has the units of kg·m^2/s^2 or newton-meters. The unit of energy is the joule, which is one newton-meter. The joule is also the unit of work.

Imagine that a weightless container filled with water of mass m is moved vertically upward a distance, z, from some reference surface (a datum). Work is done in moving the mass of water upward. This work is equal to

$$W = Fz = (mg)z \tag{5-2}$$

where

W is work (ML^2/T^2; slug-ft^2/s^2 or kg·m^2/s^2)

z is the elevation of the center of gravity of the fluid above the reference elevation (L; ft or m)

m is the mass (M; slugs or kg)

g is the acceleration of gravity (L/T^2; ft/s^2 or m/s^2)

F is a force (ML/T^2; slug-ft/s^2 or kg·m/s^2)

The mass of water has now acquired energy equal to the work done in lifting the mass. This is a potential energy, due to the position of the fluid mass with respect to the datum. E_g is *gravitational potential energy:*

$$W = E_g = mgz \tag{5-3}$$

A fluid mass has another source of potential energy owing to the **pressure** of the surrounding fluid acting upon it. Pressure is the force per unit area acting on a body:

$$P = F/A \tag{5-4}$$

where

P is the pressure (M/LT^2; slug/ft-s^2 or (kg·m/s^2)/m^2)

A is the cross-sectional area perpendicular to the direction of the force (L^2; ft^2 or m^2)

The units of pressure are pascals (Pa), or N/m^2. A N/m^2 is equal to a N·m/m^3, or J/m^3. Pressure may thus be thought of as potential energy per unit volume of fluid.

For a unit volume of fluid, the mass, m, is numerically equal to the density, ρ, since density is defined as mass per unit volume. The total energy of the unit volume of fluid is the sum of the three components—kinetic, gravitational, and fluid-pressure energy:

$$E_{tv} = \tfrac{1}{2}\rho v^2 + \rho gz + P \qquad\qquad (5-5)$$

where E_{tv} is the total energy per unit volume.

If Equation 5–5 is divided by ρ, the result is total energy per unit mass, E_{tm}:

$$E_{tm} = \frac{v^2}{2} + gz + \frac{P}{\rho} \qquad\qquad (5-6)$$

which is known as the *Bernoulli equation*. The derivation of the Bernoulli equation may be found in textbooks on fluid mechanics (Streeter 1962).

For steady flow of a frictionless, incompressible fluid along a smooth line of flow, the sum of the three components is a constant. Each term of Equation 5–6 has the units of $(L/T)^2$:

$$\frac{v^2}{2} + gz + \frac{P}{\rho} = \text{constant} \qquad\qquad (5-7)$$

Steady flow indicates that the conditions do not change with time. The density of an incompressible fluid would not change with changes in pressure. A frictionless fluid would not require energy to overcome resistance to flow. An ideal fluid would have both of these characteristics; real fluids have neither one. Real fluids are compressible and do suffer frictional flow losses; however, Equation 5–7 is useful for purposes of comparing the components of mechanical energy.

If each term of Equation 5–7 is divided by g, the following expression results:

$$\frac{v^2}{2g} + z + \frac{P}{\rho g} = \text{constant} \qquad\qquad (5-8)$$

This equation expresses all terms in units of energy per unit weight. These are J/N, or m. Thus, Equation 5–8 has the advantage of having all units in length dimensions (L). The first term of $v^2/2g$ is (m/s)^2/(m/s^2), or m; the second term, z, is already in m; and the third term, $P/\rho g$, is Pa/(kg/m^3)(m/s^2), or

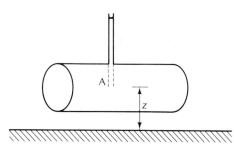

FIGURE 5.1   Piezometer measuring fluid pressure and the elevation of water.

$(N/m^2)/(kg/m^3)(m/s^2)$, which reduces to m. The sum of these three factors is the total mechanical energy per unit weight, known as the **hydraulic head,** h. This is usually measured in the field or laboratory in units of length.

5.3   HYDRAULIC HEAD

A **piezometer***** is used to measure the total energy of the fluid flowing through a pipe packed with sand, as shown in Figure 5.1. The piezometer is open at the top and bottom, and water rises in it in direct proportion to the total fluid energy at the point at which the bottom of the piezometer is open in the sand. At point A, which is at an elevation, z, above a datum, there is a fluid pressure, P. The fluid is flowing at a velocity, v. The total energy per unit mass can be found from Equation 5–6.

EXAMPLE PROBLEM

At a place where $g = 9.80$ m/s^2 the fluid pressure is 1500 N/m^2; the distance above a reference elevation is 0.75 m; and the fluid density is 1.02×10^3 kg/m^3. The fluid is moving at a velocity of 10^{-6} m/s. Find E_{tm}.

$$E_{tm} = gz + \frac{P}{\rho} + \frac{v^2}{2}$$

$$= 9.80 \text{ m/s}^2 \times 0.75 \text{ m} + \frac{1500 \text{ N/m}^2}{1.02 \times 10^3 \text{ kg/m}^3} + \frac{(10^{-6})^2}{2} \frac{\text{m}^2}{\text{s}^2}$$

$$= 7.35 \text{ m}^2/\text{s}^2 + 1.47 \text{ m}^2/\text{s}^2 + 5.0 \times 10^{-13} \text{ m}^2/\text{s}^2$$

The total energy per unit mass is 8.8 m^2/s^2. The energy is almost exclusively in the pressure and gravitational potential energy terms, which are thirteen orders of magnitude greater than the value of kinetic energy.

*****A piezometer is a small-diameter well with a very short well screen or section of slotted pipe at the end. It measures the hydraulic head at a point in the aquifer.

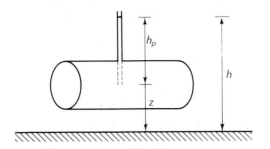

FIGURE 5.2 Total head, h, elevation head, z, and pressure head, h_p.

The preceding problem shows that the amount of energy developed as kinetic energy by flowing ground water is small. The velocity of ground water flowing in porous media under natural hydraulic gradients is very low. The example velocity of 10^{-6} m/s results in a movement of 30 m/y, which is typical for ground water.

Velocity components of energy may be safely ignored in ground-water flow because they are so much smaller than the other two terms. By dropping $v^2/2g$ from Equation 5–8, the total hydraulic head, h, is given by the formula

$$h = z + \frac{P}{\rho g} \qquad\qquad (5\text{--}9)$$

Figure 5.2 shows the components of head. The head is the total mechanical energy per unit weight of water. For a fluid at rest, the pressure at a point is equal to the weight of the overlying water per unit cross-sectional area:

$$P = \rho g h_p \qquad\qquad (5\text{--}10)$$

where h_p is the height of the water column that provides a *pressure head*. Substituting into Equation 5–9, we see that

$$h = z + h_p \qquad\qquad (5\text{--}11)$$

The total hydraulic head is equal to the sum of the elevation head and the pressure head. The elevation and pressure heads, when used in the form of Equation 5–11, correlate with energy per unit weight of water with dimensions L.

EXAMPLE PROBLEM Two points in the same confined aquifer are located on a vertical line. Point 1 is 100 m below mean sea level and point 2 is 50 m below mean sea level. The fluid pressure at point 1 is 9.0×10^5 N/m^2 and at point 2, it is 6.1×10^5 N/m^2.
Part A: Calculate the pressure and total heads at each point.

Assume that the deeper point is at zero datum. Therefore, the elevation head at point 1 is 0 and at point 2 it is 50 m.

By rearranging Equation 5–10 we can get an equation for the pressure head:

$$P = \rho g h_p$$

Therefore,

$$h_p = P/\rho g$$

Assume that $g = 9.80$ m/s^2 and $\rho = 1000$ kg/m^3. At point 1,

$$h_p = \frac{9.0 \times 10^5 \ (\text{kg·m/s}^2)/(\text{m}^2)}{1000 \ \text{kg/m}^3 \times 9.80 \ \text{m/s}^2}$$
$$= 92 \ \text{m}$$

Since total head is the sum of the elevation head and the pressure head, at point 1

$$h = h_p + z = 92 \ \text{m} + 0 \ \text{m} = 92 \ \text{m}$$

At point 2,

$$h_p = \frac{6.1 \times 10^5 (\text{kg·m/s}^2)/(\text{m}^2)}{1000 \ \text{kg/m}^3 \times 9.80 \ \text{m/s}^2}$$
$$= 62 \ \text{m}$$
$$h = 62 \ \text{m} + 50 \ \text{m} = 112 \ \text{m}$$

Part B: Does flow in the aquifer have an upward or a downward component?

Flow is downward, because the total head at 50 m is greater than the total head at 100 m, even though the pressure head at 100 m is greater.

EXAMPLE PROBLEM

The following data were collected at a nest of piezometers (several piezometers of different depths located within a few feet (1 to 2 m) of each other):

	A	B	C
Elevation at surface (m a.s.1.)	225	225	225
Depth of piezometer (m)	150	100	75
Depth to water (m below surface)	80	77	60

Part A: What is the hydraulic head at each of A, B, and C?

Hydraulic head is elevation of the water in the piezometer. It is found by subtracting the depth to water from the surface elevation.

A: 145 m   B: 148 m   C: 165 m

Part B: What is the pressure head at each of *A, B,* and *C*?

Pressure head is the height of the water in the wall above the depth of the piezometer. It is found by subtracting the depth to water from the depth of the piezometer from the surface.

A: 70 m   B: 33 m   C: 15 m

Part C: What is the elevation head in each well?

Elevation head is the height of the measuring point above the datum. In this case the datum is mean sea level and the elevation head is found by subtracting the depth of the piezometer from the surface elevation.

A: 75 m   B: 125 m   C: 150 m

Notice that the total head found in part A is the sum of the pressure head found in part B and the elevation head found in part C.

Part D: What is the vertical hydraulic gradient between the piezometers?

The hydraulic gradient is the difference in total head divided by the vertical distance between the two piezometers.

From piezometer A to piezometer B the difference in the total head is 148 m − 145 m and the vertical distance is 50 m. The hydraulic gradient is (3 m)/(50 m), or 0.06, and the direction is downward as the head in B, the shallower piezometer, is greater.

From piezometer B to piezometer C the difference in total head is 165 m − 148 m and the vertical distance is 25 m. The hydraulic gradient is (17 m)/(25 m) or 0.68. This gradient is also downward.

5.4   HEAD IN WATER OF VARIABLE DENSITY

If the salinity of ground water varies over an area or with depth, density corrections must be made to the head that is measured in a well. Recall from Equation 5–10 that the pressure at the tip of a piezometer is the height of water standing in the piezometer multiplied by the density of the water in the piezometer and the gravitational constant.

Lusczynski (1961) introduced the concept of point-water head, which is the water level in a well filled with water coming from a point in an aquifer and which is just enough to balance the pressure in the aquifer at that point. He also introduced a fresh-water head, which is the height of a column of fresh water in a well that is just sufficient to balance the pressure in the aquifer at that point. Figure 5.3 shows the head relationships in water of variable density for point-water head and fresh-water head in a confined, saline-water aquifer overlain by a fresh-water aquifer. We know that the total head is the sum of the elevation head and the pressure head. Elevation head is not dependent upon fluid density.

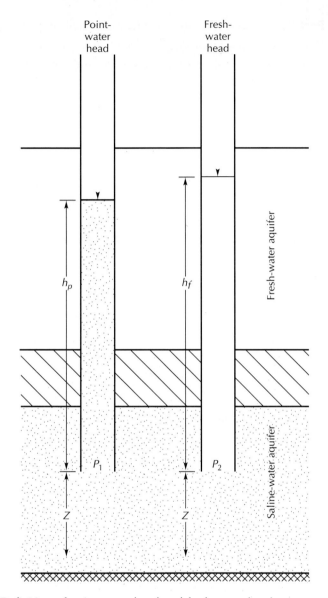

FIGURE 5.3   Definition of point-water head and fresh-water head.

Therefore point-water head is the sum of the elevation head, Z, and the point-water pressure head, h_p, and fresh-water head is the sum of elevation head, Z, and fresh-water pressure head, h_f.

   If we are in a fresh-water aquifer, all point-water heads are also fresh-water heads. However, if the density of the aquifer varies across a site, then point-water heads measured in the field must be converted to fresh-water heads before water-level maps and gradients are computed.

Consider piezometer 1 of Figure 5.3, which is filled with point water of density ρ_p. The point-water pressure head is the height of the point water in the piezometer, h_p. The pressure at P_1 may be found from

$$P_1 = \rho_p g h_p \qquad (5-12)$$

Now consider piezometer 2 of Figure 5.3, which ends in the saline aquifer but is filled with fresh water of density ρ_f; the fresh-water pressure head is h_f. The pressure at P_2 may be found from

$$P_2 = \rho_f g h_f \qquad (5-13)$$

If the two piezometers both end in the same point in the aquifer, then $P_1 = P_2$, and we can find the relationship between fresh-water pressure head and point-water pressure head.

$$\rho_p g h_p = \rho_f g h_f \qquad (5-14)$$
$$h_f = (\rho_p/\rho_f)h_p \qquad (5-15)$$

If the density of water in aquifers varies vertically, all point-water pressure heads should be converted to fresh-water pressure heads. Fresh-water heads can then be determined and used for the determination of hydraulic gradients and flow directions. For theoretical reasons equivalent fresh-water heads cannot be used to determine the hydraulic gradient in aquifers where there is a lateral variation in density. This is because the density gradient is a factor in determining lateral flow as well as the hydraulic gradient (Hubbert 1956; Hickey 1989; Oberlander 1989).

EXAMPLE PROBLEM

In Figure 5.4 there are three aquifers, a fresh-water-table aquifer and two confined saline-water aquifers, with properties as given in the following table. What are the head relationships between aquifers?

Aquifer	Water Density	Elevation Head	Point-Water Head
A	999 kg/m³	50.00 m	55.00 m
B	1040 kg/m³	31.34 m	54.67 m
C	1100 kg/m³	7.95 m	51.88 m

Field measurements of water levels in the wells give the point-water heads. The point-water pressure heads are determined by subtracting the elevation head from the point-water head. The fresh-water pressure head is then found by multiplying the point-water pressure head by the ratio of the point-water density to the fresh-water density. Finally, the fresh-water pressure head is added to the elevation head to obtain fresh-water heads.

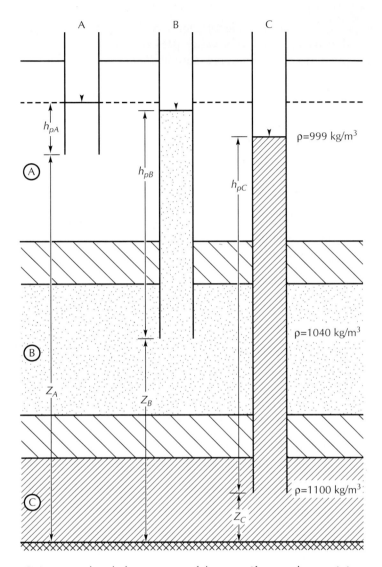

FIGURE 5.4   Point-water heads for a system of three aquifers, each containing water with a different density.

Aquifer	Point-Water Pressure Head	ρ_p/ρ_f	Fresh-Water Pressure Head	Fresh-Water Head
A	5.00 m	1.00	5.00 m	55.00 m
B	23.33 m	1.04	24.3 m	55.5 m
C	43.93 m	1.10	48.3 m	56.3 m

Note that the point-water heads indicated that the hydraulic gradient is vertically downward. However, after the point-water heads are corrected to fresh-water heads, the true hydraulic gradient is seen to be upward.

5.5   FORCE POTENTIAL AND HYDRAULIC HEAD

In Equation 5–6 we showed the total mechanical energy per unit mass to be equal to the sum of the kinetic energy, elevation energy, and pressure. This total potential energy has been termed the **force potential** and is indicated by the capital Greek letter phi, Φ (Hubbert 1940):

$$\Phi = gz + \frac{P}{\rho} = gz + \frac{\rho g h_p}{\rho} = g(z + h_p) \qquad\qquad (5\text{–}16)$$

Since $z + h_p = h$, the hydraulic head

$$\Phi = gh \qquad\qquad (5\text{–}17)$$

The force potential is the driving impetus behind ground-water flow and is equal to the product of hydraulic head and the acceleration of gravity. Both force potential and hydraulic head are potentials. Hydraulic head is energy per unit weight and force potential is energy per unit mass.

     Figure 5.5 shows a pipe filled with sand with water flowing through it from left to right. The pipe can be rotated to any inclination, with the discharge of water

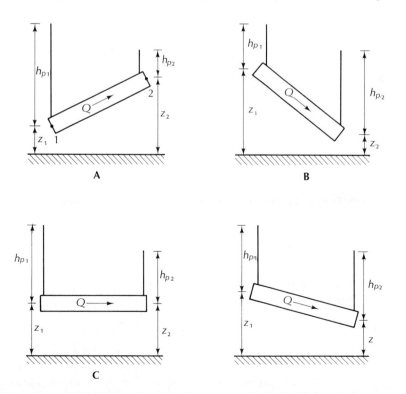

FIGURE 5.5   Apparatus to demonstrate how changing the slope of a pipe packed with sand will change the components of elevation, z, and pressure, h_p, heads. The direction of flow, Q, is indicated by the arrow.

remaining constant. In part A of the figure, the water flows from point 1 (of elevation z_1) to point 2 (of elevation z_2), z_2 being somewhat greater than z_1. In Part B, the slope has been reversed: rather than flowing uphill, the water now flows downhill. However, the fluid pressure head at point 2 (h_{p2}) is greater than at point 1 (h_{p1}). The fluid is thus moving from a region of low pressure to one of higher pressure. Clearly, neither elevation head nor pressure head alone controls ground-water motion. Part C of Figure 5.5 shows equal elevation heads, with pressure head declining in the direction of flow. Part D has equal pressure heads, but elevation head declines in the direction of flow.

In this example the total hydraulic head showed the same decrease in the direction of flow. This would be true no matter what the inclination of the pipe, so long as other factors remained constant. Since the force potential is the controlling force in ground-water flow, this demonstrates that the proportion of pressure and elevation head is not a factor.

From Figure 5.5, we see that the force potential and, hence, hydraulic head decrease in the direction of flow. As ground water moves, it encounters frictional resistance between the fluid and the porous media. The smaller the openings through which the fluid moves, the greater the friction. In overcoming the frictional resistance, some of the force potential is lost. It is transformed into heat (a lower form of energy). Thus, ground water is warmed slightly as it flows, and mechanical energy is converted to thermal energy. Under most circumstances, the resulting change in temperature is not measurable.

5.6   DARCY'S LAW

5.6.1   Darcy's Law in Terms of Force and Potential

In Section 4.5 it was shown that flow through a pipe filled with sand is proportional to the decrease in hydraulic head divided by the length of the pipe. This ratio is called the *hydraulic gradient*. It should now be apparent that the hydraulic head is the sum of the pressure head and the elevation head. Expressed in terms of hydraulic head, Darcy's law is

$$Q = -KA \frac{dh}{dl} \qquad \text{(5-18)}$$

Since the fluid potential, Φ, is equal to gh, Darcy's law can also be expressed in terms of potential as (Hubbert 1940)

$$Q = -\frac{KA}{g} \frac{d\Phi}{dl} \qquad \text{(5-19)}$$

As expressed here, Darcy's law is in a one-dimensional form, as water flows through the pipe in only one direction. In later sections, we will examine various forms of Darcy's law for two and three directions.

5.6.2 The Applicability of Darcy's Law

When a fluid at rest starts to move, it must overcome resistance to flow due to the viscosity of the fluid. Slowly moving fluids are dominated by viscous forces. There is a low energy level and the resulting fluid flow is **laminar.** In laminar flow, molecules of water follow smooth lines, called *streamlines* (Figure 5.6A).

As the velocity of flow increases, the moving fluid gains kinetic energy. Eventually, the inertial forces due to movement are more influential than the viscous forces, and the fluid particles begin to rush past each other in an erratic fashion. The result is **turbulent flow,** in which the water molecules no longer move along parallel streamlines (Figure 5.6B)

The **Reynolds number** relates the four factors that determine whether the flow will be laminar or turbulent (Streeter 1962):

$$R = \frac{\rho v d}{\mu} \qquad \textbf{(5-20)}$$

where

R is the Reynolds number, dimensionless

ρ is the fluid density (M/L^3; kg/m^3)

v is the discharge velocity (L/T; m/s)

d is the diameter of the passageway through which the fluid moves (L; m)

μ is the viscosity ($M/T\cdot L$; kg/s·m)

For open-channel or pipe flow, d is simply the hydraulic radius of the channel or the pipe diameter. In such cases, the transition from laminar to turbulent fluid flow occurs when the average velocity is such that R exceeds a value of 2000 (Streeter 1962). For a porous medium, however, it is not easy to determine the value of d. Rather than an average or characteristic pore diameter, the average grain diameter is often used.

Turbulence in ground-water flow is difficult to detect. The inception of fluid turbulent flow in ground water has been reported at a Reynolds number

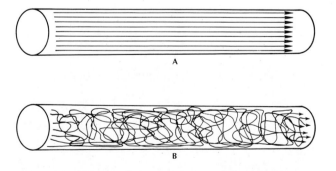

FIGURE 5.6 **A.** Flow paths of molecules of water in laminar flow. **B.** Flow paths of molecules of water in turbulent flow.

ranging from 60 (Schneebeli 1955) to 600 (Hubbert 1956). However, experimentation has shown that Darcy's law is valid only when conditions are such that the resistive forces of viscosity predominate. These conditions prevail when the Reynolds number is less than 1 to 10 (Lindquist 1933; Rose 1945a, 1945b). This means that Darcy's law applies only to very slowly moving ground waters. It is possible to have laminar ground-water flow, but under conditions such that the Reynolds number is so great as to invalidate Darcy's law. Under most natural ground-water conditions, the velocity is sufficiently low for Darcy's law to be valid. Exceptions might be areas of rock with large openings, such as solution openings and basalt flows. Likewise, areas of steep hydraulic gradients, such as the vicinity of a pumping well, might result in high velocities with a correspondingly high Reynolds number.

EXAMPLE PROBLEM

A sand aquifer has a median grain diameter of 0.050 cm. For pure water at 15°C, what is the greatest velocity for which Darcy's law is valid?

From Appendix 14,

$$\rho = 0.999 \times 10^3 \text{ kg/m}^3$$
$$\mu = 1.14 \times 10^{-2} \text{ g/s·cm}$$

Convert units to kilograms, meters, and seconds:

$$d = 0.050 \text{ cm} \times 0.01 \text{ m/cm} = 0.0005 \text{ m}$$
$$\mu = 1.14 \times 10^{-2} \text{ g/s·cm} \times 0.001 \text{ kg/g} \times 100 \text{ cm/m}$$
$$= 1.14 \times 10^{-3} \text{ kg/s·m}$$

By rearranging Equation 5–16, we can obtain a formula for velocity.

$$R = \frac{\rho v d}{\mu}$$

Therefore,

$$v = \frac{R\mu}{\rho d}$$

If R cannot exceed 1, the maximum velocity is

$$v = \frac{1 \times 1.14 \times 10^{-3} \text{ kg/s·m}}{0.999 \times 10^3 \text{ kg/m}^3 \times 0.0005 \text{ m}}$$
$$= 0.0023 \text{ m/s}$$

Darcy's law will be valid for discharge velocities equal to or less than 0.0023 m/s.

5.6.3 Specific Discharge and Average Linear Velocity

When water flows through an open channel or a pipe, the discharge, Q, is equal to the product of the velocity, v, and the cross-sectional area of flow, A:

$$Q = vA \qquad (5-21)$$

Rearrangement of Equation 5–21 yields an expression for velocity,

$$v = Q/A \qquad (5-22)$$

One can apply the same reasoning to Equation 5–18, Darcy's law, for flow through a porous medium:

$$v = \frac{Q}{A} = -K\frac{dh}{dl} \qquad (5-23)$$

A moment's reflection will reveal that this velocity is not quite the same as the velocity of water flowing through an open pipe. The discharge is measured as water coming from the pipe. In an open pipe, the cross-sectional area of flow inside the pipe is equivalent to the area of the end of the pipe. However, if the pipe is filled with sand, the open area through which water may flow is much smaller than the cross-sectional area of the pipe. The velocity of flow determined by Equation 5–23 is termed the **specific discharge,** or Darcy flux.* It is an apparent velocity, representing the velocity at which water would move through an aquifer if the aquifer were an open conduit.

The cross-sectional area of flow for a porous medium is actually much smaller than the dimensions of the aquifer. It is equal to the product of the effective porosity of the aquifer material and the physical dimensions. Water can move only through the pore spaces. Moreover, part of the pore space is occupied by stagnant water, which clings to the rock material. The effective porosity is that portion of the pore space through which saturated flow occurs.

To find the velocity at which water is actually moving, the specific discharge is divided by the effective porosity to account for the actual open space available for the flow. The result is the **seepage velocity,** or **average linear velocity**—a velocity representing the average rate at which the water moves between two points:

$$V_x = \frac{Q}{n_eA} = -\frac{Kdh}{n_edl} \qquad (5-24)$$

---

*The terms discharge velocity and Darcian velocity are synonyms for specific discharge. It would be best to avoid these as their use implies that ground water is moving at this velocity.

where

V_x is the average linear velocity (L/T; cm/s, ft/s, m/s)

n_e is the effective porosity (dimensionless)

Equation 5–24 does not take into account the factors that account for **dispersion** in flowing ground water. Dispersion is the phenomenon that results because ground water flows through different pores at different rates and various flow paths vary in length. Dispersion is discussed in detail in Section 11.6.5.

Because Equation 5–24 does not include a dispersion factor, it cannot be used to predict the average linear rate of movement of a solute front that is moving at the same rate as the flowing ground water. This is especially true for fine-grained materials, where the process of diffusion may be important in the movement of solute from an area of greater to lesser concentration. Diffusion is discussed in Section 11.6.2.

5.7   EQUATIONS OF GROUND-WATER FLOW*

5.7.1   Confined Aquifers

The flow of fluids through porous media is governed by the laws of physics. As such, it can be described by differential equations. Since the flow is a function of several variables, it is usually described by partial differential equations in which the spatial coordinates, *x, y,* and *z,* and time, *t,* are the independent variables.

In deriving the equations, the laws of conservation for mass and energy are employed. The *law of mass conservation,* or *continuity principle,* states that there can be no net change in the mass of a fluid contained in a small volume of an aquifer. Any change in mass flowing into the small volume of the aquifer must be balanced by a corresponding change in mass flux out of the volume, or a change in the mass stored in the volume, or both. The *law of conservation of energy* is also known as the *first law of thermodynamics.* It states that within any closed system there is a constant amount of energy, which can be neither lost nor increased. It can, however, change form. The *second law of thermodynamics* implies that when energy changes forms, it tends to go from a more useful form, such as mechanical energy, to a less useful form, such as heat. Based upon these principles and Darcy's law, the main equations of ground-water flow have been derived (Jacob 1940, 1950; Domenico 1972; Cooper 1966).

We will consider a very small part of the aquifer, called a *control volume.* The three sides are of lengths *dx, dy,* and *dz,* respectively. The area of the faces normal to the *x*-axis is *dydz;* the area of the faces normal to the *z*-axis is *dxdy* (Figure 5.7).

Assume the aquifer is homogeneous and isotropic. The fluid moves in only one direction through the control volume. However, the actual fluid motion

---

*The main equation of ground-water flow is derived in this section following a method used by Jacob (1940, 1950) and modified by Domenico (1972).

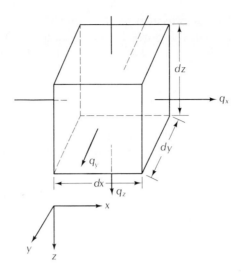

FIGURE 5.7   Control volume for flow through a confined aquifer.

can be subdivided on the basis of the components of flow parallel to the three principal axes. If q is flow per unit cross-sectional area, $\rho_w q_x$ is the portion parallel to the x-axis, etc., where ρ_w is the fluid density.

The mass flux into the control volume is $\rho_w q_x\ dydz$ along the x-axis. The mass flux[†] out of the control volume is $\rho_w q_x\ dydz\ +\ \dfrac{\partial}{\partial x}(\rho_w q_x)\ dx\ dydz$. The net accumulation in the control volume due to movement parallel to the x-axis is equal to the inflow less the outflow, or $-\dfrac{\partial}{\partial x}(\rho_w q_x)\ dx\ dydz$. Since there are flow components along all three axes, similar terms can be determined for the other two directions: $-\dfrac{\partial}{\partial y}(\rho_w q_y)\ dy\ dxdz$ and $-\dfrac{\partial}{\partial z}(\rho_w q_z)\ dz\ dxdy$. Combining these three terms yields the net total accumulation of mass in the control volume:

$$-\left(\frac{\partial}{\partial x}\,\rho_w q_x + \frac{\partial}{\partial y}\,\rho_w q_y + \frac{\partial}{\partial z}\,\rho_w q_z\right)dxdydz \qquad (5\text{--}25)$$

The volume of water in the control volume is equal to $n\ dx\ dydz$, where n is the porosity. The initial mass of the water is thus $\rho_w n\ dx\ dydz$. The volume of solid material is $(1 - n)\ dx\ dydz$. Any change in the mass of water, M, with respect to time is given by

$$\frac{\partial M}{\partial t} = \frac{\partial}{\partial t}(\rho_w n\ dxdydz) \qquad (5\text{--}26)$$

---

†Flux is a rate of flow.

As the pressure in the control volume changes, the fluid density will change, as will the porosity of the aquifer. The compressibility of water, β, is defined as the rate of change in density with a change in pressure, P:

$$\beta dP = \frac{d\rho_w}{\rho_w} \qquad (5-27)$$

The aquifer also changes in volume with a change in pressure. We will assume the only change is vertical. The aquifer compressibility, α, is given by

$$\alpha dP = \frac{d(dz)}{dz} \qquad (5-28)$$

As the aquifer compresses or expands, n will change, but the volume of solids, V_s, will be constant. Likewise, if the only deformation is in the z-direction, $d(dx)$ and $d(dy)$ will equal zero:

$$dV_s = 0 = d[(1 - n)\ dx\ dydz] \qquad (5-29)$$

Differentiation of Equation 5–29 yields

$$dz\ dn = (1 - n)d(dz) \qquad (5-30)$$

and

$$dn = \frac{(1 - n)d(dz)}{dz} \qquad (5-31)$$

The pressure, P, at a point in the aquifer is equal to $P_0 + \rho_w g h$, where P_0 is atmospheric pressure, a constant, and h is the height of a column of water above the point. Therefore, $dP = \rho_w g\ dh$, and Equations 5–27 and 5–28 become

$$d\rho_w = \rho_w \beta (\rho_w g\ dh) \qquad (5-32)$$

and

$$d(dz) = dz \alpha (\rho_w g\ dh) \qquad (5-33)$$

Equation 5–31 can be rearranged if $d(dz)$ is replaced by Equation 5–33.

$$dn = (1 - n)\alpha \rho_w g\ dh \qquad (5-34)$$

If dx and dy are constant, the equation for change of mass with time in the control volume, Equation 5–26, can be expressed as

$$\frac{\partial M}{\partial t} = \left[\rho_w n \frac{\partial(dz)}{\partial t} + \rho_w\ dz \frac{\partial n}{\partial t} + n\ dz \frac{\partial \rho_w}{\partial t} \right] dxdy \qquad (5-35)$$

Substitution of Equations 5–32, 5–33, and 5–34 into Equation 5–35 yields, after minor manipulation,

$$\frac{\partial M}{\partial t} = (\alpha\rho_w g + n\beta\rho_w g)\rho_w \, dx \, dydz \, \frac{\partial h}{\partial t} \qquad (5-36)$$

The net accumulation of material expressed as Equation 5–25 is equal to Equation 5–36, the change in mass with time:

$$-\left[\frac{\partial(q_x)}{\partial x} + \frac{\partial(q_y)}{\partial y} + \frac{\partial(q_z)}{\partial z}\right]\rho_w \, dx \, dydz = (\alpha\rho_w g + n\beta\rho_w g)\rho_w \, dx \, dydz \, \frac{\partial h}{\partial t} \qquad (5-37)$$

From Darcy's law,

$$q_x = -K\frac{\partial h}{\partial x} \qquad (5-38)$$

$$q_y = -K\frac{\partial h}{\partial y} \qquad (5-39)$$

and

$$q_z = -K\frac{\partial h}{\partial z} \qquad (5-40)$$

Substituting these into Equation 5–37 yields the main equation of flow for a confined aquifer:

$$K\left(\frac{\partial^2 h}{\partial x^2} + \frac{\partial^2 h}{\partial y^2} + \frac{\partial^2 h}{\partial z^2}\right) = (\alpha\rho_w g + n\beta\rho_w g)\frac{\partial h}{\partial t} \qquad (5-41)$$

which is a general equation for flow in three dimensions for an isotropic, homogeneous porous medium. For two-dimensional flow with no vertical components, the equation can be rearranged and terms introduced from Equations 4–31 and 4–32 for the storativity, $[S = b(\alpha\rho_w g + n\beta\rho_w g)]$, and from Equation 4–29 for the transmissivity, $(T = Kb)$, where b is the aquifer thickness:

$$\frac{\partial^2 h}{\partial x^2} + \frac{\partial^2 h}{\partial y^2} = \frac{S}{T}\frac{\partial h}{\partial t} \qquad (5-42)$$

In steady-state flow, there is no change in head with time; for example, in cases in which there is no change in the position or slope of the water table. Under such conditions, time is not one of the independent variables, and steady flow is described by the three-dimensional partial differential equation known as the **Laplace equation:**

$$\frac{\partial^2 h}{\partial x^2} + \frac{\partial^2 h}{\partial y^2} + \frac{\partial^2 h}{\partial z^2} = 0 \tag{5-43}$$

The preceding equations are based on the assumption that all flow comes from water stored in the aquifer. In the field, it is more often than not the case that significant flow is generated from leakage into the aquifer through overlying or underlying confining layers. We will consider the leakage to appear in the control volume as horizontal flow. This assumption is justified on the grounds that the conductivity of the aquifer is usually orders of magnitude greater than that of the confining layer. The law of refraction indicates that, for these conditions, flow in the confining layer will be nearly vertical if flow in the aquifer is horizontal.

The leakage rate, or rate of accumulation, is designated as e. The general equation of flow (in two dimensions, since horizontal flow was assumed) is given by

$$\frac{\partial^2 h}{\partial x^2} + \frac{\partial^2 h}{\partial y^2} + \frac{e}{T} = \frac{S}{T} \frac{\partial h}{\partial t} \tag{5-44}$$

The leakage rate can be determined from Darcy's law. If the head at the top of the aquitard is h_0 and the head in the aquifer just below the aquitard is h, the aquitard has a thickness b' and a conductivity (vertical) of K':

$$e = K' \frac{(h_0 - h)}{b'} \tag{5-45}$$

5.7.2 Unconfined Aquifers

Water is derived from storage in water-table aquifers by vertical drainage of water in the pores. This drainage results in a decline in the position of the water table near a pumping well as time progresses. In the case of a confined aquifer, although the potentiometric surface declined, the saturated thickness of the aquifer remained constant. In the case of an unconfined aquifer, the saturated thickness can change with time. Under such conditions, the ability of the aquifer to transmit water—the transmissivity—changes, as it is the product of the conductivity K and the saturated thickness h (assuming that h is measured from the horizontal base of the aquifer).

The general flow equation for two-dimensional unconfined flow is known as the **Boussinesq equation** (Boussinesq 1904):

$$\frac{\partial}{\partial x}\left(h\frac{\partial h}{\partial x}\right) + \frac{\partial}{\partial y}\left(h\frac{\partial h}{\partial y}\right) = \frac{S_y}{K}\frac{\partial h}{\partial t} \tag{5-46}$$

where S_y is specific yield. This equation is a type of differential equation that cannot be solved using calculus, except in some very specific cases. In mathematical terms, it is nonlinear.

If the drawdown in the aquifer is very small compared with the saturated thickness, the variable thickness, h, can be replaced with an average thickness, b, that is assumed to be constant over the aquifer. The Boussinesq equation can thus be linearized by this approximation to the form

$$\frac{\partial^2 h}{\partial x^2} + \frac{\partial^2 h}{\partial y^2} = \frac{S_y}{Kb} \frac{\partial h}{\partial t} \qquad\qquad \textbf{(5–47)}$$

which has the same form as Equation 5–42.

5.8   SOLUTION OF FLOW EQUATIONS

The flow of water in an aquifer can be mathematically described by Equations 5–41, 5–42, 5–43, 5–44, or 5–47. These are all partial differential equations in which the head, h, is described in terms of the variables x, y, z, and t. They are solved by means of a mathematical model consisting of the applicable governing flow equation, equations describing the hydraulic head at each of the boundaries of the aquifer, and equations describing the initial conditions of head in the aquifer.

If the aquifer is homogeneous and isotropic, and the boundaries can be described with algebraic equations, then the mathematical model can be solved by use of an analytical solution based on integral calculus. However, if the aquifer does not correspond to those conditions (e.g., a layered aquifer), then a numerical solution to the mathematical model is needed. Numerical solutions are based on the concept that the partial differential equation can be replaced by a similar equation that can be solved using arithmetic. Likewise, the equations governing initial and boundary conditions are replaced by numerical statements of these conditions. Numerical solutions are typically solved on digital computers. The use of digital-computer models is treated in Chapter 14.

5.9   GRADIENT OF HYDRAULIC HEAD

The potential energy, or force potential, Φ, of ground water consists of two parts: elevation and pressure. It is equal to the product of the acceleration of gravity and the total head (Hubbart 1940), and represents mechanical energy per unit mass:

$$\Phi = gh \qquad\qquad \textbf{(5–48)}$$

Force potential is a physical quantity. To obtain it one needs only to measure the heads in an aquifer with piezometers and multiply the results by the acceleration of gravity. If a point in an aquifer has a head of 15.1 m and the value of g is 9.81 m/s^2, then Φ is $15.1 \times 9.81 = 148.1$ m^2/s^2. For practical purposes, as g is usually constant throughout an area, most field problems are solved in terms of hydraulic head, h.

If the value of h is variable in an aquifer, a contour map may be made showing lines of equal value of h (equipotential lines). Such a map is similar to a topographic map of the land-surface elevation. In three-dimensional cases, one deals with surfaces of equal value of h (equipotential surfaces).

Figure 5.8 shows equipotential surfaces for a two-dimensional uniform flow field. The equipotential surfaces are vertical, and they form equipotential lines with a horizontal plane. The equipotential field is uniform—that is, the horizontal distance between each equipotential surface is the same. Also shown is a vector known as the gradient of h (grad h). Remember, a vector is a directed line segment, so that grad h has a magnitude and a direction. It is roughly analogous to the maximum slope of the potential field. In the notation of differential calculus,

$$\text{grad } h = \frac{dh}{ds} \qquad \qquad \textbf{(5–49)}$$

where s is the distance parallel to grad h. Grad h has a direction perpendicular to the equipotential lines.

If the potential is the same everywhere, it will be manifest in a condition such as a flat water table. In this case, grad h equals zero, since there is no slope to h. There will be no ground-water flow, since grad h must have a positive value before ground water will move.

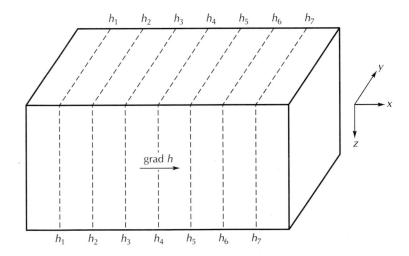

FIGURE 5.8.   Equipotential lines in a three-dimensional flow field and the gradient of h.

**5.10   RELATIONSHIP OF GROUND-WATER-FLOW
             DIRECTION TO GRAD *h***

The direction in which ground water flows is a function of the potential field and
the degree of anisotropy of the hydraulic conductivity and the orientation of the
axes of permeability with respect to grad *h*.

For isotropic aquifers, the value of K is the same in all directions. In such
aquifers, the direction of fluid flow will be parallel to grad *h*, which means that it
will also be perpendicular to the equipotential lines.

Liakopoulos (1965) presented a method for determining the direction of
ground-water flow in an anisotropic aquifer. The method assumes that there is one
plane in the ground where the hydraulic conductivity does not vary with direction
and that the axes of anisotropy are mutually perpendicular as well as being
perpendicular to the isotropic plane. For example, in many sedimentary aquifers
the hydraulic conductivity in a horizontal plane is the same in all directions, but
the horizontal hydraulic conductivity is much greater than the vertical hydraulic
conductivity.

The first step is to prepare a hydraulic-conductivity tensor ellipse. The
semiaxes of the ellipse are equal to the inverse square roots of the principal
hydraulic-conductivity values. Figure 5.9 shows the construction of a hydraulic-
conductivity tensor ellipse. Once the ellipse is constructed, the steps necessary to
find the direction of flow vis-à-vis grad *h* are also illustrated in Figure 5.9. In
general, for anisotropic media the direction of ground-water flow will not be
parallel to grad *h* and thus will not cross the equipotential lines at a right angle.
However, if one of the principal axes of hydraulic conductivity is parallel to grad
h, then the ground-water flow direction will also be parallel to grad *h*.

5.11   FLOW LINES AND FLOW NETS

A *flow line* is an imaginary line that traces the path that a particle of ground water
would follow as it flows through an aquifer. Flow lines are helpful for visualizing
the movement of ground water. In an isotropic aquifer, flow lines will cross
equipotential lines at right angles. If there is anisotropy in the plane of flow, then
the flow lines will cross the equipotential lines at an angle dictated by the degree
of anisotropy and the orientation of grad *h* to the hydraulic conductivity tensor
ellipsoid. Figure 5.10A shows equipotential lines and flow lines in an isotropic
medium and Figure 5.10B shows equipotential lines and flow lines in an anisotropic medium. It may be seen that in the isotropic medium the flow lines are parallel
to grad *h*, and in the anisotropic medium they are not.

The two-dimensional Laplace equation (5–43) for steady-flow conditions
may be solved by graphical construction of a **flow net,** which is a network of
equipotential lines and associated flow lines (Forchheimer 1914; Casagrande

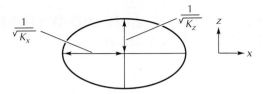

1. Construct a hydraulic-conductivity ellipse.

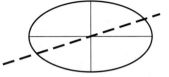

2. Draw the equipotential line as it is oriented with respect to the hydraulic-conductivity axes and passing through the origin of the ellipse.

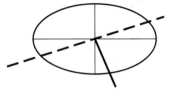

3. Draw grad *h* perpendicular to the equipotential line and starting at the origin of the ellipse.

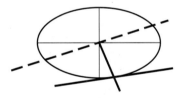

4. Draw a tangent to the ellipse at the point where grad *h* intersects the ellipse.

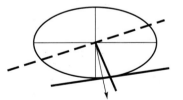

5. Draw a flow line so that it passes through the origin of the ellipse and is perpendicular to the tangent.

FIGURE 5.9 Steps in the determination of the direction of ground-water flow in an anisotropic medium using the hydraulic conductivity ellipse. Source: C. W. Fetter, *Ground Water Monitoring Review* 1, 1 (1981): 28–31. Copyright © 1981 Ground Water Publishing Co.

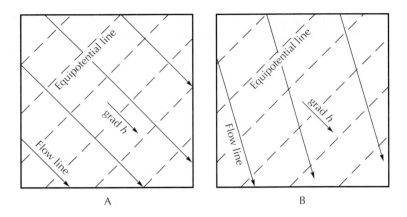

FIGURE 5.10 Relationship of flow lines to equipotential field and grad *h*. **A.** Isotropic aquifer. **B.** Anisotropic aquifer.

1940). A flow net is especially useful in anisotropic media. However, with certain transformations it can be used with anisotropic aquifers. Cedergren (1989) presents a complete discussion of the construction of flow nets, including those in anisotropic media.

The method of flow-net construction presented here is based on the following assumptions.

1. The aquifer is homogeneous.
2. The aquifer is fully saturated.
3. The aquifer is isotropic.
4. There is no change in the potential field with time.
5. The soil and water are incompressible.
6. Flow is laminar, and Darcy's law is valid.
7. All boundary conditions are known.

There are three types of boundary conditions possible. Ground water cannot pass a *no-flow boundary*. Adjacent flow lines will be parallel to a no-flow boundary, and equipotential lines will intersect it at right angles. Since the head is the same everywhere on a *constant-head boundary,* such a boundary represents an equipotential line. Flow lines will intersect a constant-head boundary at right angles and the adjacent equipotential line will be parallel. For unconfined aquifers, there is also a *water-table boundary*. The water table is neither a flow line nor an equipotential line; rather it is a line where head is known. If there is recharge or discharge across the water table, flow lines will be at an oblique angle to the water table. If there is no recharge across the water table, flow lines can be parallel to it.

A flow net is a family of equipotential lines with sufficient orthogonal flow lines drawn so that a pattern of "square" figures results. Except in cases of the most simple geometry, the figures will not truly be squares. The following steps are followed in the construction of a flow net.

1. Identify the boundary conditions.

2. Make a sketch of the boundaries to scale with the two axes of the drawing having the same scale.

3. Identify the position of known equipotential and flow-line conditions.

4. Draw a trial set of flow lines. The outer flow lines will be parallel to no-flow boundaries. The distance between adjacent flow lines should be the same at all sections of the flow field.

5. Draw a trial set of equipotential lines. Start at one end of the flow field and work toward the other. The equipotential lines should be perpendicular to flow lines. They will be parallel to constant-head boundaries and at right angles to no-flow boundaries. If there is a water-table boundary, the position of the equipotential line at the water table is based on the elevation of the water table. The equipotential lines should be spaced so that they form areas that are equidimensional. These are not necessarily squares, but should be as square as possible.

6. Erase and redraw the trial flow lines and equipotential lines until the desired flow net of orthogonal equipotential lines and flow lines is obtained.

Figure 5.11 illustrates the construction of a very simple flow net, where the pattern forms exact squares because the flow field is rectangular.

Most beginners at the art of flow-net construction will find an ample supply of paper, pencils, and erasers essential. As a check on the quality of the flow net, the diagonals of the squares can be drawn. These should form smooth curves that intersect each other at right angles. This should, of course, be done on a copy of the final product.

In addition to presenting a graphic display of the ground-water flow directions and potential distribution, the completed flow net can be used to determine the quantity of water flowing by the following formula:

$$q' = \frac{Kph}{f} \qquad\qquad (5\text{-}50)$$

where

q' is the total volume discharge per unit width of aquifer (L^3/T; ft^3/d or m^3/d)

K is the hydraulic conductivity (L/T; ft/d or m/d)

p is the number of flow paths bounded by adjacent pairs of streamlines

h is the total head loss over the length of the streamlines (L; ft or m)

f is the number of squares bounded by any two adjacent streamlines and covering the entire length of flow

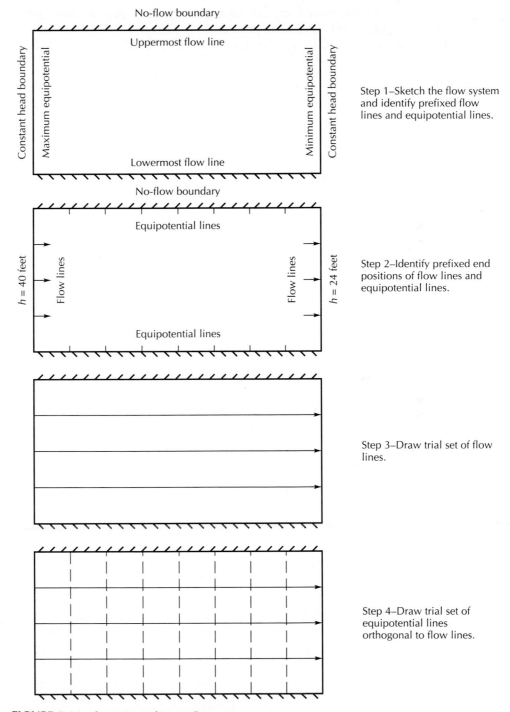

FIGURE 5.11 Steps in making a flow net.

Equation 5–50 can be used for simple flow systems with one recharge boundary and one discharge boundary. For complex systems, it is possible to find the discharge for each streamtube where $q' = (Kh)/f$. The total flow can be found by summing the flow in individual streamtubes.

EXAMPLE PROBLEM

If hydraulic conductivity is 23 ft/day, what is the discharge per unit width of the flow system in Figure 5.11?

The number of streamtubes is 4; therefore, $p = 4$. The number of equipotential drops is 8; therefore, $f = 8$. The total head loss is 40 ft − 24 ft = 16 ft. Substituting these values into Equation 5–50:

$$q' = \frac{Kph}{f}$$

$$= \frac{23 \text{ ft/d} \times 4 \times 16 \text{ ft}}{8} \times 1 \text{ ft unit width}$$

$$= 180 \text{ ft}^3/\text{d}$$

The preceding problem illustrates one of the pitfalls of two-dimensional problem solutions. It must be recognized that two-dimensional problems imply a third dimension, with an axis of symmetry perpendicular to the two-dimensional representation. The width of total flow perpendicular to this axis must be included to determine the total volume of flow. An alternative method is to state flow in terms of discharge per unit width. For an aquifer, flow might be stated in cubic meters per day per kilometer width of the aquifer (measured orthogonal to the direction of flow).

Figure 5.12 is a flow net for flow beneath a dam. The flow lines are drawn such that $6\frac{1}{2}$ flowtubes are constructed. It is possible to have a half flow tube,

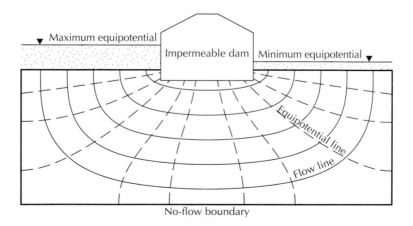

FIGURE 5.12   Flow net beneath an impermeable dam.

which may help in the drafting. This drawing shows that the "squares" are not really squares. The squares tend to get larger toward the ends of the flow lines and also to be more distorted there.

5.12   REFRACTION OF FLOW LINES

When water passes from one stratum to another stratum with a different hydraulic conductivity, the direction of the flow path will change (Hubbert 1940). Figure 5.13 shows a flowtube bounded by two flow lines. The flowtube passes from stratum 1, with a hydraulic conductivity of K_1, to stratum 2, with a hydraulic conductivity of K_2. The volume of water flowing in the streamtube in stratum 1 is Q_1, and in stratum 2 it is Q_2. The width of the flowtube in stratum 1 is a, and the width of the flowtube in stratum 2 is c. The length between adjacent equipotential lines is dl_1 in stratum 1 and dl_2 in stratum 2. There is a head loss between adjacent equipotential lines. In stratum 1, it is h_1; in stratum 2, it is h_2.

Notice in Figure 5.13 that at the boundary between the two strata, there are two triangles that have a common leg, b, at the boundary. The triangle in stratum 1 is bounded by a, b, and dl_1. The triangle in stratum 2 is bounded by c, b, and dl_2.

The flow through each streamtube is found from Darcy's law:

$$Q_1 = K_1 a \frac{dh_1}{dl_1} \quad \text{and} \quad Q_2 = K_2 c \frac{dh_2}{dl_2} \qquad (5-51)$$

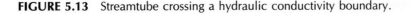

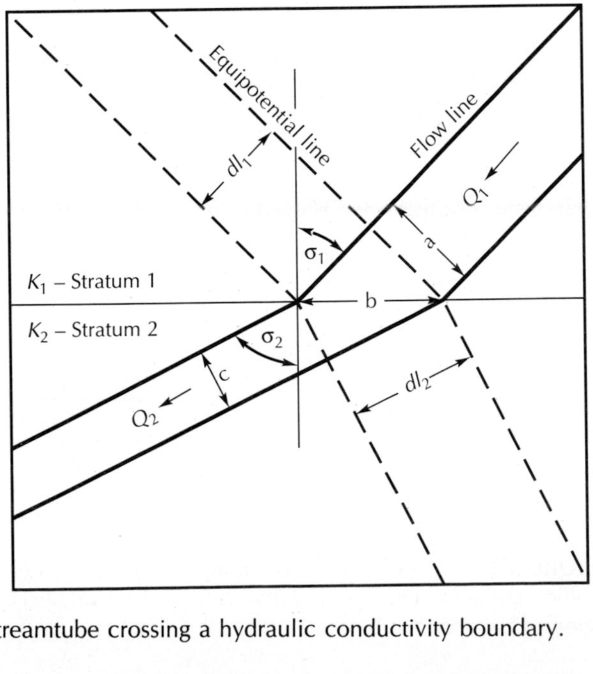

FIGURE 5.13   Streamtube crossing a hydraulic conductivity boundary.

From the principle of continuity, Q_1 must be equal to Q_2; therefore,

$$K_1 a \frac{dh_1}{dl_1} = K_2 c \frac{dh_2}{dl_2} \qquad (5-52)$$

Since the head loss between the two equipotential lines is the same in both strata, $h_1 = h_2$ and we can divide both sides of Equation 5–52 by h_1:

$$K_1 \frac{a}{dl_1} = K_2 \frac{c}{dl_2} \qquad (5-53)$$

From the geometry of the triangles, $a = b \cos \sigma_1$ and $c = b \cos \sigma_2$. Furthermore, $b/dl_1 = 1/\sin \sigma_1$ and $b/dl_2 = 1/\sin \sigma_2$. Substituting these into Equation 5–53, we obtain

$$K_1 \frac{\cos \sigma_1}{\sin \sigma_1} = K_2 \frac{\cos \sigma_2}{\sin \sigma_2} \qquad (5-54)$$

Since $\tan \sigma = (\cos \sigma)/(\sin \sigma)$, Equation 5–54 can be rewritten as the tangent law of refraction.

$$\boxed{\frac{K_1}{K_2} = \frac{\tan \sigma_1}{\tan \sigma_2}} \qquad (5-55)$$

As a consequence, the direction of refraction for flow going from a region of low conductivity to one of high conductivity will be different from that for flow going from high to low conductivity (Figure 5.14B, C). Likewise, if the streamlines are refracted, and they are perpendicular to the equipotential lines, then the equipotential lines must also be refracted. Figure 5.15 shows a portion of a flow net crossing a conductivity boundary.

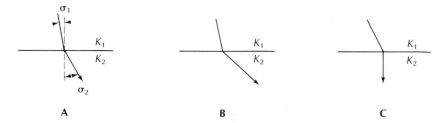

FIGURE 5.14 **A.** Refraction of a flowline crossing a conductivity boundary. **B.** Refracted flowline going from a region of low to high conductivity. **C.** Refracted flowline going from a region of high to low conductivity.

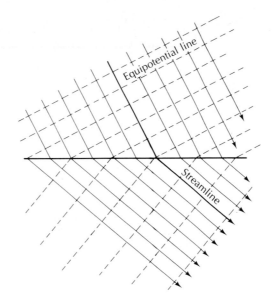

FIGURE 5.15   A flow net with flow crossing a conductivity boundary showing refraction of flowlines and equipotential lines. The hydraulic conductivity above the boundary is less than that below the boundary.

5.13   STEADY FLOW IN A CONFINED AQUIFER

If there is the steady movement of ground water in a confined aquifer, there will be a gradient or slope to the potentiometric surface of the aquifer. Likewise, we know that the water will be moving in the opposite direction of grad h. For flow of this type, Darcy's law may be used directly. In Figure 5.16, a portion of a confined aquifer of uniform thickness is shown. The potentiometric surface has a linear gradient; i.e., its two-directional projection is a straight line. There are two observation wells where the hydraulic head can be measured.

The quantity of flow per unit width, q', may be determined from Darcy's law:

$$q' = Kb \frac{dh}{dl} \tag{5-56}$$

where

K is the hydraulic conductivity (L/T; ft/d or m/d)

b is the aquifer thickness (L; ft or m)

$\dfrac{dh}{dl}$ is the slope of potentiometric surface (dimensionless)

q' is the flow per unit width (L^2/T; ft^2/d or m^2/d)

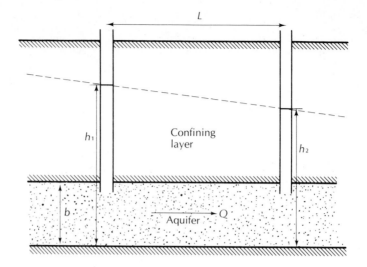

FIGURE 5.16   Steady flow through a confined aquifer of uniform thickness.

One may wish to know the head, h (L; ft or m), at some intermediate distance, x (L; ft or m), between h_1 and h_2. This may be found from the equation

$$h = h_1 - \frac{q'}{Kb}x \qquad\qquad (5\text{--}57)$$

where x is the distance from h_1.

EXAMPLE PROBLEM

A confined aquifer is 33 m thick and 7 km wide. Two observation wells are located 1.2 km apart in the direction of flow. The head in well 1 is 97.5 m and in well 2 it is 89.0 m. The hydraulic conductivity is 1.2 m/d. What is the total daily flow of water through the aquifer?

$$Q = -Kb\,\frac{dh}{dl} \times \text{width}$$

$$= 1.2 \text{ m/day} \times 33 \text{ m} \times \frac{97.5 \text{ m} - 89.0 \text{ m}}{1200 \text{ m}} \times 7000 \text{ m}$$

$$= 2000 \text{ m}^3/\text{day} \qquad \text{(to two significant digits)}$$

What is the elevation of the potentiometric surface at a point located 0.3 km from well h_1 and 0.9 km from well h_2? Discharge per unit width is (2000 m³/d)/(7000 m) = 0.29 m²/day:

$$h = h_1 - \frac{q'}{Kb}\,x$$

$$= 97.5 \text{ m} - \frac{0.29 \text{ m}^2/\text{day}}{1.2 \text{ m/d} \times 33 \text{ m}} \times 300 \text{ m}$$

$$= 97.5 \text{ m} - 2.2 \text{ m}$$

$$= 95.3 \text{ m}$$

5.14   STEADY FLOW IN AN UNCONFINED AQUIFER*

In an unconfined aquifer, the fact that the water table is also the upper boundary of the region of flow complicates flow determinations. Figure 5.17 illustrates the problem. On the left side of the figure, the saturated flow region is h_1 feet thick. On the right side, it is h_2 feet thick, which is $h_1 - h_2$ feet thinner than the left side. If there is no recharge or evaporation as the flow traverses the region, the quantity of water flowing through the left side is equal to that flowing through the right side. From Darcy's law, it is obvious that since the cross-sectional area is smaller on the right side, the hydraulic gradient must be greater. Thus, the gradient of the water table in unconfined flow is not constant; it increases in the direction of flow.

This problem was solved by Dupuit (1863), and his assumptions are known as the **Dupuit assumptions.** The assumptions are that (1) the hydraulic gradient is equal to the slope of the water table and (2) for small water-table gradients, the streamlines are horizontal and the equipotential lines are vertical. Solutions based on these assumptions have proved to be very useful in many practical problems. However, the Dupuit assumptions do not allow for a seepage face above the outflow side.

From Darcy's law,

$$q' = -Kh \frac{dh}{dx} \tag{5-58}$$

where h is the saturated thickness of the aquifer. At $x = 0$, $h = h_1$; at $x = L$, $h = h_2$.

Equation 5-58 may be set up for integration with the boundary conditions:

$$\int_0^L q' dx = -K \int_{h_1}^{h_2} h dh$$

Integration of the preceding yields

$$q'x \Big|_0^L = -K \frac{h^2}{2} \Big|_{h_1}^{h_2}$$

---

*The equations in this section are derived following methods used by Polubarinova-Kochina (1962) and Harr (1962).

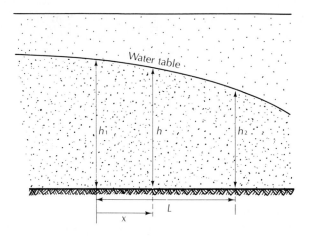

FIGURE 5.17   Steady flow through an unconfined aquifer resting on a horizontal impervious surface.

Substitution of the boundary conditions for x and h yields

$$q'L = -K\left(\frac{h_2^2}{2} - \frac{h_1^2}{2}\right)$$ **(5–59)**

Rearrangement of Equation 5–59 yields the **Dupuit equation:**

$$\boxed{q' = \frac{1}{2}K\left(\frac{h_1^2 - h_2^2}{L}\right)}$$ **(5–60)**

where

   q' is the flow per unit width (L^2/T; ft²/d or m²/day)
   K is the hydraulic conductivity (L/T; ft/d or m/day)
   h_1 is the head at the origin (L; ft or m)
   h_2 is the head at L (L; ft or m)
   L is the flow length (L; ft or m)

If we consider a small prism of the unconfined aquifer, it will have the shape of Figure 5.18. On one side it is h units high and slopes in the x-direction. Given the Dupuit assumptions, there is no flow in the z-direction. The flow in the x-direction, per unit width, is q'_x. From Darcy's law, the total flow in the x-direction through the left face of the prism is

$$q'_x dy = -K\left(h\frac{\partial h}{\partial x}\right)_x dy$$ **(5–61)**

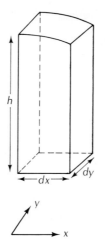

FIGURE 5.18 Control volume for flow through a prism of an unconfined aquifer with the bottom resting on a horizontal impervious surface and the top coinciding with the water table.

where dy is the width of the face of the prism. The discharge through the right face, $q'_{x\,+\,dx'}$ is

$$q'_{x\,+\,dx}\,dy = -K\left(h\frac{\partial h}{\partial x}\right)_{x\,+\,dx} dy \tag{5-62}$$

Note that $\left(h\dfrac{\partial h}{\partial x}\right)$ has different values at each face. The change in flow rate in the x-direction between the two faces is given by

$$(q'_{x\,+\,dx} - q'_x)dy = -K\frac{\partial}{\partial x}\left(h\frac{\partial h}{\partial x}\right) dx\,dy \tag{5-63}$$

Through a similar process, it can be shown that the change in the flow rate in the y-direction is

$$(q'_{y\,+\,dy} - q'_y)dx = -K\frac{\partial}{\partial y}\left(h\frac{\partial h}{\partial y}\right) dy\,dx \tag{5-64}$$

For steady flow, any change in flow through the prism must be equal to a gain or loss of water across the water table. This could be infiltration or evapotranspiration. The net addition or loss is at a rate of w, and the volume change within the initial volume is $w\,dx\,dy$ where $dx\,dy$ is the area of the surface. If w represents evapotranspiration, it will have a negative value. As the change in flow is equal to the new addition,

$$-K\frac{\partial}{\partial x}\left(h\frac{\partial h}{\partial x}\right)dx\,dy - K\frac{\partial}{\partial y}\left(h\frac{\partial h}{\partial y}\right)dy\,dx = w\,dx\,dy \tag{5-65}$$

We can simplify Equation 5–65 by dropping out $dx\,dy$ and combining the differentials:

$$-K\left(\frac{\partial^2 h^2}{\partial x^2} + \frac{\partial^2 h^2}{\partial y^2}\right) = 2w \qquad (5\text{–}66)$$

If $w = 0$, then Equation 5–66 reduces to a form of the Laplace equation:

$$\frac{\partial^2 h^2}{\partial x^2} + \frac{\partial^2 h^2}{\partial y^2} = 0 \qquad (5\text{–}67)$$

If flow is in only one direction and we align the x-axis parallel to the flow, then there is no flow in the y-direction, and Equation 5–66 becomes

$$\frac{d^2(h^2)}{dx^2} = -\frac{2w}{K} \qquad (5\text{–}68)$$

Integration of this equation yields the expression

$$h^2 = -\frac{wx^2}{K} + c_1 x + c_2 \qquad (5\text{–}69)$$

where c_1 and c_2 are constants of integration.

The following boundary conditions can be applied: at $x = 0$, $h = h_1$; at $x = L$, $h = h_2$ (Figure 5.19). By substituting these into Equation 5–69, the constants of integration can be evaluated with the following result:

$$h^2 = h_1^2 - \frac{(h_1^2 - h_2^2)x}{L} + \frac{w}{K}(L - x)x \qquad (5\text{–}70)$$

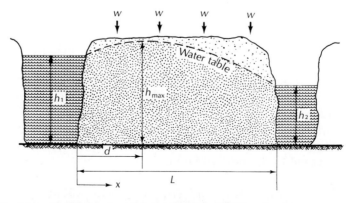

FIGURE 5.19 Unconfined flow, which is subject to infiltration or evaporation.

or

$$h = \sqrt{h_1^2 - \frac{(h_1^2 - h_2^2)x}{L} + \frac{w}{K}(L - x)x}$$

(5-71)

where

h is head at x (L; ft or m)

x is the distance from the origin (L; ft or m)

h_1 is the head at the origin (L; ft or m)

h_2 is the head at L (L; ft or m)

L is the distance from the origin at the point h_2 is measured (L; ft or m)

K is the hydraulic conductivity (L/T; ft/d or m/day)

w is the recharge rate (L/T; ft/d or m/day)

This equation can be used to find the elevation of the water table anywhere between two points located L distance apart if the saturated thickness of the aquifer is known at the two end points.

For the case in which there is no infiltration or evaporation, $w = 0$ and Equation 5–71 reduces to

$$h = \sqrt{h_1^2 - \frac{(h_1^2 - h_2^2)x}{L}}$$

(5-72)

By differentiating Equation 5–70, and because $q_x' = -Kh(dh/dx)$, it may be shown that the discharge per unit width, q_x', at any section x distance from the origin is given by

$$q_x' = \frac{K(h_1^2 - h_2^2)}{2L} - w\left(\frac{L}{2} - x\right)$$

(5-73)

where

q_x' is the flow per unit width at x (L^2/T; ft²/day or m²/day)

x is the distance from the origin (L; ft or m)

K is the hydraulic conductivity (L/T; ft/day or m/day)

h_1 is the head at the origin (L; ft or m)

h_2 is the head at L (L; ft or m)

L is the distance from the origin at the point where h_2 is measured (L; ft or m)

K is the hydraulic conductivity (L/T; ft/day or m/day)

w is the recharge rate (L/T; ft/day or m/day)

If the water table is subject to infiltration, there may be a water divide with a crest in the water table. In this case, q'_x will be zero at the water divide. If d is the distance from the origin to a water divide, then substituting $q'_x = 0$ and $x = d$ into Equation 5–72 yields

$$d = \frac{L}{2} - \frac{K}{w} \frac{(h_1^2 - h_2^2)}{2L} \qquad (5–74)$$

where

d is the distance from origin to water divide (L; ft or m)

h_1 is the head at the origin (L; ft or m)

h_2 is the head at L (L; ft or m)

L is the distance from the origin where h_2 is measured (L; ft or m)

K is the hydraulic conductivity (L/T; ft/day or m/day)

w is the recharge rate (L/T; ft/day or m/day)

Once the distance from the origin to the water divide has been found, then the elevation of the water table at the divide may be determined by substituting d for x in Equation 5–70.

$$h_{max} = \sqrt{h_1^2 - \frac{(h_1^2 - h_2^2)d}{L} - \frac{w}{K}(L - d)d} \qquad (5–75)$$

EXAMPLE PROBLEM

An unconfined aquifer has a hydraulic conductivity of 0.0020 cm/s and an effective porosity of 0.27. The aquifer is in a bed of sand with a uniform thickness of 31 m, as measured from the land surface. At well 1, the water table is 21 m below the land surface. At well 2, located some 175 m away, the water table is 23.5 m from the surface. What are (A) the discharge per unit width, (B) the average linear velocity at well 1, and (C) the water-table elevation midway between the two wells?

Part A: From Equation 5–60,

$$q' = K \frac{(h_1^2 - h_2^2)}{2L}$$

$h_1 = 31 \text{ m} - 21 \text{ m} = 10 \text{ m}$

$h_2 = 31 \text{ m} - 23.5 \text{ m} = 7.5 \text{ m}$

$L = 175 \text{ m}$

$$q' = 1.7 \text{ m/d} \times \frac{10^2 \text{ m}^2 - 7.5^2 \text{ m}^2}{2 \times 175 \text{ m}}$$

$$= 0.21 \text{ m}^2/\text{d per unit width}$$

Part B: From Equation 5–24,

$$v_x = \frac{Q}{n_e A}$$

As $Q = q' \times$ unit width and $A = h_1 \times$ unit width,

$$v_x = \frac{q'}{n_e h_1}$$

$$= \frac{0.21 \text{ m}^2/\text{d}}{0.27 \times 10 \text{ m}} = 0.08 \text{ m/day}$$

Part C: From Equation 5–71,

$$h = \sqrt{h_1^2 - (h_1^2 - h_2^2)\frac{x}{L}}$$

$$= \sqrt{(10 \text{ m})^2 - [(10 \text{ m})^2 - (7.5 \text{ m})^2]\left(\frac{87.5 \text{ m}}{175 \text{ m}}\right)}$$

$$= 8.8 \text{ m}$$

EXAMPLE PROBLEM   A canal was constructed running parallel to a river 1500 ft away. Both fully penetrate a sand aquifer with a hydraulic conductivity of 1.2 ft/d. The area is subject to rainfall of 1.8 ft/y and evaporation of 1.3 ft/y. The elevation of the water in the river is 31 ft and in the canal it is 27 ft. Determine (A) the water divide, (B) the maximum water-table elevation, (C) the daily discharge per 1000 ft into the river, and (D) the daily discharge per 1000 ft into the canal.

Part A: From Equation 5–73,

$$d = \frac{L}{2} - \frac{K}{w}\frac{(h_1^2 - h_2^2)}{2L}$$

$h_1 = 31 \text{ ft}$

$h_2 = 27 \text{ ft}$

$L = 1500 \text{ ft}$

$K = 1.2 \text{ ft/d}$

$w = 1.8 \text{ ft/y infiltration} - 1.3 \text{ ft/y evaporation}$

$\quad = 0.50 \text{ ft/y accretion}$

$\quad = 0.0014 \text{ ft/day}$

$$d = \frac{1500 \text{ ft}}{2} - \frac{1.2 \text{ ft/day}}{0.0014 \text{ ft/day}}\left(\frac{(31 \text{ ft})^2 - (27 \text{ ft})^2}{2 \times 1500 \text{ ft}}\right)$$

$\quad = 680 \text{ ft from the river}$

Part B: From Equation 5–75,

$$h_{max} = \sqrt{h_1^2 - \frac{(h_1^2 - h_2^2)d}{L} + \frac{w}{K}(L - d)d}$$

$$= \sqrt{(31\text{ft})^2 - \frac{[(31\text{ ft})^2 - (27\text{ ft})^2]\,680\text{ ft}}{1500\text{ ft}} + \frac{0.0014\text{ ft/day}}{1.2\text{ ft/day}}(1500\text{ ft} - 680\text{ ft})680\text{ ft}}$$

$$= 39\text{ ft}$$

Part C: From Equation 5–73, for $x = 0$:

$$q_x = \left[\frac{K(h_1^2 - h_2^2)}{2L} - w\left(\frac{L}{2} - x\right)\right] \times \text{width}$$

$$= \left[\frac{(1.2\text{ ft/day})[(31\text{ ft})^2 - (27\text{ ft})^2]}{2 \times 1500\text{ ft}} - (0.0014\text{ ft/day})\left(\frac{1500\text{ ft}}{2} - 0\right)\right] \times 1000\text{ ft}$$

$$q_x = -960\text{ ft}^3/\text{day}$$

The negative sign indicates that flow is in the opposite direction of x, or into the river.

Part D: From Equation 5–73,

$$x = L$$

$$q_x = \left[\frac{K(h_1^2 - h_2^2)}{2L} - w\left(\frac{L}{2} - x\right)\right] \times \text{width}$$

$$q_x = \left[\frac{(1.2\text{ ft/day})[(31\text{ ft})^2 - (27\text{ ft})^2]}{2 \times 1500\text{ ft}} - (0.0014\text{ ft/day})\left(\frac{1500\text{ ft}}{2} - 1500\text{ ft}\right)\right] \times 1000\text{ ft}$$

$$q_x = 1100\text{ ft}^3/\text{day}$$

Flow is in the direction of x, or into the canal.

NOTATION

A	Area	**dh/dl**	Hydraulic gradient
a	Width of a flowtube in the derivation of the tangent law	**dh/ds**	Grad h
		dx	Length of one side of a control volume
b	Aquifer thickness	**dy**	Length of one side of a control volume
b'	Aquitard thickness	**dz**	Length of one side of a control volume
c	Width of a flowtube in the derivation of the tangent law	**e**	Rate of vertical movement across an aquitard
d	Pore diameter	E_g	Gravitational potential energy

$\mathbf{E_k}$	Kinetic energy	$\mathbf{q'}$	Flow per unit width
$\mathbf{E_{tm}}$	Total energy per unit mass	$\mathbf{q'_x}$	Flow per unit width at location x
$\mathbf{E_{tv}}$	Total energy per unit volume	$\mathbf{q_x}$	Discharge in the x-direction
$\mathbf{f}$	Number of squares along a flowtube in a flow net	$\mathbf{q_y}$	Discharge in the y-direction
		$\mathbf{q_z}$	Discharge in the z-direction
$\mathbf{F}$	Force	$\mathbf{R}$	Reynolds number
$\mathbf{g}$	Acceleration of gravity	$\mathbf{S}$	Storage coefficient
$\mathbf{grad}\ h$	Vector representing the gradient of head	$\mathbf{S_y}$	Specific yield
$\mathbf{h}$	Head	$\mathbf{T}$	Transmissivity
$\mathbf{h_f}$	Fresh-water pressure head	$\mathbf{v}$	Velocity
$\mathbf{h_p}$	Point-water pressure head (height of point fluid in piezometer)	$\mathbf{V_x}$	Average linear velocity
		$\mathbf{V_s}$	Volume of solids
$\mathbf{K}$	Hydraulic conductivity	$\mathbf{w}$	Recharge rate to an unconfined aquifer
$\mathbf{K'}$	Vertical hydraulic conductivity of an aquitard	$\mathbf{W}$	Work
$\mathbf{L}$	Distance between h_1 and h_2 in the Dupuit equations	$\mathbf{x}$	Distance
		$\mathbf{z}$	Elevation
$\mathbf{m}$	Mass	$\mathbf{\alpha}$	Compressibility of aquifer
$\mathbf{M}$	Mass of water in the derivation of the ground-water-flow equation	$\mathbf{\beta}$	Compressibility of water
		$\mathbf{\Phi}$	Force potential
$\mathbf{n}$	Porosity	$\mathbf{\rho}$	Fluid density
$\mathbf{n_e}$	Effective porosity	$\mathbf{\rho_f}$	Density of fresh water
$\mathbf{p}$	Number of flowtubes in a flow net	$\mathbf{\rho_p}$	Density of point water
$\mathbf{P}$	Pressure	$\mathbf{\rho_w}$	Density of water
$\mathbf{P_0}$	Atmospheric pressure	$\mathbf{\mu}$	Viscosity
$\mathbf{Q}$	Discharge	$\mathbf{\sigma}$	Angle of refraction

PROBLEMS

Answers to odd-numbered problems will appear at the end of the book. Assume $g = 9.81$ m/s^2.

1. A fluid in an aquifer is 4.0 m above a reference datum, the fluid pressure is 2400 N/m^2, and the flow velocity is 1.0×10^{-5} m/s. The fluid density is 1.01×10^3 kg/m^3.

   A. What is the total energy per unit mass?

   B. What is the total energy per unit weight?

2. A fluid in an aquifer is 31.5 m above a reference datum, the fluid pressure is 3750 N/m^2, and the flow velocity is 1.35×10^{-4} m/s^2. The fluid density is 0.999×10^3 kg/m^3.

   A. What is the total energy per unit mass?

   B. What is the total energy per unit weight?

3. A piezometer is screened 273.4 m above mean sea level. The point-water pressure head in the piezometer is 23.4 m and the water in the aquifer is fresh at a temperature of 20°C.

    A. What is the total head in the aquifer at the point where the piezometer is screened?

    B. What is the fluid pressure in the aquifer at the point where the piezometer is screened?

4. A piezometer point is 23 m above mean sea level. The fluid pressure in the aquifer at that point is 6.45×106 N/m². The aquifer has fresh water at a temperature of 13°C.

    A. What is the point-water pressure head?

    B. What is the total head?

5. A piezometer in a saline water aquifer has a point-water pressure head of 15.23 m. If the water has a density of 1029 kg/m³ and is at a field temperature of 21°C, what is the equivalent fresh-water pressure head?

6. The fluid pressure in the screen of a piezometer in a saline aquifer is 4.532×10^5 N/m². The fluid density is 1073 kg/m³ and the temperature is 12°C. The elevation of the piezometer screen is 1048.54 m at surface level.

    A. Compute the point-water pressure head.

    B. Compute the fresh-water pressure head.

    C. Find the total fresh-water head.

7. A sand aquifer has a median pore diameter of 0.2 mm. The fluid density is 1.003×10^3 kg/m³ and the fluid viscosity is 1.15×10^{-3} N·s/m². If the flow rate is 0.0016 m/s, is Darcy's law valid? What is the reason for your answer?

8. An aquifer has a hydraulic conductivity of 123 ft/day and an effective porosity of 27% and is under a hydraulic gradient of 0.0003.

    A. Compute the Darcy flux.

    B. Compute the average linear velocity.

    C. The water temperature was 12°C and the mean pore diameter was 0.33 mm. Was it permissible to use Darcy's law under these circumstances? What is the reason for your answer?

9. A confined aquifer is 10 ft thick. The potentiometric surface drops 0.54 ft between two wells that are 792 ft apart. The hydraulic conductivity is 21 ft/day and the effective porosity is 0.17.

    A. How many cubic feet per day are moving through a strip of the aquifer that is 10 ft wide?

    B. What is the average linear velocity?

10. A confined aquifer is 24.5 m thick. The potentiometric surface drops 1.23 m between two wells that are 1023 m apart. If the hydraulic conductivity of the aquifer is 44 m/day, how many cubic meters of flow are moving through the aquifer per unit width?

11. An unconfined aquifer has a hydraulic conductivity of 1.7×10^{-3} cm/s. There are two observation wells 328 ft apart. Both penetrate the aquifer to the bottom. In one observation well the water stands 24.6 ft above the bottom, and in the other it is 20.0 ft above the bottom.

   A. What is the discharge per 100-ft-wide strip of the aquifer in cubic feet per day?

   B. What is the water-table elevation at a point midway between the two observation wells?

12. Refer to Figure 5.19. The hydraulic conductivity of the aquifer is 1.2 m/day. The value of h_1 is 17 m and the value of h_2 is 12 m. The distance from h_1 to h_2 is 4525 m. There is an average rate of recharge of 0.0002 m/d.

   A. What is the average discharge per unit width at $x = 0$?

   B. What is the average discharge per unit width at $x = 4525$ m?

   C. Where is the water divide located?

   D. What is the maximum height of the water table?

13. An earthen dam is constructed on an impermeable bedrock layer. It is 750 ft across (i.e., the distance from the water in the reservoir to the tailwaters below the dam is 750 ft). The average hydraulic conductivity of the material used in the dam construction is 0.23 ft/day. The water in the reservoir behind the dam is 75 ft deep and the tailwaters below the dam are 20 ft deep. Compute the volume of water that seeps from the reservoir, through the dam, and into the tailwaters per a 100-ft-wide strip of the dam in cubic feet per day.

14. Draw a flow net for seepage through the earthen dam shown in Figure 5.20. If the hydraulic conductivity of the material used in the dam is 0.22 ft/day, what is the seepage per unit width per day?

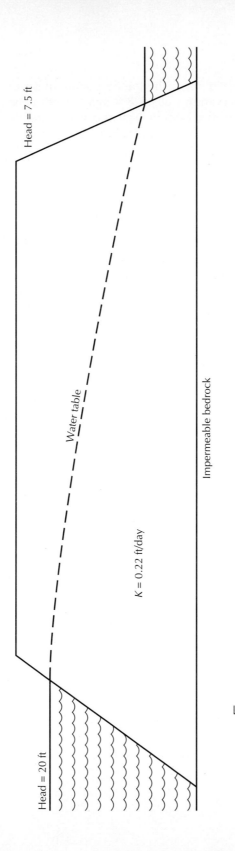

Head = 7.5 ft

Head = 20 ft

Water table

$K = 0.22$ ft/day

Impermeable bedrock

10 ft

10 ft

FIGURE 5.20 Earthen dam for Problem 14, construction of a flow net.

6 Soil Moisture and Ground-Water Recharge

There are two general methods by which water finds its way through the strata; in the one—the rock being close-textured—the water passes through fissures formed by fracture, or tubular channels formed by solution; in the other—the rock being open-textured—the water seeps through the pores, permeating the whole bed.

Requisite and Qualifying Conditions of Artesian Wells, T. C. Chamberlin, 1885

6.1   INTRODUCTION

The uppermost layer of the earth may contain a three-phase system of solid, liquid, and gaseous material. The solid phase contains mineral grains and organic matter. The organic matter represents the remains of plants and animals that are undergoing decay. The liquid phase is water containing dissolved solutes. The vapor phase includes water vapor and other gases that may not be present in the same proportions as the atmosphere (Jury, Gardner, & Gardner 1991).

The earth layer that contains the three phase system is called the *zone of aeration,* **unsaturated zone,** or *vadose zone*.

The solid phase of the vadose zone may consist of soil, which is formed by *in situ* weathering; or sediment, which has been transported from the place of weathering; or unweathered bedrock. Mineral grains may be disaggregated, for example, sand grains in a dune. Smaller grains may also be bound by organic matter to form larger units called *aggregates,* or *peds*. The peds have a specific orientation that forms the soil structure. The soil *texture* is determined by the distribution of the size fractions of mineral grains present.

6.2   POROSITY AND WATER CONTENT OF SOIL

The *porosity* of the soil is the percent of void space.

$$n = V_v/V \qquad\qquad (6-1)$$

where

n is porosity (dimensionless)

V_v is volume of the void space (L^3; cm^3 or m^3)

V is volume of the sample (L^3; cm^3 or m^3)

The *void ratio* of the soil is the ratio of the volume of the voids to the volume of the solids.

$$e = V_v/V_s \qquad\qquad (6\text{--}2)$$

where

e is void ratio (dimensionless)

V_s is volume of the solids (L^3; cm^3 or m^3)

The total volume is equal to the volume of the voids plus the volume of the solids.

$$V = V_v + V_s \qquad\qquad (6\text{--}3)$$

The void ratio is closely related to the porosity:

$$n = \frac{e}{1 + e} \qquad\qquad (6\text{--}4)$$

and

$$e = \frac{n}{1 - n} \qquad\qquad (6\text{--}5)$$

The *gravimetric water content* of the soil is the mass of the contained water divided by the mass of the solid particles (dry mass of soil):

$$\theta_g = 100(W_w/W_s) \qquad\qquad (6\text{--}6)$$

where

θ_g is the gravimetric water content (percentage)

W_w is the mass of the water in the soil (M; g or kg)

W_s is the mass of the solid particles (M; g or kg)

The *volumetric water content* of the soil is the volume of the contained water divided by the total volume of the soil:

$$\theta_v = V_w/V \qquad\qquad (6\text{--}7)$$

where

θ_v is the volumetric water content (dimensionless ratio)

V_w is the volume of the contained water (L^3; cm^3 or m^3)

The **saturation ratio** of a soil is the volume of the contained water divided by the volume of the voids:

$$R_s = V_w/V_v \tag{6-8}$$

where R_s is the saturation ratio (dimensionless ratio).

The *dry bulk density* of the soil is the mass of the soil particles (dry mass) divided by the volume of the sample:

$$\rho_b = W_s/V \tag{6-9}$$

where ρ_b is the dry bulk density (M/L^3; gm/cm^3 or kg/m^3).

The *particle density* is the mass of the mineral particles of the soil divided by the volume of the soil particles:

$$\rho_m = W_s/V_s \tag{6-10}$$

where ρ_m is the particle density (M/L^3; gm/cm^3 or kg/m^3).

The mass of water in a soil sample is equal to the product of the volumetric water content and the density of water. The mass of water is also equal to the product of the gravimetric water content and the dry bulk density of the soil:

$$\rho_w\theta_v = \rho_b\theta_g \tag{6-11}$$

Equation 6–11 can be rearranged to yield

$$\theta_v = (\rho_b/\rho_w)\theta_g \tag{6-12}$$

From Equations 6–1 and 6–3, the following relation can be obtained:

$$n = 100\left(\frac{V - V_s}{V}\right) = 100\left(1 - \frac{V_s}{V}\right) \tag{6-13}$$

From Equation 6–9, $V = W_s/\rho_b$, and from Equation 6–10, $V_s = W_s/\rho_m$. Substituting these into Equation 6–13 and dividing to eliminate W_s, we obtain

$$n = 100\left(1 - \frac{\rho_b}{\rho_m}\right) \tag{6-14}$$

The gravimetric water content of a soil sample can be measured directly by excavation of a soil sample. The volume, V, of a moist soil sample is measured and the moist mass, W_m, determined. The sample is oven-dried at 105°C until a constant mass is obtained and then the mass of the soil, W_s, is determined. The mass of the water is $W_m - W_s$. Gravimetric water content is hence $(W_m - W_s)/W_s$. To find the volumetric water content, the volume of the water must first be determined by dividing the mass of the water by the density of water, ρ_w. Volumetric water content is then determined as $[(W_m - W_s)/\rho_w]/V$.

Soil moisture can also be measured indirectly by nondestructive means. One method involves burying small blocks, called resistance cells, in which electrodes are embedded, and then passing an electrical current through the wire. The electrical resistance of the block is proportional to the moisture it contains, which, in turn, is dependent upon the soil moisture. The meter can be calibrated for the soil type and, once calibrated, can be used for repeated measurements at the same location. This type of apparatus is relatively inexpensive, but the buried resistance blocks must be left in the soil.

A more expensive device uses a source of fast neutrons enclosed in a probe. The probe is lowered into a tube in the soil. When the fast neutrons encounter hydrogen atoms in water, they become slow neutrons. The density of slow neutrons produced is a function of the amount of soil moisture. A slow neutron counter is also a part of the probe. The neutron meter must be calibrated against the known water content of one part of the soil profile. The method can be used for repeated soil-moisture measurements in the same access tube. The water content in a spherical volume of 6-in. (15-cm) radius is measured by this method.

EXAMPLE PROBLEM   A soil sample is collected in the field and placed in a container with a volume of 75.0 cm³. The mass of the soil at the natural moisture content is determined to be 150.79 g. The soil sample is then saturated with water and reweighed. The saturated mass is 153.67 g. The sample is then oven-dried to remove all the water and reweighed. The dry mass is 126.34 g. Note that masses are determined by weighing on a balance. All measurements were made at 20°C.

Part A: Determine the soil porosity.

The volume of the voids is the volume of the water at saturation. The volume of water is the mass of water divided by the density of water. The density of water at 20°C is 0.998 g/cm³. The mass of water at saturation is the saturated mass minus the dry mass.

$$W_{w(\text{saturated})} = 153.67 \text{ g} - 126.34 \text{ g} = 27.33 \text{ g}$$
$$V_{w(\text{saturated})} = (27.33 \text{ g})/(0.998 \text{ g/cm}^3) = 27.4 \text{ cm}^3$$

Porosity is $100(V_v/V)$. Since $V_v = V_{w(\text{saturated})}$,

$$n = 100(27.4/75.0) = 36.5\%$$

Part B: Determine the gravimetric water content under natural conditions.

The mass of the water is the moist mass minus the dry mass. The gravimetric water content is the ratio of the mass of the water to the dry mass of the soil.

$$W_w = 150.70 \text{ g} - 125.34 \text{ g} = 25.36 \text{ g}$$
$$\theta_g = 100(W_w/W_s)$$
$$= 100[(25.36 \text{ g})/(125.34 \text{ g})] = 20.23\%$$

Part C: Determine the volumetric water content.

The volume of the water is the weight of the water divided by the density of water.

$$V_w = (25.36 \text{ g})/(0.998 \text{ g/cm}^3)$$
$$= 25.4 \text{ cm}^3$$
$$\theta_v = V_w/V$$
$$= (25.4 \text{ cm}^3)/(75.0 \text{ cm}^3) = 0.339$$

Part D: Determine the saturation ratio.

$$S_r = V_w/V_v$$

Since V_v is equal to $V_{w(\text{saturated})}$,

$$S_r = (25.4 \text{ cm}^3)/(27.4 \text{ cm}^3) = 0.927$$

Part E: Determine the dry bulk density.

The mass of the soil particles is 126.34 g, which is the oven-dried weight. Therefore,

$$\rho_b = W_s/V$$
$$= (126.34 \text{ g})/(75.0 \text{ cm}^3) = 1.68 \text{ g/cm}^3$$

Part F: Determine the particle density.

The volume of the solids is the total volume less the volume of the voids.

$$V_s = 75.0 \text{ cm}^3 - 27.38 \text{ cm}^3 = 47.6 \text{ cm}^3$$
$$\rho_m = W_s/V_s$$
$$= (126.34 \text{ g})/(47.6 \text{ cm}^3) = 2.65 \text{ g/cm}^3$$

The experimentally determined particle density of 2.65 g/cm^3 is exactly equal to the density of quartz, which is 2.65. Quartz is a common soil mineral.

Part G: As a check on the internal consistency of the data, determine the porosity from Equation 6–14.

$$n = 100\left(1 - \frac{\rho_b}{\rho_m}\right)$$
$$= 100\left(1 - \frac{1.68 \text{ g/cm}^3}{2.65 \text{ g/cm}^3}\right)$$
$$= 36.6\%$$

The data show good internal consistency, since the porosity as determined by Equation 6–1 was 36.5%.

6.3   CAPILLARITY AND THE CAPILLARY FRINGE

If fluid pressures are measured above the water table, they will be found to be negative with respect to local atmospheric pressure. This is called a **tension.** Air may also be present in the voids above the water table. Air pressure is equal to atmospheric pressure above the water table. Water vapor is also present in the voids above the water table.

Water molecules at the water table are subject to an upward attraction due to surface tension of the air-water interface and the molecular attraction of the

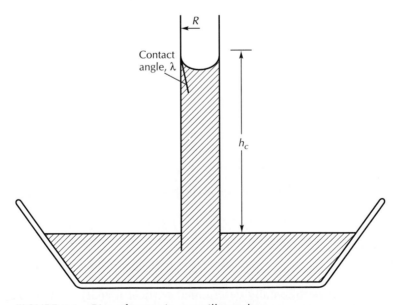

FIGURE 6.1   Rise of water in a capillary tube.

liquid and solid phases. This is known as *capillarity*—a phenomenon well studied in classical physics.

In a tube of small diameter, the free-water surface will assume a shape with the minimum surface area. The attraction of the solid for the liquid will draw the liquid up into the tube. The upward force will eventually be offset by the weight of the column of water. The water, itself, is under tension; thus, the pressure is less than atmospheric. Figure 6.1 shows the rise of a fluid in a glass capillary due to **capillary forces.** The fluid meets the glass capillary wall at a contact angle.

The rise of a fluid in a capillary tube is given by

$$h_c = \frac{2\sigma \cos \lambda}{\rho_w g R} \qquad \textbf{(6–15)}$$

where

h_c is the height of the capillary rise (L; cm or m)

σ is the surface tension of the fluid (M/T^2; g/s^2 or kg/s^2)

λ is the angle of the meniscus with the capillary tube (degrees)

ρ_w is the density of the fluid (M/L^3; g/cm^3 or kg/m^3)

g is the acceleration of gravity (L/T^2; cm/s^2 or m/s^2)

R is the radius of the capillary tube (L; cm or m)

EXAMPLE PROBLEM

Compute the rise of water in a glass capillary tube.

For water at 18°C, the surface tension is 73 g/s^2, the density is 0.999 g/cm^3, and the contact angle may be taken as 0.

$$h_c = \frac{2 \times 73 \text{ g/s}^2 \times \cos 0}{0.999 \text{ g/cm}^3 \times 980 \text{ cm/s}^2 \times R \text{ cm}} \qquad \textbf{(6–16)}$$

$$= \frac{0.15}{R} \text{ cm}$$

We can make a convenient model of soil capillaries by assuming that the pores of the soil are equivalent to a number of glass capillary tubes bundled together. The model capillary tubes are not straight but rather have a series of narrow necks separating the wider pore throats. The height of the capillary rise depends upon the widest part of the pore, where the surface attraction is the least effective. With cubic packing of spherical grains of equal diameter, it can be shown that the radius of the pore throats is equal to 0.2 times the grain diameter (Figure 6.2). For example, for a sediment with a grain diameter of 0.020 cm, the pore throat radius is 0.0040 cm, and, from Equation 6–16, the height of the capillary rise is 38 cm.

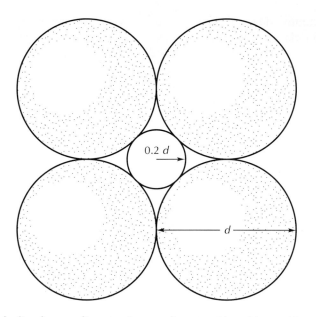

FIGURE 6.2 Idealized pore diameter in a sediment with cubic packing. The equivalent capillary tube has a radius of 0.2 the diameter of the grains.

Capillary pores in the *zone of aeration* can draw water up from the zone of saturation beneath the water table. In very fine-grained soils, this **capillary fringe** can saturate the soil above the water table. However, tensiometer readings will reveal that the head is negative, indicating that the capillary fringe is part of the vadose zone. The zone of aeration is best defined as the zone where the soil moisture is under tension.

Table 6.1 gives the mean grain size of common sediments and the range of the height of the capillary rise predicted from Equation 6–16. Because of irregularities in the size of the openings, capillary water does not rise to an even

TABLE 6.1 Height of Capillary Rise in Sediments

Sediment	Grain Diameter (cm)	Pore Radius (cm)	Capillary Rise (cm)
Fine silt	0.0008	0.0002	750
Coarse silt	0.0025	0.0005	300
Very fine sand	0.0075	0.0015	100
Fine sand	0.0150	0.003	50
Medium sand	0.03	0.006	25
Coarse sand	0.05	0.010	15
Very coarse sand	0.20	0.040	4
Fine gravel	0.50	0.100	1.5

height above the water table; rather, it forms an irregular fringe. The capillary fringe is higher in fine-grained soils than in coarse-grained ones because of the greater tensions created by the smaller pore openings. The capillary fringe can provide a means of direct evaporation of ground water if the water table is close enough to the surface that the capillary fringe reaches the ground surface. As water evaporates from the soil surface, it can be replaced by water from the zone of saturation drawn upward by capillarity.

Above the capillary fringe, there is moisture coating the solid surfaces of the fragment or rock particles. If the liquid coating becomes too thick to be held by surface tension, a droplet will pull away and be drawn downward by gravity. The fluid can also evaporate and move through the air space in the pores as water vapor.

The amount of vapor movement in the unsaturated zone is much less important than transport in the liquid form (Swartzendruber 1969). However, this might not hold true if the water content of the soil is very low or if there is a strong temperature gradient. The movement of vapor through the unsaturated zone is a function of the temperature and humidity gradients in the soil and molecular diffusion coefficients for water vapor in the soil (Ripple, Rubin, & Van Hylckama 1972).

6.4 PORE-WATER TENSION IN THE VADOSE ZONE

Fluid pressures in the unsaturated zone are negative, owing to tension of the soil-surface-water contact. The negative pressure head, ψ, is measured in the field with a **tensiometer.** This device consists of a tube that is closed at the top, with a ceramic cup at the bottom to provide a porous membrane (Figure 6.3). When the tensiometer is inserted into soil, water within it is in contact with the soil moisture through the porous membrane. The suction exerted on the water in the tensiometer can be measured with a mercury manometer, vacuum gage, or pressure transducer (Watson 1967).

6.5 SOIL MOISTURE

The water in the vadose zone that is available to growing plants is called *soil moisture,* or *soil water*. It is found in the *belt of soil moisture*. This zone extends from the land surface to the depth of plant roots. In agricultural fields this may extend downward only a meter or two. Some trees have taproots that extend downward
many meters. The top of the capillary fringe may be below the belt of soil moisture. In this case there is an **intermediate zone** between the belt of soil moisture and the capillary fringe. In other cases the belt of soil moisture may extend right to the water table and include the capillary fringe.

Soil moisture at a location varies with changes in the amount of precipitation and evapotranspiration. Figure 6.4 shows a soil-moisture budget for a

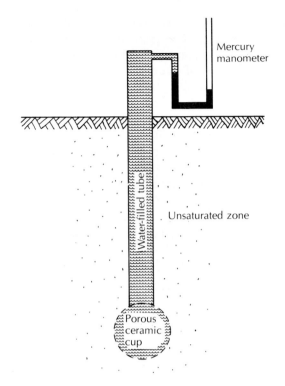

FIGURE 6.3   Porous-cup tensiometer with mercury manometer to measure soil-moisture tension.

humid area based on measured precipitation and computed potential and actual evapotranspiration. During the period of water surplus, there is moisture available for ground-water recharge and runoff. Major fluctuations are seasonal in nature. In the spring, soil moisture is high, as snowmelt and spring rains have created large amounts of water available for infiltration. At times, the top layer of soil may be completely saturated, even though lower layers are unsaturated. During these periods of very high soil moisture, ground-water recharge can occur. Moisture moves downward by gravity flow. As water is withdrawn from the soil by evapotranspiration, the soil-moisture content drops. When the soil-moisture content of a layer reaches the point at which the force of gravity acting on the water equals the surface tension, gravity drainage ceases. This soil-moisture content is the **field capacity** of the soil. Field capacity is related to specific retention but has different units. It depends upon specific retention, evaporation depth, and the unsaturated permeability characteristic curve of the soil.

The concept of field capacity is somewhat vague. Gravity drainage may take a long period of time to occur. The amount of moisture retained for a few days is much more than that retained for a long period. Table 6.2 shows the soil moisture of a silt-loam (fine-textured) soil as a function of time (Hillel 1971). It can be seen that field capacity is not a single value, but a time-dependent parameter.

Soil moisture becomes lower than the field capacity as evapotranspiration removes still more water. During summer periods, the soil often dries. Occasional

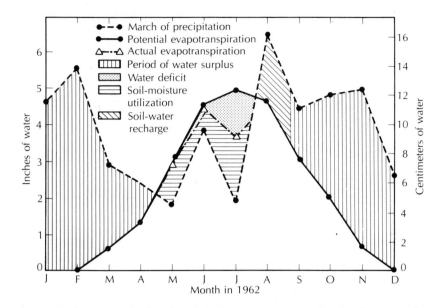

FIGURE 6.4 Soil-moisture budget for Bridgehampton, New York. The diagram is based on measured precipitation and computed potential and actual evapotranspiration. The Thornthwaite method was used for evapotranspiration computations. Source: C. W. Fetter, Jr., *Bulletin, Geological Society of America* 87 (1976): 401–6.

rainstorms may cause short-term rises in soil-moisture content, but there is generally no ground-water recharge. Exceptionally heavy summer rains, which replenish the depleted soil moisture and raise the water content above the field capacity, can create a wave of infiltrated water that courses downward through the soil-moisture zone and past the roots of thirsty plants, recharging the ground-water reservoir and thus causing the water table to rise. After fall frosts kill plants and cause deciduous trees to lose their leaves, the rate of evapotranspiration slows greatly, and soil moisture increases if rainfall and infiltration continue. Figure 6.5 illustrates a hypothetical annual cycle of soil moisture for a moderately humid area (20 to 30 in., or 50 to 75 cm per year precipitation).

TABLE 6.2 Moisture content of a silt loam as a function of time since saturation

Time (days)	θ (%)
1	20.2
7	17.5
30	15.9
60	14.7
156	13.6

Source: Hillel (1971).

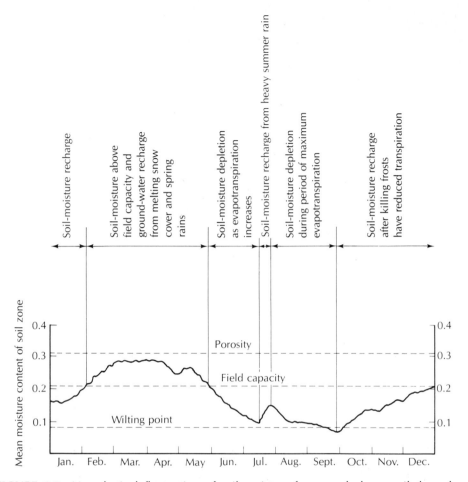

FIGURE 6.5 Hypothetical fluctuation of soil moisture for a sandy loam soil through an annual cycle in a region with a moderate amount of rainfall (20 to 30 in. (50 to 75 cm) per year) and heavy rains in the spring.

If the soil moisture drops too low, the remaining moisture is too tightly bound to the soil particles for the plant roots to withdraw it. The soil-moisture content at which this first occurs is the *wilting point*. Plants wilt and may die for lack of moisture. The wilting-point moisture content is greater for fine-textured soils owing to their greater surface area. Figure 6.6 shows typical wilting-point moisture content values for various soil types. Also shown are generalized field capacity values. The available water capacity of a soil is the difference between the field capacity and the wilting point. Water in a soil above the field will drain downward if the soil is permeable or will waterlog a fine-grained, slowly permeable soil. Brief study of Figure 6.6 shows that the available water capacity is greatest for soils of intermediate texture, e.g., loams and silt loams.

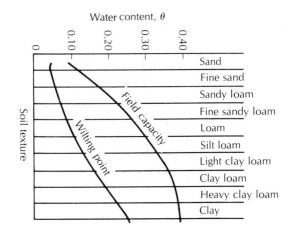

FIGURE 6.6 Water-holding properties of soils based on texture. The available water supply for a soil is the difference between field capacity and wilting point. Source: U.S. Department of Agriculture, *Yearbook*. 1955.

6.6 THEORY OF UNSATURATED FLOW

The infiltration of water into the unsaturated zone can be considered mathematically in terms of both a **gravity potential,** Z, and a **moisture potential,** ψ (Childs 1967). The moisture potential is a negative pressure due to the soil-water attraction. The moisture potential increases with decreasing amounts of soil moisture. Depending upon the soil-moisture content, either the moisture potential or the gravity potential may predominate. At moisture contents close to the specific retention, the gravity potential is greater; but when the soil is very dry, the moisture potential may be several orders of magnitude greater than the gravity potential.

When the soil is not saturated, soil moisture flows downward by gravity flow through interconnected pores that are filled with water and, to a lesser extent, as a film flowing along particle surfaces in pores that also contain air. With increasing water content, more pores fill, and the rate of downward water movement increases. Darcy's law is valid for flow in the unsaturated zone, although the *unsaturated hydraulic conductivity, $K(\theta_v)$,* is not a constant. The unsaturated hydraulic conductivity is a function of the volumetric water content, θ_v. As θ_v increases, so does $K(\theta_v)$. Figure 6.7 shows the relationship between $K(\theta_v)$ and θ_v for a clay soil. The value of the moisture potential, ψ, is also a function of θ_v, often ranging over many orders of magnitude. The relationship between moisture potential and volumetric water content is determined experimentally for a given soil. The results are graphed as a soil-water retention curve (Figure 6.8). The total potential, ϕ, in unsaturated flow is the sum of the moisture potential, ψ (θ_v), and the elevation head, Z:

$$\phi = \psi(\theta_v) + Z \qquad \qquad \textbf{(6–17)}$$

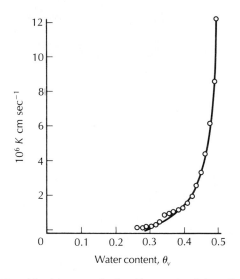

FIGURE 6.7 The relationship between hydraulic conductivity, *K*, and volumetric water content, θ_v, for a clay. Source: J. R. Philip, "Theory of Infiltration." In *Advances in Hydroscience,* vol. 5, ed. V. T. Chow. (New York: Academic Press, 1969). Used with permission.

The relationship of unsaturated hydraulic conductivity and volumetric water content is determined experimentally. A sample of the rock or sediment is placed in a container. The water content is kept constant, and the rate at which water moves through the soil is measured. The value of $K(\theta)$ can be determined by Darcy's law. Figure 6.7 is based on a number of different measurements at differing values of θ_v.

The moisture potential, ψ, is measured by a suction applied to a soil. The curve of ψ versus volumetric water content is a plot of the volumetric water content, starting with a saturated sample to which increased suction is gradually applied. A rather substantial problem occurs because of hysteresis in the volumetric water content–moisture potential relationship. The curve of ψ as a function of θ_v is different if it is determined for decreasing as opposed to increasing values of θ_v. Thus, two curves similar to Figure 6.8 are needed: one if the sediment has a particular θ_v as a result of an increase in soil-water content and another if the soil is drying. Figure 6.9 shows the impact of hysteresis for an idealized sandy soil. Therefore, to use a value of ψ, one must know the prior moisture history of the sample.

Flow in the unsaturated zone is further complicated by the fact that both *K* and ψ may change as θ_v varies, and, by its very nature, unsaturated flow involves many changes in volumetric moisture content as waves of infiltrated water pass. The flow equations are nonlinear and not subject to easy solutions (Swartzendruber 1969; Philip 1969).

The downward movement of a slug of infiltrated water is shown in Figure 6.10. During infiltration (part A), the topmost soil layer is brought up to the

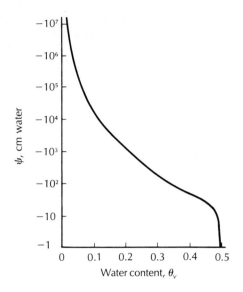

FIGURE 6.8 The relationship between moisture potential, ψ, and volumetric water content, θ_v, for the clay soil of Figure 6.7. Source: J. R. Philip, "Theory of Infiltration." In *Advances in Hydroscience*, vol. 5, ed. V. T. Chow. (New York: Academic Press, 1969). Used with permission.

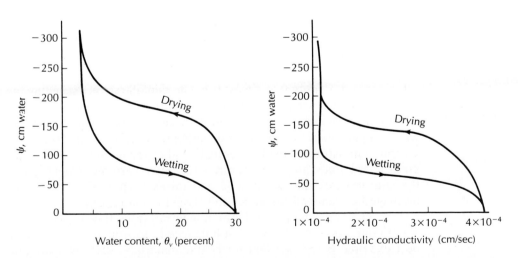

FIGURE 6.9 Idealized curves showing relationships of volumetric water content, θ_v, hydraulic conductivity, K, and soil-moisture tension head, ψ. The effect is included for wetting and drying cycles.

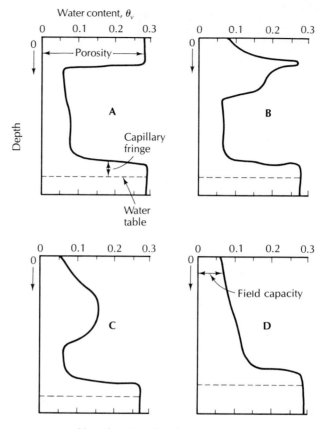

FIGURE 6.10 Moisture profiles showing the downward passage of a wave of infiltrated water. The soil is saturated at a water content of 0.29 and has a field capacity water content of 0.06. Source: Modified from Agronomy Monograph 17, "Drainage for Agriculture." 1974, pp. 359–405. Used with permission of the American Society of Agronomy.

saturated water content of almost 0.3. The rainfall rate is exceeding the infiltration capacity, so the vertical hydraulic conductivity of the soil controls the rate of downward movement. After infiltration ceases (part B), the slug of infiltrated water begins to move downward, although only a small layer remains at the saturated moisture content. Eventually, all of the downward-moving slug is at an unsaturated state, and $K(\theta_v)$ controls the downward movement (part C). Finally, the slug reaches the water table and raises it (part D). By this time, gravity drainage has reduced the upper soil layer to field capacity.

     One important consideration in unsaturated flow is that at low volumetric water contents, the relations that hold true in saturated flow may be invalid. The best example is the fact that for coarse materials, such as sand and gravel, the pores are large and drain quickly. At lower volumetric moisture contents, there may be very few saturated pores. On the other hand, finer-grained soils may have

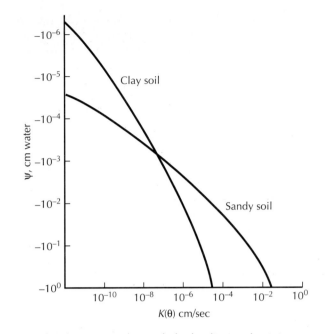

FIGURE 6.11 Typical soil-moisture-potential–hydraulic-conductivity curves for a sandy soil showing the crossover effect for increasing moisture potential (decrease in water content.)

most of the pores still saturated. Thus, at lower values of θ_v, the unsaturated hydraulic conductivity of a clay may be greater than that of a sand. A layer of sand in a fine-textured, unsaturated soil may retard downward movement of infiltrating water owing to its low unsaturated hydraulic conductivity.

Figure 6.11 shows the relationship between moisture potential and unsaturated hydraulic conductivity for a sandy soil and a clay soil. It can be seen that at low moisture potentials (high volumetric water content), the sandy soil has a greater hydraulic conductivity. However, at high (more negative) moisture potential (low water content), the clay has a greater hydraulic conductivity.

6.7   WATER-TABLE RECHARGE

When the front of infiltrating water reaches the capillary fringe, it displaces air in the pore spaces and causes the water table to rise. The capillary fringe is also higher, and the latest-arriving recharge is actually found at the top of the capillary fringe. The time of movement of infiltrating water is a function of the thickness of the unsaturated zone and the vertical unsaturated hydraulic conductivity. The presence of layers of low-permeability material, such as silts and clays, can retard the rate of recharge, even if the layers are thin. The time lag may be only a few hours in the humid regions for very coarse soils with a water table close to the surface. In arid environments, with very infrequent recharge and great depths to the water table, water may take years to pass through the unsaturated zone.

In the arid Hualapai Plateau area of northwestern Arizona, the annual recharge rate is only 0.1 in. (0.25 cm) per year. Although rainfall averages 9 to 13 in. (23 to 33 cm) per year, potential evapotranspiration is in the range of 72 to 76 in. (183 to 193 cm) per year (Huntoon 1977). Grassland regions have more precipitation and, consequently, more recharge. In the Sand Hills region of South Dakota and Nebraska, the mean annual precipitation is 18.2 in. (46.3 cm), and ground-water recharge amounts to 2.7 in. (6.9 cm) per year (Rahn & Paul 1975). On eastern Long Island, New York, precipitation averages 46 in. (116.8 cm) per year, and evapotranspiration, as computed by the Thornthwaite method, averages 22.6 in. (57.4 cm) per year. The soils are permeable, and much of the remaining 23.4 in. (59.4 cm) per year of water recharges the water table (Fetter 1976).

The rate at which water-table recharge occurs is variable, depending upon, among other things, the thickness of the unsaturated zone. Where the unsaturated zone is thinner, recharge can reach the water table first, resulting in a localized ground-water mound. The unsaturated zone is thinner in topographically low places, for example, near a lake shore or in a lowland. Soil moisture percolating through the unsaturated zone beneath upland areas takes longer so that if the water table is initially somewhat level, localized high spots can develop on the water table. Localized flow systems can develop that move water laterally from the temporary ground-water mounds toward the water table beneath the uplands, where the infiltration has not reached the water table. Eventually the water table beneath the upland area will rise owing not only to the vertical percolation of infiltrating water but also to the lateral movement from the temporary ground-water mounds. This results in the development of very complex and transitory local flow systems. In very permeable materials, the local ground-water mounds beneath the topographic lows dissipate quickly. However, in less permeable materials the local ground-water mounds could take a rather long time to dissipate. Localized flow reversals could take place with the ground water first flowing from beneath the lowland toward the upland in response to the formation of localized recharge mounds and then moving in the opposite direction as the water table beneath the upland finally is recharged by the percolating soil moisture. It would be anticipated that the flow reversals would affect only the water moving in the very top of an unsaturated flow system (Winter 1983).

On eastern Long Island, the response of the water table to recharge is fairly rapid. Figure 6.12 is a hydrograph for 1962 of a water-table well with an unsaturated zone on the order of 10 to 13 ft (3 to 4 m). This well is a few miles from the precipitation station that yielded the data for Figure 6.4. As the water table is being drained by stream and spring discharge, it will decline (1) if there is no recharge or (2) if the amount of recharge is less than ground-water

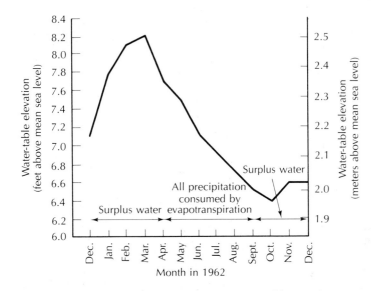

FIGURE 6.12 Hydrograph of a shallow well in a water-table aquifer in Long Island. The period of record is the same as for the soil-moisture budget diagram of Figure 6.4.

outflow. The hydrograph peaks in March and then begins a recession that lasts until October. Periods of surplus water are indicated in Figure 6.4 for January through April, and then September through December. The peak in March corresponds to snowmelt and spring rains, and the upturn in water levels by November is due to fall recharge. The response time of this water-table aquifer is more rapid than for those cases in which the unsaturated zone is many tens of feet thick.

The water table shows a seasonal fluctuation, rising during periods of recharge and falling when there is no precipitation or when evapotranspiration exceeds precipitation. Figure 6.13 is based on monthly water-level observations in a shallow well on eastern Long Island, New York. There is a seasonal fluctuation, with the peak annual water level occurring in late spring. Over the 3-y period of record there is a generally rising water table, indicating a short-term trend toward more precipitation.

If the capillary fringe reaches the land surface, direct evapotranspiration of ground water is possible. Evapotranspiration occurs primarily when the sun is shining; hence, it is at a maximum on warm, sunny days. Figure 6.14 shows the effect of direct evapotranspiration on the shallow water table on the south shore of Long Island, New York. Ground-water levels show diurnal variation in June and July when plants are actively growing and direct evapotranspiration is occurring during the daylight hours. However, in early May, before active plant growth, and in October, after plant growth has stopped, there is little evapotranspiration and no diurnal water-level variation.

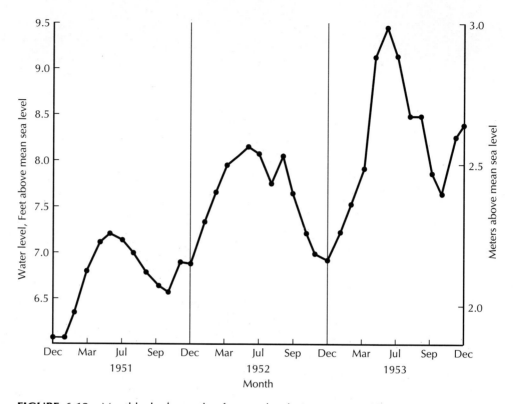

FIGURE 6.13 Monthly hydrograph of water levels in a water-table monitoring well on eastern Long Island, New York.

NOTATION

e	Void ratio	W_s	Mass of solid particles
g	Acceleration of gravity	Z	Elevation potential
h_c	Height of capillary rise	ϕ	Total potential
$K(\theta_v)$	Unsaturated hydraulic conductivity	θ_g	Gravimetric water content
n	Porosity	θ_v	Volumetric water content
R	Radius of a capillary tube	$\psi(\theta_v)$	Moisture potential
R_s	Saturation ratio	σ	Surface tension of a fluid
V	Total volume	λ	Contact angle of a fluid and a solid
V_v	Volume of the voids	ρ_b	Dry bulk density
V_s	Volume of solid particles	ρ_m	Particle density
W_m	Mass of moist soil sample	ρ_w	Fluid density
W_w	Mass of contained water		

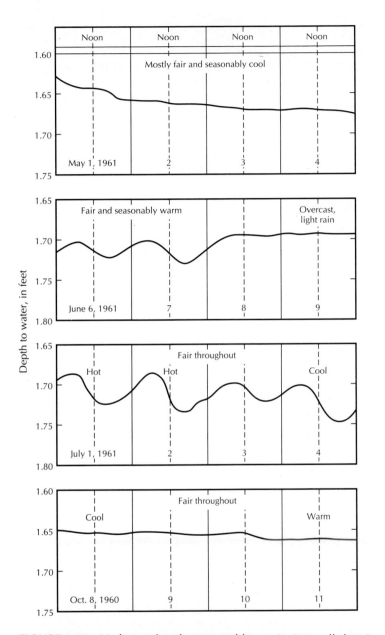

FIGURE 6.14 Hydrographs of a water-table monitoring well showing effect of discharge by evapotranspiration on the water table elevation. Source: E. J. Pluhowski & I. H. Kantrowitz, U.S. Geological Survey Water Supply Paper 1768, 1964.

PROBLEMS

Answers to odd-numbered problems will appear at the end of the book.

1. A soil sample is collected and taken to the lab. The volume of the sample is 75 cm^3. At the natural water content, the sample weighs 155.23 g. It is then saturated with water and reweighed. The saturated weight is 162.95 g. The sample is gravity-drained, and its weight is found to be 146.20 g. Finally, it is dried in an oven; the dry weight is 141.22 g. Assume the density of water is 1.00 g/cm^3. Find each of the following.

    A. Volumetric water content
    B. Gravimetric water content
    C. Saturation ratio
    D. Porosity
    E. Specific yield
    F. Specific retention
    G. Dry-bulk density
    H. Porosity computed from density

2. A soil sample is collected and taken to the lab. The volume of the sample is 75 cm^3. At the natural water content, the sample weighs 165.00 g. It is then saturated with water and reweighed. The saturated weight is 173.85 g. The sample is gravity-drained, and its weight is found to be 162.55 g. Finally, it is drained in an oven; the dry weight is 158.77 g. Assume the density of water is 1.00 g/cm^3. Find each of the following.

    A. Volumetric water content
    B. Gravimetric water content
    C. Saturation ratio
    D. Porosity
    E. Specific yield
    F. Specific retention
    G. Dry-bulk density
    H. Porosity computed from density

7 Ground-Water Flow to Wells

One must, however, note that the flow does not always remain the same. Thus when there are rains the flow is increased, for the water on the hills being in excess is more violently squeezed out. But in times of dryness the flow subsides because no additional supply of water comes into the spring. In the case of the best springs, however, the amount of flow does not contract very much. . . . It is also necessary to find the speed of the flow, for the swifter the flow the more water the spring delivers, and the slower it is, the less.

Dioptra, Hero of Alexandria (ca. A.D. 65)

7.1 INTRODUCTION

Wells are one of the most important aspects of applied hydrogeology. Water wells are used for the extraction of ground water to fill domestic, municipal, industrial, and irrigation needs. Wells have also been used to control salt-water intrusion, remove contaminated water from an aquifer, lower the water table for construction projects, relieve pressures under dams, and drain farmland.

Wells also function to inject fluids into the ground. On Long Island, New York, all ground water pumped for cooling purposes must be returned to the same aquifer by an injection well. Still another function of wells is to dispose of wastewater into isolated aquifers. Finally, as a means of ground-water management, wells are sometimes used to artificially recharge aquifers at rates greater than natural recharge.

The same theoretical considerations apply to wells that extract water and those that inject water. During well pumpage, drawdown of the head in the aquifer around the well occurs; during injection there is an increase in the head in the aquifer. From a mathematical standpoint, injection is handled by using a negative value for the pumping rate.

A **pumping cone,** or cone of depression, will form in the aquifer around a pumping well as the water level declines.

In this chapter we examine the flow to wells in two ways. We first examine how to compute the decline in the water level, or **drawdown,** around a pumping well if we know the transmissivity and storativity of the aquifer. We then see how to determine the transmissivity and storativity of an aquifer by performing an aquifer test in which a well is pumped at a constant rate and the change in drawdown is measured over time. Flow in which the head changes with time is termed **unsteady flow.**

There is a vast literature on the topic of flow to wells. In this chapter we deal only with wells that fully penetrate aquifers that are homogeneous and isotropic. Moreover, we will assume *radial symmetry*; i.e., the values of the aquifer transmissivity and storativity do not depend on the direction of flow in the aquifer. These are the most basic situations. Dawson and Istok (1991) and Kruseman and de Ridder (1991) both present exhaustive treatments of aquifer testing that include the analysis of aquifer test data from more complex situations, such as partially penetrating wells and anisotropic aquifers.

7.2   BASIC ASSUMPTIONS

In this chapter we need to make assumptions about the hydraulic conditions in the aquifer and about the pumping and observation wells. In this section we will list the basic assumptions that apply to all of the situations described in the chapter. Each situation will also have additional assumptions.

1. The aquifer is bounded on the bottom by a confining layer.
2. All geologic formations are horizontal and of infinite horizontal extent.
3. The potentiometric surface of the aquifer is horizontal prior to the start of the pumping.
4. The potentiometric surface of the aquifer is not changing with time prior to the start of the pumping.
5. All changes in the position of the potentiometric surface are due to the effect of the pumping well alone.
6. The aquifer is homogeneous and isotropic.
7. All flow is radial toward the well.
8. Ground-water flow is horizontal.
9. Darcy's law is valid.
10. Ground water has a constant density and viscosity.
11. The pumping well and the observation wells are fully penetrating; i.e., they are screened over the entire thickness of the aquifer.
12. The pumping well has an infinitesimal diameter and is 100% efficient.

7.3   COMPUTING DRAWDOWN CAUSED BY A PUMPING WELL

7.3.1   Unsteady Radial Flow

For purposes of analysis, the aquifers that we will consider will be assumed to be isotropic and homogeneous. It will be shown that solutions can be found for cases

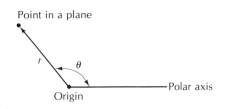

FIGURE 7.1   The use of polar coordinates to describe the position of a point in a plane. It lies a distance *r* from the origin, and the angle between the polar axis and a line connecting the point and the origin is θ.

in which the value of the horizontal conductivity is different from that of the vertical conductivity. However, we will assume that the aquifer has **radial symmetry.**

Flow toward a well has been termed **radial flow.** It moves as if along the spokes of a wagon wheel toward the hub. We can deal with radial flow by use of a coordinate system called **polar coordinates.** The position of a point in a plane is specified according to its distance and direction from a fixed point or pole. The distance is measured directly from the pole to the point in the plane. The direction is determined by the angle between the line from the pole to the point and a fixed reference line—the *polar axis*. The polar axis is usually drawn horizontally and to the right of the pole, and the angle is measured by a counterclockwise rotation from the polar axis (Figure 7.1). Only the value of the angle (θ) and the radial distance (*r*) need be specified. If the aquifer is isotropic in a horizontal plane, then flow will have radial symmetry. Under these conditions, the equation for confined flow becomes (Hantush 1964)

$$\frac{\partial^2 h}{\partial r^2} + \frac{1}{r}\frac{\partial h}{\partial r} = \frac{S}{T}\frac{\partial h}{\partial t} \qquad (7-1)$$

where

h is hydraulic head (L; ft or m)

S is storativity (dimensionless)

T is transmissivity (L^2/T); ft²/day or m²/day)

t is time (T; days)

r is radial distance from the pumping well (L; ft or m)

If there is leakage through a confining layer, or recharge to the aquifer, the flow equation is then expressed as

$$\frac{\partial^2 h}{\partial r^2} + \frac{1}{r}\frac{\partial h}{\partial r} + \frac{e}{T} = \frac{S}{T}\frac{\partial h}{\partial t} \qquad (7-2)$$

where *e* is the rate of vertical leakage (L/T; ft/day or m/day).

Solution of the flow equation for a variety of boundary values for radial flow to wells has yielded a number of extremely useful equations. The mathematics behind the basic solutions involve some rather esoteric mathematical functions. These include Laplace transforms, finite Fourier transforms, Bessel functions, and error functions.* The solutions can be used to determine the drawdown of the potentiometric surface or water table near a pumping well, if the formation characteristics are known. Conversely, if the formation characteristics of an aquifer are unknown, an aquifer test (pumping test) can be made. The well is pumped at a known rate, and the response of the potentiometric surface is measured.

Small-diameter monitoring wells used primarily for sampling ground-water quality or measuring hydraulic head can also be used to determine the hydraulic conductivity of the formation in which the well is screened. A small volume of water is instantaneously added or withdrawn from the well to create an instantaneous head change. The recovery of the water level with time is then recorded. The equations presented in this chapter can then be applied to evaluate the field data and determine the unknown aquifer characteristics.

7.3.2 Flow in a Completely Confined Aquifer

When a well is pumped in a completely confined aquifer, the water is obtained from the elastic or specific storage of the aquifer. You will recall that the elastic storage is water that is released from storage by the expansion of the water as pressure in the aquifer is reduced and by expulsion as the pore space is reduced as the aquifer compacts. The product of the specific storage, S_s, and the aquifer thickness is an aquifer parameter called *storativity*. For a confined aquifer, it is generally small (0.005 or less), and pumpage affects a relatively large area of the aquifer. Further, if there is no recharge, the area of drawdown of the potentiometric surface will expand indefinitely as the pumpage continues.

The first mathematical analysis of transient drawdown effects in a confined aquifer was published by C. V. Theis (1935). Theis made the following assumptions in addition to the basic assumptions of Section 7.2:

1. The aquifer is confined top and bottom.
2. There is no source of recharge to the aquifer.
3. The aquifer is compressible and water is released instantaneously from the aquifer as the head is lowered.
4. The well is pumped at a constant rate.

Figure 7.2 illustrates the aquifer conditions.

---

*For a review of these functions as they apply to well hydraulics, see M. S. Hantush, "Hydraulics of Wells," in *Advances in Hydroscience,* vol. 1, ed. V. T. Chow (New York, Academic Press, 1964).

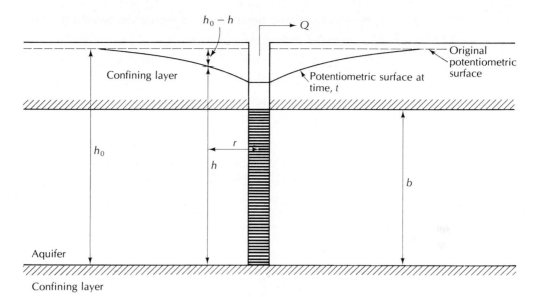

FIGURE 7.2   Fully penetrating well pumping from a confined aquifer.

Based on the preceding assumptions, an equation for the flow of water to a well was derived based on the analogy of flow of heat to a line sink. The result is the **Theis,** or **nonequilibrium, equation:***

$$h_0 - h = \frac{Q}{4\pi T} \int_u^\infty \frac{e^{-u}}{u} \, du \qquad (7\text{–}3)$$

The integral in Equation 7–3 can be replaced with an infinite series so that the Theis equation becomes:

$$h_0 - h = \frac{Q}{4\pi T} \left[-0.5772 - \ln u + u - \frac{u^2}{2 \cdot 2!} + \frac{u^3}{3 \cdot 3!} - \frac{u^4}{4 \cdot 4!} + \cdots \right] \qquad (7\text{–}4)$$

The argument u is given as

$$u = \frac{r^2 S}{4Tt} \qquad (7\text{–}5)$$

where

Q is the constant pumping rate (L^3/T; ft³/day or m³/day)

h is hydraulic head (L; ft or m)

---

*The Theis equation was actually derived by a mathematician friend of Theis, C. I. Lubin. Reportedly, Lubin declined coauthorship of the paper because he regarded his contribution as mathematically trivial.

h_0 is hydraulic head before pumping started (L; ft or m)

$h_0 - h$ is the drawdown (L; ft or m)

T is aquifer transmissivity (L^2/T; ft²/day or m²/day)

t is time since pumping began (T; days)

r is radial distance from the pumping well (L; ft or m)

S is aquifer storativity (dimensionless)

It should be noted that Q is a pumping rate and is in units of cubic feet per day or cubic meters per day. Even if the well is pumped for less than a day, the units of Q must still reflect the volume that would be pumped in a day at the rate for a partial day.

EXAMPLE PROBLEM   If a well is pumped at a constant rate for 750 min and a total of 13,500 ft³ are removed, what is the pumping rate in cubic feet per day?

$$Q = \frac{13{,}500 \text{ ft}^3}{750 \text{ min}} \times \frac{1440 \text{ min}}{\text{day}} = 25{,}900 \text{ ft}^3/\text{day}$$

EXAMPLE PROBLEM   A well is pumped at a rate of 125 gal/min for 1845 min. What is the pumping rate in cubic feet per day?

$$Q = \frac{125 \text{ gal}}{\text{min}} \times \frac{1 \text{ ft}^3}{7.48 \text{ gal}} \times \frac{1440 \text{ min}}{\text{day}} = 24{,}100 \text{ ft}^3/\text{day}$$

The infinite series term of Equation 7–4 has been called the well function and is generally designated as $W(u)$. Although the series is easy to evaluate mathematically, it has been conveniently tabulated (Wenzel 1942), and Appendix 1 contains a table of $W(u)$ for various values of u.

Using well function notation, the Theis equation is also expressed as

$$h_0 - h = \frac{Q}{4\pi T} W(u) \qquad\qquad \textbf{(7–6)}$$

EXAMPLE PROBLEM   A well is located in an aquifer with a conductivity of 14.9 m/day and a storativity of 0.0051. The aquifer is 20.1 m thick and is pumped at a rate of 2725 m³/day. What is the drawdown at a distance of 7.0 m from the well after 1 day of pumping?

$$T = Kb = 14.9 \text{ m/day} \times 20.1 \text{ m} = 299 \text{ m}^2/\text{day}$$

$$u = \frac{r^2 S}{4Tt} = \frac{(7.0 \text{ m})^2 \times 0.0051}{4 \times 300 \text{ m}^2\text{day} \times 1 \text{ day}} = 0.00021$$

From the table of $W(u)$ and u, if $u = 2 \times 10^{-4}$, $W(u) = 7.94$:

$$h_0 - h = \frac{Q}{4\pi T} W(u) = \frac{2725 \text{ m}^3/\text{day} \times 7.94}{4 \times \pi \times 299 \text{ m}^2/\text{day}} = 5.7 \text{ m}$$

7.3.3 Flow in a Leaky, Confined Aquifer

7.3.3.1 Flow Equation

Most confined aquifers are not totally isolated from sources of vertical recharge. Aquitards, either above or below the aquifer, can leak water into the aquifer if the direction of the hydraulic gradient is favorable (Figure 7.3).

The flow equation for a confined aquifer with recharge is

$$\frac{\partial^2 h}{\partial r^2} + \frac{1}{r}\frac{\partial h}{\partial r} - \frac{(h_0 - h)K'}{Tb'} = \frac{S}{T}\frac{\partial h}{\partial t} \qquad (7\text{--}7)$$

where

K' is the vertical hydraulic conductivity of the leaky layer (L/T)

b' is the thickness of the leaky layer (L)

h is the head (L)

r is the radial distance from the pumping well (L)

t is the time (T)

S is the storativity (dimensionless)

T is the transmissivity (L^2/T)

$h_0 - h$ is the drawdown (L)

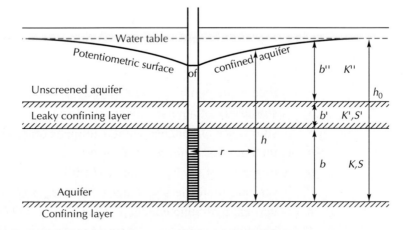

FIGURE 7.3 Fully penetrating well in an aquifer overlain by a semipermeable confining layer.

7.3.3.2   Solution for the Case When No Water Drains from the Confining Layer

The first solution to this problem was published by M. S. Hantush (1956). In this case he assumed that all the water flowing to the pumping well came from either elastic storage in the confined aquifer or leakage across the confining layer. No water is derived from elastic storage in the confining layer. To solve Equation 7–7 Hantush made the following assumptions in addition to the basic assumptions of Section 7.2.

1. The aquifer is bounded on the top by an aquitard.
2. The aquitard is overlain by an unconfined aquifer, known as the *source bed*.
3. The water table in the source bed is initially horizontal.
4. The water table in the source bed does not fall during pumping of the aquifer.
5. Ground-water flow in the aquitard is vertical.
6. The aquifer is compressible, and water drains instantaneously with a decline in head.
7. The aquitard is incompressible, so that no water is released from storage in the aquitard when the aquifer is pumped.

Assumption 4, that the water table does not decline during pumping, is difficult to attain unless there is continuous recharge to the water table aquifer. However, it can be considered to be valid when either of the following conditions is true (Neuman & Witherspoon 1969):

$$t < \frac{S'(b')^2}{10bK'} \qquad \textbf{(7–8a)}$$

or

$$b''K'' > 100bK \qquad \textbf{(7–8b)}$$

where

t is time since pumping began (T; days)

S' is storativity of aquitard (dimensionless)

S is storativity of confined aquifer (dimensionless)

b' is thickness of aquitard (L; ft or m)

b is thickness of confined aquifer (L; ft or m)

b'' is saturated thickness of water table aquifer (L; ft or m)

K' is vertical hydraulic conductivity of aquitard (L/T; ft/day or m/day)

K'' is hydraulic conductivity of water table aquifer (L/T; ft/day or m/day)

K is hydraulic conductivity of confined aquifer (L/T; ft/day or m/day)

Assumption 7, that no water is released from the aquitard, is valid under either of two conditions. Hantush (1960b) showed that the effects of water released from the aquitard are negligible if

$$t > 0.036b'S'/K' \tag{7-9}$$

Neuman and Witherspoon (1969) also showed that the assumption is valid when

$$r < 0.04b[(KS_s/K'S'_s)]^{1/2} \tag{7-10}$$

where

S_s is the specific storage of the confined aquifer ($1/L$; 1/ft or 1/m)

S'_s is the specific storage of the aquitard ($1/L$; 1/ft or 1/m)

Although the basic assumptions of Section 7.2 include an infinitesimal well diameter, the following solution is valid for any well diameter, provided that

$$t > (30r_w^2 S/T) \left[1 - (10r_w/b)^2\right] \tag{7-11}$$

and

$$r_w/(Tb'/K')^{1/2} < 0.1 \tag{7-12}$$

where

r_w is the radius of the pumping well (L; ft or m)

S is storativity of the confined aquifer (dimensionless)

T is transmissivity of the confined aquifer (L^2/T; ft²/day or m²/day)

The solution to Equation 7–7 as given by Hantush (1956, 1960b) and Hantush and Jacob (1954) is known as the **Hantush-Jacob formula** and is

$$h_0 - h = \frac{Q}{4\pi T} W(u, r/B) \tag{7-13}$$

$$u = \frac{r^2 S}{4Tt} \tag{7-14}$$

$$B = (Tb'/K')^{1/2} \tag{7-15}$$

where

Q is the pumping rate (L^3/T; ft³/day or m³/day)

$h_0 - h$ is the drawdown in the confined aquifer (L; ft or m)

T is the transmissivity of the confined aquifer (L^2/T; ft²/day or m²/day)

$W(u, r/B)$ is the leaky artesian well function (Values of this function are tabulated in Appendix 3.)

r is the distance from the pumping well to the observation well (L; ft or m)

S is the storativity of the confined aquifer (dimensionless)

t is time since pumping began (T; days)

B is leakage factor (L; ft or m)

b' is thickness of the aquitard (L; ft or m)

K' is hydraulic conductivity of the aquitard (L/T; ft/day or m/day)

The rate that water is being drawn from elastic storage in the confined aquifer, q_S (L^3/T; ft³/day or m³/day), at the specific time, t (T; days), since pumping began may be determined from

$$q_S = Q \exp(-Tt/SB^2) \qquad \textbf{(7–16)}$$

If the total discharge at time t is Q and the water drawn from storage at that time is q_S, then the rate that water is coming from leakage across the aquitard at that time, q_L, is found from

$$q_L = Q - q_S \qquad \textbf{(7–17)}$$

EXAMPLE PROBLEM   A confined aquifer is underlain by an aquiclude and overlain by an aquitard and a water-table aquifer. The following aquifer characteristics are given:

Confined aquifer: $b = 5.2$ m, $K = 0.73$ m/day, $S = 0.0035$, $T = 3.8$ m²/day
Aquitard: $b' = 1.1$ m, $K' = 5.5 \times 10^{-5}$ m/day, $S' = 0.00061$
Water table aquifer: $b'' = 25$ m, $K'' = 35$ m/day

A well that fully penetrates the aquifer has a radius of 0.15 m. If it is pumped at a rate of 28 m³/day, what is the drawdown after 1 day of pumping at the following distance from the well: 1.5 m, 5.5 m, 10 m, 25 m, 75 m, 150 m?

First, test to see if the assumptions for a leaky, confined aquifer with negligible storage in the aquitard are valid. To test to see if there will be negligible decline in the water level in the water-table aquifer, use Equation 7–8.

$$b''K'' > 100bK$$
$$25 \text{ m/day} \times 35 \text{ m} > 100 \times 5.2 \text{ m} \times 0.73 \text{ m/day}$$
$$875 \text{ m}^2/\text{day} > 380 \text{ m}^2/\text{day}$$

Therefore, the assumption is valid.

To test to see if the assumption that the contribution from storage in the aquitard is negligible, use Equation 7–9.

$$t > 0.036b'S'/K'$$

$$1 \text{ day} > (0.036 \times 1.1 \text{ m} \times 0.00081)/(5.5 \times 10^{-5} \text{ m/day})$$

$$1 \text{ day} > 0.44 \text{ day}$$

Therefore, the assumption is valid.

To test to see if the radius of the well can be considered negligible, use Equations 7–11 and 7–12.

$$t > (30r_w^2 S/T)[1 - (10r_w/b)^2]$$

$$1 \text{ day} > [30 \times (0.15 \text{ m})^2 \times (0.0035/3.8 \text{ m}^2/\text{day})][1 - (10 \times 0.15 \text{ m}/5.2 \text{ m})^2]$$

$$1 \text{ day} > (6.2 \times 10^{-4} \text{ day})(1 - 0.08)$$

$$1 \text{ day} > 5.7 \times 10^{-4} \text{ day}$$

and

$$r_w/(Tb'/K')^{1/2} < 0.1$$

$$0.15\text{m}/[(3.8 \text{ m}^2/\text{day} \times 1.1 \text{ m})/(5.5 \times 10^{-5} \text{ m/day})]^{1/2} < 0.1$$

$$(0.15 \text{ m})/(275 \text{ m}) < 0.1$$

$$5.5 \times 10^{-4} < 0.1$$

Both the preceding conditions are met, so this assumption is also true and the Hantush-Jacob equations can be used.

In order to find $W(u, r/B)$, we must first find u and r/B. As we wish to know these parameters for several values of r, we will first find them with respect to r. From Equation 7–14,

$$
\begin{aligned}
u &= (r^2 S)/(4Tt) \\
&= [(r^2 \text{ m}^2) \times 0.0035]/(4 \times 3.8 \text{ m}^2/\text{day} \times 1 \text{ day}) \\
&= (2.3 \times 10^{-4})r^2 \\
r/B &= r/(Tb'/K')^{1/2} \\
&= r\text{m}/[(3.8 \text{ m}^2/\text{day} \times 1.1 \text{ m})/(5.5 \times 10^{-5} \text{ m/day})]^{1/2} \\
&= r/275
\end{aligned}
$$

Once values of u and r/B are determined for each value of r, then the value of $W(u, r/B)$ is found in Appendix 3. The drawdown is then found from Equation 7–13.

$$
\begin{aligned}
h_0 - h &= \frac{Q}{4\pi T} W(u, r/B) \\
&= 28 \text{ m}^3/\text{day}/(4\pi \times 3.8 \text{ m}^2\text{day})W(u, r/B) \\
&= 0.59W(u, r/B) \text{ m}
\end{aligned}
$$

r	u	r/B	$W(u, r/B)$	$h_0 - h$
1.5 m	5.17×10^{-4}	5.45×10^{-3}	7.2	4.2 m
5.5 m	6.96×10^{-3}	2.00×10^{-2}	4.4	2.6 m
10 m	2.30×10^{-2}	3.64×10^{-2}	3.2	1.9 m
25 m	1.44×10^{-1}	9.09×10^{-2}	1.5	0.89 m
75 m	1.29	2.73×10^{-1}	0.17	0.10 m
150 m	5.18	5.45×10^{-1}	0.0017	0.0010 m

If the well is pumped long enough, all the water will be coming from leakage across the confining layer and none from elastic storage in the confined aquifer (Hantush & Jacob 1954). This occurs when

$$t > \frac{8b'S}{K'} \tag{7-18}$$

In this case the drawdown can be found from (Hantush & Jacob 1955):

$$h_0 - h = \frac{Q}{2\pi T} K_0(r/B) \tag{7-19}$$

where K_0 is the zero-order modified Bessel function of the second kind. A partial listing of Bessel functions is found in Appendix 5.

7.3.3.3 Solution for the Case Where Some Water Comes from Elastic Storage in the Aquitard

Examination of Equation 7–9 suggests that the contribution of water from elastic storage in the aquitard is directly proportional to the aquitard thickness and storativity and inversely proportional to the aquitard hydraulic conductivity. If the conditions of Equation 7–9 are not met, then the solution for a confined aquifer with a contribution of water from elastic storage in the aquitard must be used (Hantush 1960b; Neuman & Witherspoon 1969).

This solution is based on the assumptions of Section 7.2 as well as all of the assumptions used in Section 7.3.3.2, except the assumption that the flow contribution from the aquitard can be neglected.

This case has two solutions. During the early part of pumping, when the following condition is met, all the water will come from elastic storage in the aquifer and the aquitard. The early-time condition is

$$t > b'S'/10K' \tag{7-20}$$

The solution to Equation 7–2 is

$$h_0 - h = \frac{Q}{4\pi T} H(u, \beta) \tag{7-21}$$

where $H(u, \beta)$ is a function with values tabulated in Appendix 4 and

$$\beta = \frac{r}{4B} (S'/S)^{1/2} \tag{7-22}$$

$$B = \left(\frac{Tb'}{K'}\right)^{1/2} \tag{7-23}$$

$$u = \frac{r^2 S}{4Tt} \tag{7-24}$$

The rate of flow from storage in the main aquifer, q_s, is given by

$$q_s = .Q \exp(vt)\text{erfc}(\sqrt{vt}) \tag{7-25}$$

where

$$v = (K'/b')(S'/S^2) \ (1/T;\ 1/\text{day}) \tag{7-26}$$

and erfc is the complementary error function. The error function erf(x) is a function with tabulated values. The complementary error function erfc(x) is defined as $1 - \text{erf}(x)$. A table of the error function is given in Appendix 13.

If sufficient time elapses, the aquifer will reach equilibrium and all of the water will be coming from drainage from the overlying source bed. The time to reach this equilibrium is

$$t > \frac{8[S + (S'/3) + S'']}{[(K'/b'T) + (K''/b''T)]^{1/2}} \tag{7-27}$$

If the value of r_w/B is less than 0.01, then the solution is

$$h_0 - h = \frac{Q}{2\pi T} K_0(r/B) \tag{7-28}$$

This is the same solution as that for equilibrium conditions in the case where no water comes from elastic storage in the aquitard, since all the water comes from the source bed.

EXAMPLE PROBLEM   A confined aquifer is overlain by an aquitard and a water table aquifer. The layers have the following characteristics.

Confined aquifer: $b = 4.3$ m, $K = 1.1$ m/day, $S = 0.00053$, $T = 4.7$ m²/day
Aquitard: $b' = 7.2$ m, $K' = 5.5 \times 10^{-6}$ m/day, $S' = 0.00012$
Source bed: $b'' = 17$ m, $K'' = 87$ m/day, $S'' = 0.055$

If a well is pumped at a rate of 15 m³/day for 1.76 days, what would the drawdown be at a distance of 22 m?

First we need to test to see if the assumption that the head in the source bed will remain constant is valid. This is done with Equation 7–8:

$$b''K'' > 100bK$$
$$17 \text{ m} \times 87 \text{ m/day} > 100 \times 4.3 \text{ m} \times 1.1 \text{ m/day}$$
$$1497 \text{ m}^2/\text{day} > 473 \text{ m}^2/\text{day}$$

Therefore, this assumption is true.

We then use Equation 7–9 to see if the contribution of water from elastic storage in the aquifer needs to be considered.

$$t > 0.036 \, b'S'/K'$$
$$1.76 \text{ days} > (0.036 \times 7.2 \text{ m} \times 0.00012)/(5.5 \times 10^{-6} \text{ m/day})$$
$$1.76 \text{ days} \not> 5.66 \text{ days}$$

The conditional statement is not true; therefore, we must consider the effects of storage. We now need to see if we can use either the early time or the equilibrium equation for confined aquifers with storage in the aquitard. We will start with the late-time equation, since 1.76 days may have been enough time for equilibrium to be established. The conditional statement for equilibrium is Equation 7–27.

$$t > \frac{8[S + (S'/3) + S'']}{[(K'/b'T) + (K''/b''T)]^{1/2}}$$
$$1.76 \text{ days} > \frac{8[0.00053 + (0.00012/3) + 0.055]}{[(5.5 \times 10^{-6} \text{ m/day}/(7.2 \text{ m} \times 4.7 \text{ m}^2/\text{day})) + (87 \text{ m/day}/(17 \text{ m} \times 4.7 \text{ m}^2/\text{day}))]^{1/2}}$$
$$1.76 \text{ days} > \frac{8 \times 0.055}{[1.86 \times 10^{-4} \text{ day}^{-2} + 1.09 \text{ day}^{-2}]^{1/2}}$$
$$1.76 \text{ days} > \frac{0.44}{1.044 \text{ day}^{-1}}$$
$$1.76 \text{ days} > 0.42 \text{ days}$$

The conditional statement is true; therefore, we must use the equilibrium equation, Equation (7–28).

$$h_0 - h = \frac{Q}{4\pi T} K_0(r/B)$$
$$B = \left(\frac{Tb'}{K'}\right)^{1/2}$$
$$B = \left(\frac{4.7 \text{ m}^2/\text{day} \times 7.2 \text{ m}}{5.5 \times 10^{-6} \text{ m/day}}\right)^{1/2}$$

$$B = 2480 \text{ m}$$

$$\frac{r}{B} = \frac{22 \text{ m}}{2480 \text{ m}} = 0.009$$

From Appendix 5, $K_0(0.009) = 4.8$; thus

$$h_0 - h = \frac{15 \text{ m}^3/\text{day} \times 4.8}{4 \times \pi \times 4.7 \text{ m}^2/\text{day}}$$

$$= 1.2 \text{ m}$$

7.3.3.4 Comparison of Drawdown in Different Confined Aquifers

A plot of drawdown as a function of time for several types of confined aquifers is given in Figure 7.4. Other than at very early times, the nonleaky aquifer will decline on a straight line. As there is no recharge, drawdown per log cycle of time will remain constant.

In a leaky aquifer, the drawdown curve will initially follow the nonleaky curve. However, after a finite time interval, the lowered hydraulic head in the aquifer will induce leakage from the confining layer. As part of the well discharge is now coming through the impervious layer, the rate of decline of head will decrease. If there is storage in the confining layer, the rate of drawdown will be lower than if there were no storage. Eventually, the drawdown cone will be large

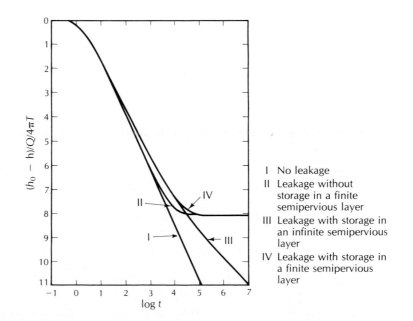

FIGURE 7.4 Plots of log of dimensionless drawdown as a function of time for an aquifer with various types of overlying confining layer. Source: M. S. Hantush, *Journal of Geophysical Research* 65 (1960): 3713–25.

enough so that the pumped water will be coming entirely from leakage through the confining layer and there will be no further drawdown with time. As water is no longer coming from storage in the leaky layer, the two curves coincide.

7.3.4   Flow in an Unconfined Aquifer

7.3.4.1   Flow Equation

The flow of water in an unconfined aquifer toward a pumping well is described by the following equation (Neuman & Witherspoon 1969):

$$K_r \frac{\partial^2 h}{\partial r^2} + \frac{K_r}{r} \frac{\partial h}{\partial r} + K_v \frac{\partial^2 h}{\partial z^2} = S_s \frac{\partial h}{\partial t} \qquad (7-29)$$

where

h is the saturated thickness of the aquifer (L; ft or m)

r is radial distance from the pumping well (L; ft or m)

z is elevation above the base of the aquifer (L; ft or m)

S_s is specific storage ($1/L$; 1/ft or 1/m)

K_r is radial hydraulic conductivity (L/T; ft/day or m/day)

K_v is vertical hydraulic conductivity (L/T; ft/day or m/day)

t is time (T; days)

A well pumping from a water-table aquifer extracts water by two mechanisms. As with confined aquifers, the decline in pressure in the aquifer yields water because of the elastic storage of the aquifer storativity (S_s). The declining water table also yields water as it drains under gravity from the sediments. This is termed specific yield (S_y). The flow equation has been solved for radial flow in compressible, unconfined aquifers under a number of different conditions and by use of a variety of mathematical gambits (Boulton & Streltsova 1975; Boulton 1954, 1955, 1963, 1973; Boulton & Pontin 1971; Streltsova 1972, 1973; Dagan 1967; Neuman 1972, 1974, 1975; Gambolati 1976). The many and various equations in these solutions can lead to confusion; however, a qualitative description of the response of a water-table well to pumping may be helpful.

There are three distinct phases of time-drawdown relations in water-table wells. We will examine the response of any typical annular region of the aquifer located a constant distance from the pumping well. Some time after pumping has begun, the pressure in the annular region will drop. As the pressure first drops, the aquifer responds by contributing a small volume of water as a result of the expansion of water and compression of the aquifer. During this time, the aquifer behaves as an artesian aquifer, and the time-drawdown data follow the Theis nonequilibrium curve for S equal to the elastic storativity of the aquifer. Flow is horizontal during this period, as the water is being derived from the entire aquifer thickness.

Following this initial phase, the water table begins to decline. Water is now being derived primarily from the gravity drainage of the aquifer, and there are both horizontal and vertical flow components. The drawdown-time relationship is a function of the ratio of horizontal-to-vertical conductivities of the aquifer, the distance to the pumping well, and the thickness of the aquifer.

As time progresses, the rate of drawdown decreases and the contribution of the particular annular region to the overall well discharge diminishes. Flow is again essentially horizontal, and the time-discharge data again follow a Theis type curve. The Theis curve now corresponds to one with a storativity equal to the specific yield of the aquifer. The importance of the vertical flow component as it affects the average drawdown is directly related to the magnitude of the ratio of the specific yield to the elastic storage coefficient (S_y/S_s). As the value of S_s approaches zero, the time duration of the first stage of drawdown also approaches zero. As S_y approaches zero, the length of the first stage increases, so that if $S_y = 0$, the aquifer behaves as an artesian aquifer of storativity S_s (Gambolati 1976).

Neuman (1972, 1973, 1974, 1987) has published a solution to Equation 7–29. There are two parts to the solution, one for the time just after pumping has begun and the water is coming from specific storage and one for much later, when the water is coming from gravity drainage with the storativity equal to the specific yield.

Neuman's solution assumes the following, in addition to the basic assumptions of Section 7.2.

1. The aquifer is unconfined.

2. The vadose zone has no influence on the drawdown.

3. Water initially pumped comes from the instantaneous release of water from elastic storage.

4. Eventually water comes from storage due to gravity drainage of interconnected pores.

5. The drawdown is negligible compared with the saturated aquifer thickness.

6. The specific yield is at least 10 times the elastic storativity.

7. The aquifer may be—but does not have to be—anisotropic with the radial hydraulic conductivity different than the vertical hydraulic conductivity.

With these assumptions Neuman's solution is

$$h_0 - h = \frac{Q}{4\pi T} W(u_A, u_B, \Gamma) \tag{7–30}$$

where $W(u_A, u_B, \Gamma)$ is the well function for the water-table aquifer, as tabulated in Appendix 6.

$$u_A = \frac{r^2 S}{4Tt} \quad \text{(for early drawdown data)} \tag{7–31}$$

$$u_B = \frac{r^2 S_y}{4Tt} \qquad \text{(for later drawdown data)} \tag{7-32}$$

$$\Gamma = \frac{r^2 K_v}{b^2 K_h} \tag{7-33}$$

where

$h_0 - h$ is the drawdown (L; ft or m)

Q is the pumping rate (L^3/T; ft^3/day or m^3/day)

T is the transmissivity (L^2/T; ft^2/day or m^2/day)

r is the radial distance from the pumping well (L; ft or m)

S is the storativity (dimensionless)

S_y is the specific yield (dimensionless)

t is the time (T; days)

K_h is the horizontal hydraulic conductivity (L/T; ft/day or m/day)

K_v is the vertical hydraulic conductivity (L/T; ft/day or m/day)

b is the initial saturated thickness of the aquifer (L; ft or m)

7.4   DETERMINING AQUIFER PARAMETERS FROM TIME-DRAWDOWN DATA

7.4.1 Introduction

In Section 7.3 we saw how to calculate the drawdown in an aquifer if we know the hydraulic parameters of the aquifer. These hydraulic parameters are usually determined by means of an **aquifer test.** In an aquifer test a well is pumped and the rate of decline of the water level in nearby observation wells is noted. The time-drawdown data are then interpreted to yield the hydraulic parameters of the aquifer.

In conducting an aquifer test, there must obviously be a pumping well. There are usually one or more observation wells. In the discussion of aquifer tests presented here, the following conditions are assumed:

1. The pumping well is screened only in the aquifer being tested.
2. All observation wells are screened only in the aquifer being tested.
3. The pumping well and the observation wells are screened throughout the entire thickness of the aquifer.

The general assumptions of Section 7.2 and the specific assumptions for each type of aquifer in Section 7.3 are also necessary.

7.4.2 Steady-State Conditions

If a well pumps long enough, the water level may reach a state of equilibrium; that is, there is no further drawdown with time. The region around the pumping well where the head has been lowered is known as the *cone of depression*. When equilibrium has been achieved, the cone of depression stops growing because it has reached a recharge boundary. These are also known as *steady-state conditions*.

The hydraulic gradient of the cone of depression causes water to flow at a constant rate from the recharge boundary to the well. The assumption of radial symmetry means that the recharge boundary has an unlikely circular geometry entered about the pumping well.

7.4.2.1 Steady Radial Flow in a Confined Aquifer

In the case of steady radial flow in a confined aquifer, the following additional assumptions are necessary:

1. The aquifer is confined top and bottom.
2. The well is pumped at a constant rate.
3. Equilibrium has been reached; i.e., there is no further change in drawdown with time.

Figure 7.5 shows a well penetrating a confined aquifer. Under steady-state conditions the rate that water is pumped from the well is equal to the rate that the aquifer transmits water to the well. This problem was first solved by G. Thiem (Thiem 1906).

From Darcy's law the flow of water through a circular section of the aquifer toward the well is the area of the circular section, $2\pi rb$, times the hydraulic conductivity times the hydraulic gradient. This can be expressed as:

$$Q = (2\pi rb)K\left(\frac{dh}{dr}\right) \qquad (7-34)$$

where

Q is the pumping rate (L^3/T)

r is the radial distance from the circular section to the well (L)

b is the aquifer thickness (L)

K is the hydraulic conductivity (L/T)

dh/dr is the hydraulic gradient (dimensionless)

Since transmissivity is the product of the aquifer thickness and the hydraulic conductivity, Equation 7–34 can also be expressed as:

$$Q = 2\pi rT\left(\frac{dh}{dr}\right) \qquad (7-35)$$

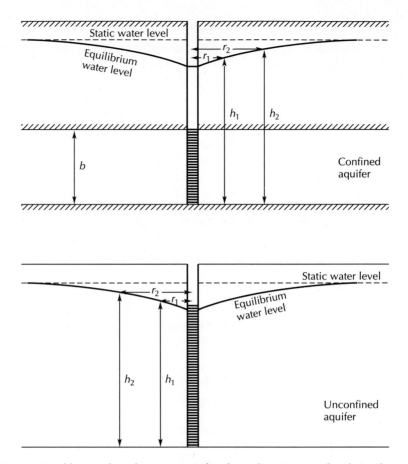

FIGURE 7.5 Equilibrium drawdown: A. confined aquifer; B. unconfined aquifer.

Equation 7–35 can be rearranged as follows:

$$dh = \frac{Q}{2\pi T}\frac{dr}{r}$$ (7–36)

If there are two observation wells, the head is h_1 at a distance r_1 from the pumping well; it is h_2 at a distance r_2. We can integrate both sides of Equation 7–36 with these boundary conditions:

$$\int_{h_1}^{h_2} dh = \frac{Q}{2\pi T}\int_{r_1}^{r_2}\frac{dr}{r}$$ (7–37)

This can be solved to yield

$$h_2 - h_1 = \frac{Q}{2\pi T} \ln\left(\frac{r_2}{r_1}\right) \qquad \textbf{(7-38)}$$

Equation 7–38 can be rearranged to yield the Thiem equation for confined aquifers:

$$\boxed{T = \frac{Q}{2\pi(h_2 - h_1)} \ln\left(\frac{r_2}{r_1}\right)} \qquad \textbf{(7-39)}$$

where

T is aquifer transmissivity (L^2/T; ft^2/day or m^2/day)

Q is pumping rate (L^3/T; ft^3/day or m^3/day)

h_1 is head at distance r_1 from the pumping well (L; ft or m)

h_2 is head at distance r_2 from the pumping well (L; ft or m)

If pumping has progressed to a point where the drawdown has stabilized, the head is determined in two observation wells, located different distances from the pumping well. The pumping rate, the heads, and the radial distances are entered into Equation 7–39, and the aquifer transmissivity is determined. Notice that there is no provision for the aquifer storativity in Equation 7–39. Under steady-state conditions there is no change in head with time and water is not coming from storage. Hence, the Thiem equation cannot be used to determine aquifer storativity.

EXAMPLE PROBLEM   A well in a confined aquifer is pumped at a rate of 220 gal/min. Measurement of drawdown in two observation wells shows that after 1270 min of pumping, no further drawdown is occurring. Well 1 is 26 ft from the pumping well and has a head of 29.34 ft above the top of the aquifer. Well 2 is 73 ft from the pumping well and has a head of 32.56 ft above the top of the aquifer. Use the Thiem equation to find the aquifer transmissivity.

We must first convert the pumping rate of 220 gal/min to an equivalent rate in cubic feet per day. We make this conversion, even though steady-state conditions were reached before a full day (1440 min) of pumping occurred.

$$220 \frac{\text{gal}}{\text{min}} \times \frac{1 \text{ ft}^3}{7.48 \text{ gal}} \times 1440 \frac{\text{min}}{\text{day}} = 42,400 \text{ ft}^3/\text{day}$$

Now we substitute the given values into Equation 7–39.

$$T = \frac{Q}{2\pi(h_2 - h_1)} \ln\left(\frac{r_2}{r_1}\right)$$

$$T = \frac{42,400 \text{ ft}^3/\text{day}}{2\pi(32.56 \text{ ft} - 29.34 \text{ ft})} \ln\left(\frac{32.56 \text{ ft}}{29.34 \text{ ft}}\right)$$

$$= \frac{42,400}{20.2} \ln(1.11) \text{ ft}^2/\text{day}$$

$$= 219 \text{ ft}^2/\text{day}$$

7.4.2.2 Steady Radial Flow in an Unconfined Aquifer

Thiem also derived an equation for steady radial flow in an unconfined aquifer. In this instance the following assumptions are needed in addition to the general assumptions of Section 7.2.

1. The aquifer is unconfined and underlain by a horizontal aquiclude.
2. The well is pumped at a constant rate.
3. Equilibrium has been reached; i.e., there is no further change in drawdown with time.

Radial flow in the unconfined aquifer is described by:

$$Q = (2\pi rh)K\left(\frac{dh}{dr}\right) \tag{7-40}$$

where

Q is the pumping rate (L^3/T)
r is the radial distance from the circular section to the well (L)
h is the saturated thickness of the aquifer (L)
K is the hydraulic conductivity (L/T)
dh/dr is the hydraulic gradient (dimensionless)

Equation 7-40 can be rearranged as follows:

$$h \, dh = \frac{Q}{2\pi K} \frac{dr}{r} \tag{7-41}$$

If there are two observation wells, the head is h_1 at a distance r_1 from the pumping well and it is h_2 at a distance r_2. We can integrate both sides of Equation 7-41 with these boundary conditions:

$$\int_{h_1}^{h_2} h \, dh = \frac{Q}{2\pi K} \int_{r_1}^{r_2} \frac{dr}{r} \tag{7-42}$$

This can be solved to yield

$$h_2^2 - h_1^2 = \frac{Q}{2\pi K} \ln\left(\frac{r_2}{r_1}\right) \qquad \text{(7-43)}$$

Equation 7–43 can be rearranged to yield the Thiem equation for an unconfined aquifer:

$$\boxed{K = \frac{Q}{2\pi(h_2^2 - h_1^2)} \ln\left(\frac{r_2}{r_1}\right)} \qquad \text{(7-44)}$$

where

K is hydraulic conductivity (L/T; ft/day or m/day)

Q is pumping rate (L^3/T; ft^3/day or m^3/day)

h_1 is head at distance r_1 from the pumping well (L; ft or m)

h_2 is head at distance r_2 from the pumping well (L; ft or m)

7.4.3 Nonequilibrium Flow Conditions

Many aquifer tests will never reach equilibrium; that is, the cone of depression will continue to grow with time. These are known as *nonequilibrium*, or *transient*, flow conditions. Analysis of the transient time-drawdown data from an observation well can be used to determine both the transmissivity and the storativity of an aquifer. If there is no observation well, the time-drawdown data from the pumping well can be used to determine aquifer transmissivity but not storativity.

7.4.3.1 Nonequilibrium Radial Flow in a Confined Aquifer—Theis Method

All assumptions are as before for radial flow in a confined aquifer. First we will recast the Theis equations to a different form.

Equation 7–6 can be rearranged as follows:

$$T = \frac{Q}{4\pi(h_0 - h)} W(u) \qquad \text{(7-45)}$$

where

T is the aquifer transmissivity (L^2/T; ft^2/day or m^2/day)

Q is the steady pumping rate (L^3/T; ft^3/day or m^3/day)

$h_0 - h$ is the drawdown (L; ft or m)

$W(u)$ is the well function of u (dimensionless)

Also, Equation 7–5 can be rearranged as:

$$S = \frac{4Tut}{r^2}$$

(7–46)

where

S is aquifer storativity (dimensionless)

t is time since pumping began (T; days)

r is radial distance from the pumping well (L; ft or m)

u is a dimensionless constant

An aquifer test consists of pumping a well at a constant rate for a period of time. The drawdown is measured as a function of time in one or more observation wells and, perhaps, the pumping well. The procedures for conducting aquifer tests are discussed in Section 7.5. The result of an aquifer test is a set of data with the drawdown given at various times after the start of pumping.

Theis developed a graphical means of solution to the Theis equations. The first step is to make a plot of $W(u)$ as a function of $1/u$ on full logarithmic paper. The type curve can be plotted by hand using the data in Appendix 1. If one has the proper graphing software, these data can also be plotted using a computer program. This graph has the shape of the cone of depression near the pumping well and is known as a **Theis type curve** or a **reverse type curve** or a nonequilibrium type curve (Figure 7.6).

Field data for drawdown, $(h_0 - h)$, as a function of time, t, at an observation well are then plotted on logarithmic paper of the same scale as the

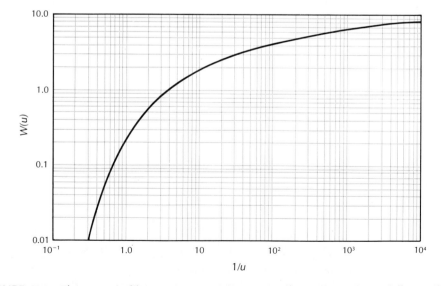

FIGURE 7.6 The nonequilibrium reverse type curve (Theis curve) for a fully confined aquifer.

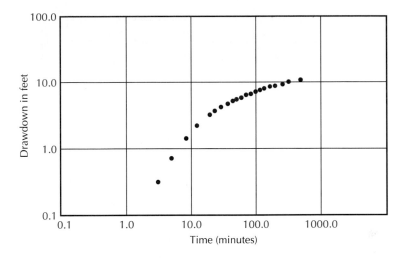

FIGURE 7.7   Field-data plot on logarithmic paper for Theis curve-matching technique.

type curve (Figure 7.7). The data points are not connected. Although the Theis equations call for time in days, it is usually more convenient to plot time in minutes, as that is the way that it is often recorded in the field. A conversion to days can be made later.

The graph paper with the type curve is taped to a light table. The graph paper with the field data is laid over the type curve, keeping the sets of axes parallel. The position of the field-data graph is adjusted until the data points overlie the type curve, with the axes of both graph sheets parallel (Figure 7.8).

A *match point* must then be selected. Any arbitrary point may be used; it does not have to be on the type curve. In fact, the intersection of the line $W(u) = 1$ and the line $1/u = 1$ is a particularly convenient match point. We need to find the value of $(h_0 - h)$ and the value of t that is on the graph paper immediately above the match point. These values do not need to correspond to one of the field data points. A pin may be pushed through the two pieces of graph paper to find the exact match point.

From the match point we obtain a set of values for $W(u)$, $1/u$, $(h_0 - h)$, and t. The reciprocal of $1/u$ is u. The value of time determined from the match point is in minutes. This must now be converted to days by dividing by 1440 min/day. The pumping rate is known, as is the radial distance from the pumping well to the observation well.

The final step in the solution of the Theis equations is to substitute the values of Q, $(h_0 - h)$, and $W(u)$ from the match point into Equation 7–45 to find the transmissivity of the aquifer. Once T is known, its value along with r and t and u from the match point can be substituted into Equation 7–46 to find aquifer storativity.

If there are several observation wells, individual graphs of drawdown as a function of time can be made and each can be analyzed separately. The values

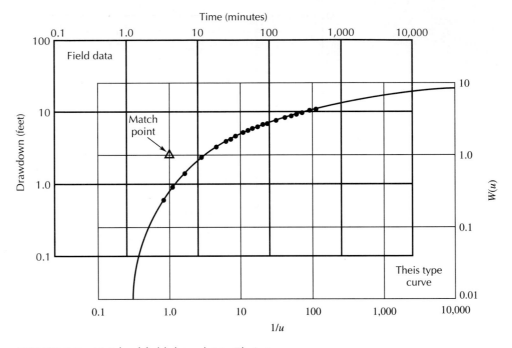

FIGURE 7.8   Match of field-data plot to Theis type curve.

for T and S determined for each observation well, as well as the value of T determined for the pumping well, can reveal how homogeneous the aquifer is.

EXAMPLE PROBLEM   A well in a confined aquifer was pumped at a rate of 220 gal/min for 500 min. The aquifer is 48 ft thick. Time-drawdown data from an observation well located 824 ft away are given in Table 7.1. Find T, K, and S.

The field data are plotted on logarithmic paper (Figure 7.7). The field data graph is then placed upon the type curve graph (Figure 7.8). The following match point is obtained:

$$W(u) = 1$$
$$1/u = 1$$
$$h_0 - h = 2.4 \text{ ft}$$
$$t = 4.1 \text{ min}$$

First, time must be converted to days.

$$t = 4.1 \text{ min} \times 1/1440 \text{ day/min} = 2.9 \times 10^{-3} \text{ day}$$

Next the pumping rate of 220 gal/min must be converted to ft^3/day.

TABLE 7.1

Time After Pumping Started (min)	Drawdown (ft)
3	0.3
5	0.7
8	1.3
12	2.1
20	3.2
24	3.6
30	4.1
38	4.7
47	5.1
50	5.3
60	5.7
70	6.1
80	6.3
90	6.7
100	7.0
130	7.5
160	8.3
200	8.5
260	9.2
320	9.7
380	10.2
500	10.9

$$220 \text{ gal/min} \times 1/7.48 \text{ ft}^3/\text{gal} \times 1440 \text{ min/day} = 42,400 \text{ ft}^3/\text{day}$$

Transmissivity is found from Equation 7–45.

$$T = \frac{Q}{4\pi(h_0 - h)} W(u)$$

$$= \frac{42,400 \text{ ft}^3/\text{day}}{4 \times \pi \times 2.4 \text{ ft}}$$

$$= 1400 \text{ ft}^2/\text{day}$$

Hydraulic conductivity is transmissivity divided by aquifer thickness.

$$K = \frac{T}{b}$$

$$= \frac{1400 \text{ ft}^2/\text{day}}{48 \text{ ft}}$$

$$= 29 \text{ ft/day}$$

Storativity is found from Equation 7–46.

$$S = \frac{4Tut}{r^2}$$

$$= \frac{4 \times 1400 \text{ ft}^2/\text{day} \times 1 \times 2.9 \times 10^{-3} \text{ day}}{824 \text{ ft} \times 824 \text{ ft}}$$

$$= 2.4 \times 10^{-5}$$

7.4.3.2 Nonequilibrium Radial Flow in a Confined Aquifer— Jacob Straight-Line Time-Drawdown Method

C. E. Jacob and H. H. Cooper (Cooper & Jacob 1946; Jacob 1950) observed that after the pumping well has been running for some time, u becomes small and the higher-power terms of the infinite series of Equation 7–4 become negligible. If u, $[(r^2S)/(4Tt)]$, < 0.05, we can ignore all higher powers of u, and Equation 7–4 can be written as:

$$T = \frac{Q}{4\pi(h_0 - h)}\left[-0.5772 - \ln\left(\frac{r^2S}{4Tt}\right)\right] \tag{7–47}$$

or

$$T = \frac{Q}{4\pi(h_0 - h)}\left[-\ln(1.78) - \ln\left(\frac{r^2S}{4Tt}\right)\right] \tag{7–48}$$

Combining natural logs we obtain

$$T = \frac{Q}{4\pi(h_0 - h)}\ln\left(\frac{4Tt}{1.78r^2S}\right) \tag{7–49}$$

We can convert this to base 10 logs and simplify:

$$T = \frac{2.3Q}{4\pi(h_0 - h)}\log\left(\frac{2.25Tt}{r^2S}\right) \tag{7–50}$$

The logarithmic Equation 7–50 will plot as a straight line on semilogarithmic paper if the limiting condition is met. This may be true for large values of t or small values of r. Thus, straight-line plots of drawdown versus time can occur after sufficient time has elapsed. In pumping tests with multiple observation wells, the closer wells will meet the conditions before the more distant ones.

In the **Jacob straight-line method,** a straight line is drawn through the field-data points and extended backward to the zero drawdown axis. It should intercept this axis at some positive value of time. This value is designated t_0. The value of the drawdown per log cycle $\Delta(h_0 - h)$ is obtained from the slope of the graph. The values of transmissivity and storativity may be found from the equations

$$T = \frac{2.3Q}{4\pi\Delta(h_0 - h)} \tag{7-51}$$

and

$$S = \frac{2.25Tt_0}{r^2} \tag{7-52}$$

where

T	is the transmissivity (L^2/T; ft²/day or m²/day)
Q	is the pumping rate (L^3/T; ft³/day or m³/day)
$\Delta(h_0 - h)$	is the drawdown per log cycle of time (L; ft or m)
S	is storativity (dimensionless)
r	is the radial distance to the well (L; ft or m)
t_0	is the time, where the straight line intersects the zero-drawdown axis (T; days)

Again, note that it is usually more convenient to plot drawdown as a function of time in minutes. The value of t_0 in minutes must be converted to days before it is used in Equation 7–52.

EXAMPLE PROBLEM   Evaluate the pumping test data of Table 7.1 by the Jacob method.

The field data are plotted on semilogarithmic paper (Figure 7.9). A straight line is fit to the later time data and extended back to the zero-drawdown axis. The value of t_0 is 5.2 min, and the drawdown per log cycle of time is 5.5 ft.

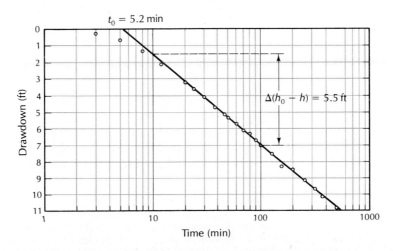

FIGURE 7.9 Jacob method of solution of pumping-test data for a fully confined aquifer. Drawdown is plotted as a function of time on semilogarithmic paper.

First, t_0 must be converted to days.

$$5.2 \text{ min} \times 1/1440 \text{ day/min} = 3.6 \times 10^{-3} \text{ day}$$

The pumping rate has already been determined to be 42,400 ft³/day. Substituting into Equation 7–51 gives

$$T = \frac{2.3Q}{4\pi\Delta(h_0 - h)}$$
$$= \frac{2.3 \times 42,400 \text{ ft}^3/\text{day}}{4 \times \pi \times 5.5 \text{ ft}}$$
$$= 1400 \text{ ft}^2/\text{day}$$

The value of T can be substituted into Equation 7–52 to find S:

$$S = \frac{2.25Tt_0}{r^2}$$
$$= \frac{2.25 \times 1400 \text{ ft}^2/\text{day} \times 3.6 \times 10^{-3} \text{ day}}{824 \text{ ft} \times 824 \text{ ft}}$$
$$= 1.7 \times 10^{-5}$$

In comparing the Jacob solution with the Theis solution in the preceding problems, we can see that the resulting answers are almost the same. As these are graphical methods of solution, there will often be a slight variation in the answers, depending upon the accuracy of the graph construction and subjective judgments in matching field data to type curves.

An aquifer test may be made even if there are no observation wells. In this case, drawdown must be measured in the pumping well. There are energy losses as the water rushes into the well, so that the head in the aquifer is higher than the water level in the pumping well. For this reason, aquifer storativity cannot be determined. However, a plot of drawdown versus time for the pumping well can be used to determine aquifer transmissivity.* Either the Theis or Jacob method can be used. It is important that the well be pumped at a constant rate, as any slight fluctuations will immediately affect the water level in the well. Likewise, drawdown data for the start of pumping are affected by the volume of water stored in the well casing. At the start of pumping, the water comes from the well casing

---

*If, because of turbulent well losses as the water enters the well, the drawdown inside the well is significantly greater than the drawdown in the formation just outside the well, use of time-drawdown data from a single well pump test will understate the formation transmissivity. This can be overcome by measuring the **recovery** of the water level in the well after the pump has been shut down. Time-recovery data can then be plotted and the aquifer transmissivity determined.

rather than from the aquifer, especially when the well diameter is large and/or the pumping rate is small. The measured drawdown data should be adjusted to compensate for this factor (Schafer 1978).

7.4.3.3 Nonequilibrium Radial Flow in a Confined Aquifer— Jacob Straight-Line Distance-Drawdown Method

If simultaneous observations are made of the drawdown in three or more observation wells, a modification of the Jacob straight-line method may be used. If drawdown is measured at the same time in several wells, it is found to vary with the distance from the pumping well in accordance with the Theis equation. Drawdown is plotted on the arithmetic scale as a function of the distance from the pumping well on the logarithmic scale. The drawdown in the wells closest to the pumping well should fall on a straight line.

A line is drawn through the data points for the closest wells. It is extended until it intercepts the zero-drawdown line. This is the distance at which the well is not affecting the water level and is designated r_0. The drawdown per log cycle is $\Delta(h_0 - h)$, as before (Figure 7.10).

Notice in Figure 7.10 the distance-drawdown data slope upward from left to right, whereas in Figure 7.9, the time-drawdown data slope downward from left to right. The greatest drawdown is close to the well, so in Figure 7.10 the drawdown becomes smaller as one goes away from the well, which is left to right on the graph. There is no drawdown before pumping starts. However, as time progresses, the drawdown increases and time increases, which is from left to right in Figure 7.9.

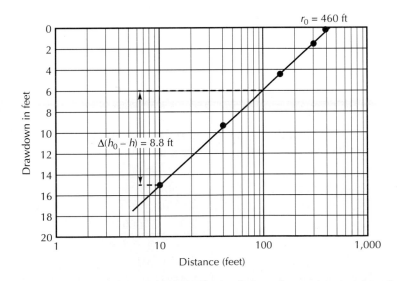

FIGURE 7.10 Variation of the Jacob method of solution of pumping-test data for a fully confined aquifer. Drawdown is plotted as a function of distance to observation well on semilogarithmic paper.

The distance drawdown formulas are:

$$T = \frac{2.3Q}{2\pi\Delta(h_0 - h)} \tag{7-53}$$

$$S = \frac{2.25Tt}{r_0^2} \tag{7-54}$$

where

T	is transmissivity (L^2/T; ft^2/day or m^2/day)
Q	is pumping rate (L^3/T; ft^3/day or m^3/day)
$\Delta(h_0 - h)$	is drawdown per log cycle of distance (L; ft or m)
S	is storativity (dimensionless)
t	is time (T; days)
r_0	is the distance at which the straight line intercepts the zero-drawdown axis (L; ft or m)

EXAMPLE PROBLEM   A well pumping at 77,000 ft^3/day has observation wells located 10, 40, 150, 300, and 400 ft away. After 0.14 days of pumping, the following drawdowns were observed:

Distance (ft)	Drawdown (ft)
10	15.1
40	9.4
150	4.4
300	1.7
400	0.25

The data are plotted in Figure 7.10. The drawdown per log cycle is 8.8 ft and r_0 is 460 ft. Find the values of T and S.

$$
\begin{aligned}
T &= \frac{2.3Q}{2\pi\Delta(h_0 - h)} \\
&= \frac{2.3 \times 77,000 \text{ ft}^3/\text{day}}{2 \times \pi \times 8.8 \text{ ft}} \\
&= 3200 \text{ ft}^2/\text{day} \\
S &= \frac{2.25Tt}{r_0^2}
\end{aligned}
$$

$$= \frac{2.25 \times 3200 \text{ ft}^2/\text{day} \times 0.14 \text{ day}}{460 \text{ ft} \times 460 \text{ ft}}$$

$$= 0.0048$$

7.4.3.4 Nonequilibrium Radial Flow in a Leaky, Confined Aquifer with No Storage in the Aquitard—Walton Graphical Method

Pumping tests may be used to determine the formation constants for both an aquifer and an overlying or underlying semipervious layer of finite thickness. Leaky artesian aquifers are also known as **semiconfined aquifers.**

W. C. Walton (1962, 1960) devised a graphical method based on type curves for $W(u,r/B)$. The type curves are plots on logarithmic paper of $W(u,r/B)$ as a function of $1/u$ for various values of $1/u$ and r/B (Figure 7.11). The type curve for $r/B = 0$ is identical to the Theis nonequilibrium reverse type curve.

Field data are plotted as drawdown versus time. The data curve is placed over the type curve with the axes parallel. The data curve should match one of the type curves for r/B; it may have to be matched to an imaginary line interpolated between two r/B lines.

The early drawdown data will tend to fall on the nonequilibrium portion of the type curve. As leakage starts to contribute to flow from the well, the drawdown will follow an r/B type curve.

A match point is picked on the graph. The coordinates on both the data plot and the type curve yield the values of $W(u,r/B)$, $1/u$, t, and $h_0 - h$. In addition, the matched type curve yields a value for r/B. These values are substituted into the Hantush-Jacob equations to find formation constants for both the aquifer and the aquitard.

$$T = \frac{Q}{4\pi(h_0 - h)} W(u, r/B) \tag{7-55}$$

$$S = \frac{4Tut}{r^2} \tag{7-56}$$

$$r/B = r/(Tb'/K')^{1/2} \tag{7-57}$$

$$K' = [Tb'(r/B)^2]/r^2 \tag{7-58}$$

where

Q is the pumping rate (L^3/T; ft^3/day or m^3/day)

T is the transmissivity of the confined aquifer (L^2/T; ft^2/day or m^2/day)

t is the time since pumping began (T; days)

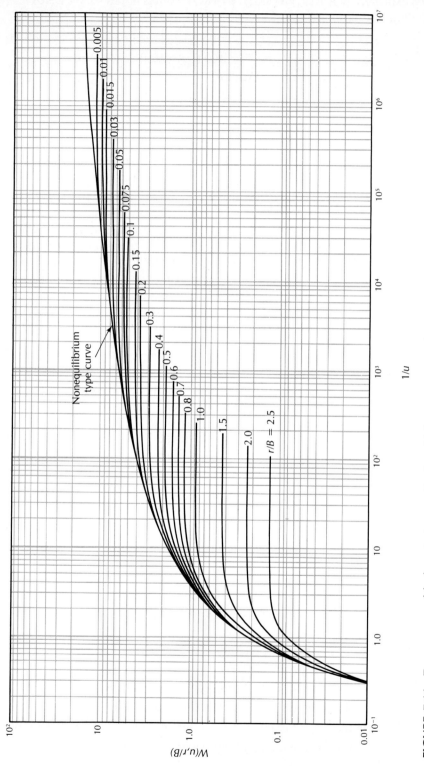

FIGURE 7.11 Type curves of leaky artesian aquifer in which no water is released from storage in the confining layer. Source: W. C. Walton, Illinois State Water Survey Bulletin 49, 1962.

S is the storativity of the confined aquifer (dimensionless)

r is the distance from the pumping well to the observation well (L; ft or m)

K' is the vertical hydraulic conductivity of the aquitard (L/T; ft/day or m/day)

b' is the thickness of the aquitard (L; ft or m)

B is the leakage factor, $(Tb'/K')^{1/2}$ (L; ft or m)

EXAMPLE PROBLEM

Walton (1960) gave time-drawdown data for an aquifer test in a well confined by a stratum of silty fine sand that was 14 ft thick (Table 7.2). Drawdown was measured in an observation well 96 ft from the pumping well, which was pumped at 25 gal/min. Using Walton's graphical method for the Hantush-Jacob formulas, find the values of T, S, and K'.

First a plot of drawdown as a function of time must be made. This is shown in Figure 7.12. The time-drawdown data are matched to an r/B curve. The match-point values are:

$$W(u, r/B) = 1.0$$
$$1/u = 10, \qquad u = 0.10$$
$$h_0 - h = 1.9 \text{ ft}$$
$$t = 33 \text{ min}$$
$$r/B = 0.22$$

Next we must convert gallons per minute to cubic feet per day and time in minutes to time in days.

TABLE 7.2

Time (min)	Drawdown (ft)
5	0.76
28	3.30
41	3.59
60	4.08
75	4.39
244	5.47
493	5.96
669	6.11
958	6.27
1129	6.40
1185	6.42

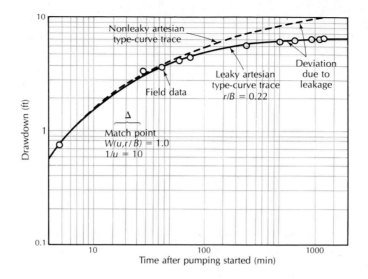

FIGURE 7.12   Field-data plot of drawdown as a function of time for a leaky, confined aquifer. Source: W. C. Walton, Illinois State Water Survey Bulletin 49, 1962.

$$Q = 25 \text{ gal/min} \times 1/7.48 \text{ ft}^3/\text{gal} \times 1440 \text{ min/day} = 4800 \text{ ft}^3/\text{day}$$
$$t = 33 \text{ min} \times 1/1440 \text{ day/min} = 0.023 \text{ day}$$

The match-point values, Q and r, are substituted into Equations 7–55, 7–56, and 7–58.

$$T = \frac{Q}{4\pi(h_0 - h)} W(u, r/B)$$
$$= \frac{4800 \text{ ft}^3/\text{day}}{4 \times \pi \times 1.9 \text{ ft}} \times 1$$
$$= 200 \text{ ft}^2/\text{day}$$
$$S = \frac{4Tut}{r^2}$$
$$= \frac{4 \times 200 \text{ ft}^2/\text{day} \times 0.10 \times 0.023 \text{ day}}{96 \text{ ft} \times 96 \text{ ft}}$$
$$= 0.00020$$
$$K' = \frac{Tb'(r/B)^2}{r^2}$$
$$= \frac{200 \text{ ft}^2/\text{day} \times 14 \text{ ft} \times 0.22^2}{96 \text{ ft} \times 96 \text{ ft}}$$
$$= 0.015 \text{ ft/day}$$

7.4.3.5 Nonequilibrium Radial Flow in a Leaky, Confined Aquifer with No Storage in the Aquitard—Hantush Inflection-Point Method

M. S. Hantush (1956) developed an alternative method that does not require the plotting and use of type curves. As can be seen from Figure 7.13, a plot of drawdown versus time on semilogarithmic paper has the shape of an elongated reverse S. At some point, the curve has an inflection. In the *Hantush inflection-point method,* the solution is based on finding this inflection point. For a pumping test with one observation well, the following procedure is used:

1. Plot the drawdown on the arithmetic scale as a function of time since pumping began on the logarithmic scale. If the test has reached equilibrium, then determine the maximum drawdown, $(h_0 - h)_{max}$. If drawdown is still occurring, extrapolate the curve to find $(h_0 - h)_{max}$.

2. The drawdown at the inflection point $(h_0 - h)_i$ is defined as being equal to one-half the maximum drawdown.

3. From the graph, determine the time, t_i, when $(h_0 - h)_i$ occurs; also graphically find the slope of the drawdown curve at the inflection point (m_i). This is generally equal to the slope of the straight portion of the drawdown curve. The slope is expressed as drawdown per log-cycle.

The following relations hold true for the inflection point:

$$u_i = \frac{r^2 S}{4 t_i T} = \frac{r}{2B} \tag{7-59}$$

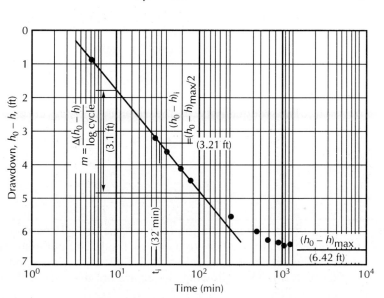

FIGURE 7.13 Plot of drawdown in a confined aquifer as a function of time on semilogarithmic paper for use in the Hantush inflection-point method of analysis.

$$m_i = \left(\frac{2.3Q}{4\pi T}\right)\exp\left(\frac{-r}{B}\right) \tag{7-60}$$

$$(h_0 - h)_i = 0.5\,(h_0 - h)_{max} = \frac{Q}{4\pi T}\,K_0\,(r/B) \tag{7-61}$$

$$B = \left(\frac{T}{K'/b'}\right)^{1/2} \tag{7-62}$$

$$f\left(\frac{r}{B}\right) = \frac{2.3(h_0 - h)_i}{m_i} = \exp\left(\frac{r}{B}\right)K_0\left(\frac{r}{B}\right) \tag{7-63}$$

where K_0 is a function with values tabulated in Appendix 5 as $K_0(x)$ and exp $(x)K_0(x)$.

From the drawdown and slope at the inflection point, the value of $f(r/B)$ may be found:

$$f(r/B) = 2.3(h_0 - h)_i/m_i \tag{7-64}$$

Knowing the value of $f(r/B)$, the function tables may be used to find the value of r/B, since $f(x) = \exp(x)K_0(x)$. Since r is known, the value of B may easily be found.

The transmissivity may be found from the relation

$$T = \frac{QK_0(r/B)}{2\pi(h_0 - h)_{max}} \tag{7-65}$$

where K_0 is a Bessel function, known as a zero-order modified Bessel function of the second kind (see Appendix 5).

The value of the storativity is found from

$$S = 4t_iT/2rB \tag{7-66}$$

The conductivity of the semipervious layer may be determined if its thickness, b', is known:

$$K' = \frac{Tb'}{B^2} \tag{7-67}$$

EXAMPLE PROBLEM Check the graphical solution for the time-drawdown data found in Table 7.2 by using the Hantush inflection-point method. The pumping rate is 4800 ft³/day, and the radial distance is 96 ft. Table 7.2 data are plotted on semilogarithmic paper (Figure 7.13). The following values are obtained from the graph:

$$(h_0 - h)_{max} = 6.42 \text{ ft}$$
$$(h_0 - h)_i = 3.21 \text{ ft}$$

$$t_i = 32 \text{ min}$$
$$m_i = 3.10 \text{ ft}$$

First convert t_i in minutes to t_i in days.

$$t_i = 32 \text{ min} \times 1/1440 \text{ day/min} = 0.022 \text{ day}$$

From Equation 7–64, $f(r/B) = 2.3(h_0 - h)_i/m_i$:

$$f(r/B) = (2.3 \times 3.21 \text{ ft})/(3.10 \text{ ft})$$
$$= 2.38$$

From Equation 7–63, we know that $f(r/B) = \exp(r/B)K_0(r/B)$; therefore, $\exp(r/B)K_0(r/B) = 2.38$. Values of this function are tabulated in Appendix 5, a portion of which is reproduced here. The value of $\exp(x)K_0(x)$, which is 2.38, falls between 2.36 and 2.68.

x	$K_0(x)$	$\exp(x)K_0(x)$
0.10	2.43	2.68
0.15	2.03	2.36

If we interpolate between these two lines, we obtain these values:

x	$K_0(x)$	$\exp(x)K_0(x)$
0.147	2.055	2.38

This gives us a value for $K_0(r/B)$ of 2.055 and for r/B of 0.147. If $r/B = 0.147$ and $r = 96$ ft, then $B = (96 \text{ ft})/(0.147) = 653$ ft.

Substituting these values into Equations 7–65, 7–66, and 7–67, we obtain the hydraulic parameters of the aquifer and aquitard.

$$T = \frac{Q}{2\pi(h_0 - h)_{max}} K_0(r/B)$$
$$= \frac{4800 \text{ ft}^3/\text{day}}{2 \times \pi \times 6.42 \text{ ft}} \times 2.055$$
$$= 240 \text{ ft}^2/\text{day}$$

$$S = \frac{4t_i T}{2rB}$$
$$= \frac{4 \times 0.022 \text{ day} \times 240 \text{ ft}^2/\text{day}}{2 \times 96 \text{ ft} \times 653 \text{ ft}}$$
$$= 0.00017$$

$$K' = \frac{Tb'}{B^2}$$

$$= \frac{240 \text{ ft}^2/\text{day} \times 14 \text{ ft}}{653 \text{ ft} \times 653 \text{ ft}}$$

$$= 0.0079 \text{ ft/day}$$

We can compare the parameters obtained by the two methods:

	T	S	K'
Walton graphical	200 ft²/day	0.00020	0.015 ft/day
Hantush inflection point	240 ft²/day	0.00017	0.0079 ft/day

The values for T and S are reasonably close, being within 20%; however, the value of K' is off by nearly 50%.

7.4.4 Nonequilibrium Radial Flow in a Leaky Aquifer with Storage in the Aquitard

A type-curve method can also be used for the leaky, confined aquifer with storage in the confining layer. A set of type curves on logarithmic paper are prepared from the tabulated values of $H(u, \beta)$ (Figure 7.14). On logarithmic paper of similar

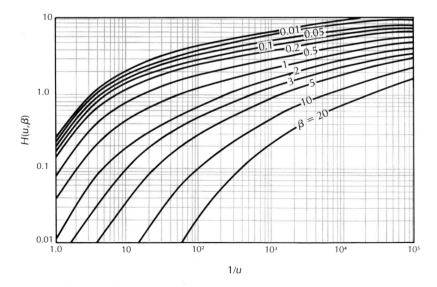

FIGURE 7.14 Type curves for a well in an aquifer confined by a leaky layer that releases water from storage. Source: Data from M. S. Hantush, *Journal of Geophysical Research* 65 (1960): 3713–25; type curves from W. C. Walton, *Groundwater Resource Evaluation* (New York: McGraw-Hill, 1970).

scale, drawdown is plotted against time. The field-data sheet is overlain on the type curves with the $H(u, \beta)$-axis parallel to the drawdown axis, and it is matched to a β-curve. A match point is selected and values of $H(u, \beta)$, $1/u$, drawdown, and time are obtained. The value of the curve is also noted.

From the match-point value of $H(u, \beta)$ and $h_0 - h$, the value of T is found from:

$$T = \frac{Q}{4\pi(h_0 - h)} H(u, \beta) \qquad (7-68)$$

and S may be found from the match-point values of t, $1/u$, the measured value of r, and the computed value of T:

$$S = \frac{4Tut}{r^2} \qquad (7-69)$$

The value of β can be used to compute the product $K'S'$:

$$\beta^2 = \frac{r^2 S'}{16B^2 \, S} \qquad (7-70)$$

$$B^2 = T/(K'/b') \qquad (7-71)$$

Combining Equations 7–70 and 7–71,

$$K'S' = \frac{16\beta^2 Tb'S}{r^2} \qquad (7-72)$$

If one of the values, either K' or S', is known, the value of the other can be found.

7.4.5 Nonequilibrium Radial Flow in an Unconfined Aquifer

A graphical method for analysis of an aquifer test in an unconfined aquifer has also been developed (Neuman 1975; Streltsova 1976a; Walton 1979). This analysis is based upon the general assumptions of Section 7.2 and the specific assumptions of Section 7.3.4.

The flow equation for unconfined aquifers is

$$T = \frac{Q}{4\pi(h_0 - h)} W(u_A, u_B, \Gamma) \qquad (7-73)$$

where $W(u_A, u_B, \Gamma)$ is the well function for the water-table aquifer and

$$S = \frac{4Tu_A \, t}{r^2} \qquad \text{(for early drawdown data)} \qquad (7-74)$$

$$S_y = \frac{4Tu_B\, t}{r^2} \qquad \text{(for later drawdown data)} \qquad \textbf{(7–75)}$$

$$\Gamma = \frac{r^2 K_v}{b^2 K_h} \qquad\qquad\qquad\qquad\quad \textbf{(7–76)}$$

where

$h_0 - h$ is the drawdown (L; ft or m)

Q is the pumping rate (L^3/T; ft³/day or m³/day)

T is the transmissivity (L^2/T; ft²/day or m²/day)

r is the radial distance from the pumping well (L; ft or m)

S is the storativity (dimensionless)

S_y is the specific yield (dimensionless)

t is the time (T; day)

K_h is the horizontal hydraulic conductivity (L/T; ft/day or m/day)

K_v is the vertical hydraulic conductivity (L/T; ft/day or m/day)

b is the initial saturated thickness of the aquifer (L; ft or m)

Two sets of type curves are used. Type-A curves are good for early drawdown data, when instantaneous release of water from storage is occurring. As time elapses, the effects of gravity drainage and vertical flow cause deviations from the nonequilibrium type curve, which is accounted for in the family of Type-A curves. The Type-B curves are used for late drawdown data, when effects of gravity drainage are becoming smaller. The Type-B curves end on a Theis curve. Figure 7.15 shows the two sets of type curves for fully penetrating wells. Values of $W(u_A, \Gamma)$ and $W(u_B, \Gamma)$ are found in Appendix 6.

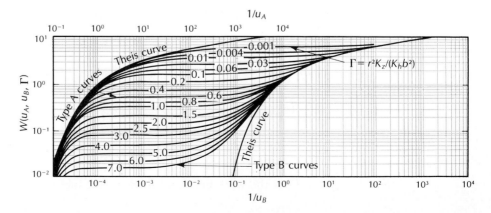

FIGURE 7.15 Type curves for drawdown data from fully penetrating wells in an unconfined aquifer. Source: S. P. Neuman, *Water Resources Research* 11 (1975):329–42.

The type curves are then used to evaluate the field data for time and drawdown, which are plotted on logarithmic paper of the same scale as the type curve. The following procedure can be used (Gambolati, 1976):

1. Superpose the early time-drawdown data on the Type-A curves. The axes of the graph papers should be parallel and the data matched to the curve with the best fit. At any match point, the values of $W(u_A, \Gamma)$, $1/u_A$, t, and $h_0 - h$ are determined. The value of Γ comes from the type curve. The value of T is found using these values and Equation 7–73. The storativity is found from Equation 7–74.

2. The latest drawdown data are then superposed on the Type-B curve for the Γ-value of the previously matched Type-A curve. A new set of match points is determined. The value of T calculated from Equation 7–73 should be approximately equal to that computed from the Type-A curve. Equation 7–75 can be used to compute the specific yield.

3. The value of the horizontal hydraulic conductivity can be determined from

$$K_h = T/b \qquad (7-77)$$

4. The value of the vertical hydraulic conductivity can also be computed using the Γ-value of the matched type curve. Rearrangement of Equation 7–70 yields the following formula:

$$K_v = \frac{\Gamma b^2 K_h}{r^2} \qquad (7-78)$$

The preceding analysis is based upon a very low value of drawdown compared with the saturated thickness of the aquifer. If the drawdown is substantial, some authorities (Neuman 1974) suggest that the drawdown data be corrected. The corrected drawdown $(h_0 - h)'$ is found from the relation (Hantush 1956)

$$(h_0 - h)' = (h_0 - h) - [(h_0 - h)^2/2h_0] \qquad (7-79)$$

This is normally necessary only for the later time-drawdown data; if $h_0 - h$ is small compared with h_0, correction will not be needed.

EXAMPLE PROBLEM

A well pumping at 1000 gal/min fully penetrates an unconfined aquifer with an initial saturated thickness of 100 ft. Time-drawdown data for a well located 200 ft away are plotted on log paper (Figure 7.16). Find T, S_y, S, K_h, and K_v.

The value of the pumping rate is

$$Q = 1000 \text{ gal/min} \times 1/7.48 \text{ ft}^3/\text{gal} \times 1440 \text{ min/day} = 1.9 \times 10^5 \text{ ft}^3/\text{day}$$

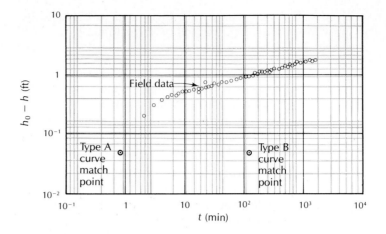

FIGURE 7.16   Field data for Example Problem of analysis of aquifer test for an unconfined aquifer.

The early time-drawdown data fit best on the Type-A curve for $\Gamma = 0.1$. The selected match point is $W(u_A, \Gamma) = 0.1$, $1/u_A = 1.0$, $h_0 - h = 0.041$ ft, and $t = 0.9$ min. The value of u_A is 1.0 and $t = 6.25 \times 10^{-4}$ day. From Equation 7–73,

$$T = \frac{Q}{4\pi(h_0 - h)} W(u_A, \Gamma)$$

$$= \frac{1.9 \times 10^5 \text{ ft}^3/\text{day} \times 0.1}{4 \times \pi \times 0.041 \text{ ft}}$$

$$= 3.7 \times 10^4 \text{ ft}^2/\text{day}$$

The storativity value is found from Equation 7–74:

$$S = \frac{4Tu_A t}{r^2}$$

$$= \frac{4 \times 3.7 \times 10^4 \text{ ft}^2/\text{day} \times 6.25 \times 10^{-4} \text{ day} \times 1.0}{(200 \text{ ft})^2}$$

$$= 0.0023$$

The later time-drawdown data are now matched to the Type-B curve for $\Gamma = 0.1$. With the axes of the two sheets of graph paper parallel, the selected match point has values of $W(u_B, \Gamma) = 0.1$, $1/u_B = 10$, $h_0 - h = 0.043$ ft, and $t = 128$ min. The value of u_B is 0.1 and $t = 0.089$ days. From Equation 7–73,

$$T = \frac{Q}{4\pi(h_0 - h)} W(u_B, \Gamma)$$

$$T = \frac{1.9 \times 10^5 \text{ ft}^3/\text{day} \times 0.1}{4 \times \pi \times 0.043 \text{ ft}}$$

$$= 3.5 \times 10^4 \text{ ft}^2/\text{day}$$

The value of the specific yield can be determined from Equation 7–75:

$$S_y = \frac{4Tu_B t}{r^2}$$

$$= \frac{4 \times 3.5 \times 10^4 \text{ ft}^2/\text{day} \times 0.089 \text{ days} \times 0.1}{(200 \text{ ft})^2}$$

$$= 0.031$$

The value of the horizontal hydraulic conductivity can be found from Equation 7–77. The average of T is 3.6×10^4 ft^2 per day:

$$K_h = T/b$$

$$= (3.6 \times 10^4 \text{ ft}^2/\text{day})/100 \text{ ft}$$

$$= 360 \text{ ft/day}$$

and the value of K_v is determined using Equation 7–78:

$$K_v = \frac{\Gamma b^2 K_h}{r^2}$$

$$= \frac{0.1 \times (100 \text{ ft})^2 \times 360 \text{ ft/day}}{(200 \text{ ft})^2}$$

$$= 9.0 \text{ ft/day}$$

7.4.6 Effect of Partial Penetration of Wells

In many cases, the open hole or well screen of a pumping well does not extend from the top to the bottom of an aquifer. In all of the cases considered thus far, we assumed the well to penetrate the entire thickness of the aquifer. This caused flow in the aquifer to be essentially horizontal. However, the flow toward a partially penetrating well will be three-dimensional owing to the vertical flow components (Figure 7.17). In addition, if the aquifer is anisotropic, the value of the vertical conductivity, K_v, and the horizontal conductivity, K_h, are important. This will affect both the amount of water pumped from the well and the potential field caused by drawdown.

   If an observation well completely penetrates an aquifer, or if a partially penetrating well is located more than $1.5b\sqrt{K_h/K_v}$ distance units from the pumped well, the effect of a partially penetrating pumping well is negligible (Hantush 1964). The drawdown is described by either the Hantush-Jacob formula for leaky aquifers or the Theis formula for nonleaky aquifers.

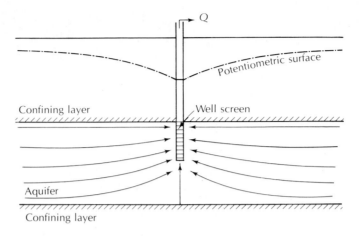

FIGURE 7.17   Flow lines toward a partially penetrating well in a confined aquifer.

If, however, the pumping well is partially penetrating, and the observation wells are also partially penetrating and located closer to the pumping well than $1.5b\sqrt{K_h/K_v}$ distance units, then the drawdown formula is different and quite complex. Hantush (1964) makes the following observations about the drawdown in observation wells in such cases:

1. If two observation wells equidistant from the pumping well are screened in different parts of the aquifer, the time-drawdown curves may be different.

2. Depending upon the length and relative position of observation-well screens, it is possible for a more distant well to have a greater drawdown than a closer well.

3. The effects of partial penetration produce a time-drawdown curve similar in shape to one produced when there is a downward leakage from storage through a thick, semipervious layer.

4. Partial-penetration effects may produce a time-drawdown curve that resembles the effect of a recharge boundary, a fully penetrating well in either a sloping water-table aquifer or an aquifer of nonuniform thickness.

The preceding observations suggest that the hydrogeologist must always use extreme care in collecting and interpreting pumping-test data. Test drilling should be used to delineate the aquifer system before the pumping test. If a pumping test is to be made on a partially penetrating well, fully penetrating observation wells or wells located at a distance of more than $1.5b\sqrt{K_h/K_v}$ would be desirable. The analysis of partial-penetrating effects is beyond the scope of this book. The hydrogeologist who must deal with this problem can use the *Hantush partial-penetration method* (Hantush 1961). Using partially penetrating wells, it is also possible to run pumping tests that determine the value of both horizontal and vertical conductivity of the confining layer and the storativity of both the aquifer and the confining layer (Hantush, 1960a, 1966; Weeks 1969; Boulton & Streltsova, 1975).

The pumping-test solutions given in the preceding sections are all based on some simplifying assumptions that may not be valid in certain circumstances. For leaky artesian aquifer flow, the assumption that the storage in the aquifer is negligible may be erroneous (Hantush 1960b; Neuman & Witherspoon 1969). The effect of storage in the confining layer occurs early in the pumping test, as leakage water is being furnished from storage in the confining layer. With increasing time, more of the leakage is being contributed by the aquifer above the leaky, confining layer, and the amount of water from storage diminishes. If analysis of pumping-test data for a leaky artesian aquifer does not follow an r/B-curve for early drawdown data, significant flow from storage should be suspected and an alternative analysis made (Hantush 1960b; Neuman & Witherspoon 1969; Boulton & Streltsova 1975; Streltsova 1976b; Gambolati 1977).

The Theis solution may overestimate the value of the storativity of shallow elastic aquifers. If the ratio of the average depth of a confined aquifer to its thickness is less than 0.5, the overestimation can be as great as 40 percent owing to three-dimensional consolidation of the aquifer (Neuman & Witherspoon 1969). If the average depth-to-thickness ratio is 1.0 or more, then the Theis solution is valid even for deforming aquifers.

Neuman (1974) points out that the effects of partial penetration of a well pumping from a water-table aquifer can be minimized if the observation well fully penetrates the saturated thickness of the aquifer. Under these conditions the time-drawdown curve for observation wells located at distances greater than $r = b/(K_v/K_h)^{1/2}$ and for times greater than $t = S_y/r^2 T$ will follow the late-time Theis curve. The time-drawdown curve for observation wells located at distances less than $r = 0.03b/(K_v/K_r)^{1/2}$ and for times less than $t = Sr^2/T$ will follow the early-time Theis curve. Thus wells located some distance from the partially penetrating observation well can be utilized with late-time-drawdown data to find T and S_y, and close wells can be utilized with early-time-drawdown data to find T and S. Intermediate-distance wells are of no use in the analysis of an aquifer test performed on a partially penetrating well in an unconfined aquifer.

7.5 SLUG TESTS

7.5.1 Determination of Aquifer Parameters with Slug Tests

Aquifer tests are expensive to conduct, both in terms of the cost of installation of the pumping well and observation well(s) and also in terms of the personnel time needed. Moreover, if the water in the aquifer is contaminated, it may be necessary to treat it prior to disposal. This may not be feasible.

In some field investigations the practicing hydrogeologist may need to know the hydraulic conductivity of low-permeability materials. This is especially likely in site studies for areas of potential waste storage or disposal. The earth materials in these areas may have a hydraulic conductivity that is too low to conduct a pumping test.

As an alternative to an aquifer test, a **slug** or **bail-down test** can be performed in a small-diameter monitoring well. This type of test can be used to determine the hydraulic conductivity of the formation in the immediate vicinity of a monitoring well. A known volume of water is quickly drawn from or added to the monitoring well. The rate at which the water level falls or rises is measured. These data are then analyzed by an appropriate method.

7.5.2   Cooper-Bredehoeft-Papadopulos Method for a Confined Aquifer

In the Cooper-Bredehoeft-Papadopulos method a confined aquifer is completely penetrated by an open borehole or a well screen. A known volume of water is poured into the well in a time frame of a few seconds. This causes the water to rise to a maximum height, H_0 (L; ft, m, or cm), above the initial head. The excess head will then decay as the water drains from the well into the formation. As the water level falls, the height of the water level above the original elevation, H (L; ft, m, or cm), is measured with respect to time, t (T; s). See Figure 7.18.

We do not need to pour water into the well. The water in the well can be displaced by rapidly lowering a solid piece of pipe, or slug, on a line into the well. The volume of the slug will displace an equal volume of water, which must flow into the formation. Also, the well can be bailed to remove a volume of water, in which case the water level will rise rather than fall.

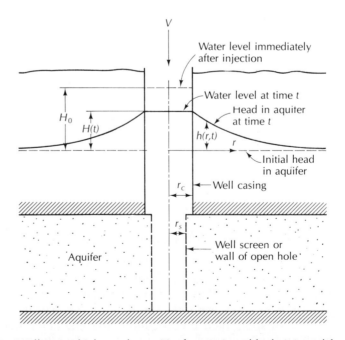

FIGURE 7.18   Well into which a volume, V, of water is suddenly injected for a slug test of a confined aquifer. Source: H. H. Cooper, Jr., J. D. Bredehoeft, & I. S. Papadopulos, *Water Resources Research* 3 (1967): 263–9.

A plot of the ratio of the measured head to the head after injection (H/H_0) is made as a function of time. The ratio H/H_0 is on the arithmetic scale, and time is on the logarithmic scale of semilogarithmic paper. The ratio H/H_0 is equal to a defined function:

$$H/H_0 = F(\eta, \mu) \tag{7-80}$$

where

$$\eta = Tt/r_c^2 \tag{7-81}$$

and

$$\mu = r_s^2 S/r_c^2 \tag{7-82}$$

If time is measured in seconds and all dimensions are in centimeters, then the value of T will be in cm^2/s. If time is in seconds and all dimensions are in feet, then T will be in ft^2/s.

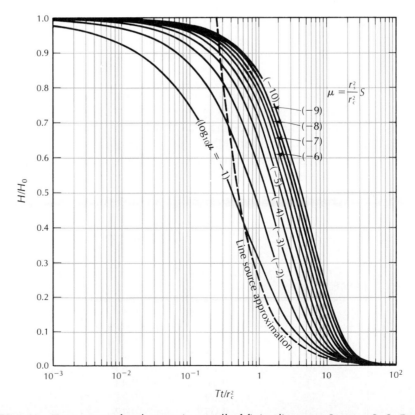

FIGURE 7.19 Type curves for slug test in a well of finite diameter. Source: S. S. Papadopulos, J. D. Bredehoeft, and H. H. Cooper, Jr., *Water Resources Research* 9 (1973): 1087–89.

$F(\eta, \mu)$ is a function, the values of which are given in tabulated form in Appendix 2 (Papadopulos, Bredehoeft, & Cooper 1973). The tabulated values are plotted as a series of type curves in Figure 7.19. The field data should be plotted on the same-scale semilogarithmic paper as the type curves.

The field-data curve is placed over the type curves with the arithmetic axis coincident. That is, the value of $H/H_0 = 1$ for the field data lies on the horizontal axis of 1.0 on the type curve. The data are matched to the type curve (μ), which has the same curvature. The vertical time-axis, t_1, which overlays the vertical axis for $Tt/r_c^2 = 1.0$, is selected. The transmissivity is found from

$$T = \frac{1.0r_c^2}{t_1} \tag{7-83}$$

The value of storativity can be found from the value of the μ-curve for the field data. Since $\mu = (r_s^2/r_c^2)S$,

$$S = (r_c^2\mu)/r_s^2 \tag{7-84}$$

For small values of μ, however, the curves are often very similar; therefore, in matching the field data, the question of which μ-value to use is often encountered. The use of this method to estimate storativity should be approached with caution. Likewise, the value of T that is determined is representative of the formation only in the immediate vicinity of the test hole.

EXAMPLE PROBLEM

A casing with a radius of 7.6 cm is installed through a confining layer. A screen with a radius of 5.1 cm is installed in a formation with a thickness of 5 m. A slug of water is injected, raising the water level 0.42 m. The decline of the head is given in Table 7.3. Find the values of T, K, and S.

A plot of H/H_0 as a function of t is made on semilogarithmic paper (Figure 7.20). It is overlain on the type curve (Figure 7.19). At the axis for $Tt/r^2 = 1.0$, the value t_1 is 13 s.

$$T = \frac{1.0r_c^2}{t_1} = \frac{1.0 \times (7.6 \text{ cm})^2}{13 \text{ s}} = 4.4 \text{ cm}^2/\text{s}$$

TABLE 7.3

Time (s)	H (m)	H/H_0
2	0.37	0.88
5	0.34	0.81
10	0.27	0.64
21	0.18	0.43
46	0.09	0.21
70	0.05	0.12
100	0.02	0.05

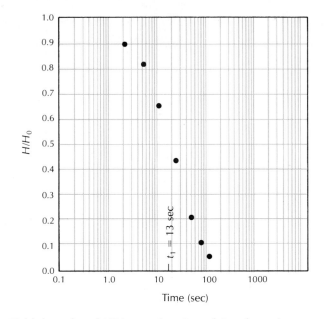

FIGURE 7.20 Field-data plot of H/H_0 as a function of time for a slug-test analysis.

$$K = T/b$$
$$= 4.4 \text{ cm}^2/\text{s}/500 \text{ cm}$$
$$= 8.8 \times 10^{-3} \text{ cm/s}$$

The μ-curve is 10^{-3}. With $r_c = 7.6$ cm and $r_s = 5.1$ cm, we find

$$S = (\mu r_c^2)/r_s^2$$
$$= (10^{-3} \times 7.6^2 \text{ cm}^2)/(5.1 \text{ cm}^2)^2$$
$$= 2.2 \times 10^{-3}$$

7.5.3 Hvorslev Slug Test Method

In many cases piezometers, or auger holes, are installed that do not fully penetrate an aquifer. They are generally installed at a specific depth to monitor head and sample ground-water quality. A very convenient method exists to use these piezometers to determine the hydraulic conductivity of the formation in which the screen is installed. This is the **Hvorslev method** (Hvorslev 1951).

Figure 7.21 shows the geometry of a piezometer installed in an aquifer. In the case of a piezometer installed into a low-permeability unit, special attention must be paid to the method of construction. In many such cases, a hole is augered into a clay unit, the well and well screen are lowered into the open borehole, and gravel is used to fill the open annulus between the well screen and the wall of the open hole. Under such conditions the radius of the well screen, *R*, is the radius of

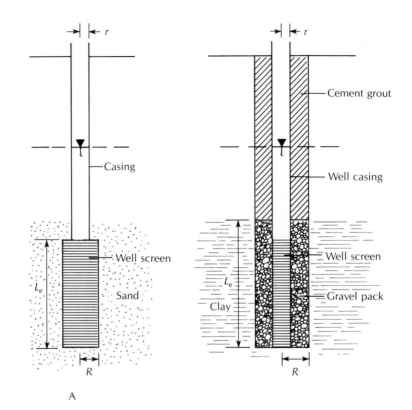

FIGURE 7.21 Piezometer geometry for Hvorslev method. Note that for a piezometer installed in a low-permeability unit the value R is the radius of the highest permeable zone that includes the gravel pack zone and L is the length of the gravel pack zone.

the bored hole and the length of the well screen, L, is the length of the gravel pack. (The gravel pack would typically be extended one to several feet above the well screen and the remainder of the open hole backfilled with some type of grout.)

Naturally the Hvorslev method can be applied only below the water table. Water is either added to the well casing or withdrawn by bailing out the casing with a special tool called a bailer. It is possible to induce the water column in the well casing to rise by rapidly lowering a solid piece of metal, called a slug, into the well and submerging it below the original water surface, or static water level. This is equivalent to adding a volume of water to the well equal to the volume of the slug. The water level in the well is measured prior to the time that the slug is lowered. Immediately after the slug is lowered, the water level in the well is again measured. The water levels are measured at timed intervals as the water level falls back toward the static water level. The height to which the water level rises above the static water level immediately upon lowering the slug is h_0. The height of the water level above the static water level some time, t, after the slug is lowered is h. The data are plotted by computing the ratio h/h_0 and plotting that versus time

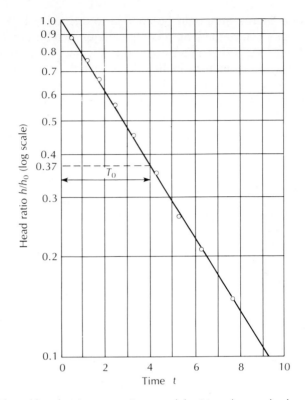

FIGURE 7.22   Plot of head ratio versus time used for Hvorslev method.

on semilogarithmic paper, as shown in Figure 7.22. The time-drawdown data should plot on a straight line.

If the length of the piezometer is more than 8 times the radius of the well screen ($L_e/R > 8$), the following formula applies:

$$K = \frac{r^2 \ln(L/R)}{2 L_e T_0} \tag{7–85}$$

where

K is hydraulic conductivity (L/T; ft/day, m/day, or cm/s)

r is the radius of the well casing (L; ft, m, or cm)

R is the radius of the well screen (L; ft, m, or cm)

L_e is the length of the well screen (L; ft, m, or cm)

T_0 is the time it takes for the water level to rise or fall to 37 percent of the initial change (Figure 7.22) (T; day or s)

Equation 7–85 is but one of many formulas presented by Hvorslev for differing piezometer geometry and aquifer conditions. However, it is one that is quite useful and could be applied to unconfined conditions for most piezometer designs where the length is typically quite a bit greater than the radius of the well screen. For other conditions, the original paper should be consulted.

Piezometers in highly permeable material may recover to the original water level in 30 s or less. In such cases, it is necessary to use a pressure transducer to record pressure changes in the well as the water level changes along with automatic electronic signal recording equipment. Commercial equipment of this type can measure the water level every second and automatically record it. Piezometers in low-permeability clays may take hours to days to recover. In these instances the water levels may be easily measured with a steel tape.

EXAMPLE PROBLEM

A slug test is performed by lowering a metal slug into a piezometer that is screened in a coarse sand. The inside diameter of both the well screen and the well casing is 2 in. The well screen is 10 ft in length. A pressure transducer was used to record the water level every second for the first 10 s and less frequently thereafter. The following data were obtained when the slug was rapidly pulled from the piezometer:

Elapsed time (s)	Depth to Water (ft)	Change in Water Level h (ft)	h/h_0
Static Level	13.99		
0	14.87	0.88 (h_0)	1.000
1	14.59	0.60	0.682
2	14.37	0.38	0.432
3	14.20	0.21	0.239
4	14.11	0.12	0.136
5	14.05	0.06	0.068
6	14.03	0.04	0.045
7	14.01	0.02	0.023
8	14.00	0.01	0.011
9	13.99	0.00	0.000

Figure 7.23 contains a plot of the data. The time for the head to rise to 37 percent of the initial change is 1.8 s (T_0). The following values are obtained from the geometry of the piezometer:

$$r = 0.083 \text{ ft}$$
$$R = 0.083 \text{ ft}$$
$$L_e = 10 \text{ ft}$$

The ratio L_e/R is 120.5, which is more than 8, so that Equation 7–85 can be used:

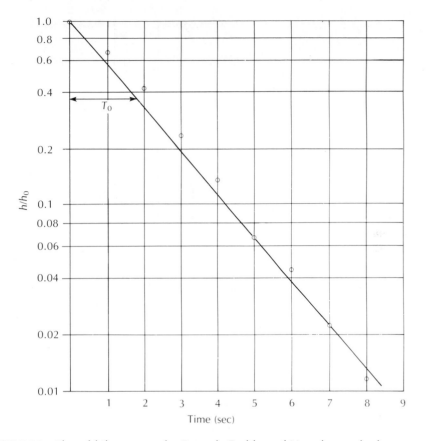

FIGURE 7.23 Plot of h/h_0 versus t for Example Problem of Hvorslev method.

$$K = \frac{r^2 \ln(L_e/R)}{2LT_0}$$

$$= \frac{(0.083 \text{ ft})^2 \times \ln(10 \text{ ft}/0.083 \text{ ft})}{2 \times 10 \text{ ft} \times 1.8 \text{ s}}$$

$$= 9.17 \times 10^{-4} \text{ ft/s} \times 8.64 \times 10^4 \text{ s/day}$$

$$= 79 \text{ ft/day}$$

7.5.4 Bouwer and Rice Slug-Test Method

A very useful slug-test method was developed by H. Bouwer (Bouwer & Rice 1976; Bouwer 1989). This test can be performed on open boreholes or screened wells. The wells can be fully or partially penetrating. Although the test was originally developed for unconfined aquifers, it can be used in confined aquifers if the top of the well screen is some distance below the bottom of the confining layer.

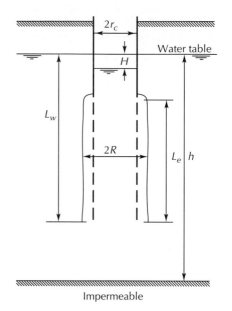

FIGURE 7.24 Geometry and symbols for a slug test on a partially penetrating screened well in an unconfined aquifer with a gravel pack around the screen. Source: Herman Bouwer, *Ground Water* 27 (1989): 304–309. Used with permission. © 1989, Ground Water Publishing Company.

The geometry of the borehole for the Bouwer and Rice slug test is shown in Figure 7.24. In this diagram the parameter r_c is the radius of the well casing in which the water level is rising and R is the radius of the gravel pack or developed area around the well screen.

The Bouwer and Rice equation is:

$$K = \frac{r_c^2 \ln(R_e/R)}{2L_e} \frac{1}{t} \ln\frac{H_t}{H_0} \qquad \textbf{(7–86)}$$

where

K is hydraulic conductivity (L/T; ft/day, m/day, or cm/sec)

r_c is the radius of the well casing (L; ft, m, or cm)

R is the radius of the gravel envelope (L; ft, m, or cm)

R_e is the effective radial distance over which head is dissipated (L; ft, m, or cm) (This is also the distance away from the well over which the average value of K is being measured.)

L_e is the length of the screen or open section of the well through which water can enter (L; ft, m, or cm)

H_0 is the drawdown at time $t = 0$ (L; ft, m, or cm)

H_t is the drawdown at time $t = t$ (L; ft, m, or cm)

t is the time since $H = H_0$ (T; day or s)

The effective distance over which the induced head is dissipated, R_e, is also the distance away from the well that the average value of K is being measured. However, there is no way to know what the value of R_e is for a given well. Bouwer (Bouwer & Rice 1976; Bouwer 1989) has presented a method of estimating the dimensionless ratio $\ln(R_e/R)$ found in Equation 7–86.

If L_w is less than h, the saturated thickness of the aquifer, then

$$\ln \frac{R_e}{R} = \left[\frac{1.1}{\ln(L_w/R)} + \frac{A + B \ln[(h - L_w)/R]}{L_e/r_e} \right]^{-1} \qquad (7\text{–}87)$$

If L_w is equal to h, then

$$\ln \frac{R_e}{R} = \left[\frac{1.1}{\ln(L_w/R)} + \frac{C}{L_e/R} \right]^{-1} \qquad (7\text{–}88)$$

where A, B, and C are dimensionless numbers that can be found from Figure 7.25, where they are plotted as a function of L_e/R.

The value of H_t as a function of t is plotted on semilogarithmic paper, with H_t on the logarithmic axis. The data pairs will fall on a straight line from small

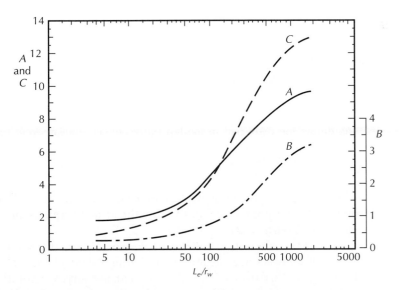

FIGURE 7.25 Dimensionless parameters A, B, and C plotted as a function of L_e/r_w. These parameters are used in the determination of $\ln(R_e/R)$ in Equations 7–87 and 7–88. Source: Herman Bouwer, *Ground Water* 27 (1989): 304–309. Used with permission. © 1989, Ground Water Publishing Company.

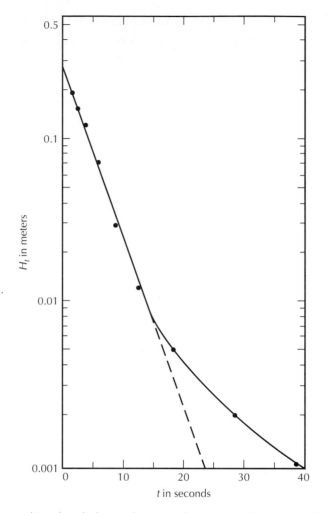

FIGURE 7.26   Head in a borehole as a function of time. Note that the data form on a straight line during the early part of the test but eventually deviate from the straight line. Source: Herman Bouwer, *Ground Water* 27 (1989): 304–309. Used with permission. © 1989, Ground Water Publishing Company.

values of time and large values of head. As the head dissipates and the time increases, the points may not follow the straight line. A plot of H_t versus time is shown in Figure 7.26.

The value of $(1/t)\ln(H_0/H_t)$ may be obtained from a figure such as Figure 7.26. Two points are picked on the straight-line portion of the graph. Although Equation 7–86 calls for H_0 and H_t, we can use any two points on the graph. At one point we have values H_1 and t_1 and at the other point we have H_2 and t_2. Under these conditions $(1/t)\ln(H_0/H_t) = (1/(t_2 - t_1))\ln(H_1/H_2)$.

In some cases the plot of head versus time will yield a curve with two straight-line segments. This case is illustrated in Figure 7.27. Such a situation

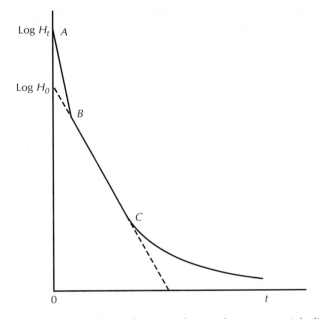

FIGURE 7.27 Head in a borehole as a function of time where two straight-line segments are formed during the early part of the test but eventually deviate from the straight line. Source: Herman Bouwer, *Ground Water* 27 (1989): 304–309. Used with permission. © 1989, Ground Water Publishing Company.

occurs when the water in the gravel pack drains rapidly into the well. Once the water level in the gravel pack equals the water level in the well, then the second straight-line segment forms. This reflects the hydraulic conductivity of the undisturbed aquifer. If a double straight line forms, the second segment should be used to find $(1/t)\ln(H_0/H_t)$. If the water level in the well is not lowered to the level of the gravel pack or if the gravel pack is so permeable that it will drain at the same rate that the water level is lowered, then the double straight-line effect should not develop.

The Bouwer and Rice slug test was developed as a bail-down test; that is, the water level is lowered by bailing or pumping so that water flows from the aquifer into the well. The method can also be used when water is added to the well and heads in the casing fall, provided that the static water level is above the well screen or open borehole. If this is not the case, then water would drain from the well into the vadose zone as well as the saturated aquifer. This would result in an overestimate of the permeability of the saturated aquifer.

Some studies have been performed that compared results of slug-test data analyzed by two or more methods. In one study in highly permeable sands (Fetter 1985), comparable results were obtained by using either the Bouwer and Rice method (Bouwer & Rice 1976) or the Hvorslev method (Hvorslev 1951). Another study, performed on a fractured glacial till (Herzog & Morse 1984), compared the results of the Cooper-Bredehoeft-Papadopulos method (Cooper, Bredehoeft, &

Papadopulos 1967; Papadopulos, Bredehoeft, & Cooper 1973) and the Nguyen and Pinder method (Nguyen & Pinder 1984) and found both methods yielded similar results. Methods may vary in the type of wells for which they are applicable and the length of time over which the data may need to be collected.

7.6 ESTIMATING AQUIFER TRANSMISSIVITY FROM SPECIFIC CAPACITY DATA

There is a large body of data on the specific capacity of water wells. Aquifer tests are usually not performed when production wells are installed. However, the well is normally test pumped for at least a few hours and the yield and maximum drawdown are recorded. The yield of the well divided by the drawdown is called the **specific capacity.** Typical units are ft^3/day/ft of drawdown or m^3/day/m of drawdown.

Theis (1963) proposed a way of estimating the transmissivity of an aquifer from the specific capacity of a well. For a confined aquifer, one can rearrange Equation 7–50 to yield:

$$T = \frac{Q}{(h_0 - h)} \frac{2.3}{4\pi} \log \frac{2.25Tt}{r^2S} \qquad \textbf{(7–89)}$$

where

$Q/(h_0 - h)$ is the specific capacity of the well (L^2/T; ft^3/day/ft or m^3/day/m)

t is the period of pumping (T; day)

r is the radius of the pumping well (L; ft or m)

T is aquifer transmissivity (L^2/T; ft^2/day or m^2/day)

S is aquifer storativity (dimensionless)

Notice that T appears in both the arithmetic and the logarithmic portions of Equation 7–89. In order to solve this equation for T when $Q/(h_0 - h)$ is known, we must make an initial guess of the value of T, substitute it into the equation, and solve for $Q/(h_0 - h)$. The value of T is then adjusted until the calculated value of $Q/(h_0 - h)$ is reasonably close to the measured value.

This procedure has two flaws. First, we must estimate a value for S. This means that the estimate of T is based on an estimate of S—a somewhat dubious situation. Second, the assumption is made that the well is 100% efficient. In the real world in which the specific capacity is measured, the well will not be 100% efficient; that is, drawdown within the well will be greater than the drawdown in the aquifer just outside the well. This is due to turbulent friction losses as the water passes into the well.

Bradbury and Rothschild (1985) developed a computer program to estimate T from specific-capacity data. They utilized a method of correcting the observed specific capacity for well losses and introduced a partial penetration correction factor to Equation 7–89. They found a reasonable agreement between T determined by aquifer tests and T estimated from specific-capacity data using this method on 5 wells in fractured dolomite and 11 wells in sandy outwash.

Razack and Huntley (1991) utilized a different approach in studying the relationship between T and specific capacity in an alluvial ground-water basin in Morocco. They analyzed 215 data pairs where transmissivity and specific capacity were known for a well. Application of Equation 7–89 generally underpredicted the actual specific capacity due to the turbulent well losses. They were able to find an empirical relationship between the two parameters, which had a correlation coefficient of 0.63. This relationship can be expressed as

$$T = 15.3\left(\frac{Q}{h_0 - h}\right)^{0.67} \tag{7–90a}$$

where

$\quad T \qquad$ is transmissivity (m^2/day)

$\quad Q \qquad$ is pumping rate (m^3/day)

$\quad h_0 - h$ is drawdown (m)

or

$$T = 33.6\left(\frac{Q}{h_0 - h}\right)^{0.67} \tag{7–90b}$$

where

$\quad T \qquad$ is transmissivity (ft^2/day)

$\quad Q \qquad$ is pumping rate (ft^3/day)

$\quad h_0 - h$ is drawdown (ft)

Well efficiency can be estimated for wells where an aquifer test has been performed so that values of T and S are known. The theoretical drawdown at a distance equal to the radius of the pumping well would be the water level in a well that is 100% efficient. This value can be compared to the drawdown that is measured inside the well to find the well efficiency.

7.7   INTERSECTING PUMPING CONES AND WELL INTERFERENCE

The cases we have considered thus far have involved only one well pumping from an aquifer. However, there are often several wells tapping the same aquifer, resulting in intersecting pumping cones. At any given point in a confined aquifer, the total drawdown is the sum of the individual drawdowns for each well. Because the Laplace equation is linear, the superposition of drawdown effects is found by

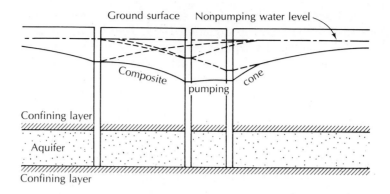
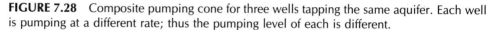

FIGURE 7.28 Composite pumping cone for three wells tapping the same aquifer. Each well is pumping at a different rate; thus the pumping level of each is different.

simple addition. In Figure 7.28, the well interference for a multiple-aquifer well field is presented graphically. Linear superposition is valid only for confined aquifers, in which the value of the transmissivity does not change with drawdown. In water-table aquifers, if the drawdown is significant in relation to the total saturated thickness, the use of linear superposition will result in a predicted composite drawdown that is less than the actual composite drawdown. As a decrease in saturated thickness reduces the transmissivity, the multiple-well system will result in a composite hydraulic gradient greater than that of an equivalent confined system in order to compensate for a reduced value of aquifer transmissivity.

In designing well-field layouts, it is necessary to take into account well interference. The level of the water in the well during pumping determines the length of pipe necessary to carry water to the surface. The characteristics of the well pump and the horsepower requirements of the motor also depend upon the depth to the pumping level. If wells are spaced too closely together, the amount of well interference could be excessive. Aligning wells parallel to a line source of recharge, such as a river, would result in less well interference than would a perpendicular configuration.

7.8   EFFECT OF HYDROGEOLOGIC BOUNDARIES

If a well is not located in an aquifer of infinite areal extent, as is the case with all real wells in real aquifers, the drawdown cone will extend until either the well is supplied by vertical recharge or a hydrogeologic boundary is reached. A hydrogeologic boundary could be the edge of the aquifer, a region of recharge to a fully confined artesian aquifer, or a source of recharge, such as a stream or lake.

Boundaries are considered to be either recharge or barrier boundaries. A **recharge boundary** is a region in which the aquifer is replenished. A **barrier boundary** is an edge of the aquifer, where it terminates, either by thinning or abutting a low-permeability formation, or has been eroded away.

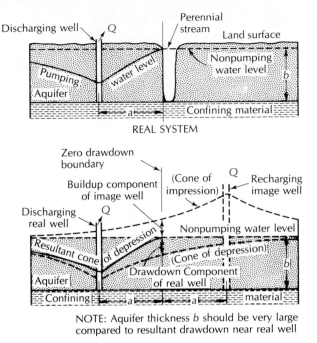

REAL SYSTEM

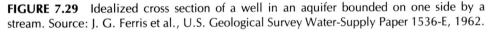

NOTE: Aquifer thickness b should be very large
compared to resultant drawdown near real well

HYDRAULIC COUNTERPART OF REAL SYSTEM

FIGURE 7.29 Idealized cross section of a well in an aquifer bounded on one side by a stream. Source: J. G. Ferris et al., U.S. Geological Survey Water-Supply Paper 1536-E, 1962.

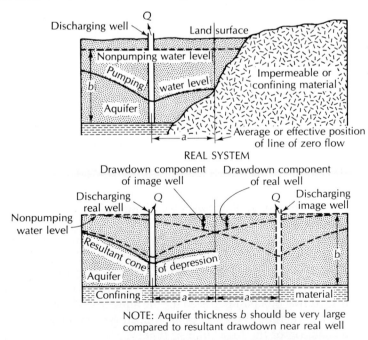

NOTE: Aquifer thickness b should be very large
compared to resultant drawdown near real well

HYDRAULIC COUNTERPART OF REAL SYSTEM

FIGURE 7.30 Idealized cross section of a well in an aquifer bounded on one side by an impermeable boundary. Source: J. G. Ferris et al., U.S. Geological Survey Water-Supply Paper 1536-E, 1962.

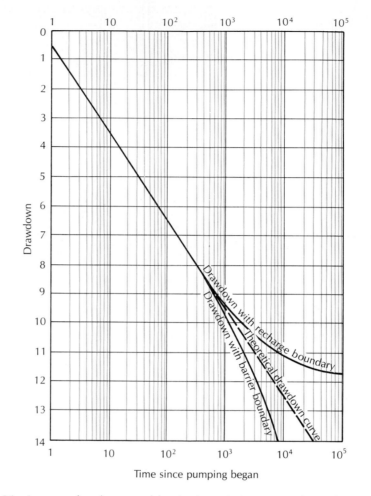

FIGURE 7.31 Impact of recharge and barrier boundaries on semilogarithmic drawdown-time curves.

Figure 7.29 shows a well bounded by a recharge boundary. The recharge boundary can be simulated by a recharging **image well** located an equivalent distance away from the recharge boundary but on the opposite side. Figure 7.30 indicates the presence of a barrier boundary. The barrier boundary is simulated by a discharging image well located an equivalent distance away from the boundary but on the opposite side. Boundaries have the most dramatic impact on the drawdown of a pumped well for the aquifer with no source of vertical recharge. As the well withdraws water only from storage in the aquifer, drawdown proceeds as a function of the logarithm of time.

Figure 7.31 shows a theoretical straight-line plot of drawdown as a function of time on semilogarithmic paper. The effect of a recharge boundary is to retard the rate of drawdown. Change in drawdown can become zero if the well comes to be supplied entirely with recharged water. The effect of a barrier to flow in some region of the aquifer is to accelerate the drawdown rate. The water level declines faster than the theoretical straight line.

7.9 AQUIFER-TEST DESIGN

Adequate design and execution of an aquifer test involves considerable planning and attention to detail. An understanding of fundamental well hydraulics is necessary, not only for the interpretation of data, but also for the experimental design by which valid and usable data are obtained. The purpose of the pumping test must be established first. Determining the yield of a new well involves simply pumping the well. This type of test, as it is generally conducted, yields only the scantiest information about the aquifer itself. With careful planning, the pumping-well test can yield data to compute the aquifer transmissivity. It can also indicate the general type of aquifer.

If a test well has been drilled prior to the installation of a production well, a reasonable conjecture can be made as to the probability the well will be unconfined, semiconfined, or confined. However, the presence or absence of recharge or barrier boundaries may not be known. Indeed, this is one of the reasons to perform a long-term pumping test. If one makes a semilogarithmic plot of drawdown versus time (e.g., Figure 7.31) one can inspect it to see if the pumping level of the well stabilizes. If this occurs, this means that there is a source of recharge, either vertically by leakage across a semiconfining layer or horizontally from a recharge boundary. If the water level falls faster than the theoretical drawdown curve, then the presence of a barrier boundary must be considered.

The amount of information gained from an aquifer test expands greatly if one or more observation wells are involved in addition to the pumping well. Both transmissivity and storativity of the aquifer can be determined, as can the vertical hydraulic conductivity of any overlying semipervious layers. More eloquent tests can be used to determine the value of the vertical anisotropy of the formation. Radial anisotropy and recharge or barrier boundaries can also be detected.

7.9.1 Single-Well Aquifer Tests

The basics of a single-well aquifer test are also applicable to aquifer tests involving multiple wells. The first step is to determine the location of the well to be drilled. This is best done on the basis of detailed exploration using geological, geophysical, and perhaps aerial photo techniques. However, the location of the well is often dictated by economic or engineering factors. If economic or engineering factors predominate, the hydrogeologist should determine if there is a reasonable chance of success based on the known hydrogeology of the site.

A test well may be bored as the first step, or the production well may be drilled immediately. The geologist should make a log of the geologic formations encountered. The water level in the drilled hole should be recorded as a function of the depth of the hole; however, this might not be possible if certain drilling techniques such as rotary and reverse rotary are used. Based on the test hole and selected borehole geophysical studies, the hydrogeologist can determine the depth and thickness of potential aquifer zones. An aquifer is selected, and a test or permanent well is installed. If at all feasible, the well should be open throughout the entire thickness of the aquifer. The physical dimensions of the well should be recorded, along with the depth, thickness, and type of aquifer. A description of the aquifer material should be included. An inventory of nearby wells should be made, and it should be determined whether any other wells will be running when an aquifer test is planned. Intersecting cones of depression during an aquifer test should be avoided.

A pump is installed in the completed well. The pump, engine, wiring, piping, and assorted equipment must be reliable. If a pumping test is terminated owing to mechanical failure prior to the planned time, the data may not be sufficient. The discharge pipe from the pump must be equipped with a valve to control the volume of flow and with a means of measuring the flow.

Small pumping rates may be measured by means of a water meter in the line or by filling a container of known volume in a measured time. These methods are generally useful for flows of 100 gal/min (6 L/s) or less. For larger flows, a common method is to use a circular-orifice weir on the discharge pipe (Johnson Division, Universal Oil Products Co. 1966). Generally, the well-drilling contractor furnishes the equipment and makes the discharge measurements. The hydrogeologist should always check the apparatus and measurements.

There must be a means of conveying the pumped water away from the test site. This is especially true for shallow, unconfined aquifers, where the water could recharge the aquifer and render any pumping-test results useless. If a pumping test runs for several days, the quantity of pumped water that must be conveyed from the site can be considerable.

A means to measure the water level in the well before as well as during pumping must be available. A steel tape, air line, or electrical tape can serve this function (Johnson Division, Universal Oil Products Co. 1966).

The aquifer tests we have studied all have been based on a constant discharge rate. In reality, the water level in the well falls with pumping, the pumping lift increases, and the discharge of the pump tends to decline. To avoid this, the valve on the discharge pipe should be partially closed to restrict the initial discharge. During the course of the test, the valve can be opened as necessary to keep the pumping rate constant. There should be no more than a 10 percent variation in rate during the test, with a smaller pumping-rate variation if possible. After construction, a new well is usually developed by on-and-off pumping, which causes the water to surge back and forth through the well screen or open hole. This increases the yield of the well by washing out fine particles and mud used for drilling. The well should be fully developed before pumping tests are made.

It is necessary to select a measuring point on the well to serve as a fixed reference for water-level measurements. The measuring point should be marked, and its elevation measured and recorded. Prior to the pumping test, the water level should be measured. Other production wells that may be nearby should either be shut down for the duration of the test or pumped at a constant rate. It is difficult to correct pumping-test data for the effect of wells starting and stopping. For the most accurate results, the nonpumping water elevation should be measured several times before pumping begins. Ground-water levels may have a long-term trend of rising or falling. They may also be affected by tides or changes in the barometric pressure. If the static level is found to fluctuate, then detailed pretest measurements must be made for at least twice the expected length of the pumping test. If a long-term linear trend is observed, the drawdown observed during the pumping test must be corrected. The corrected drawdown is the difference between the measured depth to water and the projected static level based on the long-term trend. Tidal fluctuations and fluctuations due to barometric pressure require measurements of the water level in the well and either the tide or air pressure prior to pumping, in order to establish a relationship between them. Measurements of tides or air pressure must then be made during the pumping test to find a corrected static level at the time of each drawdown measurement.

Prior to the test, the discharge of the well should be measured and a planned pumping rate selected. The valve should be preset for this rate. When the pump is started for the test, the valve is adjusted to yield the desired pumping rate. Periodic measurements of discharge and corrections using the valve should be made about every half hour. The time and measurement of pumping rate should be recorded in the field notes.

Water-level readings in the well commence after one minute of pumping. The usual procedure is to record the depth of water below the measuring point and the time of measurement. Computation of drawdown is made later. On the order of ten readings per log cycle of time are made. The first reading could be at 1 min, then at 1½ min, and then at every minute up to 10. Between 10 and 20 min, readings are taken every 2 min; then every 5 min between 20 min and 1 h; and every 10 min between 1 and 2 h. After 2 h have elapsed, the recommended 10 readings per log cycle of time are made. The most work occurs during the first minutes of the pumping test.

There must be some advance planning for the length of time the test is to proceed. For an artesian or leaky artesian aquifer, the test may last 24 h or less. This is often sufficient to delineate values for the formation constants and to determine whether there are any recharge or barrier boundaries. Water-table wells must be pumped for a sufficient duration to preclude any significant effects of vertical flow near the well. The time period is a function of the distance from the pumping well, the conductivity of the formation, and the degree of anisotropy. The time is naturally greater the farther the distance from the pumping well, the lower the horizontal conductivity, and the greater the anisotropy. Pumping tests of unconfined aquifers normally run for several days to several weeks.

Periodically during the pumping test, a sample of the water is collected for chemical analysis. A series of samples will reveal any trend in chemical or bacteriological quality with continued pumping. It is possible to predict final well quality fairly accurately, even on the basis of chemical analysis of water from temporary test wells (Fetter 1975).

Following the collection of time and drawdown data in the pumped well, an analysis is made using the appropriate type curves. It should be remembered that with only one well, the valve of T, but not of S, can be determined. In another computation usually made at the end of the pumping test, the well yield is divided by the maximum drawdown to obtain the specific capacity, which is widely used as an index of the capacity of the well.

7.9.2   Aquifer Tests with Observation Wells

If it is feasible, drawdown should be measured in one or more nonpumping wells situated close to the pumping well. These observation wells are often constructed especially for the pumping tests. However, domestic-supply wells, abandoned wells, other wells in a well field, or other wells under construction are sometimes used. The use of one or more observation wells can enable the hydrogeologist to compute the storativity or specific yield of the aquifer. Under some circumstances, the anisotropy of the pumping well and the leakage factor for leaky, confined aquifers may be determined. In most cases, the hydrogeologist will be able to employ only one observation well—especially if the aquifer is deep and the area is undeveloped, with no existing wells available.

The selection of the location of the single observation well is critical. It should be located at a radial distance such that the time-drawdown data collected during the planned pumping period will fall on a type curve of unique curvature. If the curvature is too flat, selection of the proper type curve is difficult. After a test boring is made, the hydrogeologist should know the type of aquifer system to be confronted. The formation characteristics are then estimated, and using the correct formula with the planned pumping rate, time-drawdown curves for several hypothetical observation wells at different distances are plotted. On the basis of these curves, the location of the single observation well is selected. As a general rule, it will be closer for a water-table well than for a confined well.

If there are two observation wells, the second should be in a radial line with the first, but at ten times the distance. If there are more than two wells, they should form two or more radial lines from the pumping well. This will indicate any radial anisotropy in the aquifer. A map should be carefully made in the field showing the relative locations of the pumping well and all observation wells. The distance from the pumping well to each observation well should be measured with a steel tape.

Ideally, observation wells should fully penetrate the aquifer, so that they measure the average head in the formation at that location.* Plastic (PVC) pipe with slots cut out in the zone of the aquifer serves as a suitable observation well. The annular space between the drilled hole and the plastic pipe can be backfilled with coarse sand. If the well goes through a confining layer, the hole must be backfilled with clay and tamped to prevent leakage around the pipe.

---

*Observation wells should also not be screened in aquifers other than the one in which the pumping well is also screened. The author has seen cases where very expensive pumping tests yielded meaningless results because either the pumping well or the observation well was screened in more than one aquifer. Time-drawdown data can be obtained from such misbegotten wells, and misguided hydrogeologists have attempted to evaluate them. The results, however, are garbage.

If the observation well has a short well screen, then the screen should be placed such that the head it measures is representative of the average head in the formation at that location. For a confined aquifer, it should be at the middepth of the stratum. For a water-table aquifer, it should be one-third of the distance from the static water table to the bottom of the aquifer.

The observation well should have a rapid response to changes in the water level in the aquifer. One way to test response time is to pour water into the observation well. The induced head should drain away in a fairly short time, usually in a few hours or less in most aquifers. If the observation well does not show a good response, an effort should be made to unclog it. Pouring water in the well may clear it. If the water level is within a few meters of the surface, a plunger on a stick can be used to surge the well to clear it.

Prior to the pumping test, a measuring point should be chosen for each observation well. Usually this is the top of the casing. The elevation of each measuring point should be determined. The depth to the static water level should also be measured prior to pumping. This will be useful in mapping the potentiometric surface.

Depth-to-water measurements are made in the observation wells on the same schedule as the pumping-well measurements. For the first minutes, this probably will necessitate one observer for each well. After 20 min, when the readings go to a frequency of every 10 min, fewer people will be needed. After 2½ h, readings are made only every 100 min, so that one observer usually can do everything. Also, at the start of the test, one or two additional people should be on hand to measure pump discharge and adjust the flow.

After the end of the scheduled pumping test, recovery measurements can be made in the wells. The water levels will recover at the same rate they fall. In some cases, the drawdown data are affected by uncontrolled variations in the pumping rate. This does not affect the recovery rate. The flow rate for recovery data is equal to the mean discharge for the entire pumping period. In using recovery data, the difference between the water level at the end of pumping and that after a given time since pumping stopped is plotted as a function of the time since pumping stopped. The standard methods of well-test analysis are used. Recovery measurements are a standard part of the aquifer test. In many cases, the recovery data prove to be more useful than the drawdown data, for example, if there was a short period when the well shut down during the test or if the rate of pumping was extremely variable during the drawdown phase of the test.

Butler (1990) observed that in analysis of aquifer-test data from confined aquifers, the hydrogeologist will usually analyze the data by either or both the Theis curve-matching method and the Cooper-Jacob straight-line approximation. If the aquifer is nonuniform, as real aquifers typically are, these two approaches can yield differing results. The Theis curve-matching method is influenced heavily by the hydraulic properties of the aquifer in close proximity to the pumping well. On the other hand, the Cooper-Jacob approximation emphasizes the hydraulic properties of the aquifer near the edge of the cone of depression, which may be far from the pumping well.

The difference, if any, in the results of the two methods is a function of the degree of nonuniformity of the aquifer and the distance from the pumping well to the observation wells. The further the distance of the observation well is from the pumping well, the smaller the difference in results that is observed for a given degree of nonuniformity.

If we are using an aquifer test to predict the long-term drawdown of a pumping well, then the most appropriate method is the Theis curve-matching technique because it reflects conditions near the well. However, if we are using an aquifer test to predict aquifer yield, then the Cooper-Jacob method should be used, since this tests the properties of the aquifer over a greater distance from the pumping well.

COMPUTER NOTES

There are several programs on the market that enable us easily to find hydraulic parameters for an aquifer from aquifer-test data. One of the more popular programs is AQTESOLV, which is an interactive, menu-driven program that can solve for the following types of aquifers:

- ☐ Confined aquifers, using the Theis and Cooper-Jacob nonequilibrium methods.
- ☐ Leaky, confined aquifers, using Hantush's nonequilibrium methods with or without storage in the leaky confining layers.
- ☐ Unconfined aquifers, using Neuman's nonequilibrium method.
- ☐ Slug tests on unconfined aquifers, using the Bouwer and Rice method.
- ☐ Slug tests on confined aquifers, using the Cooper, Bredehoeft, and Papadopolus method.
- ☐ Aquifer tests in fractured rock.
- ☐ Large-diameter well solutions for confined and leaky, confined aquifers.
- ☐ Water-level recovery data after an aquifer test.

AQTESOLV is menu-driven, with pull-down menus that are accessed by moving the highlight bar with the arrow keys and then pressing the enter key when the desired menu function is highlighted.

In order to enter a new data set and analyze it with AQTESOLV, we can follow these steps:

1. Access the **Data Set Manager** from the program control menu.
2. Access **Create new data set** from the Data Set Manager menu.
3. When the programs asks, Do you wish to initialize the data set?, type Y and press the Enter key.

4. The **Input Options** menu will appear. Press the Enter key to select the **title for the data set** option. Type in a title and then press the Enter key.

5. Move the highlight bar to the type of aquifer that is being tested and press Return. You can now enter data. You must use consistent units for data. For example, select either meters or feet and minutes. Pumping rates are in ft^3/min or m^3/min, distance and drawdown are in m or ft, and time is in min. Computed Computed transmissivities are in m^2/min or ft^2/min. To convert these computed values to m^2/day or ft^2/day, multiply by 1440 min/day.

Enter the pumping rate, the distance to the observation well, the radius of the casing, and the effective radius of the well, which is the radius of the gravel pack if it is a gravel-packed well, or the radius of the well screen if it is a naturally developed well. Use the Up- and Down-arrow keys to move from one data entry box to another and press the Return key when all data are entered.

6. Move the highlight bar to the "Observation well measurement" line and press the Enter key. You will wish to enter the data from the keyboard, so press the Enter key with that command highlighted.

7. Enter the time (in min) and the drawdown (in ft or m). Leave the value 1 in the column marked "weight." After each data entry, move the highlight box with the Arrow keys. When all the data are entered, press the Esc key three times. At this point you have the option of saving the data set on a disk. It is not necessary to do this. Press N for no and then press the Enter key. You will now have the option of saving the output file on a disk. This is also not necessary. Press N and then the Enter key.

8. Press the Esc key to leave the Data Set Manager menu. Move the highlight bar on the Program Control menu to the type of solution desired and press the Enter key. You now have the choice of a solution methodology. Select that method with the highlight bar and press the Enter key. The Parameter Estimation Control menu will appear. There are a number of default values that the program automatically computes; although these can be changed, you don't need to do so. Move the highlight bar to Graph Data—Match Curve and press Enter to access the Plot Options menu. Now select Plot Graph and press the Enter key. A time-drawdown plot will appear on the screen, along with a type curve that the program has matched to the data. Computed values of T and S are given. If you wish, you can move the type curve with the Arrow keys to see if you can make a better fit than the program. As the type curve is moved, the values of T and S are updated.

9. If you wish to see if a different type of analysis would give a better match, press the Esc key several times to get to the Solution Type menu and select a different method of solution. Then proceed as before.

There is a free student version of AQTESOLV on the computer disk that accompanies this book. This version of the program uses the Theis solution for confined aquifers; the Hantush-Jacob solution for leaky, confined aquifers; and the Bouwer and Rice method for slug tests. You might try to use this program to check your answers to some of the programs in this chapter.

If you wish to obtain additional information about this program or to purchase the fully configured model, contact:

Geraghty & Miller, Inc.
Modeling Group
10700 Parkridge Boulevard, Suite 600
Reston, VA 22091
(703) 758-1200

NOTATION

b	Thickness of confined aquifer	r	Radial distance to pumping well
b'	Thickness of aquitard	r_c	Radius of a piezometer casing
b''	Saturated thickness of water-table aquifer	r_0	Intercept of distance drawdown line with zero drawdown axis
B	$(Tb'/K)^{1/2}$ (leakage factor)	r_s	Radius of a piezometer screen
e	Rate of vertical leakage	r_w	Radius of pumping well
h	Hydraulic head	R	Radius of screen or gravel pack
h_0	Head prior to pumping	R_e	Radial distance over which slug test head is dissipated
$h_0 - h$	Drawdown		
$\Delta(h_0 - h)$	Drawdown per log cycle of time	S	Storativity
H_0	Water level at start of a slug test	S'	Storativity of the confining layer
H	Water level during a slug test	S_s	Specific storage of confining layer
K	Hydraulic conductivity	S_y	Specific yield
K'	Hydraulic conductivity of aquitard	t	Time
K''	Hydraulic conductivity of water-table aquifer	t_0	Intercept of time-drawdown line with zero drawdown axis
$K_0(x)$	Zero-order modified Bessel function of the second kind	T	Transmissivity
		T_0	Time that it takes for water level to fall to 37 percent of original value
K_r	Radial hydraulic conductivity	u	$(r^2S)/(4Tt)$
K_v	Vertical hydraulic conductivity		
L_e	Effective length of a well screen	$W(u)$	Well function of u
Q	Pumping rate	z	Elevation
q_L	Rate that water leaks across an aquitard into a leaky, confined aquifer	β	$(r/4B)((S'/S)^{1/2})$
		ν	$(K'/b')(S'/S^2)$
q_s	Rate at which water is withdrawn from elastic storage in a leaky, confined aquifer	η	Tt/r_c^2
		μ	r_s^2S/r_c^2

PROBLEMS

Answers to odd-numbered problems will appear at the end of the book.

Type curves will be necessary for the solution of many of these problems. Type curves can be constructed from the data in the appendices, although this is laborious. Type curves have been published for a number of aquifer tests on confined aquifers. The curves were derived by, among others, the Theis method, the two methods for leaky artesian aquifers given in this chapter, and the Cooper-Bredehoeft-Papadopulos method. (J. E. Reed, "Type Curves for Selected Problems of Flow to Wells in Confined Aquifers," in *Techniques of Water-Resources Investigations of the United States Geological Survey,* Book 3, Chapter B3, 1980. This is available from the U.S. Government Printing Office, Washington, D.C., or Scientific Publications Company, P.O. Box 23041, Washington, D.C. 20026-3041.)

The published type curves use 3×5 cycle logarithmic graph paper with 1.85 in. for each log cycle, such as Keuffel and Esser Co. 46 7522, and semilogarithmic graph paper with 2.00 in. per log cycle, such as Keuffel and Esser Co. 46 6213. The Jacob straight-line methods will require 4-cycle semilogarithmic paper such as Keuffel and Esser Co. 46 6013.

Instructors may request a copy of the solution manual for *Applied Hydrogeology* from their Macmillan representative. Type curves for these problems are contained therein.

1. A community is installing a new well in a regionally confined aquifer with a transmissivity of 2675 ft²/day and a storativity of 0.0002. The planned pumping rate is 750 gal/min. There are several nearby wells tapping the same aquifer, and the hydrogeologist in charge needs to know if the new well will cause significant interference with these wells. Compute the theoretical drawdown caused by the new well after 30 days of continuous pumping at the following distances: 50, 150, 250, 500, 1000, 3000, 6000, and 10,000 ft.

2. A well that is screened in a confined aquifer is to be pumped at a rate of 125,000 ft³/day for 30 days. If the aquifer transmissivity is 4675 ft²/day, and the storativity is 0.005, what is the drawdown at distances of 50, 150, 250, 500, 1000, 3000, 5000, and 10,000 ft?

3. A. Plot the distance-drawdown data from problem 1 on semilog paper.
   B. If the pumping well from problem 1 has a radius of 1 ft, and the observed drawdown in the pumping well is 122 ft, what is the efficiency of the well?

4. Plot the distance-drawdown data from problem 2 on semilog paper. If the pumping well has a radius of 1 ft, and the observed drawdown in the pumping well is 48 ft, what is the efficiency of the well?

5. If the aquifer in problem 1 is not fully confined, but is overlain by a 10-ft-thick confining layer with a vertical hydraulic conductivity of 0.16 ft/day and no storativity, what would the drawdown values be after 30 days of pumping at 750 gal/min at the indicated distances?

6. If the aquifer described in problem 2 is not fully confined, but is overlain by a 5.0-ft-thick leaky, confining layer with a vertical hydraulic conductivity of 0.019 ft^2/day, what would the drawdown values be after 30 days of pumping at 125,000 ft^3/day at the indicated distances?

7. With reference to the well and aquifer system in problem 1, compute the drawdown at a distance of 250 ft at the following times: 1, 2, 5, 10, 15, 30, and 60 min; 2, 5, and 12 h; and 1, 5, 10, 20, and 30 days.

8. With reference to the well and aquifer system in problem 6, compute the drawdown at a distance of 100 ft from the well at the following times: 1, 2, 5, 10, 15, 30, and 60 min; 2, 5, and 12 h; and 1, 5, 10, 20, and 30 days.

9. Plot the time-drawdown data from problem 7 on semilog paper.

10. Plot the time-drawdown data from problem 8 on semilog paper. How is this plot different from that of problem 9?

11. The following data are from a pumping test where a well was pumped at a rate of 200 gallons per minute. Drawdown as shown below was measured in an observation well 250 ft away from the pumped well. The geologist's log of the well is

0–23 ft	Glacial till, brown, clayey
23–77 ft	Dolomite, fractured
77–182 ft	Shale, black, dense
182–217 ft	Sandstone, well-cemented, coarse
217–221 ft	Shale, gray, limy

Elapsed Time (min)	Drawdown (ft)
0	0.00
1	0.66
1.5	0.87
2.0	0.99
2.5	1.11
3.0	1.21
4.0	1.36
5	1.49
6	1.59
8	1.75
10	1.86
12	1.97
14	2.08
18	2.20
24	2.36
30	2.49
40	2.65
50	2.78
60	2.88

Elapsed Time (min)	Drawdown (ft)
80	3.04
100	3.16
120	3.28
150	3.42
180	3.51
210	3.61
240	3.67

A steel well casing was cemented to a depth of 182 ft and the well was extended as an open boring past that point.

A. Plot the time-drawdown data on 3 × 5 cycle logarithmic paper. Use the Theis type curve to find the aquifer transmissivity and storativity. Compute the average hydraulic conductivity.

B. Replot the data on 4-cycle semilogarithmic paper. Use the Jacob straight-line method to find the aquifer transmissivity and storativity.

12. A test well was drilled to a total depth of 117 ft with the following geologist's log:

0–73 ft	Coarse sand
73–82 ft	Clayey sand
82–117 ft	Coarse sand
117 ft	Crystalline bedrock

The depth to water was 55 ft. The test well was screened from 82 to 117 ft. It was pumped at a rate of 560 gal/min. Drawdown was measured in an observation well that was also screened from 82 to 117 ft and was located 82 ft away from the pumping well. The following time-drawdown data were obtained:

Elapsed Time (min)	Drawdown (ft)
0	0.00
1	0.90
2	2.15
3	3.05
4	3.64
5	4.07
6	4.52
7	4.74
8	5.02
9	5.21
10	5.53

Elapsed Time (min)	Drawdown (ft)
15	5.72
20	5.97
30	6.12
40	6.20
50	6.25
60	6.27
90	6.29
120	6.29

A. Plot the time-drawdown data on 3×5 cycle logarithmic paper. Compute the value of the storativity and transmissivity of the aquifer using the graphical method for leaky aquifers. Find the vertical hydraulic conductivity of the confining layer.

B. Compute the value of aquifer storativity and transmissivity of the aquifer using the Hantush inflection-point method.

13. A slug test was made with a piezometer that had a casing radius of 2.54 cm and a screen of radius 2.54 cm. A slug of 4000 cm³ of water was injected; this raised the water level by 197.3 cm. The well completely penetrated a confined stratum that was 2.3 m thick. The decline in head with time is given in the following chart:

Elapsed Time (s)	Head (cm)
0	197.3
1	185.4
2	178.6
3	173.6
5	167.7
7	158.8
10	147.0
13	140.0
17	129.2
22	118.4
32	99.6
53	74.0
84	51.3
119	35.5
170	23.3
245	15.2
400	8.7
800	4.3

Plot the data on semilogarithmic paper and find the aquifer transmissivity and storativity using the Cooper-Bredehoeft-Papadopulos method.

14. A well in a water-table aquifer was pumped at a rate of 873 m^3/day. Drawdown was measured in a fully penetrating observation well located 90 m away. The following data were obtained (Kruseman & deRitter 1991):

Time (min)	Drawdown (m)	Time (min)	Drawdown (m)	Time (min)	Drawdown (m)
0	0	18	0.098	300	0.173
1.17	0.004	21	0.103	370	0.173
1.34	0.009	26	0.110	430	0.179
1.7	0.015	31	0.115	485	0.183
2.5	0.030	41	0.128	665	0.182
4.0	0.047	51	0.133	1340	0.200
5.0	0.054	65	0.141	1490	0.203
6.0	0.061	85	0.146	1520	0.204
7.5	0.068	115	0.161		
9.0	0.064	175	0.161		
14	0.090	260	0.172		

Find the transmissivity, storativity, and specific yield of the aquifer.

15. A well that pumps at a constant rate of 50,000 ft^3/day has achieved equilibrium so that there is no change in the drawdown with time. (The cone of depression has expanded to include a recharge zone equal to the amount of water being pumped.) The well taps a confined aquifer that is 24 ft thick. An observation well 125 ft away has a head of 273 ft above sea level; another observation well 315 ft away has a head of 291 ft. Compute the value of aquifer transmissivity using the Theim equation.

8 Regional Ground-Water Flow

Again, most springs are in the neighborhood of mountains and high grounds, whereas if we except rivers, water rarely appears in the plains. For mountains and high ground suspended over the country like a saturated sponge, make the water ooze out and trickle together in minute quantities but in many places. They also receive a great deal of water falling as rain.

Meteorologica, Aristotle (384−322 B.C.)

8.1 INTRODUCTION

In the zone of actively flowing ground water, the water moves through the porous media under the influence of the fluid potential. This movement is a three-dimensional phenomenon, yet we are usually forced to represent it on a two-dimensional medium. In the diagrams in this chapter, the reader will have to imagine the implied third dimension. We will start by examining steady flow through isotropic, homogeneous media and then include the effects of nonhomogeneity and anisotropy.

Flow nets will be used to illustrate the various regional flow patterns. These are a means of portraying the solution to the Laplace equation (5–43), which governs steady flow. The various solutions will represent differing conditions of hydraulic conductivity and aquifer geometry. This type of flow net is constructed by drawing flow lines on a potential field. The potential fields are solutions to a mathematical model of the aquifer systems. Laplace's equation was solved either analytically (Tóth 1962, 1963) or numerically (Freeze & Witherspoon 1966, 1967) with different boundary conditions. One of the most critical boundary conditions is the shape of the water table or potentiometric surface. The flow lines are drawn to illustrate some of the possible flow paths.

8.2 STEADY REGIONAL GROUND-WATER FLOW IN UNCONFINED AQUIFERS

8.2.1 Recharge and Discharge Areas

In unconfined aquifers, some characteristics are common to most **recharge areas;** likewise, most **discharge areas** have some common denominators. Recharge areas are usually in topographical high places; discharge areas are located in topographic lows. In the recharge areas, there is often a rather deep unsaturated zone between the water table and the land surface. Conversely, the water table is found either close to or at the land surface in discharge areas.

Flow lines on a flow net tend to diverge from recharge areas and converge toward discharge areas. This convergence will not occur if the discharge zone is large, such as a coastline. A water-table contour map can often be used to locate ground-water recharge and discharge areas. Flow lines can be drawn on the basis of ground-water contours, crossing them at right angles if the aquifer is isotropic.

In the field, vegetation and surface water can sometimes be used to locate discharge areas. There may be some physical manifestation of the discharging ground water, which can take the form of a spring, seep, lake, or stream. The presence of vegetation common to wet soils may be indicative of discharge areas. In arid regions, ground water may be discharged as evaporation or transpiration. In such cases, a thicker-than-normal cover of vegetation or a salt deposit may indicate a discharge area.

The aforementioned physical manifestations sometimes betoken a ground-water discharge area—but not always. In nonhomogeneous materials, a low-permeability layer may, for example, form a perched aquifer, which could result in a wetland or pond. A thorough evaluation of all hydrogeologic information should always be made.

Many field studies conducted in humid regions note that the water table in unconfined aquifers usually has the same general shape as the surface topography. This is not surprising, since recharge taking place in topographical high areas has a greater potential energy than recharge in lower areas. This greater energy is reflected in the higher elevations of the water table in those locations. This generalization may not be true in arid regions.

8.2.2 Ground-Water Flow Patterns in Homogeneous Aquifers

A descriptive model of regional steady-state ground-water flow in an unconfined aquifer was first published by M. K. Hubbert in his pioneering paper, "The Theory of Ground-Water Motion" (Hubbert 1940). Figure 8.1 was first published in that paper. This figure is a cross section of a homogeneous aquifer with the water table rising in a hill between two valleys. Equipotential lines are shown as dashed lines, and flow lines are solid. The diagram does not extend to the bottom of the aquifer, so the flow lines continue below the bottom of Figure 8.1.

Hubbert showed the equipotential lines extending horizontally above the water table, reflecting the elevation head. Below the water table the equipotential lines are curvilinear, reflecting the sum of the elevation and pressure heads.

Figure 8.1 shows that below the hill, where the water table is high, ground-water flow is downward, into the aquifer. This is a recharge zone. However, ground-water flow is upward beneath the valleys, indicating that discharge zones are present there. The valley bottoms are concentrated areas of ground-water discharge into the streams, with the flow lines converging toward

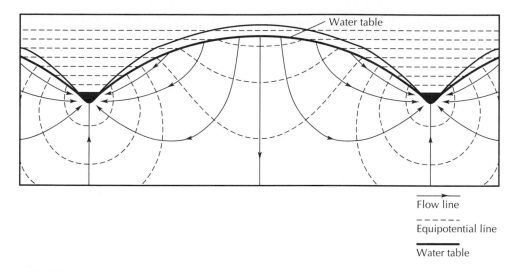

FIGURE 8.1 Cross-sectional flow net in an isotropic, homogeneous aquifer. The aquifer is much deeper than the diagram. Source: M. K. Hubbert, *Journal of Geology* 48, no. 8 (1940): 795–944. Used with permission of the University of Chicago Press.

them. The crest of the water table is a ground-water divide, with flow on either side going in opposite directions.

The water level in a piezometer will extend to the elevation represented by the equipotential at the bottom end. Figure 8.2 shows several piezometers superimposed on the flow net of Figure 8.1. Head values have also been assigned to the equipotential lines on Figure 8.2.

Piezometers C, E, and F all end on the 50-m equipotential line. The water level in each piezometer will rise to the same elevation, which represents a head

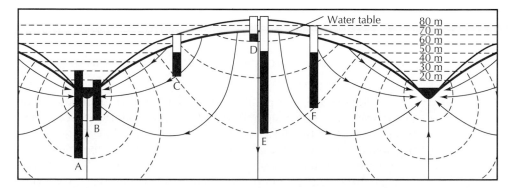

FIGURE 8.2 Piezometers superimposed on Figure 8.1. The water level in the piezometer will rise to the elevation of the hydraulic head, which is represented by the equipotential line at the open end of the piezometer. Source: Modified from M. K. Hubbert, *Journal of Geology* 48, no. 8 (1940): 795–944. Used with permission of the University of Chicago Press.

of 50 m above a datum. Notice that the length of the water column in each piezometer is different, because the 50-m equipotential line is at a different depth at each location.

Piezometers A and B are in essentially the same location, with A being deeper than B. Piezometer A has a head of 30 m, whereas piezometer B has a head of 20 m. The hydraulic gradient at this location is upward, indicating that the valley is a discharge zone. Likewise, piezometers D and E are at the same location, with D being shallow and E being deep. The head in piezometer D is 70 m; in E, it is 50 m. The gradient here is downward, so this is a ground-water recharge zone. Notice that, although the head in D is greater than the head in E, the water column in piezometer D is much shorter. The important factor is the elevation of the water level in the piezometer, not the length of the water column.

Figure 8.1 was a conceptual model based on Hubbert's mathematical intuition. The next major advancement in the understanding of regional flow systems came in 1962 when J. Tóth found an analytical solution to the Laplace equation. For the case where the water table has a linear slope, his solution was (Tóth 1962):

$$h = g\left(z_0 + \frac{\tan \alpha L}{2}\right) - \frac{4g \tan \alpha L}{\pi^2} \cdot \sum_{m=0}^{\infty} \frac{\cos[(2m + 1)\pi x/L]\cosh[(2m + 1)\pi z/L]}{(2m + 1)^2\cosh[(2m + 1)\pi z_0/L]}$$

(8–1)

where

h is the head (L)

g is the gravitational constant (L/T^2)

z_0 is the elevation of the water table at its lowest point above the bottom of the aquifer (L)

z is the elevation of the water table above the bottom of the aquifer (L)

$\tan \alpha$ is the slope of the water table

L is the total length of the flow system (L)

x is the horizontal distance from the place where the water table is at its lowest elevation (L)

Figure 8.3 is a graphical representation of Tóth's solution for the case where there is a linear slope to the water table and the depth of the ground-water basin is one-half the flow length. The base of the flow system is an impermeable boundary, so that flow cannot cross it. The sides of the system are also no-flow boundaries because of hydrodynamic considerations. The solution to Equation 8–1 gave the distribution of hydraulic head. Based on this distribution, the flow lines were drawn at right angles, since the aquifer was assumed to be isotropic and homogeneous.

In the ground-water basin represented by Figure 8.3, ground water is discharged primarily by evapotranspiration or diffused springs on the lower slopes

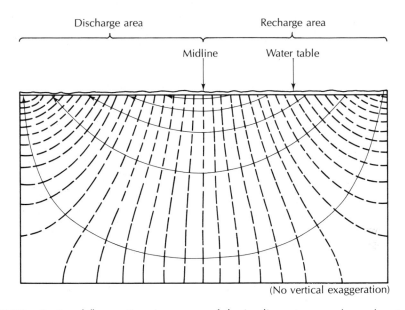

(No vertical exaggeration)

FIGURE 8.3   Regional flow pattern in an area of sloping linear topography and water table. The flow pattern is symmetrical about the midline. Source: J. A. Tóth, *Journal of Geophysical Research* 67 (1962): 4375–87.

of the basin. This flow system does not have converging flow lines as shown in Figure 8.1 because there is not a place where ground-water discharge is concentrated. The entire upper part of the ground-water basin is a ground-water recharge basin, and the entire lower part is a ground-water discharge area. In recharge areas the angle between the water table and flow lines is oblique and points upstream. At the midline, or hinge point, the flow lines are parallel to the water table. In the discharge area, the flow lines are again oblique with the water table but are pointing downstream.

In order to solve the Laplace equation by analytical means, the boundary conditions must be specified. In 1963 Tóth published another solution to the Laplace equation, this time for an undulating water table, which was described by a sine-wave function. The results of this analysis provided hydrogeologists with great insight into ground-water flow systems (Tóth 1963).

If the surface topography has well-defined local relief, a series of *local ground-water flow systems* can form in humid regions. This is due to the fact that the topographic relief causes undulations in the water table. A local ground-water flow system has the recharge area at a topographic high spot and its discharge area at an adjacent topographic low. Figure 8.4A shows a flow net of a ground-water drainage basin with a series of local flow systems. The basin depth is one-twentieth of the basin length from the regional ground-water divide to the lowest part of the basin.

If the basin depth-to-width ratio increases, other flow systems may also develop. *Intermediate flow systems* have at least one local flow system between

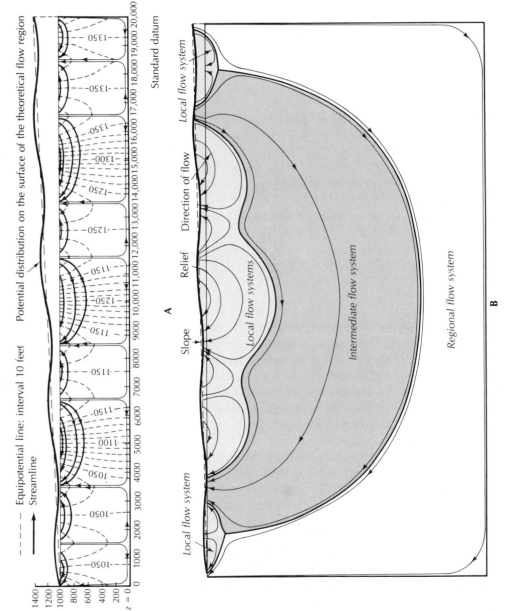

- - - - Equipotential line: interval 10 feet   Potential distribution on the surface of the theoretical flow region

→ Streamline

A

Topographic elevation and head of water above standard datum (ft)

FIGURE 8.4 The effect of increased basin depth is shown on these two figures. In Part A, the basin depth/length ratio is 1:20; in Part B, it is 1:2. The shallow basin has only local flow systems, whereas the deep basin has local, intermediate, and regional flow systems. The water-table configuration is the same for both basins. Source: J. A. Tóth, *Journal of Geophysical Research* 68 (1963): 4795–4811.

their recharge and discharge areas. *Regional flow systems* have the recharge area in the basin divide and the discharge area at the valley bottom (Figure 8.4B). Depending upon the drainage basin topography and the basin-shape geometry, flow systems may have regional; local; local and intermediate; or local, intermediate, and regional components.

In addition to the influence of the drainage basin depth/length ratio, it has been shown that the more pronounced the relief of the undulating water table, the deeper the local flow systems extend. In some basins, both local and regional flow systems may exist, while in other basins with a similar depth/length ratio but with a more pronounced water-table relief, only deep local flow systems develop. This is illustrated in Figure 8.5A, which has local and intermediate flow systems, and in part B of the figure, where the more pronounced relief of the water table has resulted in the exclusive formation of local flow systems.

One of the features of complex flow systems is the presence of **stagnation points** in the flow field (Tóth 1963). At a stagnation point, the magnitudes of the vectors in the flow field are equal but opposite in direction and cancel each other. The value of the hydraulic potential is higher at the stagnation point than at any part of the surrounding region. Ground-water flow paths diverge around stagnation points, which are found at the juncture of local and regional flow systems. Figure 8.6 illustrates the potential field and flow paths at a stagnation point. Stagnation points can exist in materials that are completely isotropic and homogeneous.

It has been suggested that "dead cells" or stagnation points might be appropriate areas in which to inject waste fluids for permanent disposal (Maxey 1968). Diffusion would then be the only physical mechanism to disperse the fluid. However, if the waste fluid were not of the exact density and temperature as the native ground water, the original potential field might be disrupted, causing flow in the area of the stagnation point and resultant movement of the waste fluid. Likewise, ground-water pumpage could change the potential field, shifting or eliminating the locations of stagnation points.

If regional flow systems develop, the flow paths are long compared with those of local flow systems (Tóth 1963). In aquifers composed of soluble rock material, the degree of mineralization is a function of both the initial chemistry of the water and the length of time it is in contact with the aquifer (Back & Hanshaw 1970). Referring back to Figure 8.4B, we see the boundaries of local, intermediate, and regional flow systems for a deep aquifer with an undulating water table. The surface area where recharge to the regional flow system takes place is quite small in relation to the volume of water stored in that region of the aquifer. The water moves slowly and circulates deeply within the aquifer, as the flow paths are long. At the point of discharge, the water from the regional flow system is likely to have relatively high mineralization and an elevated temperature due to the geothermal gradient. (The temperature of the earth increases with depth at a more or less constant rate of 2.5°C per 100 m of depth.)

Local flow systems are shallower, with short flow paths. The size of the recharge area is much greater with respect to the volume of water in the aquifer. Thus, water has a shorter contact time with the rocks and is potentially mineral-

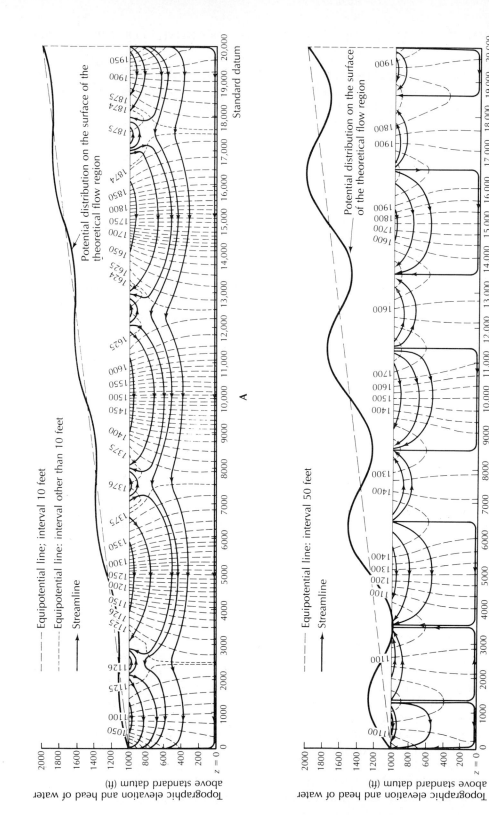

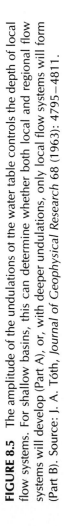

FIGURE 8.5 The amplitude of the undulations of the water table controls the depth of local flow systems. For shallow basins, this can determine whether both local and regional flow systems will develop (Part A), or, with deeper undulations, only local flow systems will form (Part B). Source: J. A. Tóth, *Journal of Geophysical Research* 68 (1963): 4795–4811.

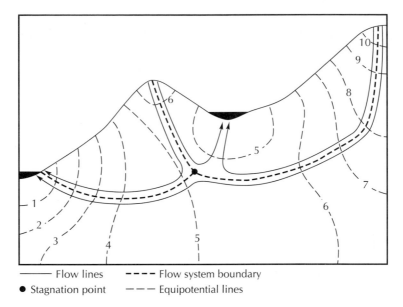

FIGURE 8.6   The potential field and flow lines in the vicinity of a stagnation point, which will develop at the intersection of three flow systems.

ized to a lesser degree than that of the regional system. The temperature of water discharging from the local flow systems is close to the mean annual air temperature. Local flow systems are areas of rapid circulation of ground water; therefore, ground water in these systems is much more active in the hydrologic cycle than ground water in regional flow systems (Tóth 1963). Spring discharge of local flow systems is closely related to recharge of precipitation and shows wide fluctuations (Sartz, Curtis, & Tolsted 1977). Intermediate flow systems have properties falling between those of local and regional flow systems.

If a flow system has extended areas with a flat water table, the potential is the same in all parts of the field. Neither local nor regional flow systems can develop, and the ground water is stagnant. Evapotranspiration is the only method of ground-water discharge. Ground water under such conditions is likely to be highly mineralized owing to a long contact time with the aquifer rocks and the concentration of dissolved salts by evaporation.

8.2.3   Effect of Buried Lenses

Tóth (1962) also showed why piezometers sometimes yield water levels that are apparently anomalous with respect to the expected regional flow pattern. A set of piezometers at various depths may show a water elevation equal to the water table for a shallow well, a lower water elevation for a piezometer of intermediate depth, and then a water elevation equal to the water-table elevation for a deep well. Geologic logs of these piezometers might show the shallow one to end in a fine, silty sand; the one of intermediate depth to end in coarse sand; and the deepest

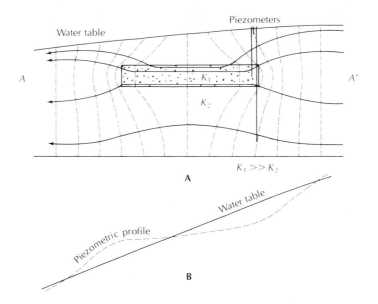

FIGURE 8.7   A. Equipotential field and flow lines in a region where a high-conductivity body is buried in a lower-conductivity aquifer. **B.** The water table and the potentiometric profile of a line of piezometers, each ending at the same elevation along line A-A' of Part A.

one to end in the fine, silty sand. Figure 8.7A shows a cross section of the potential distribution where a body of material of high hydraulic conductivity is surrounded by material with a lower conductivity. The high-conductivity zone acts as a conduit for flow, attracting water from much of the aquifer. The result is that the potential field bends away from the high-conductivity zone on either end. Flow will thus converge toward the high-conductivity zone on the upstream end and diverge away from it on the downstream side.

A line of piezometers of equal depth, extending to line A-A', would have a potentiometric profile that would differ from the water-table profile (Figure 8.7B). Upstream of the midpoint of the high-conductivity layer, the potentiometric profile would be lower than the water table. It would cross the water table at the midpoint and be higher than the water table below the midpoint. Such a profile would not occur in a homogeneous aquifer.

If a lens of low-permeability material is buried in an aquifer, it acts as a partial barrier to ground-water flow. The ground-water streamlines diverge around the lens. While a modicum of the flow is carried in the low-conductivity layer, the majority of the flow tends to be in the aquifer.

8.2.4   Nonhomogeneous and Anisotropic Aquifers

Analytical solutions to the Laplace equation are limited to homogeneous, isotropic aquifers with easily described boundary and initial conditions. Most natural aquifers are anisotropic and nonhomogeneous with complex boundary conditions. R. A. Freeze and P. A. Witherspoon developed a numerical method of solving the

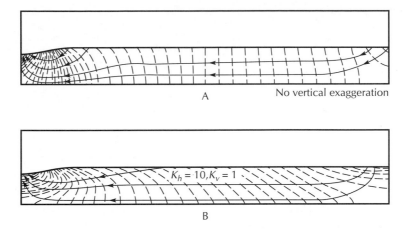

A   No vertical exaggeration

$K_h = 10, K_v = 1$

B

FIGURE 8.8 Ground-water flow in an aquifer with concentrated discharge. **A.** Flow in an isotropic aquifer. **B.** Flow in an anisotropic aquifer with the horizontal hydraulic conductivity ten times the vertical. Source: R. A. Freeze & P. A. Witherspoon, *Water Resources Research* 3 (1967): 623–34. Copyright by the American Geophysical Union.

Laplace equation that could be applied to virtually any aquifer (Freeze & Witherspoon 1966, 1967). These types of numerical solutions are evaluated by the use of a computer.

In Tóth's model with a linear slope to the water table, the discharge was dispersed over half of the ground-water basin. By selecting a water table with two linear slopes, one being much steeper and covering only a small portion of the lower part of the ground-water basin, Freeze and Witherspoon were able to show the effect of concentrated ground water discharge. Figure 8.8A shows the potential distribution and flow lines for a ground-water basin with concentrated discharge for an isotropic and homogeneous aquifer.

There is considerable evidence that many aquifers are anisotropic. For deposits as uniform as glacial outwash, the horizontal permeability may be two to twenty times as great as the vertical (Weeks 1969). Figure 8.8B shows the potential distribution in an aquifer in which horizontal conductivity is 10 times as great as the vertical. In anisotropic aquifers, the flow lines do not cross the equipotential lines at right angles; the correct angles can be obtained graphically (Liakopoulos 1965). Compared with an isotropic medium, the vertical components of flow are more pronounced in the anisotropic aquifer. The greatest variation in the potential field occurs at the extreme ends of the ground-water basin.

Layered aquifers are especially prevalent in sedimentary basins, with individual hydrostatigraphic* units having different hydraulic conductivities. If a

*Hydrostratigraphic units comprise geologic units grouped together on the basis of similar hydraulic conductivity. Several geologic formations may be grouped into a single aquifer, for example. A single geologic formation may be divided into both aquifers and confining units.

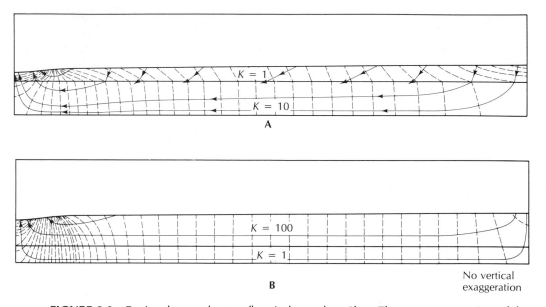

FIGURE 8.9   Regional ground-water flow in layered aquifers. The greater proportion of the flow occurs in the layer with higher hydraulic conductivity. Source: R. A. Freeze & P. A. Witherspoon, *Water Resources Research* 3 (1967): 623–34.

lower formation has a substantially higher conductivity than the surface layer, it acts as the major conduit of flow (Freeze & Witherspoon 1967). Figure 8.9A shows the potential distribution in a layered aquifer when the lower unit has a conductivity ten times that of the upper. Flow in the lower unit is horizontal, while the upper unit has vertical flow components in the recharge and discharge areas. As the difference in conductivity between the upper and lower layer increases, the components of vertical flow in the upper unit increase as more of the flow is carried in the lower unit.

If a high-conductivity layer overlies a unit of substantially lower conductivity, the potential field is very similar to that of an isotropic aquifer. Most of the flow is carried in the upper, more conductive, layer, as Figure 8.9B illustrates. The potential field of Figure 8.9B is quite similar to that of Figure 8.8A, which is homogeneous.

Aquifers that are overlain by a layer of substantially lower hydraulic conductivity are confined. The hydraulic gradient is generally greater in the confining bed than in the aquifer. Since the frictional resistance to flow is so much greater in the confining layer, most of the available energy of the potential field is dissipated there.

Confined aquifers may be either sloping or flat. If the aquifer crops out near a topographic high, substantial recharge takes place in the outcrop area. In the sloping aquifer shown in Figure 8.10A, the confining layer retards flow. Wells drilled through it to the underlying aquifer would yield artesian flow. Flow lines

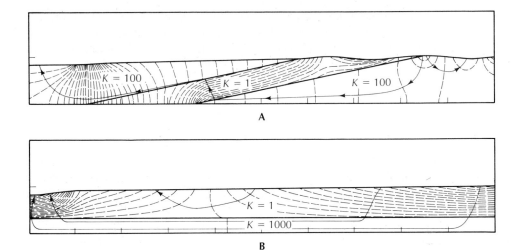

FIGURE 8.10   Regional ground-water flow in confined aquifers: **A.** Aquifer confined by a sloping confining layer. **B.** Aquifer confined by a flat-lying confining layer. Source: R. A. Freeze & P. A. Witherspoon, *Water Resources Research* 3 (1967): 623–34.

refract as they cross the confining layer. Discharge of the regional flow system is concentrated in the valley bottom.

The flat-lying confined aquifer in Figure 8.10B does not have an outcrop area. Recharge to the aquifer occurs by downward flow through the confining layer. Almost all of the energy of the potential field is consumed as flow moves through the confining layer in recharge and discharge areas. Only one equipotential line crosses the aquifer. The volume of water flowing through the buried aquifer is less than it would be if the aquifer cropped out in the recharge area. This is due to the use of much of the available potential energy in forcing recharge through the low-conductivity layer. If a well were drilled in the discharge area, artesian conditions would occur.

Unless the confined aquifer is capped by a completely impermeable layer, there will be some discharge from the aquifer in the form of upward leakage in the area of upward hydraulic gradient. Many confined aquifers have heads above the land surface when the first wells are drilled (Weidman & Schultz 1915). The amount of upward flow occurring through the confining beds is typically small, and the water does not circulate rapidly. Ground-water withdrawals in many confined regional aquifers have lowered the potentiometric head. This actually increases the rate of lateral ground-water flow in the aquifer, as the hydraulic gradient between the recharge area and the well-field area is increased.

8.3   TRANSIENT FLOW IN REGIONAL GROUND-WATER SYSTEMS

The systems we have considered have been in a state of **dynamic equilibrium.** The amount of water recharging the aquifer is balanced by an equal amount of natural discharge, and the potential field is more or less constant. If a well field is established in the ground-water basin, the withdrawal of well water increases the discharge from the system, disrupting the equilibrium. Thus, a new equilibrium must be established (Theis 1938).

In the case of an unconfined aquifer, the water table around the well field will be drawn down. As the discharge exceeds the recharge, the difference comes from gravity drainage of ground water stored in the aquifer. The cone of depression around the well field will slowly expand until it affects the flow system enough to create a new equilibrium condition. This will occur when the area of the cone of depression is large enough to intercept sufficient aquifer recharge to supply the well discharge. This will reduce natural discharge somewhere else, and a new condition of dynamic equilibrium will be reached. Should the rate of withdrawal be so great that the cone of depression reaches the boundaries of the aquifer without intercepting sufficient recharge, the aquifer will not reach equilibrium and eventually could be drained.

In confined and leaky, confined aquifers, pumping will reduce the heads near the wells. As a result, the potentiometric surface will decline. The cone of depression will expand rapidly owing to the small value of storativity of confined aquifers. Initially, the water being pumped comes from storage in the aquifer. The cone of depression in a leaky, confined aquifer will stabilize when enough downward leakage is induced to balance pumpage. This, of course, will upset the natural equilibrium in the overlying aquifer that is furnishing the water.

In a confined aquifer, the cone of depression will grow until it reaches either the recharge area of the aquifer, or the discharge area, or both. The resulting change in the potential field will induce either increased recharge, or decreased natural discharge, or both. If this is sufficient to balance recharge and discharge, the aquifer will again be in dynamic equilibrium. If not, the water levels will continue to decline.

8.4   NONCYCLICAL GROUND WATER

There is a certain amount of water in the ground that is not encompassed by the hydrologic cycle. When sediments are deposited, water is present in the pores. The same may be true for undersea volcanic rocks. Later geologic events may bury the sediment or rock and its contained pore water. Water buried with the rock is termed **fossil water** (White 1957a). Interstitial water that was not buried with the rock but has been out of contact with the atmosphere for an appreciable part of a geologic period is called **connate water** (White 1957a).

Magmatic water is associated with a magma. It may be in part **juvenile water,** having never before circulated in the hydrologic cycle (White 1957b). However, most magmatic water comes from the recycling of connate or fossil water. Magmatic water can re-enter the hydrologic cycle through volcanic eruptions or thermal springs.

8.5   SPRINGS

Springs have played a role in the settlement pattern of many lands, where they have served as a local water supply. Mineralized and thermal springs have been thought to have therapeutic value. The importance of springs is evident from the many localities named for the springs found there (e.g., Tarpon Springs, Florida; Palm Springs, California; Hot Springs, Arkansas; Steamboat Springs, Colorado).

A spring may have a discharge that is fairly constant, or the discharge may vary. Springs can be permanent or ephemeral. The water may contain dissolved minerals of many different types or certain dissolved gases or petroleum. The temperature of the water may be close to the mean annual air temperature or be lower or higher—even boiling. Flow may range from a barely perceptible seepage to 1000 ft^3 (30 m^3) or more per second.

Topographic low spots provide the simplest mechanism for the formation of springs. **Depression springs** are formed when the water table reaches the surface (Bryan 1919). The change in topography creates a corresponding undulation in the water-table configuration. A local flow system is thus created, with a spring formed at the local discharge zone (Figure 8.11A).

Where permeable rock units overlie rocks of much lower permeability, a **contact spring** may result (Bryan 1919). A lithologic contact is often marked by a line of springs, which may be either in the main water table or in a perched water table. It is not necessary for the underlying layer to be impermeable, merely that the difference in hydraulic conductivity be great enough to preclude transmission of all of the water that is moving through the upper horizon (Figure 8.11B).

A classic occurrence of contact springs is found along the eastern side of Chuska Mountain, New Mexico. A sandstone cliff rises 197 to 492 ft (60 to 150 m) above a terrace composed of shale, which also underlies the sandstone. More than 30 springs are found at the foot of the cliff at the contact of the sandstone and shale (Gregory 1916). One of the most spectacular series of springs in the world is in the Snake River Canyon below Shoshone Falls in Idaho. Along a 40-mi (64-km) reach of the canyon, there are 11 springs with a discharge of more than 100 ft^3 (2.8 m^3) per second. The springs issue from permeable basalt flows; total spring flow is about 5000 ft^3/s (140 m^3/s) in this reach of the Snake River (Meinzer 1927).

Faulting may also create a geologic control favoring spring formation. A faulted rock unit that is impermeable may be emplaced adjacent to an aquifer. This can form a regional boundary to ground-water movement and force water in the aquifer to discharge as a **fault spring** (Figure 8.11C).

Some of the largest springs are found in areas of limestone bedrock. In such areas, the runoff may be carried in part or totally as subterranean flow. It may be either diffused flow in pores and fractures in the rock or channelized flow in caverns. Springs may be found where a cavern is connected to a shaft that rises to the surface. Many of the famous springs of Florida cover an area of several acres in which water rises to the surface through sinkholes (Figure 8.11D). The water in these **sinkhole springs** is under artesian pressure and comes from the

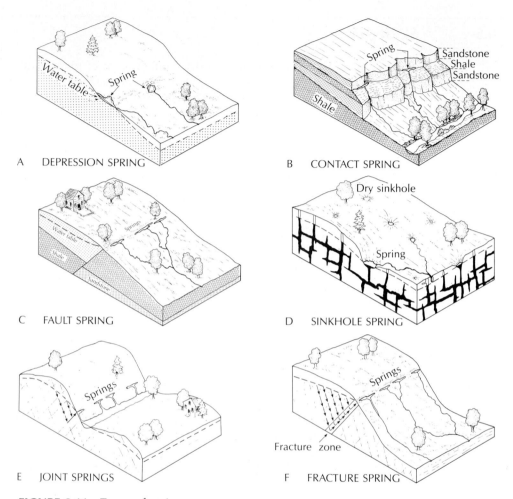

A   DEPRESSION SPRING

B   CONTACT SPRING

C   FAULT SPRING

D   SINKHOLE SPRING

E   JOINT SPRINGS

F   FRACTURE SPRING

FIGURE 8.11   Types of springs.

principal artesian aquifer, or Floridan aquifer, which underlies Florida (Stringfield 1966). This aquifer is in Tertiary-age limestones.

    Joint springs or **fracture springs** may occur from the existence of jointed or permeable fault zones in low-permeability rock. Water movement through such rock is principally through fractures, and springs can form where these fractures intersect the land surface at low elevations (Figure 8.11E, F).

    Springs in limestone terrane can be interconnected to topographic depressions caused by collapsed caverns (sinkholes) at higher elevations. Water level in the sinkholes may rise and fall owing to variations in runoff (Brook 1977). Discharge of these springs, known as *karst springs,* may correspond with the elevation of water in the sinkholes.

8.6   GEOLOGY OF REGIONAL FLOW SYSTEMS

Several case studies illustrate different types of regional flow systems.

CASE STUDY: REGIONAL FLOW SYSTEMS IN THE GREAT BASIN

The Basin and Range Province contains a number of topographically closed basins. These intermontane basins are characterized by an accumulation of relatively permeable clastic sediments. The mountains surrounding the basins are composed of bedrock, which also underlies the basins at depth. The hydraulic conductivity of the bedrock types is extremely variable (Mifflin 1968).

Annual precipitation is greatest in the mountains and least in the valleys (Eakin 1966). Below an elevation of 6000 ft (1800 m), annual precipitation is less than 8 in. (20 cm). This is almost all evaporated, with virtually no recharge of ground water. Above 9000 ft (2750 m), there may be more than 20 in. (50 cm) of precipitation, with up to 5 in. (13 cm) of ground-water recharge. The areas of greatest precipitation and recharge are in topographic highs, which are good recharge zones.

Those mountain areas formed by crystalline rocks or low-permeability sedimentary rocks have near-surface permeability due to fracturing. Such mountain areas have many small springs and perennial streams, as the ground water is discharged by local flow systems (Maxey 1968). Mountains underlain by highly permeable carbonate rocks are generally dry. The ground water appears as the discharge of relatively large springs at the foot of the mountains or in the intermontane valleys. In the areas of carbonate aquifers, the water table is relatively flat and may extend with a regional slope beneath topographic divides (Maxey 1968; Mifflin 1968; Eakin 1966).

In the White River area of southeastern Nevada, a regional interbasin ground-water flow system has been identified (Eakin 1966). There are thirteen topographic basins: seven of them are closed; the other six were drained by the White River during the Pleistocene. The mountains are 8000 to 10,000 ft (2450 to 3050 m) high, with the valley bottoms 2000 to 4000 ft (600 to 1220 m) lower. The principal water-bearing units are Paleozoic limestone and dolomites, up to 30,000 ft (9150 m) thick. There are some volcanic rocks (tuffs and welded tuffs) that can form locally perched aquifers. The valleys are filled with Tertiary-age clastic sedimentary rocks and evaporites.

Ground water is discharged by means of several large springs. The flow of Muddy River Springs, which is the largest, is highly uniform, suggesting a regional flow system as the source (Eakin & Moore, 1964). A longitudinal profile of the area, which shows both the topography and the potentiometric surface, reveals that the regional hydraulic gradient is unaffected by crossing topographic divides (Figure 8.12).

The amount of ground-water recharge is much greater than discharge in the topographically higher basins. The water balance is reversed for the topographically lower basins where the large springs are located. However, the regional water budget is balanced when all thirteen of the basins are included (Eakin, Price, & Harrill 1976).

In the Great Basin area, local flow systems have small drainage areas and short flow paths within the same topographic basin. Springs have a wide fluctuation in discharge. The temperature of the spring water is about the same as mean annual air temperature, and the dissolved ion content is relatively low. Regional flow systems have long flow paths, which often cross basin divides. The drainage basin area is large. Springs are large,

FIGURE 8.12 Flow paths and longitudinal profile of a regional ground-water flow system in Nevada. Source: T. E. Eakin, *Water Resources Research* 2 (1966): 251–71.

with fairly constant discharge, elevated temperatures, and higher concentrations of dissolved salts (Maxey 1968).

Another large ground-water flow system has been delineated in southern Nevada at the Nevada Test Site (Winograd & Thordarson 1975). The area has typical basin and range topography. Rocks range in age from Recent sediments in the valleys to Precambrian. A basement unit of Cambrian and Precambrian siltstones and quartzites is a regional confining layer. Above this layer is a major regional aquifer—the lower carbonate aquifer of Lower Paleozoic age. There are solution openings in the aquifer, but the hydraulic conductivity is due primarily to fractures. The saturated thickness ranges from a hundred to several thousand feet or more. There are several caves in the outcrop area. One of them, Devils Hole, is partially filled with water and extends vertically more than 300 ft (100 m) below the water table. The lower carbonate aquifer feeds many large springs.

No other aquifer has a regional extent, although several are important locally. There is a regional confining layer that overlies the lower carbonate aquifer. This is the welded tuff confining layer of Tertiary age. The clay-rich beds of this layer restrict ground-water movement. The three regional units control ground-water flow. The welded tuff confining layer permits slow downward movement of water to the lower carbonate aquifer in recharge zones and upward movement in discharge zones. The lower carbonate aquifer underlies most of the valleys and ridges.

In some of the upland valleys, such as Yucca Flats and Frenchman Flats, the water table lies from 700 to 2000 ft (210 to 610 m) below the valley floor (Figure 8.13). Water drains from these valleys downward to the lower carbonate aquifer. Water flows laterally in this unit until it reaches the Ash Meadows area of the Amargosa Desert Basin. Running across Ash Meadows is a normal fault with a displacement of at least 500 ft (160 m). This has formed a hydraulic barrier that impedes the ground-water flow. The result is a line of springs discharging at the outcrop of the lower carbonate aquifer. The average discharge of all the springs is 10,000 gal/min (630 L/s).

The inferred ground-water drainage basin contains 10 intermontane valleys and is on the order of 4500 mi^2 (11,650 km^2) in area. The Ash Meadows Basin may be hydrologically interconnected with other intermontane valleys from which ground-water flow may be obtained. There may also be underflow past the spring line at Ash Meadows. This spring line feeds into the central Amargosa Desert. The lack of wells and the extremely complex structure of the area make exact delineation of flow systems difficult.

CASE STUDY: REGIONAL FLOW SYSTEMS IN THE COASTAL ZONE OF THE SOUTHEASTERN UNITED STATES

Ground-water flow regimes in humid regions characteristically have a much thinner unsaturated zone than those in arid regions. The saturated zone is close to the surface in recharge areas as well as in discharge areas. Because of the greater amounts of precipitation, the volume of recharge is much higher, and proportionally more water circulates through ground-water flow systems.

The coastal zone of the southeastern United States is underlain by extensive aquifers contained in Tertiary- and Quarternary-age limestones, forming some of the most productive aquifers in the United States (Stringfield 1966; Stringfield & LeGrand 1960; Miller 1986). Sediments on the coastal plain dip seaward, and the units thicken as the coast is approached. Figure 8.14 shows the isopach map of the Cenozoic sediments and sedimentary rocks of the region. Starting at a feather-edge near the interior border of the Atlantic

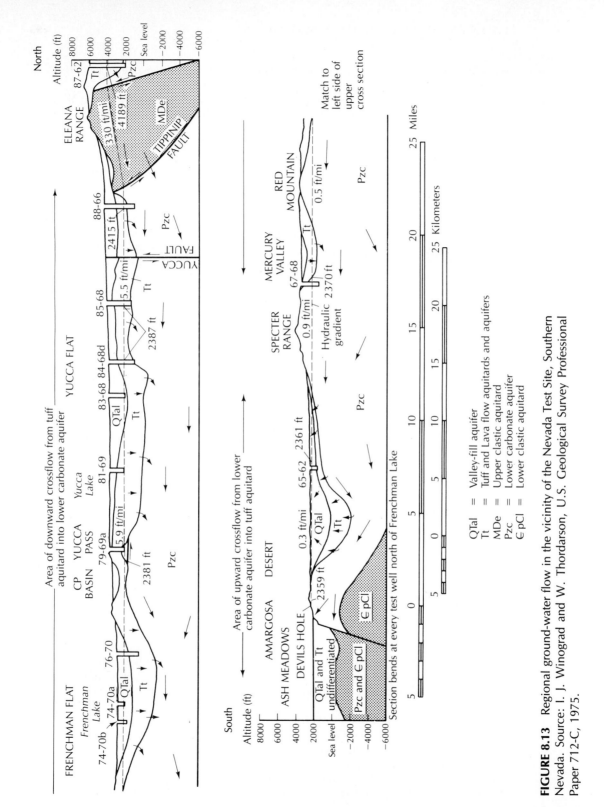

FIGURE 8.13 Regional ground-water flow in the vicinity of the Nevada Test Site, Southern Nevada. Source: I. J. Winograd and W. Thordarson, U.S. Geological Survey Professional Paper 712-C, 1975.

QTal = Valley-fill aquifer
Tt = Tuff and Lava flow aquitards and aquifers
MDe = Upper clastic aquitard
Pzc = Lower carbonate aquifer
€pCl = Lower clastic aquitard

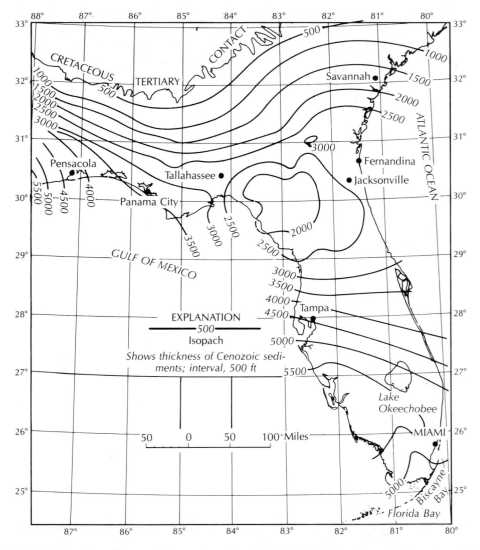

FIGURE 8.14 Isopach map of Cenozoic sediments in Florida, Georgia, and adjacent parts of Alabama and South Carolina. Source: V. T. Stringfield, U.S. Geological Survey Professional Paper 517, 1966.

Coastal Plain, they thicken to more than 5500 ft (1680 m) in southwestern Florida. The structure is disrupted by the Ocala uplift north of Tampa, resulting in a thinning of the sediments.

There are a number of hydrostratigraphic units in the region. There is a surficial aquifer system, an upper confining unit, the upper Floridan aquifer, a middle confining unit, the lower Floridan aquifer, and a lower confining unit (Miller 1986; Bush & Johnson 1986).

Most of the region has permeable materials at or near the surface. These are predominantly sand, but also include gravel, sandy limestone, and limestone. In most places the sands are relatively thin and are termed the *surficial aquifer.* In southwest Florida the surficial material consists of highly permeable sands and limestones that are as much as 200 ft (60 m) thick and occur in a wedge shape that thins to the northwest. It has been termed the *Biscayne aquifer.* In the western part of the panhandle of Florida the surficial material consists of a thick deposit of sand and gravel termed the *sand and gravel aquifer.* The surficial aquifer system is mostly unconfined and receives recharge directly from precipitation. It is underlain by either the upper confining unit or, if that unit is missing, by the upper Floridan aquifer. The surficial aquifer materials are Pleistocene to Recent in age. Figure 8.15 shows the surficial hydrostratigraphic units of the area.

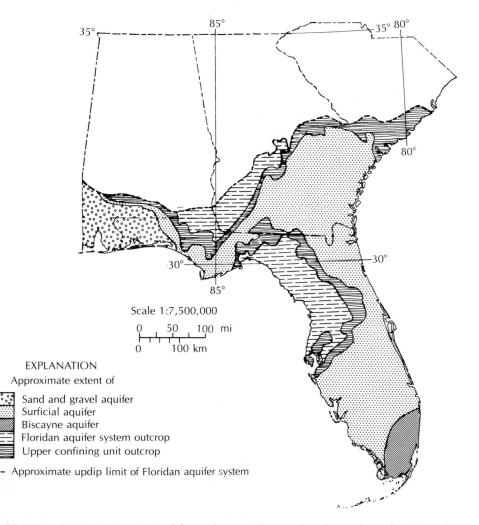

Scale 1:7,500,000

EXPLANATION
Approximate extent of

Sand and gravel aquifer
Surficial aquifer
Biscayne aquifer
Floridan aquifer system outcrop
Upper confining unit outcrop

----- Approximate updip limit of Floridan aquifer system

FIGURE 8.15 Approximate extent of the surface aquifer, sand-and-gravel aquifer, Biscayne aquifer, and outcrop area of the upper confining unit in the southeastern United States. Source: J. A. Miller, "Hydrogeologic Framework of the Floridan Aquifer System in Florida and Parts of George, Alabama, and South Carolina." U.S. Geological Survey Professional Paper 1403-B:B41, 1986.

The *upper confining layer* consists of low-permeability rocks that are primarily clastic. It basically consists of the Hawthorn formation of Miocene age, although in some areas the lower Miocene Tampa Limestone and the Oligocene Suwannee Limestone are included if the permeability of these limestones is comparatively low. The thickness and lithology of this unit are variable and hence the degree to which it retards vertical flow differs from place to place. In some areas where it is thin the Hawthorn has been breached by sinkholes so that direct pathways for water movement between the surficial deposits and the Floridan aquifer are present. In some places it is so thick, as much as 500 ft (150 m), that it directs virtually all of the water in the surficial aquifer laterally toward the coastline. The Hawthorn Formation has been removed by erosion in the area of the Ocala uplift of Florida so that the underlying Floridan aquifer system forms the surface material (Figure 8.15).

The *Floridan aquifer system* is quite thick and consists of Eocene- to Miocene-age limestones of different permeability. The units contained in the Floridan aquifer system consist of the early Miocene Tampa Limestone, where it is permeable; the Oligocene Suwannee Limestone; and the Eocene Ocala Limestone, Avon Park Formation, and Oldsmar Formation. The hydraulic conductivity of the rocks of the Floridan aquifer ranges from one to several orders of magnitude greater than that of the overlying confining layer. The Floridan aquifer system is unconfined in areas where the confining layer is missing and is semiconfined or confined elsewhere, depending upon the thickness and vertical hydraulic conductivity of the Hawthorn Formation. It is recharged in upland areas either directly, where the Hawthorn Formation is missing or is breached by sinkholes, or by leakage through the Hawthorn Formation. Water discharging from the Floridan aquifer forms some of the major springs of Florida. Water also discharges as subsea outflow in coastal areas. The Floridan aquifer is subdivided into an upper aquifer unit, a middle confining layer, and a lower aquifer unit (Figure 8.16). The *middle confining layer* is discontinuous and is represented by several different geologic units. In places the *upper Floridan aquifer* and the *lower Floridan aquifer* are joined into one unit. The upper aquifer unit is generally highly transmissive, although there is a complex relationship between the transmissivity of the limestone units and their postdepositional erosional history. In areas where the upper Floridan aquifer is unconfined, transmissivities of as much as 1×10^6 ft^2/day (9×10^4 m^2/day) occur because of sinkholes and caves. Where the aquifer is confined by a thick zone of the Hawthorn Formation, the transmissivity is much less, being as little as 5×10^4 ft^2/day (5×10^3 m^2/day) in south Florida. The upper Floridan aquifer typically has good water quality and is extensively used as a water supply. The lower Floridan aquifer frequently contains saline water. It has some very highly transmissive areas. The *Boulder zone* of South Florida is a paleokarst zone developed in early Eocene rocks. It contains saline water and is extensively used for disposal of treated municipal sewage.

The Floridan aquifer is bounded on the bottom by a *lower confining unit.* This is a low-permeability unit that consists of the Cedar Keys Formation, which contains massive anhydrite beds as well as other units with a much lower hydraulic conductivity than the Floridan aquifer.

Precipitation in Florida averages about 55 in. (140 cm) annually with about 39 in. (100 cm) of evaporation (Parker 1975). The 16-in. (40-cm) difference represents the recharge to aquifers and overland flow into streams. The vast majority of the available water, 14 in. (35 cm), flows into surface water, with only 1 to 2 in. (3 to 5 cm) recharging the Floridan aquifer. The major recharge areas are discernible as the areas of high elevation on the potentiometric map (Figure 8.17). The Floridan aquifer has two major hydraulic regions separated by a hydrologic divide (Parker 1975). This is shown as a dashed line in Figure 8.17.

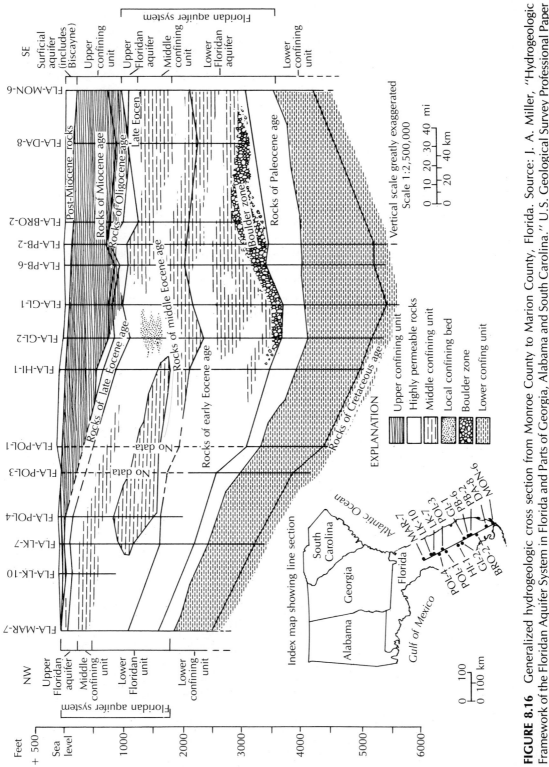

FIGURE 8.16 Generalized hydrogeologic cross section from Monroe County to Marion County, Florida. Source: J. A. Miller, "Hydrogeologic Framework of the Floridan Aquifer System in Florida and Parts of Georgia, Alabama and South Carolina." U.S. Geological Survey Professional Paper 1403-B:B78, 1986.

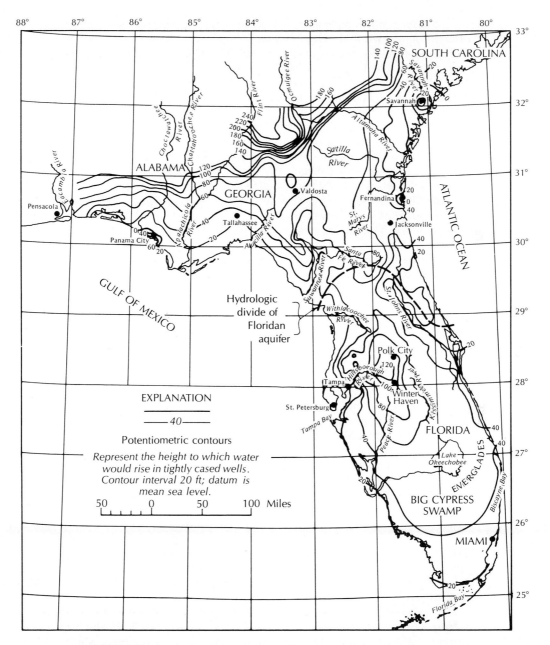

FIGURE 8.17 Potentiometric surface of water in the principal artesian aquifer of the southeastern United States. Source: V. T. Stringfield, U.S. Geological Survey Professional Paper 517, 1966.

The northern portion of the Floridan aquifer is recharged in the outcrop area of central Georgia, where the elevations are 100 to 230 ft (30 to 70 m) above sea level. Recharge to the aquifer also occurs in a region extending from north of Valdosta, Georgia, to east of Gainesville, Florida. This is an area of high topography with many sinkhole lakes. The sinkholes breach the Hawthorn Formation, a confining layer, so recharge occurs through the sinkhole lakes into the underlying Floridan aquifer. Water flows from the recharge areas, following the regional hydraulic gradient, and discharges along the coastlines. There are also lesser amounts of recharge from downward leakage through the Hawthorn Formation over much of Florida.

The southern portion of the Floridan aquifer is hydraulically separated from the north by the potentiometric divide. The Withlacoochee River and the St. Johns River act as ground-water drains to lower the potentiometric surface and isolate the southern aquifer. In the Polk City area, north of Winter Haven, the potentiometric surface is as much as 115 ft (35 m) above sea level (Stewart et al. 1971). This is another area in which sinkhole lakes as deep as 200 ft (60 m) are present and can serve as recharge conduits.

The general circulation of ground water in the Floridan aquifer is downward in the recharge areas and then laterally toward the coasts. To the west, discharge is by upward leakage through the Hawthorn Formation into the Gulf of Mexico or into coastal springs located in sinkholes. Flow eastward can reach the Atlantic Ocean floor, where some units of the Floridan aquifer crop out. There is also discharge into the rivers that drain the coastal area. Major rivers, such as the Suwannee and St. Johns in northern Florida, have a profound effect on the potentiometric surface, as they are regions of major discharge from the aquifer.

Zones of high-salinity ground water are found in the lower parts of the Floridan aquifer. These are naturally occurring, but heavy pumping of overlying fresh water is causing an upconing of the saline water. In a number of areas in the southeastern Coastal Plain, the principal artesian aquifer has been heavily pumped, with a subsequent lowering of the potentiometric surface. In an area east of Tampa, the potentiometric surface declined by as much as 60 ft (18 m) from 1949 to 1969 (Stewart 1971). In the Savannah area, the cone of depression resulting from ground-water pumping has reversed the regional hydraulic gradient and has caused salt-water encroachment (Counts & Donsky, 1963; Bush, Miller, & Maslia 1986). This is becoming common throughout much of the southeastern Coastal Plain.

The estimated predevelopment discharge of water from the Floridan aquifer system is shown in Figure 8.18. Water from the aquifer discharges into known springs as well as lakes and streams in areas where the aquifer is unconfined or loosely confined. As of 1980 the total discharge from some 300 known springs was about 13,000 ft^3/s (370 m^3/s) with an additional 7000 ft^3/s (200 m^3/s) estimated discharge into the lakes and streams. The remainder of the water from the Floridan aquifer, about 2500 ft^3/s (70 m^3/s), diffused upward across the upper confining layer into surface aquifers.

Development of the Floridan aquifer system has resulted in the withdrawal of some 4000 ft^3/s (110 m^3/s) of ground water by pumping. Overall, the discharge from the Floridan aquifer system is now estimated to be 24,100 ft^3/s (680 m^3/s), which is 12 percent more than under predevelopment conditions. Figure 8.19 shows the estimated 1980 discharge from the Floridan aquifer. The creation of cones of depression in the potentiometric surface of the Floridan aquifer has enlarged the recharge area and resulted in increased recharge. Long-term regional water-level declines of more than 10 ft (3 m) have occurred in the Savannah–Hilton Head Island area, Brunswick-Jacksonville-Fernandina beach area, Fort Walton beach area, and west-central Florida. Most of the discharge is still through the major springs, with some 18,000 ft^3/s (510 m^3/s) of flow into known springs and surface-

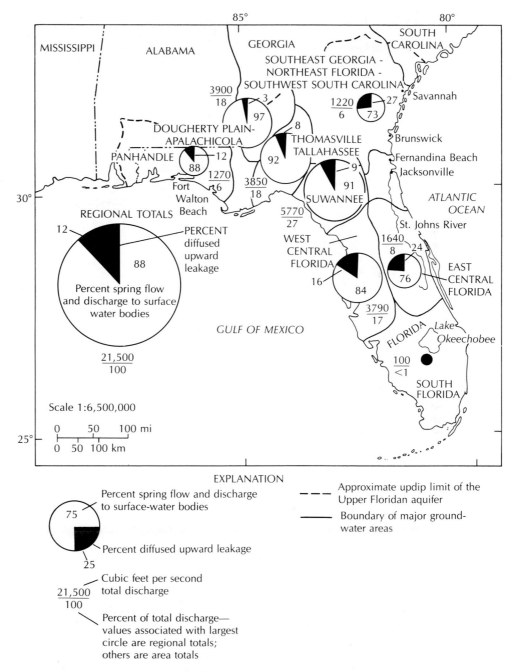

FIGURE 8.18 Estimated predevelopment discharge for major ground-water areas of the upper Floridan aquifer. Source: P. W. Bush & R. H. Johnson, "Floridan Regional Aquifer-System Study." In *Regional Aquifer-System Analysis Program of the U.S. Geological Survey, Summary of Projects, 1978-84.* U.S. Geological Survey Circular 1002 (1986): 17–29.

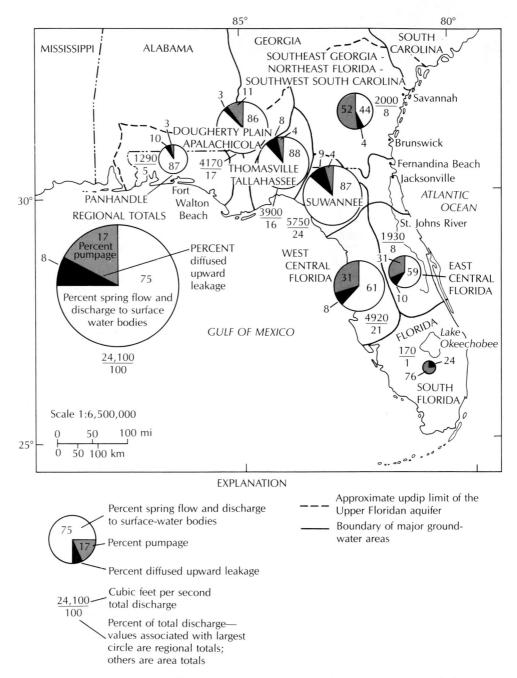

FIGURE 8.19 Estimated 1980 discharge from major ground-water areas of the upper Floridan aquifer. Source: P. W. Bush & R. H. Johnson, "Floridan Regional Aquifer-System Study." In *Regional Aquifer-System Analysis Program of the U.S. Geological Survey, Summary of Projects, 1978-84*. U.S. Geological Survey Circular 1002 (1986): 17–29.

water bodies. There still remains a large amount of water in the Floridan aquifer that could be developed, especially in southeast Georgia and north-central Florida (Bush & Johnson 1986). However, future development could be hindered by salt-water encroachment, which is also a threat to those areas that have already developed deep cones of depression (Bush, Miller, & Maslia 1986).

CASE STUDY: REGIONAL FLOW SYSTEM OF THE HIGH PLAINS AQUIFER

The *High Plains aquifer* occurs in the states of Colorado, Kansas, Nebraska, New Mexico, Oklahoma, South Dakota, Texas, and Wyoming. It is a water-table aquifer system consisting primarily of near-surface sand and gravel deposits that were formed as alluvium deposited by streams draining from the Rocky Mountains, which lie to the west. The *Ogalalla Formation,* which consists of alluvium, is the principal hydrogeologic unit; but the aquifer system also includes the fractured and thus permeable zones of the Brule Formation, a massive siltstone, and the Arikaree Group, a sandstone. Dune sands of Quaternary age are also an important part of the High Plains aquifer, especially in Nebraska. Figure 8.20 is a geologic map showing the principal geologic units of the aquifer system (Gutentag et al. 1984; Weeks 1986).

The High Plains aquifer is recharged by rainfall, which varies from 16 in. (41 cm) annually in the western part of the aquifer to 28 in. (71 cm) in the extreme eastern part. However, potential evapotranspiration greatly exceeds the available precipitation. Class A pan evaporation ranges from 60 in. (152 cm) in the northern part of the aquifer to 105 in. (267 cm) in the southern. As a result, net annual recharge to the aquifer system is as low as 0.024 in. (0.061 cm) per year in Texas to 6 in. (15 cm) per year in south-central Kansas. Recharge rates are greatest in areas of surficial sand dunes. Except in dune areas the long-term annual recharge is 1 in. (2.5 cm) per year or less (Gutentag et al. 1984).

The hydrogeologic characteristics of the High Plains aquifer materials vary widely because of the differing grain-size distribution of the sediments. The specific yield varies from less than 5 to more than 30 percent and averages 15 percent. Hydraulic conductivity varies from 25 to 300 ft/day (7.6 to 91 m/day) and averages 60 ft/day (18 m/day). There is a complex areal variability of the hydrogeologic characteristics because of the fluvial depositional history of the formations that comprise the aquifer (Figure 8.21).

The water table slopes from west to east (Figure 8.22). This is due to the regional topographic slope in this direction. Ground water is flowing from west to east at an average rate of about 1 ft/day (0.3 m/day). It discharges naturally into streams and springs as well as by evapotranspiration.

The saturated thickness of the aquifer ranges from zero at the western edge to about 1000 ft (305 m) in west-central Nebraska and averages 200 ft (60 m). Prior to development there were about 3.42 billion acre-feet (4.22×10^{12} m³) of drainable water in storage in the aquifer system. Starting in the late nineteenth century, the aquifer was tapped by wells for irrigation. In 1978 there were some 170,000 wells pumping 23 million acre-feet (2.84×10^{10} m³) of water annually (Gutentag et al. 1984). In some areas the amount of annual pumpage is from 2 to 100 times greater than the annual recharge. This has resulted in declines in the water table of more than 100 ft (30 m) (Figure 8.23).

The volume of water in storage in the aquifer has decreased by about 166 million acre-feet (2.05×10^{11} m³) per year, with most of the decline occurring in Kansas and Texas. Some 3.25 billion acre-feet (4×10^{12} m³) of drainable water still remain in the

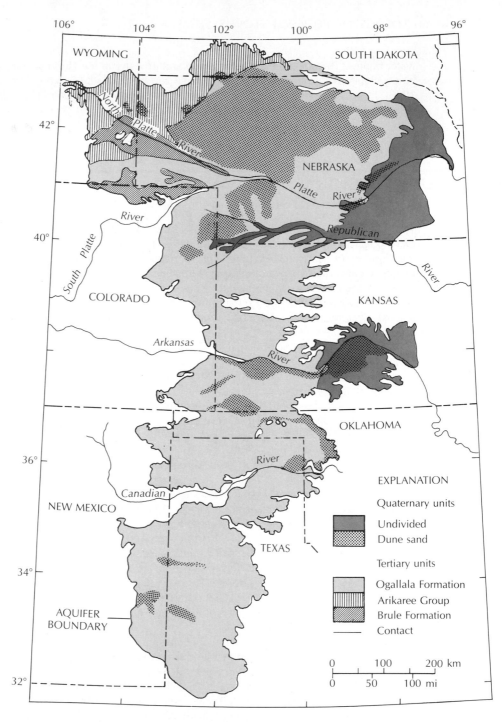

FIGURE 8.20 Principal geologic units of the High Plains aquifer. Source: E. D. Gutentag, F. J. Heimes, N. C. Krothe, R. R. Luckey, & J. B. Weeks, U.S. Geological Survey Professional Paper 1400-B, 1984.

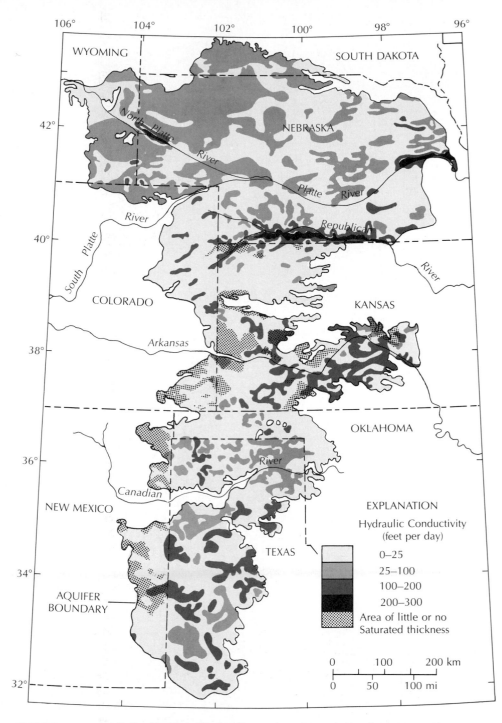

FIGURE 8.21 Areal distribution of hydraulic conductivity in the High Plains aquifer. Source: E. D. Gutentag, F. J. Heimes, N. C. Krothe, R. R. Luckey, & J. B. Weeks, U.S. Geological Survey Professional Paper 1400-B, 1984.

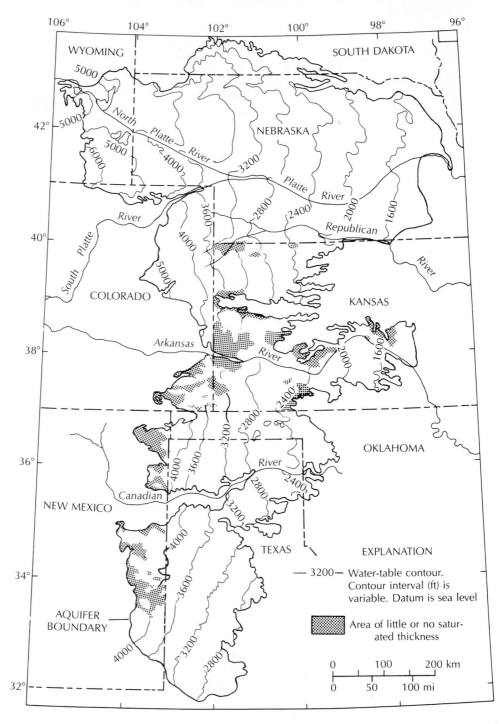

FIGURE 8.22 Water table in the High Plains aquifer, 1980. Source: E. D. Gutentag, F. J. Heimes, N. C. Krothe, R. R. Luckey, & J. B. Weeks, U.S. Geological Survey Professional Paper 1400-B, 1984.

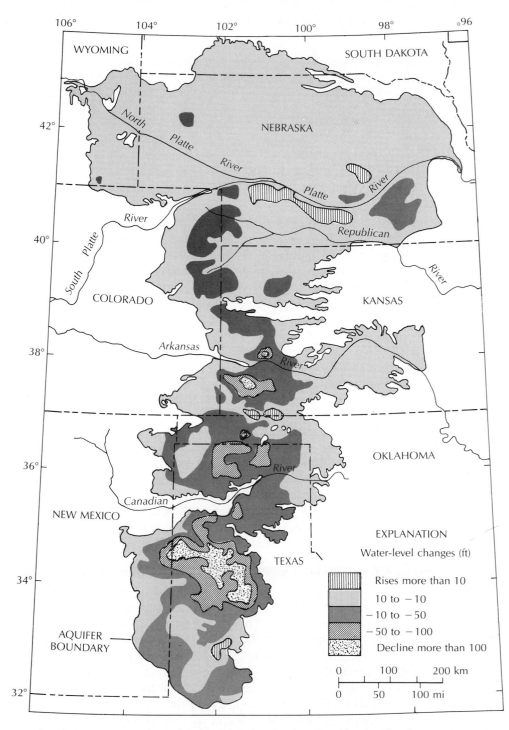

FIGURE 8.23 Water-level changes in the High Plains aquifer, predevelopment to 1980. Source: E. D. Gutentag, F. J. Heimes, N. C. Krothe, R. R. Luckey, & J. B. Weeks, U.S. Geological Survey Professional Paper 1400-B, 1984.

aquifer system, but the costs of obtaining it increase as the depth to water increases. In some places, the natural quality of the water precludes its use for drinking water, although it is generally usable for irrigation. Careful management of this resource will help to extend the period of time that it will continue to supply water for irrigation in those areas of low recharge and high use.

8.7   INTERACTIONS OF GROUND WATER AND LAKES OR WETLANDS

One of the important aspects of the hydrology of wetlands and lakes is their interaction with ground water. This interaction plays an important role in determining the water budget for a lake or a wetland (Winter 1977; Siegel 1988a, 1988b). Lakes may be classified hydrogeologically on the basis of domination of the annual hydrologic budget by surface water or ground water. Surface-water-dominated lakes typically have both inflow and outflow streams, whereas seepage lakes are ground-water dominated (Born, Smith, & Stephenson, 1979). Siegel (1988a) points out some wetlands are recharge areas, where excess water recharges the water table. Other wetlands are maintained by ground-water discharge. For example, bogs can be ground-water-recharge areas, whereas fens are typically supported by ground water discharge (Siegel 1988b).

There is a continuum from wetlands to lakes, starting with a wet meadow, which has saturated surface soils but no standing water, and ending with a deep lake. Thus the hydrologic relationships of ground water and lakes can also be applied to ground water and wetlands. Indeed, many lakes are ringed by wetlands.

Figure 8.24 shows several possible relationships between ground water, surface water, and lakes or wetlands. All the lakes or wetlands receive direct precipitation and lose water by evaporation. In most cases, neither of these will dominate the water balance. Figure 8.24A shows a seepage lake. It is ground-water dominated, being fed by ground-water inflow on one side and drained by ground-water outflow on the other. The lake shown in Figure 8.24B is also ground-water dominated, being fed by ground water and drained by stream flow. Some lakes may be controlled by neither ground water nor surface water. The lake shown in Figure 8.24C is fed by both stream flow and ground-water inflow. However, the amount of stream flow is low enough that it can be drained entirely by ground-water outflow. The lake in Figure 8.24D is surface-water dominated, being fed by streams and also drained by streams. The ground-water inflow and outflow are minimal. Figure 8.24E represents a reservoir created behind a dam. The water level created by the reservoir is much greater than the adjacent ground-water level. Water can be lost by seepage along both sides and the dam. A terminal lake is shown in Figure 8.24F. It is fed by both streams and ground-water inflow on all sides. Water is discharged only by evaporation, which dominates the water balance.

In a field study of the ground-water regime of permanent lakes in hummocky moraine of western Canada, both local and intermediate flow systems were identified (Meyboom 1967). The lakes were seepage-type (no surface outlet) and

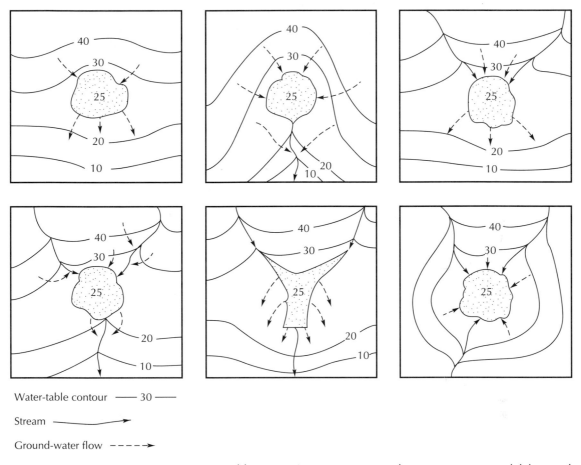

Water-table contour ——— 30 ———

Stream

Ground-water flow - - - - ➤

FIGURE 8.24 Some possible interactions among ground water, streams, and lakes and wetlands.

received inflow from either local or local plus regional flow systems. Figure 8.25 shows the water table and two lakes. Interlake areas are recharge areas, with the higher-elevation lake receiving water from local flow systems; the lower lake is fed by both local and regional flow systems.

In early spring, the water table was high, and most lakes were like those of Figure 8.25A. However, as the water table fell during the dry season, the ground-water divide disappeared between some lakes. These lakes became flow-through lakes, with ground water flowing in on one side and out on the other. The upper lake in Figure 8.25B illustrates this condition.

The flow conditions close to each lake were complicated by a growth of willow trees around the lake. Willows are *phreatophytes*—plants that use large amounts of ground water. During the growing season, the water table may be locally depressed below lake level beneath the phreatophyte fringe (Figure 8.25C). This can cause outflow at the margins of the lake, still leaving inflow beneath the

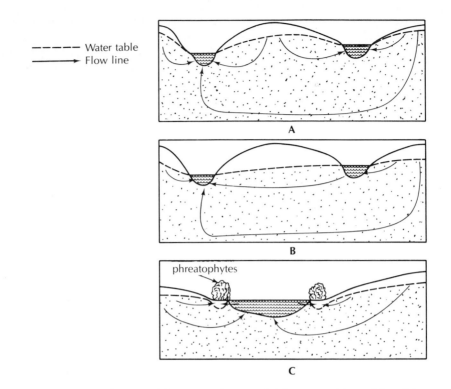

- - - - - Water table
———→ Flow line

FIGURE 8.25 Ground-water–lake interactions: **A.** High water table and interlake ground-water divide. **B.** Low water table and no interlake divide. **C.** Depressed water table due to fringe of phreatophytes. Source: Redrawn from P. Meyboom, *Journal of Hydrology* 5 (1967): 117–42. Used with permission of Elsevier Scientific Publishing Company, Amsterdam.

bottom. Phreatophyte water usage is diurnal, with resulting fluctuations in the seepage balance of the lake (Meyboom 1967).

A very interesting aspect of lake hydrology has been revealed through the numerical modeling of ground-water–lake systems. If the water table is higher than the lake level on all sides of a seepage lake, ground water will seep into the lake from all sides. Figure 8.26 shows a cross section through a seepage lake, with the water table higher than the lake surface. Potential distribution is based on a two-dimensional steady-state numerical model (Winter 1976). There is a stagnation zone beneath the lake, indicating both local and regional ground-water flow. Upward seepage takes place throughout the lake bottom. As long as the stagnation zone is present, the lake will not lose water through the bottom. However, should an aquifer or high-conductivity layer underlie the lake, the stagnation zone could be eliminated (Winter 1976). Without the stagnation zone, the lake will lose water through part or all of the bottom, even with the presence of a water-table mound downslope (Figure 8.27). Flow patterns of considerable complexity can form in multiple-lake systems, with some lakes having inflow through the bottom and some outseepage (Figure 8.28).

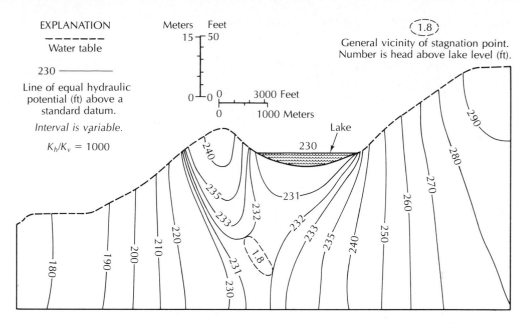

FIGURE 8.26 Hydrogeologic cross section showing head distribution in a one-lake system with a homogeneous, anisotropic aquifer system. Results are based on a two-dimensional, steady-state, numerical-simulation model. Source: T. C. Winter, U.S. Geological Survey Professional Paper 1001, 1976.

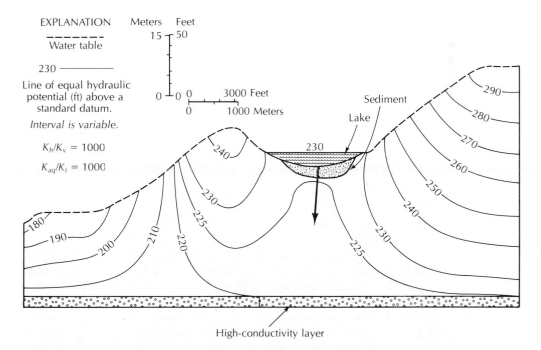

FIGURE 8.27 Hydrogeologic cross section showing head distribution in a one-lake system with a layered aquifer system. The high-conductivity layer has a conductivity 1000 times as great as the low-conductivity layer. The lake loses water to the aquifer. Source: T. C. Winter, U.S. Geological Survey Professional Paper 1001, 1976.

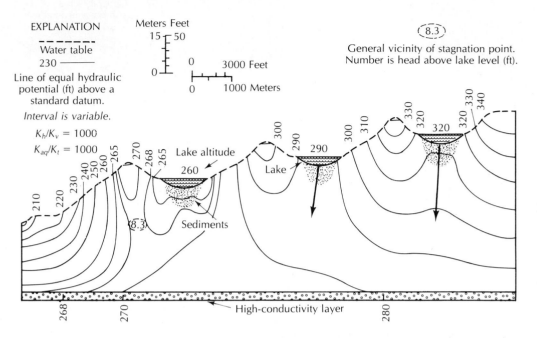

FIGURE 8.28 Hydrogeologic cross section through a three-lake system with a complex aquifer. Local and regional ground-water flow systems are present. Source: T. C. Winter, U.S. Geological Survey Professional Paper 1001, 1976.

Three-dimensional numerical analysis of ground-water–lake systems has shown that the stagnation zone, if present, is beneath the down-gradient side of the lake and follows the lakeshore (Winter 1978). The area of outseepage, if present, can be estimated from the three-dimensional model. The one-lake system in Figure 8.29 is shown in top view and two cross sections. Outseepage occurs through the center of the lake, while seepage into the lake takes place all around the edges.

Whether or not a stagnation zone will develop is largely determined by the following factors: (a) the height of adjacent water-table mounds relative to the lake level, (b) the position and conductivity of highly conductive layers in the ground-water system, (c) the ratio of horizontal to vertical hydraulic conductivity, (d) the regional slope of the water table, and (e) the lake depth (Winter 1976).

As the downgradient ground-water mound height increases, the less is the likelihood that the lake will leak. If the ratio of horizontal to vertical hydraulic conductivity is low (less than 100), the model studies have indicated that stagnation points typically are present. As the ratio increases, so does the likelihood of the stagnation point disappearing and outseepage developing. Also, the presence of high-conductivity layers increases the possibility of outseepage, as does the presence of deeper lakes. Flow systems with a low regional water-table gradient, as well as thin aquifers, are more likely to have only local flow systems—hence, lakes that do not leak if there is a down-gradient water-table divide.

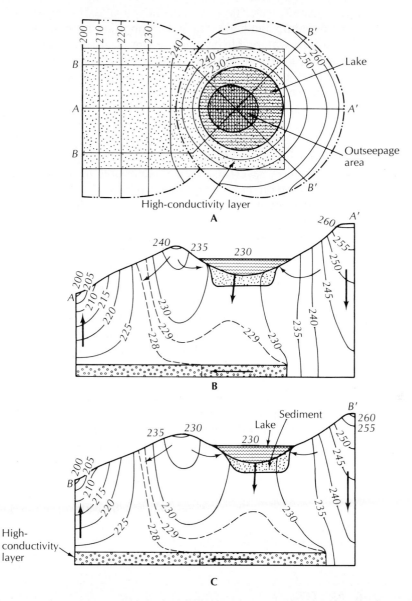

FIGURE 8.29 Three-dimensional analysis showing the area of outseepage in the center of the lake. Seepage into the lake occurs over the rest of the bottom. Areal extent of the high-conductivity layer is indicated by stipples. Source: T. C. Winter, *Water Resources Research* 14, no. 2 (1978): 245–54.

Numerical modeling studies have shown that, of all of the factors that affect ground-water flow systems near a lake, the water-table configuration is the most dynamic and hence the most important to monitor during field studies. The slope of the water table beyond the immediate drainage basin of a lake has a strong impact on lake-bed seepage. This is particularly true for the slope between the ground-water mound on a lake's down-gradient side and a regional discharge area. The steeper the slope, the greater the likelihood that lake-bed seepage will occur (Winter 1981).

Because of the complex nature of ground-water recharge, localized ground-water mounds may develop temporarily. These typically form in areas where the unsaturated zone is thin, for example, near lakes. They can create temporary flow reversals. Thus it is possible for ground-water mounds to temporarily develop on the down-gradient side of a lake so that lakes that normally have outseepage could undergo transient metamorphosis to lakes with no outseepage (Winter 1983).

In field studies of ground-water–lake interactions, there is no substitute for piezometers in defining the flow field. There must be several shallow piezometers all around the lake in order to define the water-table configuration (Winter 1976, 1978). It is necessary to determine whether a ground-water divide exists on the down-gradient side of a lake or between two lakes. If a divide exists, then the maximum water-table elevation must be found. The existence of a stagnation point can be determined by placing a nest of closely spaced piezometers of different depths below the shoreline on the downslope side of the lake. If the head at all depths is greater than the elevation of the lake surface, then a stagnation point is indicated.

In order to calculate the seepage rate into a lake, the seepage per unit area can be measured at the sediment–lake interface (Lee 1976). This is done with a device made of half an oil drum, with the open part pushed into the sediments. The rate of flux across the sediments is measured. Inflow generally occurs in the littoral zone of a lake (Winter 1978; Lee 1976). Outseepage in lakes with no down-gradient water-table divide can occur in both the down-gradient littoral zone and the deeper part of the lake. If the lake has a down-gradient water-table divide and there still is outseepage, it will be through the deeper part of the lake bottom. This area typically contains low-conductivity sediments, limiting the rate of water loss (Winter 1976).

COMPUTER NOTES

FLOWNET is a finite-difference computer program that models vertical slices through the earth, similar to the diagrams in this chapter. Chapter 14 explains what a finite-difference model is. It is a very simple program to use, yet it is very powerful. This is a steady-state model in which the potentiometric head along one or more of its boundaries is entered. The output is a flow net that illustrates the potential distribution within the earth as well as some illustrative flow paths. The model has an animation feature that allows the user to compare the relative rates of ground-water flow along different flow paths. Both isotropic and anisotropic aquifers can be modeled, and it is very simple to account for heterogeneities.

Input to the model is the head distribution along its open boundaries and the horizontal and vertical hydraulic conductivity and porosity for each cell of the model. The model consists of cells in a matrix of vertical columns and horizontal rows. The model can have up to about 5000 cells—for example, 100 columns and 50 rows.

On the disk that accompanies this book, there is a free student version of FLOWNET called FLOWNETD. It is the same as the full program, except that it is limited to a maximum of 200 cells, with 3 to 20 columns and 3 to 13 rows. The program is interactive—that is, you enter the data as the program is running. Directions on how to load and run the program are in Appendix 15. Once the program is running, the instructions are on the screen. You turn from one page in the program to the next with the Page Up and Page Down keys on the keyboard. If you simply page through all seven pages of the program, there is a test case of Tóth's model (Figure 8.3). Once the flow net is on the screen, press Page Up once; then use the Down-arrow key to move the little arrow to the line marked "Time steps are not drawn." Press the Right-arrow key, and the line will then read: "Time steps are drawn." Now press the Page Down key, and the flow net will reappear with little dots on the flow lines. Press the space bar and the little dots will begin to move. Use the + key to make them move faster and the − key to make them move more slowly. The geometry of the ground-water basin can be defined on page 3 by selecting the number of rows and columns and the length and height of the model. It is a good idea to keep the model to true scale by having the ratio of the number of columns to rows the same as the ratio of length to height. The default model has closed boundaries on all sides but the top. It is possible to have the left and right sides open to flow, but heads need to be specified along any open boundary. The heads are specified on page 4. They can be positive or negative with respect to an arbitrary datum. These fixed heads are displayed at a relative scale along each open boundary. Conductivity values are entered on page 5. The matrix on this page represents the rows and columns. There is a blinking cursor that can be moved with the Arrow keys. If you wish to put in a layer of different hydraulic conductivity, move the cursor to the point where the layer starts. Then type any letter except (lowercase) x—for example, b. You must now enter hydraulic conductivity values and porosity for the new symbol, using the Down-arrow key to move from one value to the next. After you enter the porosity and press the Down-arrow key, the program will return to the hydraulic conductivity matrix. The blinking cursor is at the same place. Type b to access the newly defined symbol. Move the cursor with the Arrow keys to encompass the area of the different conductivity layer. This can be more than one row thick. When you have entered the position of the layer, press x to return the cursor to the default value. Using this method you can enter several layers with different conductivity values. Once the conductivity values are entered, press Page Down twice to obtain a solution.

As an exercise, start with the default (Tóth model) solution, and then place a horizontal lens with a K_h and K_v equal to 100 and effective porosity equal to 0.3. It is a very good exercise to use this program to reproduce the figures in this chapter.

FLOWNET is available for purchase for $200.00 (U.S.) from:

> Dr. C. J. Hemker
> Geohydroloog
> Elandsgracht 83
> 1016 TR Amsterdam
> The Netherlands

NOTATION

h Head

g Gravitational constant

L Length of flow system

x Horizontal distance from lowest point in the water table

z_0 Lowest water-table elevation

z   Water-table elevation

tan α Slope of the water table

PROBLEMS

1. Make a copy of Figures 8.26, 8.27, and 8.28. Draw sufficient flow lines on them to illustrate the regional flow patterns. Even though these aquifers are anisotropic, make the flow lines cross the equipotentials at right angles.

2. Based on the flow fields shown on Figures 8.3 and 8.7, draw a flow net on Figure 8.30.

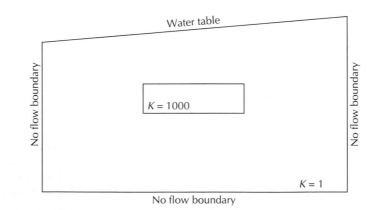

FIGURE 8.30   Diagram for Problem 2.

3. Answer the following questions based on Figure 8.31.
    A. Fill in the heads at the locations labeled on the diagram.

Location	Elevation Head	Pressure Head	Total Head
A			
B			
C			
D			
E			
F			

    B. Find one place on Figure 8.31 where recharge is occurring, and label it *R*.
    C. Find one place on Figure 8.31 where discharge is occurring, and label it *D*.
    D. Draw in flow lines on Figure 8.31 starting at points *X*, *Y*, and *Z*.

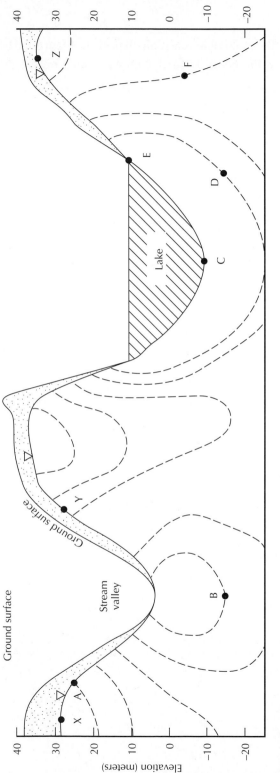

FIGURE 8.31 Diagram for Problem 3. The dashed lines are equipotentials. Heads are in meters.

9 Geology of Ground-Water Occurrence

Let us not forget in this connection that every stream of water whenever it comes from a higher point and flows to a delivery tank through a short length of pipe, not only comes up to its measure but yields, moreover, a surplus; but whenever it comes from a low point, that is, under a less head, and is conducted a tolerable long distance, it will actually shrink in measure by the resistance of its own conduit.
De aquis urbis Romae, libri II, Sextus Julius Frontinus (ca. A.D. 35–104)

9.1 INTRODUCTION

Ground water inevitably occurs in geological formations. Knowledge of how these earth materials formed and the changes they have undergone is vital to the hydrogeologist. The earth is basically heterogeneous, and geological training is prerequisite to understanding the distribution of geological materials of varying hydraulic conductivity and porosity.

A ground-water study of an area necessarily includes a review of previous geologic studies. In an area of limited geologic knowledge, a detailed geologic study must be made along with the ground-water study. The hydrogeologist is concerned with the distribution of earth materials as they affect the porosity and hydraulic conductivity of the earth. The results of a geologic study conducted by a ground-water geologist may well differ substantively from a study in the same area by a paleontologist or a petrologist. The methods employed by each would be very similar, however. Rock outcrops would be examined, certain properties noted, and the results mapped. Test borings and geophysical methods would be used to examine the subsurface. Earth materials would be grouped according to similar properties. The hydrogeologist most often begins with the proper formation names that have been assigned to an area. But these formations may then be grouped or split according to hydraulic characteristics rather than by lithology or fossil species present. For example, some bedding planes in a limestone may have been enlarged by solution, so as to transmit significant amounts of water. A hydrogeologic study might identify these on the basis of outcrop and subsurface data and map their locations. Other geological studies might not even note their occurrence.

A locality almost always has a variety of geologic materials resulting from different processes. For example, the Keweenaw Peninsula of Michigan has Precambrian bedrock famous for copper mineralization in the Portage Lakes Lava Series and Copper Harbor Conglomerate. Some water wells obtain a small supply of water from fractures in the dense bedrock, which has a very low primary permeability. One community obtains a supply from an old mine adit originally constructed to intercept seepage into the lower working levels of the mine. Sediments deposited as glaciofluvial deposits and as deltaic or long-shore deposits in Lake Superior (especially during the Pleistocene, when lake levels were higher) also serve as aquifers for community water supplies (Doonan, Hendrickson, & Byerlay, 1970). A hydrogeological study of ground-water exploration in this area must examine a number of different geologic settings. A study focused on bedrock aquifers might locate wells yielding 1 to 15 gal/min (0.06 to 1 L/s) from fractured bedrock. Sand deposited by long-shore currents along Lake Superior can yield up to 100 gal/min (6.3 L/s) to wells. If present, this is a preferred alternative to bedrock wells.

The hydrogeologist should consider all alternatives before selecting one or more sites for detailed exploration. The preliminary survey usually consists of examination of air photos, topographic and geologic maps, and logs and reports of existing wells and then a "walking survey" of the area. This is followed by detailed study using methods such as test borings, fracture-trace analyses, and geophysical surveys. If the results of the detailed study are favorable, one or more test wells are usually installed and pumping tests made.

Hydrogeologic studies are also made to evaluate the suitability of a site for such uses as sanitary landfills, land treatment of wastewater, seepage ponds for wastewater disposal and spray irrigation of wastewater, or nuclear power plant siting. The basic methods of study and evaluation are similar. However, the target geologic terrane will differ according to the application. Landfills are usually placed in areas of low-permeability material in order to minimize the possibility of ground-water pollution. On the other hand, seepage ponds for artificial aquifer recharge would be located in as coarse a material as possible. While this chapter will focus on ground-water as a resource, the basic principles of occurrence are the same, regardless of the application.

9.2   UNCONSOLIDATED AQUIFERS

Materials ranging in texture from fine sand to coarse gravel are capable of being developed into a water-supply well. Material that is well sorted and free from silt and clay is best. The hydraulic conductivities of some deposits of unconsolidated sands and gravels are among the highest of any earth materials. The filterlike nature of many fine- to medium-grain sediments removes such particulate matter as bacteria and viruses, so that epidemiological water quality is usually good. However, dissolved contaminants such as nitrate, chloride, and tetrahydrofuran can travel through most host sediment and rock with little attenuation other than dilution. Unconsolidated materials are also often close to a source of recharge, such as a stream or lake. Shallow unconsolidated aquifers are in the region of rapid circulation of water—usually in local flow systems.

Unconsolidated deposits range in thickness up to tens of thousands of feet in sedimentary basins such as the Gulf of Mexico. Local well-construction regulations often call for about 30 ft (10 m) of surface casing to prevent surface water or very shallow ground water from entering a well. An unconsolidated deposit must extend more than 30 ft deep to be useful for a water supply. Even for a domestic well, at least 3 to 6 ft (1 to 2 m) of slotted casing, well screen, or a well point is needed below the minimum casing depth.

9.2.1 Glaciated Terrane

The midcontinental area of North America has been covered a number of times in the past by great sheets of glacier ice. The moving ice eroded and deposited as it waxed and waned across the land surface. As a result, deposits of glacial drift from less than a few feet to scores of feet thick mantle the bedrock. This covering of glacial drift is a potential source of ground water, although the hydrogeology of glaciated areas can be very complex (Figure 9.1).

Glacially related sediments have a wide range of hydraulic conductivity values. Materials carried by the glaciers ranged in size from clay to large boulders. The mode of deposition determined how well the sediment was size-sorted. Glacial till was deposited directly from the glacial ice without significant sorting by running water. As a result, it can contain particles of any size. If the ice held a wide range of particle sizes, so will the till it deposited. Generally the till will have a low hydraulic conductivity, especially if it is clay-rich. Mountain glaciers, in particular, have deposited some very coarse till materials.

Several studies have resulted in published and unpublished hydraulic conductivity values of glacial till deposits (Norris 1963; Sharp 1984; Hendry 1982; Prudic 1982; Herzog & Morse 1984; Gordon & Huebner 1983). Permeability of glacial tills depends upon the clay content, mode of deposition, and degree of weathering. *Ablation till* is deposited as the ice melts and is typically slightly water-sorted; this should make it more permeable than *basal till,* which is laid down beneath the ice. In one study of the Vandalia Till in Illinois, the hydraulic conductivity was determined by slug tests (Cooper et al. method) (Herzog & Morse 1984). The unaltered basal till had a mean hydraulic conductivity of 1.35×10^{-7} cm/s; the basal till that was weathered and fractured had a mean hydraulic conductivity of 5.86×10^{-6} cm/s; and the weathered ablation till had a mean hydraulic conductivity of 3.81×10^{-5} cm/s. In New York State a homogeneous till was tested that contained 50% clay, 27% silt, 10% sand, and 13% gravel (Prudic 1982). The mean hydraulic conductivity as determined from slug tests (Hvorslev method) was 2×10^{-8} cm/s and ranged from 1×10^{-8} to 3×10^{-8} cm/s. Falling-head permeameter tests on the same till had an average vertical hydraulic conductivity of 6.2×10^{-8} cm/s and horizontal hydraulic conductivity of 3.8×10^{-8} cm/s. A till that consisted of an upper, weathered zone and an unweathered lower zone was studied in Alberta (Hendry 1982). The unweathered till had a hydraulic conductivity of about 10^{-8} cm/s. The weathered till had both small-scale

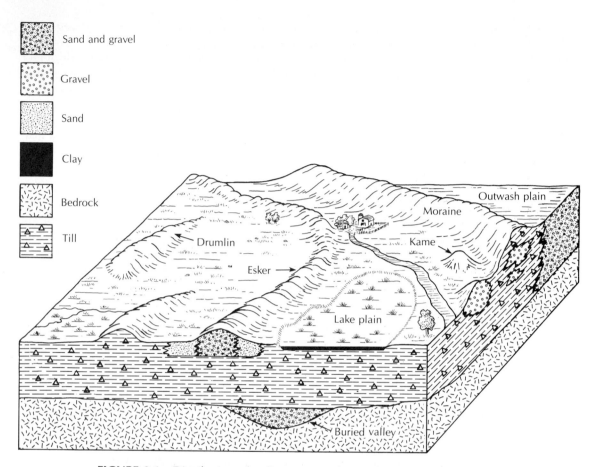

Sand and gravel

Gravel

Sand

Clay

Bedrock

Till

Outwash plain

Moraine

Drumlin

Kame

Esker

Lake plain

Buried valley

FIGURE 9.1   Distribution of sediments in a glaciated terrane.

fractures with a hydraulic conductivity of 10^{-7} cm/s and large-scale fractures with a hydraulic conductivity of 10^{-5} cm/s. In Wisconsin a till that averaged 83 percent clay and silt had a mean hydraulic conductivity of 3×10^{-8} cm/s on the basis of laboratory tests (Gordon & Huebner 1983). However, slug tests in the field yielded a hydraulic conductivity of 1×10^{-5} cm/s. This till is highly fractured and the matrix conductivity as revealed by the permeameter test is much less than the bulk conductivity, which is measured by the slug test. Other reports of till permeability indicate that it ranges from as little as 1.9×10^{-9} to 2.3×10^{-3} cm/s (Norris 1963; Sharp 1984). In most cases, if both laboratory and field tests are performed on the same till, the field test indicates one to three orders of magnitude more permeability. The reason for this is that the field tests measure the properties of a larger sample of the material, which may have fractures or sand and silt seams with higher conductivity values than the clay matrix (Stephenson, Flemming, & Mickelson 1988).

In a study of ground-water flow in a clayey till in Ontario, it was found that the till was fractured near the surface. Even with the fracturing, the bulk hydraulic conductivity of the till as determined by slug tests using the Hvorslev method was in the range of 10^{-7} to 10^{-8} cm/s. Active weathering along the fractures was noted to a depth of 5.7 m in test pits, and tritium dating of ground water showed that water to depths as great as 12 m was no older than 40 y. This suggests that water is actively circulating through the fractures to depths as great as 12 m (Ruland, Cherry, & Feenstra, 1991).

In a study in central Iowa it was found that the hydraulic conductivity of a weathered till was 3×10^{-4} to 5×10^{-4} cm/s, which was three orders of magnitude greater than the unweathered till. This till was quite sandy, having an average of 48% sand, 37% silt, and 15% clay (Jones, Lemar, & Tsai 1992).

Glacial-lacustrine sediments have hydraulic conductivity values similar to those of tills. Lake Michigan bottom sediments were shown to have a hydraulic conductivity ranging from 2×10^{-5} to 5×10^{-8} cm/s (Bradbury & Taylor 1984). Fine-grained glacial-lacustrine sediments in southern Indiana had lab permeability values ranging from 3.3×10^{-6} to 2.7×10^{-8} cm/s and averaged 5.9×10^{-7} cm/s (Fetter 1985). A sandy lacustrine layer at the same site had a higher hydraulic conductivity of 3.3×10^{-4} cm/s.

Glacial drift may also have been well sorted by running water—usually the meltwater from the glacier. Such sediments, called glacial outwash, can be seen forming in the braided stream deposits of meltwater streams draining modern mountain glaciers. The coarse gravel is close to the terminus, sand is deposited farther downstream, and the silt and clay are carried away to be deposited far downstream in lakes or the ocean. Coarser, water-sorted glacial materials can have very high hydraulic conductivities. The general range for these sediments is 2.8 to 2835 ft/day (10^{-3} to 1 cm/s). The materials with lower hydraulic conductivities are not as well sorted. Some silty outwash materials have hydraulic conductivities as low as 2.8×10^{-2} ft/day (10^{-5} cm/s). Well-sorted glacial deposits may be found associated with recognizable geomorphic features such as moulin kames, kame terraces, and eskers. However, these features are topographically high and, except for the base, may be unsaturated (Figure 9.1). If they are buried, then they can be excellent sources of water. Such buried gravel deposits can sometimes be located by electrical resistivity methods. Outwash plains and valley-train deposits are not as obvious geomorphic features, although they are often found near terminal moraines. Outwash deposits are usually well-sorted sand or gravel or mixtures of both. Some glacial outwash materials are tens of feet thick and make prolific aquifers. The city of Tacoma, Washington, obtains ground water from outwash deposits in the valley of the North Fork of the Green River. Six production wells were tested at rates of 7500 to 8600 gal/min (470 to 540 L/s) with total drawdowns of 1.9 to 7.3 ft (0.58 to 2.23 m). *Specific capacities** as high as 4400 gal/m/ft (0.9 m²/s) were obtained (Carr 1976).

A key mark of glaciated terrane is the variability of hydraulic conductivity. A single test boring might reveal sequences of glacial deposits with a variation

---

*The specific capacity of a well is the yield divided by the drawdown of the water level from the nonpumping level. The units are thus flow-distance or m³/s/m (m²/s) or gal/min/ft.; 1 gal/min/ft = 2.07×10^{-4} m²/s.

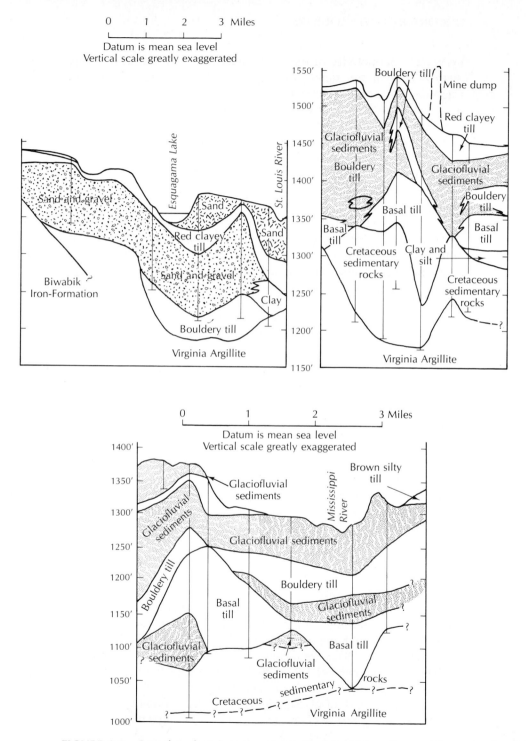

FIGURE 9.2 Complex glacial stratigraphy in the Mesabi Iron Range, Minnesota. Sand and gravel and glaciofluvial sediments are potential aquifers. Source: T. C. Winter, U.S. Geological Survey Water-Supply Paper 2029-A, 1973.

of hydraulic conductivity of 280 ft/day (10^{-1} cm/s) for well-sorted glacial sands to 0.003 ft/day (10^{-7} cm/s) for interbedded tills or clays. Figure 9.2 shows glacial stratigraphic cross sections for three areas in the Mesabi Iron Range of Minnesota with a high potential for ground-water development (Winter 1973).

During the Pleistocene, meltwaters from continental glaciers flowed across much of the North American landscape. This was true even in areas south of the limit of glaciation. These rivers carried large volumes of sediment. Well-sorted glaciofluvial sediments can provide excellent supplies of ground water.

The pre-Pleistocene landscape of the midcontinent was a bedrock erosional surface. Deeply incised rivers drained the land with a well-developed drainage network. The glacial sculpting of the North American landscape resulted in many changes of the preglacial drainage patterns. Many of the deep bedrock valleys were filled with sediment. Layers of till, lacustrine silts, and clays alternate with glaciofluvial deposits. Modern rivers follow the courses of some of the buried channels, whereas other former channels lie inconspicuously beneath the farmland of the Midwest. The courses of buried bedrock valleys can be determined by use of a number of remote-sensing and geophysical methods along with test drilling. In one study in Kansas (Denne et al. 1984) tonal patterns on springtime *LANDSAT* imagery and winter/summer anomalies in soil temperature were used along with seismic refraction, earth resistivity, and gravity measurements to define the channels of two bedrock valleys that had been filled with glacial drift. Test drilling was used to confirm the results of the remote techniques.

Hydraulic barriers may be present in Pleistocene buried valley aquifers. These barriers may be parallel to or across the trend of the buried valley aquifer. Their presence may be determined by aquifer test analysis as well as evaluation of hydraulic head relationships found in piezometers (Shaver & Pusc 1992).

CASE STUDY: HYDROGEOLOGY OF A BURIED VALLEY AQUIFER AT DAYTON, OHIO

There is a classic buried valley running beneath the city of Dayton, Ohio. One of the factors promoting the growth of Dayton has been the ready availability of a source of high-quality ground water (Norris & Spieker 1966). Permeable layers of glacial drift in the bedrock valley furnish water in large quantities to wells. In turn, these are recharged by infiltration of precipitation, as well as by water from the Miami River and its tributaries.

The bedrock in the area is the Richmond Shale of Ordovician age. During the Tertiary, an erosional surface developed, which was cleft by deeply incised rivers. During the late Tertiary, the main river draining the area was the Teays. This drainage system cut a number of valleys in the bedrock. Early Pleistocene glaciation dammed the rivers, so that lacustrine silts are found filling many parts of the Teays system. During the Kansan-Illinoian interglacial period, a radically different drainage system, the Deep Stage, prevailed in southwestern Ohio. Bedrock valleys were deeply entrenched during this time. The Deep Stage Valley passed through the present site of Dayton. Glacial processes of Illinoian and Wisconsinan age filled the Deep Stage Valley with layers of till and outwash. The character of the sediment is quite heterogeneous, as revealed by a test hole in the bedrock valley

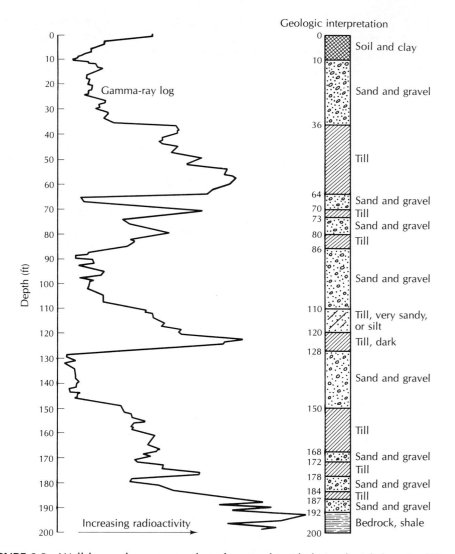

FIGURE 9.3 Well log and gamma-ray log of uncased test hole in glacial deposits filling a buried bedrock valley south of Dayton, Ohio. Source: S. E. Norris & A. M. Spieker, U.S. Geological Survey Water-Supply Paper 1808, 1966.

south of Dayton (Figure 9.3). The valley-train gravel aquifers alternate with confining till sheets. The aquifer layers beneath the till sheets are recharged where the till is missing because of either nondeposition or river-channel erosion. A cross section of the valley just south of Dayton shows the upper and lower aquifers separated by a discontinuous till sheet (Figure 9.4). The potentiometric surface in the lower aquifer lies below the bed of the Miami River. The potentiometric surface rises sharply near the river when the discharge and stage of the Miami River rise. This indicates a good hydraulic connection between the river and the aquifer.

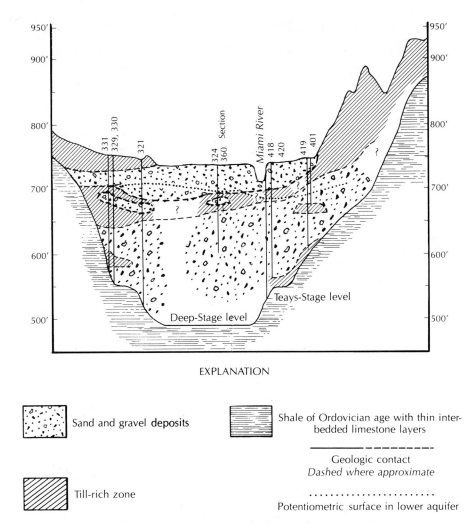

EXPLANATION

Sand and gravel **deposits**

Shale of Ordovician age with thin interbedded limestone layers

——————— – – – – – – – –
Geologic contact
Dashed where approximate

Till-rich zone

· ·
Potentiometric surface in lower aquifer

FIGURE 9.4 Cross section of buried bedrock valley at Dayton, Ohio, showing upper
(water-table) aquifer and lower (confined) aquifer. Source: Modified from S. E. Norris & A. M.
Spieker, U.S. Geological Survey Water-Supply Paper 1808, 1966.

The Rohrers Island Well Field is located on an island in the Mad River, a tributary
of the Miami River. The upper sand and gravel aquifer yields up to 90 million gallons per
day (3940 L/s). The upper aquifer, which is up to 65 ft (20 m) thick, is artificially recharged with river water that floods onto about 20 ac (8 ha) of infiltration lagoons during
periods when the river turbidity is low. The lagoon bottoms are annually cleaned of silt
and clay. This is a classic example of induced stream infiltration used as a water resource
management technique. Virtually all of the water pumped from this aquifer comes indirectly from the river. The filtration through the sediments reduces turbidity and removes
pathogens.

9.2.2 Alluvial Valleys

Flowing rivers deposit sediment, generally termed **alluvium.** During periods of flooding, alluvium is deposited in the channel as well as in the floodplain. As the flood peak passes, flow velocities start to drop, the energy available to transport sediment decreases, and deposition begins. Coarse gravel is deposited in the stream channel, sand and fine gravel form natural levees along the banks, and silt and clay come to rest on the floodplain. Point bars are formed by deposition on the inside of a bend in the river. Point-bar formation is not limited to floods.

The alluvium may be reworked by a meandering stream, even during quiescent periods for the river. As the channel swings back and forth across the floodplain, point-bar deposits of sand and coarse gravel are left behind. If the stream is aggrading, the general land level subsiding, or both, the alluvial deposits will thicken with time (Figure 9.5).

Rivers draining glaciers, either modern or Pleistocene, have very heavy sediment loads. Braided streams and gravel bars in the river are typical and connote the aggrading nature of such rivers. Thick deposits of sand, gravel, or mixtures of both are formed. Downcutting by a river through previously deposited sediment can form terraces on the sides of the lower-stage floodplain. Many modern rivers, even some outside of glaciated regions, have terraces formed of Pleistocene-age gravel deposits.

Alluvial valleys can be excellent sources of water. There are zones of gravel in old channel and point-bar deposits with a very high conductivity. The task of the hydrogeologist is to find these gravel zones, since silt and clay floodplain deposits may also be present, obscuring their locations. A knowledge of fluvial processes is helpful in this regard. Surface geophysical surveys can often be used to locate sand and gravel deposits where they are surrounded by cohesive

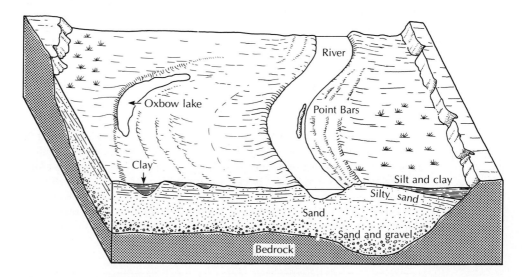

FIGURE 9.5   Distribution of alluvium in a flowing river.

sediments. Test drilling is usually necessary to confirm the preliminary conclusions of the hydrogeologist and geophysicist.

In evaluating the water-yielding potential of stream-terrace deposits, it should be kept in mind that erosion, as well as deposition, can result in terrace formation. If a terrace represents an erosional surface on bedrock, it is not underlain by potentially permeable alluvium.

9.2.3 Alluvium in Tectonic Valleys

Many major valley systems are products of tectonic activity rather than of fluvial or glacial erosion. During mountain-building episodes, the uplift of mountain masses will result in intermontane basins being formed. Fault-block valleys can also be created by down-dropping of large crustal pieces along faults. Erosion of the mountains creates sediment that is carried into the valleys, forming talus slopes, alluvial fans, and alluvial and lacustrine deposits. These sediments can be very coarse, with high hydraulic conductivities—alluvial fan gravels and channel deposits, for example. Lacustrine clays, on the other hand, can be fine, with low conductivity. Gravel aquifers confined by lacustrine sediment are quite typical of such basins.

Intermontane valleys of the interior basins of southeastern California and the Great Basin area of Nevada and Utah are typically tectonic. The unconsolidated sediments in the basins may be part of either local or regional flow systems (Figure 9.6). In regions that are semiarid to arid, ground water is recharged by precipitation in the mountains. Bedrock beneath the mountains may receive direct recharge and then feed the valley-fill sediments. Streams originating in the

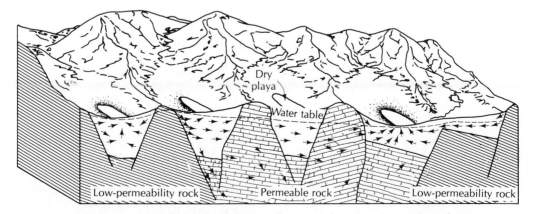

FIGURE 9.6 Common ground-water flow systems in tectonic valleys filled with sediment. Basins bounded by impermeable rock may form local or single-valley flow systems. If the interbasin rock is permeable, regional flow systems may form. In closed basins, ground water discharges into playas, from which it is discharged by evaporation and transpiration by phreatophytes. Source: Modified from T. E. Eakin, D. Price, & J. R. Harrill, U.S. Geological Survey Professional Paper 813-G, 1976.

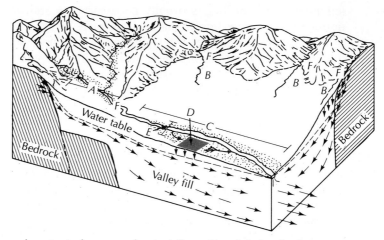

A, Gaining reach, net gain from ground-water inflow, although in localized areas stream may recharge wet meadows along floodplain. Hydraulic continuity is maintained between stream and ground-water reservoir. Pumping can affect streamflow by inducing stream recharge or by diverting ground-water inflow that would have contributed to streamflow.

B, Minor tributary streams, may be perennial in the mountains but become losing ephemeral streams on the alluvial fans. Pumping will not affect the flow of these streams because hydraulic continuity is not maintained between streams and the principal groundwater reservoir. These streams are the only ones present in arid basins.

C, Losing reach, net loss in flow due to surface-water diversions and seepage to ground water. Local sections may lose or gain depending on hydraulic gradient between stream and ground-water reservoir. Gradient may reverse during certain times of the year. Hydraulic continuity is maintained between stream and groundwater reservoir. Pumping can affect streamflow by inducing recharge or by diverting irrigation return flows.

D, Irrigated area, some return flow from irrigation water recharges ground water.

E, Floodplain, hydrologic regimen of this area dominated by the river. Water table fluctuates in response to changes in river stage and diversions. Area commonly covered by phreatophytes (shown by random dot patterns).

F, Approximate point of maximum stream flow.

FIGURE 9.7   Ground-water–surface-water relationships in valley-fill aquifers located in arid and semiarid climates. Source: T. E. Eakin, D. Price, & J. R. Harrill, U.S. Geological Survey Professional Paper 813-G, 1976.

mountains may also lose water to the alluvium when the flow goes across the valley bottoms (Figure 9.7).

The water table is generally closer to the surface in the lower elevations of intermontane basins than it is at the edges next to the mountains, so wells in the former locations would have a smaller pumping lift—hence, lower energy use. The intermontane deposits typically slope downward from the mountain flanks. Beneath the upper parts of alluvial fans, the depth to ground water may be hundreds of feet. Surface streams flowing onto high alluvial fans may disappear as they lose water to the coarse sediment.

In the Great Basin region of the western United States, saturated valley-fill deposits over 1000 ft (300 m) thick are common in the large valleys (Eakin,

Price, & Harrill 1976). Artesian conditions are often found in the lower elevations of these basins. Pleistocene lacustrine clays near the surface overlying coarse alluvium create this situation.

In tectonic valleys, ground-water outflow can occur by transpiration, evaporation from surface water or saturated soils, spring discharge, and/or underflow into adjacent basins. Water pumped and used consumptively for irrigation is also an outflow. If the basin is topographically closed, there is no stream discharge. In such a case, evapotranspiration and underflow are the natural drainage methods.

Wells in tectonic valleys should be located where the aquifer material is coarse, the depth to water is not great, and a source of recharge water is available. These criteria are often met in well fields located near modern rivers. However, there may not be any surface streams near areas where wells are needed. Artesian aquifers overlain by thick lacustrine deposits may be too deep to find by surficial geophysical methods, such as electrical resistivity. Test drilling may be the only recourse in such cases, albeit an expensive one. Ground-water quality in tectonic valleys can have significant spatial variation. In general, ground water that is not actively circulating may be high in dissolved solids. Shallow ground water, with a high rate of direct evaporation, may also have high salinity. In the Great Basin area, individual wells have been developed with yields of up to 8600 gal/min (540 L/s) with specific capacities of up to 3000 gal/min/ft (0.6 m^2/s). These are prodigious wells. The average yield is lower but still substantial at 1000 gal/min (65 L/s) (Eakin, Price, & Harrill 1976).

CASE STUDY: TECTONIC VALLEYS—SAN BERNARDINO AREA

The San Bernardino area is in the upper Santa Ana Valley of the coastal area of southern California (Dutcher & Garrett, 1963). The valley is tectonic in origin and bounded on the north, east, and south by mountains of consolidated rocks. The alluvial valley is subdivided into separate ground-water basins by faults in the alluvial materials, which are barriers to ground-water movement. The Bunker Hill Basin has mountains on three sides and the San Jacinto Fault on the fourth side (Figure 9.8).

The alluvium in the basin consists of Pleistocene-age deposits overlain by Recent alluvium, river-channel deposits, and dune sands. The alluvium is from 690 to 1400 ft (210 to 425 m) thick at the center of the basin. The older deposits consist of alluvial fans, terrace deposits, and stream channels. The deposits can be highly permeable, but facies of low permeability exist and form confining layers. Faulting cuts the older alluvium, and the faults are generally barriers to ground-water movement. They may impede ground-water flow by (1) offsetting of gravel beds against clay layers, (2) folding of impermeable beds upward along the fault, (3) cementation by carbonate formation, and (4) formation of clayey gouge.

Recent alluvium includes highly permeable river-channel deposits through which much of the ground-water recharge occurs. They underlie active and abandoned stream channels and are above the water table, for the most part. The Recent floodplain material is less permeable than the stream channels. It is 50 to 100 ft (15 to 30 m) thick and is not known to be faulted.

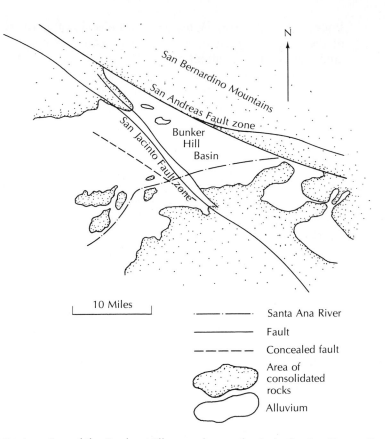

FIGURE 9.8 Location of the Bunker Hill ground-water basin in the San Bernardino area of southern California. Source: Modified from L. C. Dutcher & A. A. Garrett, U.S. Geological Survey Water-Supply Paper 1419, 1963.

The lower parts of the valley floor receive 12 to 17 in. (30 to 43 cm) of precipitation per year. Owing to orographic effects, precipitation in the San Bernardino Mountains to the east is as much as 28 in. (71 cm). Runoff from the mountains flows into the Santa Ana River and its tributaries. In the upper reaches of the basin, the river is a losing stream. Ground-water recharge by floodwater takes place through the permeable river-channel deposits.

A cross section of the Bunker Hill area is shown in Figure 9.9. Ground-water recharge takes place in the upper parts of the basin and flows toward lower elevations. Well hydrographs on the upstream side of the San Jacinto Fault show that the head in a shallow well is lower than in deeper wells, indicating upward flow. In the San Bernardino area, the younger alluvium is confining, and when wells were first drilled in the area they were flowing.

At one time, considerable ground-water outflow occurred from the Bunker Hill Basin in the area where the Santa Ana River crosses the San Jacinto Fault Zone (Colton Narrows). Although the fault zone does not offset the Recent alluvium, it forces deeper water to move upward. Except for Recent river-channel deposits beneath the Santa Ana River, the surficial deposits are, for the most part, impermeable. At the turn of the century, large

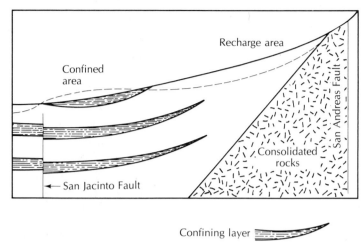

FIGURE 9.9 Cross section through the unconsolidated deposits of the Bunker Hill ground-water basin. The section is along the course of the Santa Ana River, which provides recharge to the aquifers. Source: Modified from L. C. Dutcher & A. A. Garrett, U.S. Geological Survey Water-Supply Paper 1419, 1963.

springs made this area marshy, and ponds were present. Heavy ground-water withdrawals for irrigation have lowered the water table and reduced the amount of outflow. Figure 9.10 shows the water-table contours where the Santa Ana River crosses the San Jacinto Fault as they existed in 1939. Ground-water contours northeast (upstream) of the fault indicate that ground water is moving toward the surface. Well hydrographs in the area also indicate upward ground-water movement. The water-table contours on either side of the fault indicate a very steep gradient across the fault zone.

By the 1970s the basin had undergone extensive ground-water overdrafts. The water level in the swampy area along the Santa Ana River dropped 50 to 150 ft (15 to 46 m) below the land surface. This drained the wetland, and the land became dry. Extensive urban development occurred, with the site of the former swamp becoming downtown San Bernardino.

A computer model of the ground-water basin was developed (Hardt & Hutchinson 1978). This model predicted that if ground-water pumpage were to be reduced or natural recharge were to increase, ground-water levels could rise. In the first edition of this book the author pointed out that this would be especially severe in the San Jacinto Fault area, which is the point of natural ground-water discharge from the Bunker Hill Basin, and indicated that "Abandoned and forgotten wells could begin to flow again. Basements would be flooded and soils waterlogged." This prediction came true. By 1982 several years of above-average precipitation in the Bunker Hill Basin resulted in ground-water levels that were high enough to cause severe problems. Basements were flooded; abandoned wells began to flow (one beneath a building split the foundation); and springs at the land surface made a new water hazard in the fifth fairway of the San Bernardino Golf Club (Hoffman 1983). An even greater fear was that the high ground-water levels might result in the failure of building foundations owing to soil liquefaction during an earthquake (Hoffman 1983).

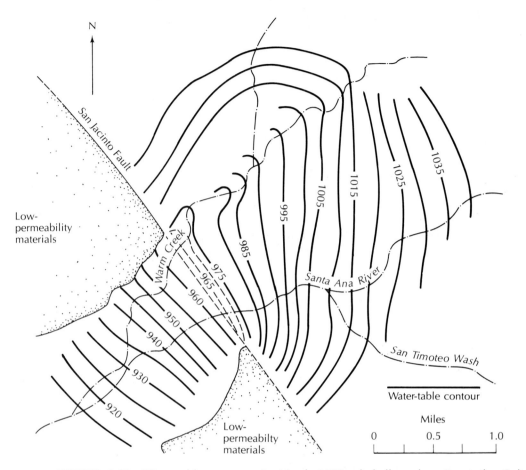

FIGURE 9.10 Water-table contours in March 1939 of shallow deposits at the Colton Narrows area, where the Santa Ana River crosses the San Jacinto Fault. The contours of 970 and 965 are shown, but the exact positions were not known. Source: L. C. Dutcher & A. A. Garrett, U.S. Geological Survey Water-Supply Paper 1419, 1963.

The U.S. Geological Survey made a detailed study of the potential for liquefaction and found that an earthquake on the San Jacinto Fault with a magnitude of 7.0 on the Richter scale might trigger liquefaction under much of greater San Bernardino if the water table was within 30 ft (10 m) of the land surface (McGreevy 1987). Meanwhile the municipal authorities had made plans to install dewatering wells to lower the water table in the downtown area. About the time that the USGS report was released—and before the dewatering wells could be installed—nature did an about-face. The wet period ended and ground-water levels began to fall (Sears 1987). Within a few years Southern California was in the grip of a severe drought. Between 1986 and 1990 ground-water levels in the San Bernardino area declined as much as 67 ft, and high ground-water problems were a thing of the past (Atwood 1990)—that is, until the next wet period.

9.3   LITHIFIED SEDIMENTARY ROCKS

Clastic sedimentary rocks are typically composed of silicate, carbonate, or clay minerals. Chemically precipitated sedimentary rocks are primarily limestone, dolomite, salt, or gypsum. Coal and lignite can also be considered to be sedimentary deposits, with the original sediments being organic matter.

      Sedimentary rock sequences were formed with younger beds laid down upon older ones. The original sediments may have been subaqueous or terrestrial. Rarely do sedimentary rocks occur as a single unit; there is typically a sequence of many beds. The original layered sequence may be undisturbed or it may be extensively folded and faulted.

CASE STUDY: SANDSTONE AQUIFER OF NORTHEASTERN ILLINOIS–SOUTHEASTERN WISCONSIN

The sandstone or deep aquifer of northeastern Illinois–southeastern Wisconsin comprises formations of Cambrian and Ordovician age, primarily sandstones and dolomites. The formations strike north-south and dip to the east. The aquifer is confined by the Maquoketa Shale (Figure 9.11) and makes a classic confined artesian system in clastic rock. The aqui-

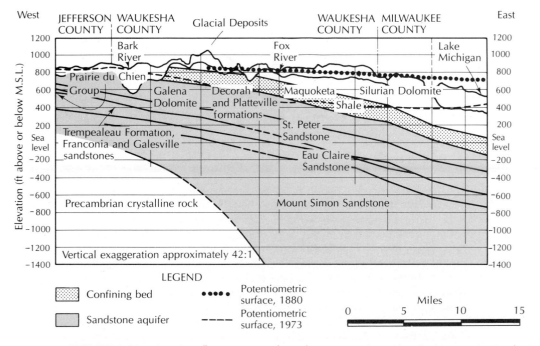

FIGURE 9.11 Artesian flow system of southeastern Wisconsin. Source: U.S. Geological Survey.

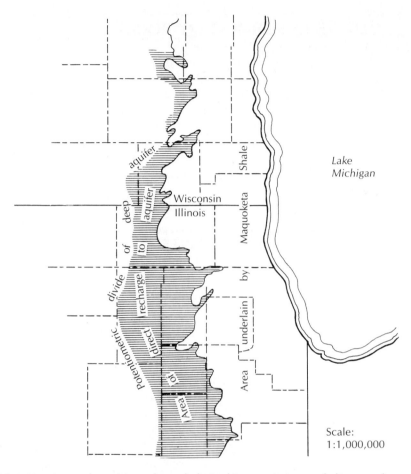

FIGURE 9.12 Area where Maquoketa shale is thin or missing and direct recharge to the Cambrian-Ordovician aquifer can occur. Source: C. W. Fetter, Jr., *Ground Water* 19 (1981): 201–13.

fer extends from the area north of Milwaukee, Wisconsin, to south and west of the Chicago, Illinois, area (Young 1976; Fetter 1981a; Visocky, Sherrill, & Cartwright, 1985).

The Maquoketa Shale has been eroded away in the western part of the area so that the upper formations of the sandstone aquifer can be recharged directly (Figure 9.12). The average direct recharge rate to the sandstone aquifer in Illinois is estimated to be 20,400 gal/day/mi^2 (30 m^3/day/km^2) (Visocky, Sherrill, & Cartwright, 1985).

In the area where the aquifer is confined by the Maquoketa Shale the amount of recharge across the shale is determined by the thickness and vertical hydraulic conductivity of the shale and the hydraulic gradient across it.

Before the sandstone aquifer was developed, there was a small amount of recharge in the western part of the region, with lateral flow to the east and upward flow from the deep aquifers to the overlying shallow aquifer in the vicinity of Lake Michigan. Figure 9.13A shows the potentiometric surface of this aquifer system prior to development. When wells were first drilled in eastern Wisconsin the artesian head in the sandstone aquifer was

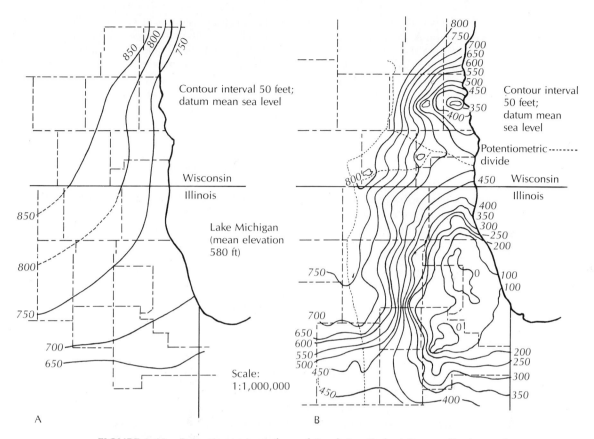

FIGURE 9.13 Potentiometric surface of Cambrian-Ordovician aquifer in southeastern Wisconsin and northeastern Illinois. **A.** About 1865–1880. **B.** In 1973. Source: C. W. Fetter, Jr., *Ground Water* 19 (1981): 201–13.

130 to 200 ft (40 to 60 m) above the level of Lake Michigan. As water was pumped from the deep sandstone aquifer, the potentiometric surface fell and the upward gradient in the eastern part of the aquifer disappeared. Eventually, substantial downward hydraulic gradients were developed, and the amount of downward recharge across the Maquoketa Shale increased. The maximum rate of recharge across the shale layer in Illinois is estimated to be about 3000 gal/day/mi² (4.4 m³/day/km²) in Illinois (Visocky, Sherrill, & Cartwright 1985) and 4200 gal/day/mi² (6.1 m³/day/km²) in Wisconsin (Fetter 1981a).

The practical sustained yield of the sandstone aquifer in Illinois has been estimated to be 65 million gallons (246,000 cubic meters) per day (Visocky, Sherrill, & Cartwright 1985). This consists of 50 million gallons (189,240 m³) a day of direct recharge, 12 million gallons (45,400 m³) per day of downward leakage across the Maquoketa Shale and 3 million gallons (11,360 m³) per day of upward leakage from deeper fresh-water aquifers (Mt. Simon aquifer). In Wisconsin the practical sustained aquifer yield has been estimated to be 34 million gallons (128,700 m³) per day—25 million gallons (94,600 m³) per day of direct recharge and 9 million gallons (34,100 m³) per day of recharge across the shale layer (Fetter 1981a).

Since 1958 the ground-water withdrawals from this aquifer system in Illinois have exceeded the practical sustained yield and in Wisconsin the pumpage first exceeded the practical sustained yield about 1980. As a result of the ground-water withdrawals, the piezometric pressure in the aquifer has declined and the potentiometric surface has declined. Figure 9.13B shows the potentiometric surface as it existed in 1973. Deep pumping cones had developed in the Milwaukee and Chicago areas. A ground-water divide formed in southern Wisconsin, with flow north of the divide going toward Milwaukee and flow south of the divide going toward Chicago.

The potentiometric surface in the Milwaukee area has fallen more than 400 ft (120 m); in the Chicago region it has declined to below sea level, a distance of more than 900 ft (275 m) (Young et al. 1986). The potentiometric surface in parts of northeastern Illinois has fallen below the Maquoketa Shale so that the aquifer, which was formerly confined, is now unconfined. In some areas, parts of the Galena-Platteville unit have been dewatered along with the upper part of the Ancell unit. The great decline in the water levels indicates that much of the water that has been withdrawn has come from storage. Unless ground-water withdrawals are reduced to the amount of recharge in this circumstance, water levels will continue to decline until the aquifers can no longer transmit water from the recharge area to the well fields. Even if the pumpage were to be reduced to the practical sustained yield, ground-water levels would not recover to their former positions.

9.3.1 Complex Stratigraphy

The fact that a formation may change in lithology from one locality to another accounts for some of the difficulties associated with studies of sedimentary rock units. For example, a sandstone may grade into a shaly sandstone or siltstone, yet still have the same fossil fauna assemblage and the same stratigraphic nomenclature. The Eau Claire Formation of Cambrian age is a sandstone in east-central Wisconsin, but it grades into a shale and siltstone in northern Illinois. In Wisconsin, it has high conductivity and is an aquifer; in Illinois it becomes a confining layer (Walton 1965; Schicht, Adams, & Stall 1976).

Complex stratigraphy can be a very real hindrance to ground-water exploration. On the Hualapai Plateau of northwestern Arizona, the best potential aquifer is the stratigraphic sequence of sedimentary rocks in the Rampart Cave Member of the Muav Limestone Formation (Twenter 1962). Springs discharge from this member into the Grand Canyon along the flanks of the plateau. The Bright Angel Shale stratigraphically underlies the Rampart Cave Member of the Muav Limestone, causing the ground water to be perched (Figure 9.14). However, because these units were being deposited in a transgressing sea, the various beds are interfingering. As a result, at some localities the Bright Angel Shale can be both above and below the Rampart Cave Member (Figure 9.15). Drillers in the area have had unsuccessful wells because they stopped drilling when the Bright Angel Shale was first reached (Huntoon 1977). Unfortunately, the target formation had not yet been penetrated. A well drilled at Location *A* of Figure 9.15 would not hit Bright Angel Shale until the Rampart Cave Member was penetrated. However, at location *B*, three shale members alternating with limestone would have to be penetrated before the Rampart Cave Member was reached. The area of these interfingering members extends for tens of miles. The hydrogeologist

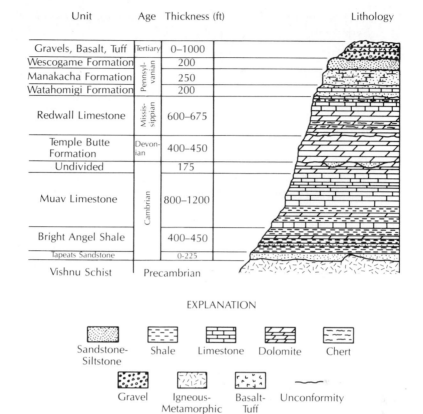

Unit	Age	Thickness (ft)	Lithology
Gravels, Basalt, Tuff	Tertiary	0–1000	
Wescogame Formation	Pennsylvanian	200	
Manakacha Formation	Pennsylvanian	250	
Watahomigi Formation	Pennsylvanian	200	
Redwall Limestone	Mississippian	600–675	
Temple Butte Formation	Devonian	400–450	
Undivided		175	
Muav Limestone	Cambrian	800–1200	
Bright Angel Shale	Cambrian	400–450	
Tapeats Sandstone	Cambrian	0–225	
Vishnu Schist	Precambrian		

EXPLANATION

Sandstone-Siltstone   Shale   Limestone   Dolomite   Chert

Gravel   Igneous-Metamorphic   Basalt-Tuff   Unconformity

FIGURE 9.14 Stratigraphy of the Grand Canyon area. Source: P. W. Huntoon, *Ground Water* 15 (1977): 426–33.

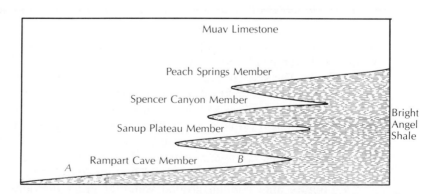

FIGURE 9.15 Interfingering of sedimentary rock units of the Hualapai Plateau area. Source: Adapted from P. W. Huntoon, *Ground Water* 15 (1977): 426–33.

working in areas of such complex stratigraphy must be aware of the possible hydrogeologic consequences.

9.3.2 Folds and Faults

Folding and faulting of sedimentary rocks can create very complex hydrogeologic systems, in which determination of the locations of recharge and discharge zones and flow systems is confounded. Not only must the hydrogeologist determine the hydraulic characteristics of rock units and measure groundwater levels in wells to determine flow systems, but detailed geology must also be evaluated. In most cases, the basic geologic structure will have already been determined; however, logs of test wells and borings must be reconciled with the pre-existing geologic knowledge.

Fault zones can act either as barriers to ground-water flow or as ground-water conduits, depending upon the nature of the material in the fault zone. If the fault zone consists of finely ground rock and clay (gouge), the material may have a very low hydraulic conductivity. Significant differences in ground-water levels can occur across such faults (see Figures 9.9 and 9.10). Impounding faults can occur in unconsolidated materials with clay present, as well as in sedimentary rocks where interbedded shales, which normally would not hinder lateral ground-water flow, can be smeared along the fault by drag folds. **Clastic dikes** are intrusions of sediment that are forced into rock fractures. If they are clay-rich, they can act as ground-water barriers in either sediments or in a lithified sedimentary rock. Clastic dikes are known to occur in alluvial sediments, glacial outwash, and lithified sedimentary rock.

Faults in consolidated rock units can act either as pathways for water movement or as flow barriers. If there has been little displacement along the fault, then the fault is more likely to develop fracture permeability because there is less opportunity for the formation of soft, ground-up rock, called **gouge,** to form between the moving surfaces (Huntoon 1986). Fault gouge can have a matrix of rock breccia encased in clay and can have a wider range of permeability (Snipes et al. 1986). Faults in poorly consolidated rocks in South Carolina had greater permeability than those in well-consolidated rocks (Snipes et al. 1986).

If the fault zone has a high porosity and hydraulic conductivity, it can serve as a conduit for ground-water movement. Springs discharging into the Colorado River in Marble Canyon are controlled by a vertical fault zone, the Fence Fault. The springs discharge where the faults intersect the river. The fault zones provide for vertical movement of recharging ground water from the land surface as well as lateral movement toward the river. The geochemistry of the spring water indicates that some of the water discharging on one side of the river originated in the ground-water basin on the opposite side of the river, indicating the fault zone was conducting some ground-water flow beneath the river even though it is a regional discharge zone (Huntoon 1981).

Faults may contain ground water at great pressures at depths where tunnels or mines may be constructed. One of the dangers of hard-rock tunneling is the possibility of breaching an unexpected fault zone. Damaging and dangerous flooding can occur if the fault contains ground water with a high hydraulic head. In Utah a well being drilled through an anticlinal structure that in an unfaulted state creates a regional confining layer encountered an exceptionally high-permeability zone created by normal faults and associated extensional joints that imprinted joint permeability on the brittle rocks. The well was being drilled to initiate solution mining of an abandoned underground potash mine. Unfortunately, ground water under high pressure in the fracture zone rushed into the well boring when the mine level was reached and flooded it with mineralized water (Huntoon 1986).

Overthrust faulting can create conditions in which a rock, normally found as an impermeable basement unit, is overlying the sedimentary rock units, typically a ground-water source. In such a case, the hydrogeologist might recommend drilling through the "basement" rocks to attempt to obtain a ground-water supply from younger sedimentary units, provided there is an opportunity for recharge to occur and the water is not known to have a high mineral content (Figure 9.16).

The major artesian basins of Wyoming consist of sedimentary rock units bounded by major thrust faults. The fault-severed margins of the basins have good permeability, adequate recharge, and good-quality water in the sedimentary rock aquifers of the hanging wall. However, the foot-wall rocks receive little recharge, have poor-quality water, and have permeabilities often many orders of magnitude lower than those of the adjacent hanging wall segments of the same formation (Huntoon 1985).

Folding can affect the hydrogeology of sedimentary rocks in several ways. The most obvious is the creation of confined aquifers at the centers of synclines. The nature of the fold will affect the availability of water. A tight, deeply plunging fold might carry the aquifer too deep beneath the surface to be economically developed. Deeply circulating ground water is also typically

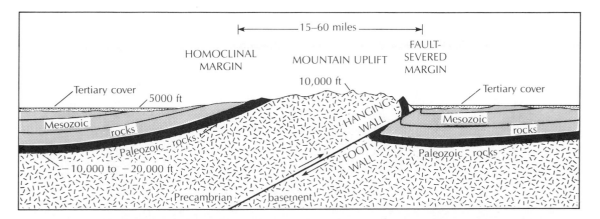

FIGURE 9.16 Schematic cross section through a typical Wyoming mountain uplift showing the style of deformation that results in approximately equal percentages of fault severed and homoclinal basin perimeters. Source: P. W. Huntoon, *Ground Water* 23 (1985): 176–81.

warmed by the geothermal gradient and may be highly mineralized. A broad, gentle fold can create a relatively shallow, confined aquifer that extends over a large area. This might be a good source of water if sufficient recharge can occur through the confining layer or if the aquifer can transmit enough water from areas where the confining layer is absent.

Another effect of folding is to create a series of outcrops of soluble rock, such as limestone, alternating with rock units that are not as permeable. Smaller streams flowing across the limestone might sink at the upper end, only to reappear at the lower outcrop. The type of trellis drainage that can develop on folded rocks is shown in Figure 9.17. Surface streams follow the strike of rock outcrops, usually along fault or fracture traces. In folded sedimentary rocks with solutional conduits in carbonate units, ground-water flow may be along the conduits that parallel the strike of the fold and not down the dip.

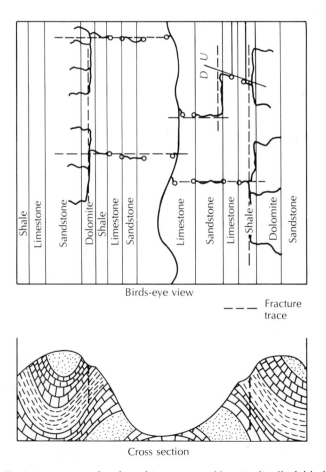

FIGURE 9.17 Drainage pattern developed in an area of longitudinally folded rock strata: **A.** Top view. **B.** Cross section.

In areas of homoclinal folds, the outcrop areas usually have bands of sedimentary rocks, with resistant rocks forming ridges and more easily erodable rocks forming valleys. The ridges may create ground-water divides, with aquifers outcropping in the valleys. The outcrop area of an aquifer will have local water-table flow systems with relatively large amounts of water circulating. These areas also serve as the recharge zones for the more distal parts of the aquifer, which are downdip in the basin and are confined. There is a limited amount of natural discharge from the confined portions of the aquifer; this is typically upward leakage into overlying beds with lower hydraulic head. Because of poor ground-water circulation, the confined portions of the aquifer may have low hydraulic conductivity and poor water quality (LeGrand & Pettyjohn 1981).

Complex folded and faulted sedimentary rock units are a challenge to the hydrogeologist. Competency in geology is necessary in order to construct geologic cross sections based on well logs, drill-core samples, and outcrops. Regional flow systems can be controlled by the structural and stratigraphic relations of the confining beds and aquifer units. In addition, distribution of hydraulic potential must be determined, very often from limited data.

9.3.3 Clastic Sedimentary Rocks

Hydraulic conductivity of clastic sedimentary rocks, based on primary permeability, is a function of the grain size, shape, and sorting of the original sediment. The same factors that affect the permeability and porosity of loose sediments also are important in sedimentary rocks. **Cementation,** in which parts of the voids are filled with precipitated material such as silica, calcite, or iron oxide, can reduce the original porosity. Solution of the original material may occur during and after diagenesis, resulting in an increase in porosity.

Consolidated rocks also contain secondary porosity and permeability due to fracturing. Microfractures may add very little to the original hydraulic characteristics; however, major fracture zones may have localized hydraulic conductivities several orders of magnitude greater than that of the unfractured rock. Fracturing can occur through several geologic processes. Rock at depth is under great pressure owing to the weight of the overburden. As uplift and erosion bring the consolidated rock to the surface, it expands as the pressure is reduced. The expansion can cause fracturing of the rock, with the majority of the expansion fractures occurring within about 300 ft (100 m) of the surface. Vertical fractures carry recharging precipitation downward and provide a very important function in bypassing low-permeability layers near the surface. Wells located in surface fracture zones are generally highly successful.

Fracturing may also be associated with tectonic activity. Rock deformed by faulting or folding may fracture when it is subjected to tension or compression. Such activity can take place at substantial depths; thus, secondary permeability is not strictly a near-surface phenomenon. However, great pressure on the deeper fractures does not permit them to be as open (have as great a porosity) as shallow fractures.

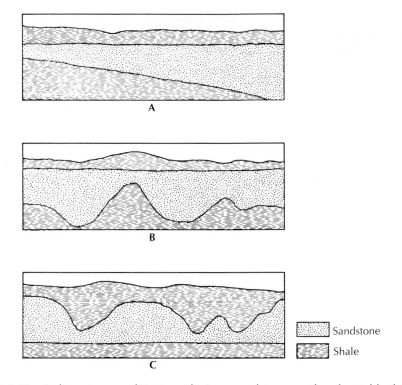

Sandstone

Shale

FIGURE 9.18 Sedimentary conditions producing a sandstone aquifer of variable thickness: **A.** Sandstone deposited in a sedimentary basin. **B.** Sandstone deposited uncomformably over an erosional surface. **C.** Surface of sandstone dissected by erosion prior to deposition of overlying beds.

The yield to wells is proportional to the transmissivity of the aquifer. This, in turn, is proportional to the aquifer thickness if the hydraulic conductivity is uniform throughout the aquifer. Sedimentary aquifers were deposited in sedimentary basins in which units gradually thicken. Variable thickness of a sedimentary aquifer may also be due to the deposition of the aquifer material over an eroded surface with high relief or a dissection of the top of the aquifer after deposition (Figure 9.18). Higher well yields will be obtained from thicker sections of the aquifer. The relationship between the specific capacity of wells and the uncased thickness of two sandstone aquifers in northern Illinois is shown in Figure 9.19.

Wells in sandstone aquifers should be located in such a manner as to penetrate the maximum saturated thickness of the aquifer. If one area of the aquifer is known to have a higher hydraulic conductivity than other areas, the combination of hydraulic conductivity and thickness should be considered in order to locate the well in the area of greatest aquifer transmissivity.

The yield of sandstone wells can sometimes be increased by the detonation of explosives in the uncased hole. The shots are generally located opposite the most permeable zones of the sandstone. The loosened rock and sand is bailed

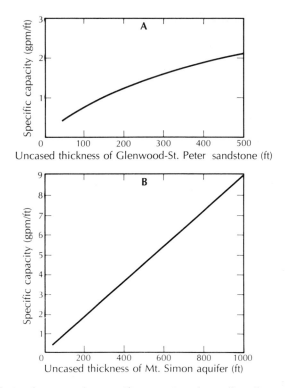

FIGURE 9.19 Relation between the specific capacity of a well (gallons per minute of yield per foot of drawdown) and the uncased thickness of the sandstone aquifer: **A.** Glenwood–St. Peter Sandstone. **B.** Mt. Simon Sandstone. Both of northern Illinois. Source: W. C. Walton & S. Csallany, Illinois State Water Survey Report of Investigation 43, 1962.

from the well prior to the installation of the pump. The shooting process has two effects on the borehole: it enlarges the diameter of the well in the permeable zones and also breaks off the surface of the sandstone, which may have been clogged by fine material during drilling. Fractures near the well may be opened all the way to the borehole by shooting. Old wells may be rehabilitated by shooting if the yield has decreased owing to mineral deposition on the well face. Shooting has increased the specific capacities of sandstone wells in northern Illinois by an average of 22 to 38%, depending upon the formation (Walton & Csallany 1962).

9.3.4 Carbonate Rocks

The primary porosity of limestone and dolomite is variable. If the rock is clastic, the primary porosity can be high. Chemically precipitated rocks can have a very low porosity and permeability if they are crystalline. Bedding planes can be zones of high primary porosity and permeability.

Limestone and (to a much lesser extent) dolomite are soluble in water that is mildly acidic. In general, if the water is unsaturated with respect to calcite or

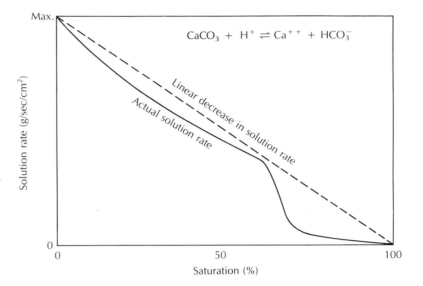

FIGURE 9.20 Solution rate vs. degree of saturation. Instead of decreasing linearly, the solution rate drops sharply to a low level at 65–90% saturation. Source: A. N. Palmer, *Journal of Geological Education* 32 (1984): 247–53.

dolomite, it will dissolve the mineral until it reaches about 99+% saturation with respect to calcite (Plummer, Wigley, & Parkhurst 1978; Palmer 1984). The rate of solution is linear with respect to increasing solute concentration until somewhere between 65% and 90% saturation, at which value the rate decreases dramatically. Figure 9.20 shows the general nature of the solution rate as a function of degree of saturation.

Massive chemically precipitated limestones can have very low primary porosity and permeability. Secondary permeability in carbonate aquifers is due to the solutional enlargement of bedding planes, fractures, and faults (Ford & Ewers 1978). The rate of solution is a function of the amount of ground water moving through the system and the degree of saturation (with respect to the particular carbonate rock present) but it is nearly independent of the velocity of flow (Palmer 1984). The width of the initial fracture is one of the factors controlling how long the flow path is until the water reaches 99+% saturation and dissolution ceases (Palmer 1984).

Initially, more ground water flows through the larger fractures and bedding planes, which have a greater hydraulic conductivity. These become enlarged with respect to lesser fractures; hence, even more water flows through them. Solution mechanisms of carbonate rocks favor the development of larger openings at the expense of smaller ones. Carbonate aquifers can be highly anisotropic and nonhomogeneous if water moves only through fractures and bedding planes that have been preferentially enlarged. Water entering the carbonate rock is typically unsaturated. As it flows through the aquifer, it approaches saturation, and dissolution slows and finally ceases. It has been shown experimentally that solution passages form from the recharge area to the discharge area and that, as

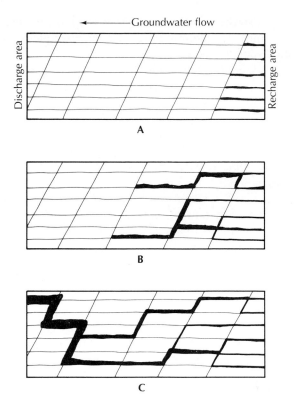

FIGURE 9.21 Growth of a carbonate aquifer drainage system starting in the recharge area and growing toward the discharge area. **A.** At first, most joints in the recharge area undergo solution enlargement. **B.** As the solution passages grow, they join and become fewer. **C.** Eventually, one outlet appears at the discharge zone.

they follow fracture patterns, many smaller solution openings join to form fewer but larger ones (Ewers et al. 1978) (Figure 9.21). Eventually, many passages join to form one outlet. Greater ground-water movement—hence, solution—takes place along the intersection of two joints or a joint and a bedding plane. Ground water moving along a bedding plane tends to follow the strike of intersecting joints.

A second mode for the entry of unsaturated water into a carbonate aquifer occurs near valley bottoms. In **karst*** regions, flow in valleys with permanent streams is usually discharged from carbonate aquifers recharged beneath highlands. Water tables in many karst areas are almost flat owing to the high hydraulic conductivity. Floodwaters from surface streams can enter the carbonate aquifers and reverse the normal flow. If the floodwaters are unsaturated with respect to the mineral in the aquifer, solution will occur (White 1969).

---

*Karst is a term applied to topography formed over limestone, dolomite, or gypsum; and characterized by sinkholes, caverns, and lack of surface streams.

Swallow holes, or shafts leading from surface streams, can carry surface water underground into caverns. Swallow holes can drain an entire stream or only a small portion of one.

Geochemical studies have shown that there are two types of ground water found in complex carbonate aquifer systems (Shuster & White 1971). The joints and bedding planes that are not enlarged by solution contain water that is saturated with respect to calcite (or dolomite). Because of the low hydraulic conductivity of these openings, this water mass moves slowly. Another mass of water, generally undersaturated, moves more rapidly through well-defined solution channels close to the water table. It is this second body that forms the passageways.

Cave systems can be formed above, at, or below the water table. They form when free-flowing water enlarges a fracture or bedding plane sufficiently for non-Darcian flow to occur. This can be above the water table if a surface stream enters the ground in the unsaturated zone **(vadose cave),** below the water table if the joint or bedding plane through which flow is occurring dips below the water table **(phreatic cave),** or at the water table itself **(water-table cave)** (Ford & Ewers 1978). The pattern of cave passages is controlled by the pattern and density of the joints and/or bedding planes in the carbonate rock (Ford & Ewers 1978). Figure 9.22 shows the influence of fissure density and orientation on cave formation. With widely spaced fissures, the cave can develop below the potentiometric surface because the fissure pattern is too coarse to allow the cave development to parallel the water table (Figure 9.22A and B). If the fissure density is great enough, cave development can occur along the water table (Figure 9.22C and D). Vertical shafts can form in the vadose zone by undersaturated infiltrating water trickling down the rock surface (Brucker, Hess, & White 1972). Some caves that are presently dry were formed at or below the water table when the regional water table was higher. The regional base level of a karst region is typically a large river. If the river is downcutting, the regional water table will be lowering. The result will be a series of dry caves at different elevations, each formed when the regional water table was at a different level.

Carbonate aquifers show a very wide range of hydrologic characteristics. There are, to be sure, a number of "underground rivers" where a surface stream disappears and flows through caves as open channel flow. At the other extreme, some carbonate aquifers behave almost like a homogeneous, isotropic porous medium. Most lie between these extremes.

Three conceptual models for carbonate aquifers have been proposed (White 1969). *Diffuse-flow carbonate aquifers* have had little solutional activity directed toward opening large channels; these are to some extent homogeneous. *Free-flow carbonate aquifers* receive diffused recharge but have well-developed solution channels along which most flow occurs. Ground-water flow in free-flow aquifers is controlled by the orientation of the bedding planes and fractures that determine the locations of solutional conduits, but not by any confining beds. *Confined-flow carbonate aquifers* have solution openings in the carbonate units, but low-permeability noncarbonate beds exert control over the direction of ground-water movement.

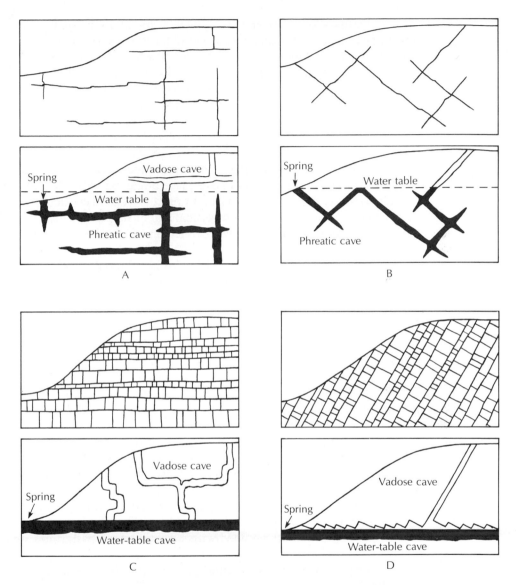

FIGURE 9.22 Effects of fissure density and orientation on the development of caverns. Source: Modified from D. C. Ford & R. O. Ewers, *Canadian Journal of Earth Science* 15 (1978).

Diffuse-flow aquifers are typically found in dolomitic rocks or shaly limestones, neither of which is easily soluble. Water movement is along joints and bedding planes that have been only modestly affected by solution. Moving ground water is not concentrated in certain zones in the aquifer and, if caves are present, they are small and not interconnected. Discharge is likely to be through a number of small springs and seeps. The Silurian-age dolomite aquifer of the Door Peninsula of Wisconsin is an example. Well tests have shown that the horizontal flow of water is along seven different bedding planes in the dolomite. Vertical recharge is through fractures. The bedding-plane zones can be identified by borehole geophysical means (caliper logs) and correlated across several miles (Sherrill 1978). Because water movement takes place along broad bedding planes, the yield of wells is fairly constant from place to place. Wells in vertical fractures have a higher yield, as they possess both vertical and horizontal conductivity. The water table in diffuse-flow aquifers is well defined and can rise to a substantial elevation above the regional base level.

Free-flow carbonate aquifers have substantial development of solution passages. Not only are many joints and bedding planes enlarged, but some have formed large conduits. Although all the openings are saturated, the vast majority of flow occurs in the large channels; the flow behaves hydraulically as pipe flow. Velocities are similar to those of surface streams. Flow is turbulent, and the stream may carry a sediment load—as suspended material, bedload, and suspended bedload. Water quality is similar to that of surface water, and the regional discharge may occur through a few large springs. Because of the rapid drainage, the water table is nearly flat, having only a small elevation above the regional base level. The very low hydraulic gradient indicates that diffused flow through the unenlarged joints and fractures is exceedingly slow. Recharge to the subterranean drainage system is rapid, as water drains quickly through the vadose zone. The water level in the open pipe network may rise rapidly in a recharge event (Fox & Rushton 1976). The spring discharge will also increase in response to the amount of recharge, so that the spring hydrograph may resemble the flood peak of a surface stream. The water levels in the open-pipe network will also fall rapidly as the water drains. Caving expeditions have been known to end tragically when a "dry" cave passage became filled with surcharged water during a rapid recharge event.

The depth of major solution openings below base level is probably less than 200 ft (60 m), unless artesian flow conditions are present (White 1969). However, in areas where the regional base level was formerly at a lower level (for example, where a buried bedrock valley is present), cavern development may have taken place graded to that base level. In the coastal aquifers of the southeastern United States, the drilling fluid may suddenly drain from a well being drilled when it is 300 ft (100 m) or more deep. Cavernous zones found at these depths are well below the present water table. They formed when mean sea level and the regional water table were lower during the Pleistocene. The development of a sinkhole in Hernando County, Florida, was initiated by drilling in the Suwannee Limestone. The drilling fluid was lost several times, and the drill-bit would drop through small caverns. At 200 ft (62 m), the drill broke into a cavern and, within 10 min, a large depression had formed, with the drill rig sinking into the ground and the drillers narrowly escaping the same fate. The present-day water level is close to the land surface (Boatwright & Ailman 1975).

Sinking surface streams may also feed the pipe-flow network of a karst region. Lost River of southern Indiana is a typical headwater surface stream flowing across a thick clay layer formed as a weathering residuum on the St. Louis and St. Genevieve limestones. Where the clay thins, a karst landscape is present, and Lost River sinks beneath an abandoned surface channel. It appears as a large spring some miles away. There is no surface drainage in the karst region other than some ephemeral streams flowing into sinkholes (Ruhe 1977).

If the carbonate rock beneath the uplands between regional drainage systems is capped by a clastic rock, karst landforms will not form. Dry caves in the uplands capped with clastics are less likely to have collapsed than caves in areas that are not capped. Recharge to the phreatic zone occurs through vertical shafts located at the edge of the caprock outcrop (Brucker, Hess, & White 1972). These shafts, which may be as large as 30 ft (10 m) in diameter and more than 300 ft (100 m) deep, extend only to the water table. Water flows from them through horizontally oriented drains.

The central Kentucky karst region, including the Mammoth Cave area, is a capped carbonate aquifer system in some areas (Brown 1966; White 1970). A cross section through the Mammoth Cave Plateau is shown in Figure 9.23. The plateau is capped by the Big Clifty Sandstone Member of the Golconda Formation, with cavern development in the underlying limestone formations, including the Girkin, St. Genevieve, and St. Louis limestones. Contact springs are found at the margins of the top of the plateau, as there is a thin shale layer at the top of the Girkin Formation. Recharge to the main carbonate rock aquifer takes place through vertical shafts formed in the plateau where the shale layer is absent. Karst drainage in the Pennyroyal Plain also contributes to the regional water table. Drainage is to the Green River through large springs, such as the River Styx outlet and Echo River outlet. These streams may also be seen underground where they flow in cave passages. Wells in the area draw water from the Big Clifty Sandstone, which yields enough for domestic supplies. There are some perched water bodies in the limestone, but the amount of water in storage is limited. A more permanent supply is reached if the well penetrates the regional water table, below the level of the Green River. The water level in these deep wells can rise 25 ft (8 m) in a few hours during heavy rains and then can fall almost as fast (Brown 1966).

If a carbonate rock is confined by strata of low hydraulic conductivity, those strata may control the rate and direction of ground-water flow (White 1969). Such confined systems may have ground water flowing to great depths, with solution openings that are much deeper than those found in free-flow aquifers. Flow is not localized, and a greater density of joint solution takes place. The cavern formation of the Pahasapa Limestone of the Black Hills is apparently of this type, as water flows through the equivalent Madison Limestone aquifer eastward from a recharge area at the eastern side of the Black Hills. It has been

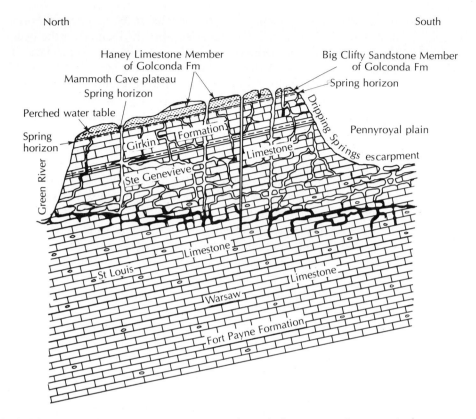

FIGURE 9.23 Diagrammatic cross section through the Mammoth Cave Plateau. Ground-water flow in the carbonate aquifer is from south to north. Source: R. F. Brown, U.S. Geological Survey Water-Supply Paper 1837, 1966.

suggested that because of the low gradient (0.00022) of the potentiometric surface, the Madison Limestone is highly permeable, owing to solution openings, for at least 130 mi (200 km) east of the Black Hills (Swenson 1968).

In the preceding discussion of karst hydrology, it was assumed that the various types of carbonate rock aquifers were isolated; however, this may not actually be the case. Highly soluble carbonate rock may be adjacent to a shaly carbonate unit with only slight solubility. The slope and position of the water table is a reliable indicator of the relative hydraulic conductivity of different carbonate rock units. In general, the water table will have a steeper gradient in rocks of lower hydraulic conductivity (LeGrand & Stringfield 1971). This may be due to either a change in lithology or the degree of solution enlargement of joints. Figure 9.24 illustrates some conditions that might result in a change in the water-table gradient. In Part A, the hilltop is capped by sandstone, with a low-permeability shale between the sandstone and the underlying limestone. Only a limited amount of recharge occurs through the shale. A spring horizon exists at the sandstone-shale contact, with small streams flowing across the shale outcrop area, only to

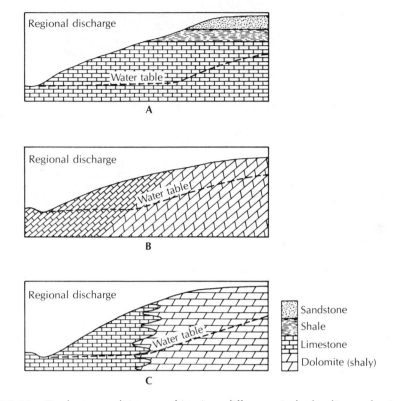

FIGURE 9.24 Geologic conditions resulting in a difference in hydraulic conductivity and, hence, a difference in the water-table gradient.

sink into the limestone terrane. The much greater amount of water circulating through the limestone in the area where the shale is absent has created a highly permeable, cavernous unit. Beneath the caprock, the limestone is less dissolved, owing to lower ground-water recharge. Because of the lower hydraulic conductivity, the water-table gradient is steeper. This type of situation has been reported in the central Kentucky karst, with the hydraulic gradient beneath the caprock near the drainage divide being much steeper (0.01) than that near the Green River (0.0005) (Cushman 1967). If an area has two rock units, one of which is more soluble, the more soluble rock may develop large solution passages and, hence, have greater conductivity. Parts B and C of Figure 9.24 illustrate two situations in which the upland is underlain by shaly dolomite and the lowland by cavernous limestone. The difference in rock solubility creates a change in hydraulic gradient.

The general concept of a water table in free-flowing karstic regions may be different from the model water table found in sandstones or sand-and-gravel aquifers. Because of the extremely high conductivity of some limestones, the water table can occur far beneath the land surface in mountains. Water can "perch" in solution depressions above the main water table. The level of free-flowing streams in caves is controlled by the regional water table, and the streams

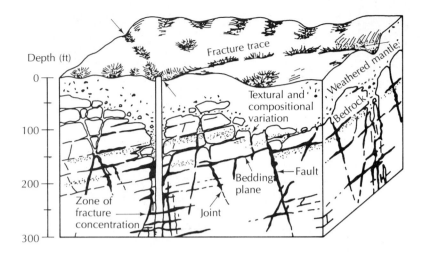

Depth (ft)

Fracture trace

Weathered mantle

Bedrock

Textural and compositional variation

Fault

Bedding plane

Joint

Zone of fracture concentration

FIGURE 9.25 Concentration of ground water along zones of fracture concentrations in carbonate rock. Wells that do not intercept an enlarged fracture or a bedding plane may be dry, thus indicating a discontinuous water table. Source: L. H. Lattman & R. R. Parizek. *Journal of Hydrology* 2 (1964): 73–91. Used with permission of Elsevier Scientific Publishing Company, Amsterdam.

can have losing and gaining reaches, just as surface streams. Finally, because the solution of carbonate rock can be isolated along such features as fracture zones, the water table may be discontinuous. Wells drilled between fractures may not have any water in them, whereas nearby wells of the same depth may be in a fracture and therefore measure a water table (Figure 9.25).

The selection of well locations in carbonate terrane is one of the great challenges for the hydrogeologist. As the porosity and permeability may be localized, it is necessary to find the zones of high hydraulic conductivity. One of the most productive approaches to the task is the use of **fracture traces** (Lattman & Parizek 1964; Parizek 1976). Fracture traces (up to 1 mi, or 1.5 km) and lineaments (1 to 100 mi, or 1.5 to 150 km) are found in all types of geologic terrane. As they represent the surface expression of nearly vertical zones of fracture concentrations, they are often areas with hydraulic conductivity 10 to 1000 times that of adjacent rocks. The fracture zones are from 6 to 65 ft (2 to 20 m) wide and have surface expressions such as swales and sags in the land surface; vegetation differences, due to variations in soil moisture and depth to the water table; alignment of vegetation type; straight stream and valley segments; and alignment of sinkholes in karst. The surface features can reveal fracture traces covered by up to 300 ft (100 m) of residual or transported soils. Fracture traces are found over carbonate rocks, siltstones, sandstones, and crystalline rocks.

Because of differential solution in carbonate rocks, if a fracture zone has somewhat higher conductivity than that of the unfractured rock, flowing ground water will eventually create a much larger conductivity difference. Wells located in a fracture trace, or especially at the intersection of two fracture traces, have a statistically significant greater yield than wells not located on a fracture trace

(Siddiqui & Parizek 1971). The same relationship is apparently true for wells located on lineaments, as opposed to those not on lineaments (Parizek 1976).

In areas where topography is influenced by structure, valleys may form along fracture traces. In central Pennsylvania 60% to 90% of a valley may be underlain by fracture traces (Siddiqui & Parizek 1971). Under such conditions, valley bottoms are good places to prospect for ground water. In the Valley and Ridge Province of Pennsylvania, the valleys are structural, and wells in valley bottoms are statistically more productive. In Illinois, just the reverse is true. For the shallow dolomite aquifer, the yields of wells in bedrock uplands are greater than those in bedrock valleys (Csallany & Walton 1963).

In central Pennsylvania, the yield of wells drilled into anticlines is greater than the yield of wells drilled into synclines. However, the proximity to a fault trace and rock type are more significant than structure in determining yield (Siddiqui & Parizek 1971). In other karst areas, synclines have been noted as major water producers (LaMoureaux & Powell 1960). The relation between structure and well yield is not clear; thus, local experience must be used as a guide.

One of the integral parts of carbonate terrane hydrogeology is the **regolith**—the layer of soil and weathered rock above bedrock. The regolith can be composed of weathering residuum of insoluble minerals remaining after solution of the carbonate minerals. This is typically reddish in color, owing to iron oxides, and contains a high proportion of clay minerals. The regolith may also include transported materials, such as glacial drift. If recharge must first pass through a low-permeability regolith, the rapid response of a carbonate aquifer will be reduced or eliminated. As the regolith slowly releases water from storage, spring discharge from areas overlain by a thick regolith will be more constant than if the regolith were absent. The regolith may also be a local aquifer. The weathering residuum of the Highland Rim area of Tennessee contains localized zones of chert, which can yield water in small amounts (Marcher, Bingham, Lounsbury 1964; Zurawski 1978).

9.3.5 Coal and Lignite

Coal contains bedding planes cut by fractures that are termed **cleat.** Cleat is similar to joint sets in other rock. It is formed as a response to local or regional folding of the coal. There are typically two trends of cleat—normal to the bedding planes and cutting each other at about a 90° angle (Stone & Snoeberger 1977). Coal is often an aquifer and yields water from the cleat and bedding. The quality of water from coal aquifers is variable and sometimes can be poor. Such coals are typically anisotropic, with the maximum hydraulic conductivity oriented along the face cleat, which develops perpendicular to the axis of folding.

There is not a great deal of information on the hydraulic characteristics of coals. The maximum hydraulic conductivity of the Felix No. 2 coal of the Wasatch Formation was determined to be 0.88 ft/day (3.1×10^{-4} cm/s), with a storativity of 1.2×10^{-3} (Stone & Snoeberger 1977). The mean hydraulic conductivity of lignite from four sites in western North Dakota is 1.1 ft/day (3.9×10^{-4} cm/s) (Rehm, Groenewold, & Moran 1978). The Sawyer-A coal aquifer (Fort

Union Formation, Montana) exhibits horizontal anisotropy, with a maximum hydraulic conductivity of 3.3 ft/day (1.2×10^{-3} cm/s) and a minimum hydraulic conductivity of 0.85 ft/day (3.0×10^{-4} cm/s). The storativity was 3.4×10^{-4}. The Anderson coal aquifer of the same formation had a maximum horizontal hydraulic conductivity of 0.66 ft/day (2.3×10^{-4} cm/s) and a minimum value of 0.23 ft/day (8.1×10^{-5} cm/s) (Stoner 1981).

Thick coal beds, such as those of the Powder River Basin of Wyoming, may be important regional aquifers. Well yields of 10 to 100 gal/min (0.6 to 6 L/s) are possible from these coals (Keefer & Hadley 1976). Conflicts arise between water supply from coal aquifers and energy development. Strip mining takes place at the outcrop areas of the coal. If these are recharge areas for the coal aquifer, mine dewatering will adversely affect the potentiometric level in the coal aquifer. If the infiltration capacity of the area is reduced by mining, spoil disposal, or the like, this can also reduce the available recharge to the aquifer.

9.4   IGNEOUS AND METAMORPHIC ROCKS

9.4.1   Intrusive Igneous and Metamorphic Rocks

Intrusive igneous and highly metamorphosed crystalline rocks generally have very little, if any, primary porosity. In order for ground water to occur, there must be openings developed through fracturing, faulting, or weathering. Fractures can be developed by tectonic movements, pressure relief due to erosion of overburden rock, loading and unloading of glaciation, shrinking during cooling of the rock mass, and the compression and tensional forces caused by regional tectonic stresses.

In general, the amount of fracturing in crystalline rocks decreases with depth (Davis & Turk 1964). However, two deep test wells drilled in northern Illinois as exploratory holes for a possible pumped hydroelectric storage project have shown this is not always the case (Coates et al. 1983; Haimson & Doe 1983; Daniels, Olhoeft, & Scott 1983; Coutre, Steitz, & Steindler 1983). These wells were drilled in an area of Paleozoic bedrock overlying crystalline bedrock comprised of biotite granite. Crystalline rocks were penetrated from a depth of 2179 to 5443 ft (664 to 1669 m) in one hole and 2179 to 5273 ft (664 to 1607 m) in the other. Even at these great depths, fractures were found in the crystalline rock. Porosity ranged from 1.42% to 2.15% and intrinsic permeability, determined from in situ packer tests,* from 10^{-4} darcy in a more highly fractured zone near the top of the crystalline basement to 10^{-8} darcy in an area with few fractures. Brines were found in the fractures of the rock at these depths. Brines have also been found in mines at depths in excess of 3000 ft (900 m) in a number of areas of the Precambrian shield of North America (Brace 1980). As crystalline rocks are a

---

*A **packer test** is performed in an open borehole through rock. Inflatable seals, called packers, are used to isolate a particular segment of the open borehole. A pump test or a slug test is then performed on the isolated segment of the borehole.

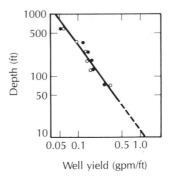

FIGURE 9.26   Yields of wells in crystalline rock in the eastern United States. Open circles represent grouped mean yields of granite rock wells and black dots represent grouped mean yields for schist wells. Source: S. N. Davis & L. J. Turk, *Ground Water* 2 (1964): 6–11.

potential medium for the construction of mined repositories for high-level nuclear waste, the presence of fracture permeability and porosity at great depths is significant.

One study of crystalline rock wells in the eastern United States has shown that the yield, expressed in gallons per minute divided by the depth of the saturated zone penetrated by the well, decreases rapidly with depth (Davis & Turk 1964) (Figure 9.26). However, jointed crystalline rock in the Piedmont of the eastern United States is known to be fractured to depths of 500 ft (150 m) (Stewart 1962). In Wisconsin, fracture zones in crystalline rock are known to be present and can be delineated from air photos. One well drilled on a fracture trace to a depth of 353 ft (108 m) was test-pumped for 12 h at 200 gal/min (13 L/s) with a drawdown of 134 ft (41 m); another well drilled to a depth of 400 ft (122 m) had a yield of 80 gal/min (5 L/s) (Socha 1983).

Chemical weathering of crystalline rock can produce a weathering product called **saprolite.** This material has porosities of 40% to 50% and a specific yield of 15% to 30%. It acts as a reservoir, storing infiltrated water and releasing it to wells intersecting fractures in the underlying crystalline rock (Welby 1984).

The probability of obtaining a high-yield well in crystalline rock areas can be maximized if drilling takes place in an area where fractures are localized. It has been observed that zones of high conductivity in crystalline rock areas underlie linear sags in the surface topography (LeGrand 1962). Such sags are the surface features that overlie major zones of fracture concentration. These show as fracture traces and lineaments on aerial and satellite photographs (Lattman & Parizek 1964). If, in drilling a water-supply well in a crystalline rock area, sufficient water is not encountered in the first 300 ft (100 m) of drilling, in most situations other than where deep tectonic fracturing is suspected a new location should be sought rather than drilling any deeper. Because most fractures are vertical, or nearly so, an angled borehole will be more likely to intersect fractures and create a successful well. Well yields in some areas of crystalline rock are greater when the wells are located on valley bottoms (LeGrand 1954). Many of the valley bottoms probably developed along fracture traces.

Limitations on the use of angled boreholes in fractured rock include stability problems, especially blocks of rock breaking off and lodging in the borehole, and lower potential drawdown than in a vertical borehole of the same length (Banks 1992).

9.4.2   Volcanic Rocks

Because volcanic rocks crystallize at the surface, they can retain porosity associated with lava-flow features and pyroclastic deposition. Hydraulic conductivity of volcanic rocks such as lava flows and cinder beds is typically quite high. However, ash beds, intrusive dikes, and sills may have a much lower hydraulic conductivity. Younger basalt flows tend to have greater conductivity than older ones. Flow features such as clinker zones and gas vesicles in *aa* (a type of lava) and lava tubes and gas vesicles in *pahoehoe* (another lava type), as well as vertical contraction joints and surface irregularities and stream gravels buried between successive flows, contribute to overall conductivity (Peterson 1972). Some of the most productive aquifers are located in basalt flows, as is described in the following case studies.

CASE STUDY: VOLCANIC PLATEAUS—COLUMBIA RIVER BASALTS

The Columbia Plateau area of Washington, Oregon, and Idaho consists of a very thick sequence of Miocene-age basaltic lava flows that erupted from fissures. The lava was very fluid: individual flows are 150 to 500 ft (50 to 150 m) thick, and some can be traced for 125 mi (200 km). For the most part, the basalt flows are either flat-lying or gently tilted. In some places, the basalts have been folded by later tectonic activity. The basalt flows of the Columbia River Group of east-central Washington are as thick as 10,000 ft (3000 m), although in most areas a thickness of 4600 ft (1400 m) is more typical (Luzier & Skrvian 1975). The basalt flows dip at angles of 1° to 2° from northwest to southeast. Water occurs in distinct zones, probably related to interflow boundaries.

A number of irrigation wells have been drilled in the uppermost 1000 ft (300 m) of the basalt flows. Typical well yields of 1000 to 2000 gal/min (60 to 120 L/s) are obtained from confined aquifers located at a depth of between 500 and 1000 ft (150 and 300 m). Shallower aquifers are not as productive, but have a hydraulic head up to 100 ft (30 m) greater than those of the deeper aquifers. Uncased well bores are draining water from the shallower aquifers into the deeper ones (Newcomb 1972). Recharge to the aquifer systems comes from precipitation and loss from ephemeral streams. Precipitation—hence, recharge—increases to the north and east. As these are also topographical high areas, regional ground-water flow is away from them, to the southwest. Age determinations based on carbon-14 dating of ground water of the basalts range from modern to as old as 32,000 years B.P. Age appears to increase with depth, but relationships are obscured by mixing of water of different ages coming from different aquifer zones (Newcomb 1972).

The classic conceptual model of a series of confined aquifers corresponding to individual basalt flows might not be valid in all parts of the Columbia Plateau. Vertical joints form in basalt as it shrinks during cooling. These can provide substantial vertical conductivity, producing unconfined aquifer conditions in thick sequences of lava flows (Foxworthy 1983). In Horse Heaven Plateau, Washington, there is reported to be an unconfined aquifer consisting of the Saddle Mountains Basalt and the upper part of the underlying Wanapum Basalt. A deeper basalt, the Grande Ronde, is confined by a saprolite layer, which formed by weathering of a basalt flow (Brown 1979).

CASE STUDY: VOLCANIC DOMES—HAWAIIAN ISLANDS

In contrast to the very large area of similar geology of the Columbia Plateau Basalts, lava flows associated with some volcanic eruptions can have very heterogeneous aquifer systems. The Hawaiian Islands provide the classic example (Peterson 1972; Cox 1954; Visher & Mink 1964; Takasaki 1978).

Each of the Hawaiian Islands consists of shield volcanoes forming from one to five volcanic domes. Each dome is composed of thousands of individual basaltic lava flows coming from either craters or fissures. Lava flows cooling above sea level are thin bedded, highly fractured, or composed of vesicular and very permeable basalt. Those cooled under water are more massive and less permeable. However, owing to lowered sea levels during the Pleistocene and isostatic sinking of the islands, highly permeable, air-cooled basalt is found below present-day sea level. Interbedded with the lava flows are ash beds, which have a lower permeability. In the zone in which lava flows originated, igneous dikes with low porosity and permeability cut across the lava beds. The original dome structure may be partially eroded. Sediments have accumulated along some of the coastal areas. These coastal plain sediments consist of both terrestrial and marine deposits. The cross section of Figure 9.27 is through an idealized Hawaiian volcanic dome, with both the original and eroded states illustrated.

The Hawaiian Islands are typical examples of oceanic islands where fresh ground water is underlain by salty ground water. Because fresh water is less dense than salty water, the fresh ground water beneath an oceanic island can be thought of as "floating" as a thin lens in the salty ground water. The fresh ground water grades into salty ground water in a zone of mixing.

Ground water in the Hawaiian Islands is contained in the highly permeable basalt flows. It is recharged by rainfall, which can average as much as 20.8 ft (635 cm) per year on Oahu. In the interior of the islands, the basalt flows are isolated by cross-cutting igne-

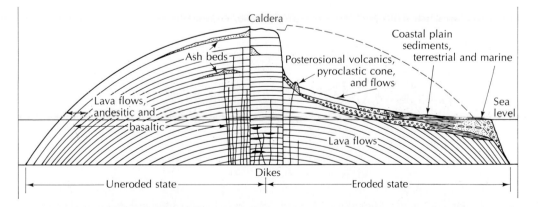

FIGURE 9.27 Geologic structure of an idealized Hawaiian volcanic dome. Source: D. C. Cox, *Hawaiian Planters Record* 54 (1954). Used with permission.

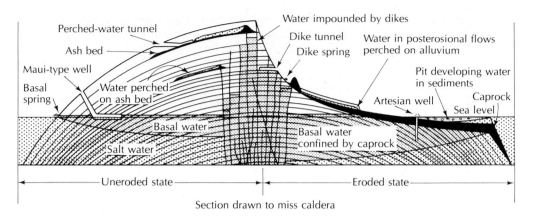

FIGURE 9.28 Occurrence and development of ground water in an idealized Hawaiian volcanic dome. Source: D. C. Cox, *Hawaiian Planters Record* 54 (1954). Used with permission.

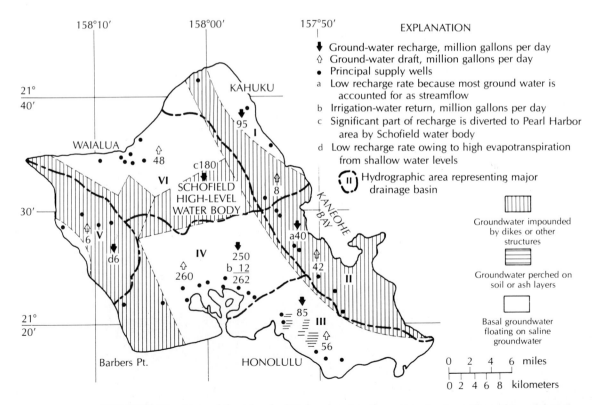

FIGURE 9.29 Map of the island of Oahu showing the approximate outline of ground-water reservoirs, recharge, 1975 draft, and principal supply wells by hydrographic areas representing major drainage basins. Source: K. J. Takasaki, U.S. Geological Survey Professional Paper 813-M, 1978.

ous dikes, which are ground-water dams. The ground water is trapped at a high elevation behind these dams. High-level ground water is also found as perched water in lava beds overlying low-permeability ash beds. If the infiltration is not trapped on an ash bed or behind a dike dam, it moves downward to the basal ground-water body, which is a fresh-water lens in dynamic balance with salty ground water. The basal ground water may be unconfined or, in areas of coastal plain sediments, may be confined by the low-permeability sediments locally termed *caprock.* A cross section of the occurrence of ground water is shown in Figure 9.28. A bird's-eye view of the island of Oahu indicates areas of high-level water bodies impounded by dikes (Figure 9.29).

Springs issue from ash-bed perched aquifers and also from dike-dammed water bodies. Some of these are 300 ft (100 m) or more above sea level. High-level water is developed by tunnels into the tops of ash beds or penetrating dike dams. Most ground water is developed from the basal ground-water body. Unconfined basal water is collected in horizontal skimming tunnels, called *Maui tunnels,* which are at sea level and slightly below. These skimming tunnels can develop water where the fresh-water lens is very thin and conventional wells would pump brackish or salt water. Maui tunnels are capable of producing up to 3.1×10^4 gal/min (2×10^3 L/s), although in most cases the yield is much less. Where the basal water is confined by coastal plain sediments, conventional wells are used for ground-water development.

9.5 GROUND WATER IN PERMAFROST REGIONS

In polar latitudes and high mountains, the mean annual temperatures may be sufficiently low for the ground to be at a temperature below 0°C. If this temperature persists for two or more years, the condition is known as **permafrost** (Cederstrom, Johnson, & Subitzky 1953; Brandon 1965; Williams 1970). During the summer, warm temperatures may cause the upper 3 to 6 ft (1 to 2 m) of the soil or rock to thaw. This is called the *active layer,* but underneath it the ground may be frozen to depths to 1300 ft (400 m).

The magnitude of the annual temperature fluctuation of the soil is greatest at the surface and diminishes with depth until, at some depth, there is zero annual temperature amplitude. The depth at which the maximum annual soil temperature is 0°C is the *permafrost table.* In some areas, the permafrost may occur in layers, with zones of unfrozen ground between them. This condition is usually the result of past climatic events, and the permafrost distribution is not congruent with the present climatic and thermal regime. The local depth of permafrost is a function of the geothermal gradient and the mean annual air temperature (Terzagki 1950).

The insulating cover of glacier ice and large lakes may prevent the formation of permafrost. Permafrost is likewise not present beneath the ocean. At higher elevations and on north-facing slopes, where the mean temperature is lower, permafrost may be thicker (Figure 9.30A). Lakes and streams also affect the permafrost. Shallow lakes—6 ft (2 m) deep or less—freeze to the bottom in winter and have little effect on the permafrost table, but the permafrost may be warmer; hence, not as thick. Deeper lakes are unfrozen at the bottom and have an insulating effect. Small deep lakes create a saucer-shaped depression in the permafrost table, which, in turn, creates an upward indentation in the bottom of the permafrost (Lachenbruch et al. 1962). Permafrost is absent beneath large deep

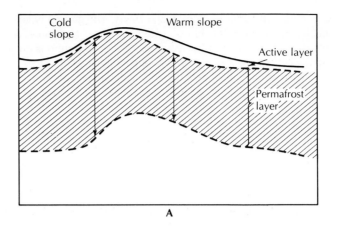

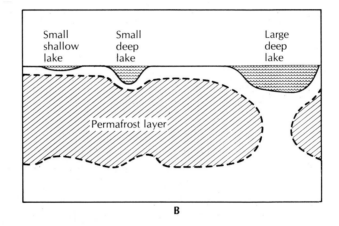

FIGURE 9.30   A. Effect of topography on the permafrost layer with the thinnest layer found beneath a warm, south-facing slope. **B.** Effect of lakes on the permafrost layer.

lakes—even in the continuous permafrost zone (Figure 9.30B). The effects of large rivers on distribution of permafrost are similar to those of large deep lakes.

The permafrost table creates perched water in the active layer. This results in poorly drained soils and the typical muskeg and marsh vegetation of tundra regions. Deeper aquifers are recharged only in the absence of permafrost, as the permafrost layer acts as a confining layer.

Water below the permafrost layer is confined. The potentiometric surface may be in the permafrost layer or even above land surface (Williams 1970). Subpermafrost water may discharge into large rivers and lakes, beneath which unfrozen conduits are open. Saline ground water may exist beneath permafrost, but it is typically not actively circulating (Williams 1970). Discharge of ground water at the surface, especially in winter, may result in the development of sheets of surface ice or large conical hills called *pingos*. Pingos have ice cores and are formed by the upward arch of the ground surface due to hydraulic pressure of ground water confined by permafrost.

Alluvial river valleys are good sources of ground water in permafrost regions. The permafrost beneath and along the river may be thin, or absent in places, and alluvial gravel deposits are good aquifers. Large rivers will have more unfrozen ground—hence, more available water—than smaller tributaries and headwater streams.

In the far northern parts of Alaska, Canada, and the Soviet Union, permafrost is present nearly everywhere. It is in this continuous permafrost region that maximum permafrost depths are found. To the south, the permafrost layer is discontinuous. It may be up to 600 ft (180 m) thick locally, but elsewhere there could be unfrozen ground. In the continuous permafrost regions of Alaska, alluvial valleys of large rivers may be the only water available. As the active layer freezes in winter, baseflow to even the largest rivers may be reduced to zero. Subpermafrost ground water is typically saline or brackish. In the discontinuous permafrost region, the permafrost is thinner, and the areas free of permafrost are much more extensive in alluvial river valleys.

Alluvial fans are found in Alaska at the margins of many of the mountain ranges. The fans are composed of glaciofluvial deposits, which tend to be coarse sand and gravel. Ground water can be obtained from these deposits, either below the permafrost or near rivers or lakes, where the permafrost may be thin or absent. In the more southerly alluvial fans, the water table may be below the permafrost layer. In this case, the permafrost has little impact on the hydrogeology other than preventing recharge to areas where it is present.

The distribution of permafrost in consolidated rocks is similar to that in unconsolidated deposits. If a rock unit has significant hydraulic conductivity in both the horizontal and vertical directions, ground-water hydrology should be similar to that of unconsolidated deposits. In highly anisotropic aquifers, even discontinuous permafrost bodies could act as ground-water dams, preventing horizontal flow and significantly reducing or eliminating vertical recharge. For example, ground water in fracture zones is replenished by downward recharge. If the fracture zone were covered by a patch of permafrost, recharge would be difficult, even if the fracture zone extended below the permafrost layer.

CASE STUDY: ALLUVIAL AQUIFERS—FAIRBANKS, ALASKA

In the area of Fairbanks, Alaska, the Chena and Tanaua rivers have alluvial deposits up to 800 ft (240 m) thick. The alluvium consists of interbedded gravel, sand, and silt. The floodplains are interspersed with terraces from 3 to 25 ft (1 to 8 m) in height. The distribution of permafrost in the area is irregular. Permafrost is absent or nearly so in the alluvium beneath the river channels. Thin permafrost may be found under islands. The youngest, low-terrace deposits are underlain by unfrozen alluvium, with some isolated permafrost bodies up to 80 ft (24 m) thick. Higher terraces have more continuous permafrost, which can be 200 ft (60 m) deep. The older terraces have nearly continuous permafrost up to 280 ft (85 m) deep (Williams 1970). Permafrost thickness increases beneath progressively older terraces. Wells in the alluvial valley obtain water from unfrozen areas or, if there is permafrost, from either above or below it (Cederstrom 1963).

9.6   GROUND WATER IN DESERT AREAS

Desert areas receive 10 in. (25 cm) of precipitation or less each year. Cold deserts, such as Antarctica, can have great accumulations of water. In warm deserts, however, the potential evapotranspiration may be many times the annual precipitation. Under such conditions, there is often virtually no local ground-water recharge. Yet, some warm deserts can have large volumes of fresh water stored beneath them.

Both bedrock and unconsolidated aquifers are known in desert areas. As the dry climate promotes mechanical weathering of rock, the unconsolidated materials are often rather coarse and permeable. Alluvial fans and talus slopes at the bases of mountains and escarpments can be productive. In South Yemen on the Arabian peninsula, ground water is obtained from shallow wells in alluvial materials (Cederstrom 1971). Recharge occurs during flooding that accompanies the infrequent precipitation.

Vast bedrock aquifers are known to occur in the sedimentary basins of Egypt, Jordan, and Saudi Arabia (Harshbarger 1968). The Sahara Desert is underlain by the Nubian aquifer, a sandstone up to 3000 ft (900 m) thick. Several younger aquifers overlie it. The ground water in the Nubian Sandstone is confined, and there are large initial heads. High-capacity wells now tap this aquifer. Several major aquifer systems, primarily sandstone, also underlie the Arabian peninsula. Carbonate aquifers may not have high-solution permeability owing to the lack of actively circulating ground water. Those that are permeable are a result of primary permeability or solution permeability developed during moister periods of the past.

Recharge to aquifers can occur through runoff from adjacent mountains, which receive relatively more water. However, under much of North Africa and Arabia, the ground water is old, exceeding 35,000 radiocarbon years B.P. (Lloyd & Farag 1978). It has been shown by model studies that the water in these aquifers was probably recharged during wet climatic periods during the Pleistocene. Under such conditions, ground water is mined from the aquifers, as there is no modern recharge.

Arid zones may have significant water-quality problems. The slow circulation of ground water results in mineralization. In addition, evaporation of ground water from discharge areas results in salt deposition, with consequent high salinity in the soil and shallow ground water.

9.7   COASTAL PLAIN AQUIFERS

Coastal plains are found on all of the continents, with the exception of Antarctica. They are regional features, bounded on the continental side by highlands and seaward by a coastline. Coastal plains exist in areas of both stable basement rock as well as in those areas where the basement is sinking. The coastal plain may include large areas of former sea floor, and the geology of the coastal zone may be very similar to that of the adjacent continental shelf.

Sediment and sedimentary rocks of coastal plains were formed as either terrestrial or marine deposits. The terrestrial deposits tend to be landward and the marine deposits seaward, although fluctuating sea levels have resulted in alternating continental and marine strata—units deposited at a given time grade from continental to marine. Individual units have a variety of shapes, although the overall sequence usually thickens seaward. Coastal plains almost always contain Quaternary sediments; many also contain Tertiary- and Mesozoic-age deposits as well. The older rock units tend to crop out on the landward side, whereas younger, Quarternary and Recent, rock and sediments are found at the surface near the coast.

The coastal plain sediments may be unconsolidated or lithified, with no relationship to age. The Magothy aquifer of the northeastern United States is an unconsolidated sand of Cretaceous age, while the Biscayne aquifer of southeastern Florida consists of Pleistocene-age marine limestone. Aquifers typically are continental sands, gravels, and sandstones or marine sands or limestones. Confining beds consist of marine and continental silts and clays.

Typically, alternating and interfingering facies of different lithology are encountered. Some units are thick and extend for hundreds of miles; others are often not traceable for more than a few miles. One or more aquifers may be found at most locations. Baton Rouge, Louisiana, has 10 aquifers in the first 3000 ft (900 m) of sediments (Kazmann 1970). Some of the world's most prolific aquifers are found in coastal plain deposits. A notable example is the Atlantic and Gulf coastal plain of North America (Miller 1986; Grubb 1986; Wait et al. 1986; Meisler et al. 1986).

The sediments of coastal plain areas were deposited either adjacent to or in shallow marine waters. Thus, pore water was originally saline. Fluctuating sea levels during the Pleistocene inundated many areas that are now land. As a result, saline waters occupied many contemporary fresh-water coastal plain aquifers in the not-too-distant geologic past. Relative sea level was up to 300 ft (90 m) lower during the last Wisconsinan glaciation than present sea level, uncovering more of the continental shelf than is now exposed. During this period, saline water was flushed from inland aquifers to considerable depths.

Because of the seaward-sloping nature of coastal plain strata, deep aquifers are recharged in inland areas. Fresh water flows down-gradient and then discharges by several mechanisms to the coastal waters. The amount of aquifer flushing that has occurred is a function of the volume of aquifer recharge and the amount of fresh water than can escape down-gradient via the available mechanisms. Saline water has been flushed to a depth of 5900 ft (1800 m) in Karnes County, Texas (McGuinness 1963), and 3495 ft (1070 m) below sea level in St. Tammany Parish, Louisiana (Rollo 1960), although this is unusual. Fresh water is found at depths of 1000 ft (300 m) or more below sea level in many areas of the coastal plain aquifer (McGuinness 1963). Fresh water has been found to occur in deep, confined coastal plain sediments many miles at sea. In a deep test well on Nantucket Island, located 40 mi (64 km) off the New England coast, fresh water

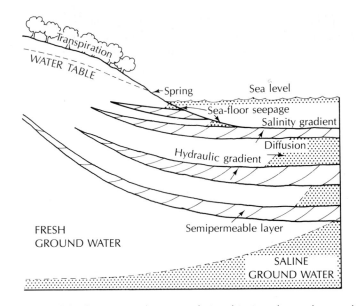

FIGURE 9.31   Typical fresh-water—salt-water relationship in a layered coastal aquifer.

was found in confined aquifers at depths of 730 to 820 ft (223 to 250 m) and 900 to 930 ft (274 to 283 m). These are Cretaceous-age sands and are confined by clays (Kohout et al. 1977). They represent aquifers that were recharged during the Pleistocene, when sea level was lower (Kohout et al. 1977; Collins 1978).

Fresh water can discharge from a coastal aquifer via several natural mechanisms: (1) evapotranspiration, (2) direct seepage into springs, streams, tidal water, and the ocean floor, (3) mixing with saline ground water in a zone of diffusion, (4) flow across a semipermeable layer under the influence of a hydraulic gradient, and (5) flow across a semipermeable layer due to osmotic pressure caused by a salinity gradient. Examples of the various discharge mechanisms in a coastal area are shown in Figure 9.31. Mechanisms (1) and (2) are very efficient in discharging water from unconfined aquifers. Fresh-water springs in the sea bottom occur in unconfined aquifers or confined aquifers where the confining layer is breached. Deep confined aquifers utilize only the last three methods—1, 2, and 3—which are not as efficient. Assuming that all aquifers have equal transmissivities, the deeper the aquifer, the less fresh water it can discharge because of the reduction in the number and efficiency of discharge methods. Confined coastal aquifers can contain fresh water, even though overlying aquifers are salty.

On Long Island, New York, there is an unconfined aquifer and two deep confined aquifers, all recharged by precipitation at the center of the island. Most of the fresh ground water discharges through the unconfined aquifer, with much lower volumes flowing in the confined aquifers. Wells drilled in a barrier island 28,600 ft (8700 m) from the coastline show fresh water the full depth of both confined aquifers, with 5 to 10 ft (1.5 to 3 m) of artesian head in the upper confined aquifer and 20 ft (6 m) in the lower. This is a distance of at least 11 to 12 mi (18 to 19 km) from the closest part of the recharge area for the confined aquifers (Pluhowski & Kantrowitz 1964).

The most characteristic type of water-quality degradation occurring in coastal plain aquifers is saline-water intrusion. Sources of saline water are found as connate water below inland fresh-water aquifers, as subsurface sea water below island aquifers, on the seaward edges of coastal aquifers, and as surface tidal waters in natural estuaries and artificial canals.

The shape and position of the boundary between saline ground water and fresh ground water is a function of the volume of fresh water discharging from the aquifer (excluding intrinsic aquifer characteristics). Any action that changes the volume of fresh-water discharge results in a consequent change in the salt-water–fresh-water boundary. It should be noted that minor fluctuations in the boundary position occur with tidal actions and seasonal and annual changes in the amount of fresh-water discharge. For this reason, the boundary is in a state of quasi-equilibrium. Natural changes in the equilibrium position can result from long-term changes in climatic patterns or the position of relative sea level.

Human action that results in saline ground water entering a fresh-water aquifer is termed **saline-water encroachment.** It occurs as a result of a diversion of fresh water that previously had discharged from a coastal aquifer. Salt-water encroachment can be either *active* or *passive* (Fetter 1973).

Passive saline-water encroachment occurs when some fresh water has been diverted from the aquifer—yet the hydraulic gradient in the aquifer is still sloping toward the salt-water–fresh-water boundary. In this case, the boundary will slowly shift landward until it reaches an equilibrium position based on the new discharge conditions. Passive saline-water encroachment is taking place today in many coastal plain aquifers where ground-water resources are being developed. It acts slowly and, in some areas, it may take hundreds of years for the boundary to move a significant distance (Figure 9.32).

The consequences of active saline-water encroachment are considerably more severe, as the natural hydraulic gradient has been reversed and fresh water is actually moving away from the salt-water–fresh-water boundary (Figure 9.33). This occurrence is due mainly to the concentrated withdrawals of ground water creating a deep cone of depression. The boundary zone moves much more rapidly than it does during passive saline-water encroachment. Furthermore, it will not stop until it has reached the low point of the hydraulic gradient: the center of pumping. It is this type of rapid encroachment that destroyed the aquifers beneath Brooklyn, New York, in the 1930s, when the water table was lowered 30 to 50 ft (9 to 15 m) below sea level.

Both types of saline-water encroachment can occur in areas of inland connate saline waters as well as in coastal areas. In Baton Rouge, Louisiana, heavy industrial pumpage has lowered the potentiometric surface in the "600-ft sand" aquifer. This has caused connate saline ground water in the aquifer to move northward toward the well field (Kazmann 1970).

Saline-water encroachment is also a serious problem in the Miami area of Florida. The unconfined Biscayne aquifer is the sole source of ground water for this area, as the Floridan aquifer is salty here. The encroachment began with the 1907 installation of inland drainage canals to lower the water table. This diverted

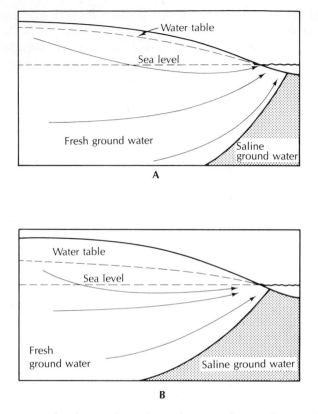

FIGURE 9.32 **A.** Unconfined coastal aquifer under natural ground-water discharge conditions. **B.** Passive saline-water encroachment due to a general lowering of the water table. Flow in the fresh-water zone is still seaward.

fresh water that had been flowing through the Biscayne aquifer to Biscayne Bay. Passive saline-water encroachment resulted as the salt-water–fresh-water boundary sought a new equilibrium position. Because of the high transmissivity of the aquifer, saline water occupied many areas of the Biscayne aquifer beneath Miami in only a few years (Parker et al. 1955). Seawater also moved up the canals toward the aquifer during high tides. Dams have now been built across the canals in order to keep salt water from traveling up them. Water impounded behind the dams is fresh, adding to the available recharge to the Biscayne aquifer. This fresh water had formerly been lost to the sea during flooding. New canals must be constructed with salinity-control dams.

9.8 FRESH-WATER–SALINE-WATER RELATIONS

9.8.1 Coastal Aquifers

We have assumed to this point that the content of dissolved solids of ground water is so low that it does not affect the physics of flow. However, if fresh ground water

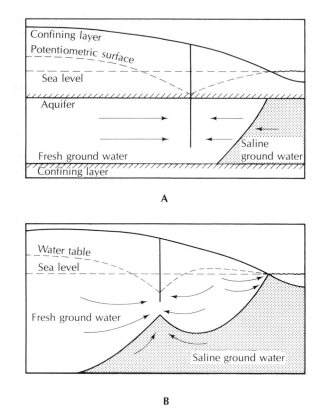

FIGURE 9.33 **A.** Active saline-water encroachment in a confined aquifer with the potentiometric surface below sea level. **B.** Active saline-water encroachment in an unconfined aquifer with the water table drawn below sea level.

is adjacent to saline ground water, the difference in density between the two fluids becomes very important. Due to the difference in dissolved solids, the density of the saline water, ρ_s, is greater than the density of fresh water, ρ_w. Salt water is found adjacent to fresh water in inland areas, often in the same aquifer, as well as in oceanic coastal areas and oceanic islands. Highly saline water in inland aquifers could be either trapped from the time of formation of the rock unit (connate water) or occur through mineralization due to stagnant flow conditions. At coastal locations, the fresh ground water beneath land is discharging near the coast and mixing with saline ground water beneath the sea floor.

Fresh ground water usually grades into saline water with a steady increase in the content of dissolved solids. In some situations, the contact may be quite sharp; that is, a very thin zone of mixed water. The mixture of fresh water and salt water yields the zone in which there is a salinity gradient. If the aquifer is subject to hydraulic head fluctuations caused by tides, the zone of mixed water will be enlarged. In unconfined coastal aquifers, there is ground-water flow occurring in both the fresh zone and the saline zone (Cooper 1959). Fresh water is flowing upward to discharge near the shoreline, and there is a cyclic flow in the

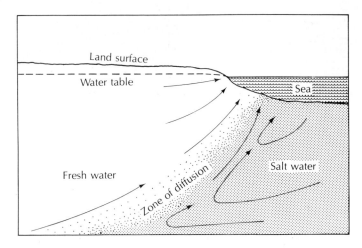

FIGURE 9.34   Circulation of fresh and saline ground water at a zone of diffusion in a coastal aquifer. Source: H. H. Cooper, Jr., U.S. Geological Survey Circular 1613-C, 1964.

salty water near the interface (Figure 9.34). We will make the simplifying assumption that there is a sharp interface between fresh water and saline water. Although the salt-water interface problem can be studied using dispersion and mass-transport theory (Bredehoeft & Pinder 1973; Segol, Pinder, & Gray 1975), the mathematical treatment is beyond the scope of this book. The zone of dispersion is often thin with respect to the overall thickness of the fresh-water lens. Likewise, we will consider only the steady-state case. Solutions have been developed for moving interface problems (Bredehoeft & Pinder 1973; Segol, Pinder, & Gray 1975; Anderson 1976; Collins & Gelhar 1971; Pinder & Cooper 1970), but they are too complex to be considered here.

A number of scientists have made significant contributions to the study of the saline-water–fresh-water interface in coastal aquifers. Studies by W. Baydon-Ghyben (1888–1889) and A. Herzberg (1901) in the late nineteenth century have been widely cited and have given rise to the Ghyben-Herzberg principle, which we will now discuss. However, their work was antedated by more than half a century by an American, Joseph DuCommun (1828), who clearly made the same observations in 1828. Unfortunately, DuCommun is not given just credit in the literature.

These early observers noted that in unconfined coastal aquifers the depth to which fresh water extends below sea level is approximately 40 times the height of the water table above sea level. The (misnamed) **Ghyben-Herzberg principle** states that

$$z_{(x,y)} = \frac{\rho_w}{\rho_s - \rho_w} h_{(x,y)} \tag{9–1}$$

where

$z_{(x,y)}$ is the depth to the salt-water interface below sea level at location (x,y) (L; ft or m)

$h_{(x,y)}$ is the elevation of the water table above sea level at point (x,y) (L; ft or m)

ρ_w is the density of fresh water (M/L^3; g/cm^3)

ρ_s is the density of salt water (M/L^3; g/cm^3)

The application of this principle is limited to situations in which both the fresh water and salt water are static.

EXAMPLE PROBLEM

If $\rho_w = 1.000$ g/cm^3 and $\rho_s = 1.025$ g/cm^3, what is the ratio of $z_{(x,y)}$ to $h_{(x,y)}$?

$$z_{(x,y)} = \frac{\rho_w}{\rho_s - \rho_w} h_{(x,y)}$$

$$= \frac{1.000 \text{ g/cm}^3}{1.025 \text{ g/cm}^3 - 1.000 \text{ g/cm}^3} h_{(x,y)}$$

$$= 40 h_{(x,y)}$$

Figure 9.35 illustrates the Ghyben-Herzberg principle for an unconfined coastal aquifer. Hubbert (1940) pointed out that $h_{(x,y)}$ should actually be the

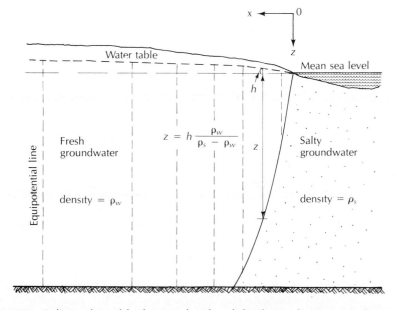

FIGURE 9.35 Relationship of fresh-water head and depth to salt-water interface.

hydraulic head at the interface at point (x,y). However, for thin aquifers with a large vertical extent, the Dupuit assumption that equipotential lines are vertical can be made, so that the hydraulic head at the salt-water interface is equal to the elevation of the water table at that location (Fetter 1972b). In a study of the salt-water interface on eastern Long Island, it was shown that even at the coastline, where the greatest deviations from the Dupuit assumptions could be expected, there was almost no difference in the position of the interface as computed from potential theory versus Dupuit flow (Figure 9.36).

Flow in coastal aquifers can be described by means of the Dupuit equations in combination with the Ghyben-Herzberg principle. The steady flow of ground water is given by the partial differential equation (Fetter 1972b)

$$\frac{\partial^2 h^2}{\partial x^2} + \frac{\partial^2 h^2}{\partial y^2} = \frac{-2w}{K\,(1 + g)} \tag{9-2}$$

where

w is the recharge to the aquifer (L/T; ft/day, m/day)

K is the hydraulic conductivity (L/T; ft/day, m/day)

G is equal to $\dfrac{\rho_w}{\rho_s - \rho_w}$ (dimensionless)

Should the value of the depth to the salt-water interface, as computed by Equation 9–1, exceed the depth of the aquifer, then the salt-water wedge is

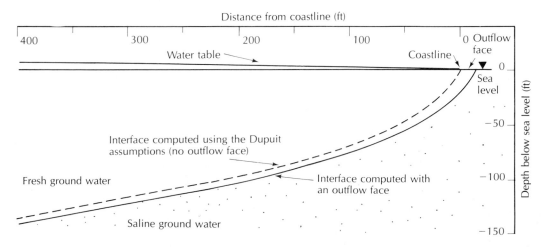

FIGURE 9.36 Comparison of the position of the salt-water interface on eastern Long Island as computed using the Dupuit assumptions with that computed using an outflow face. Source: C. W. Fetter, Jr., *Water Resources* 8 (1972): 1307–14.

missing. This is the case on the left side of Figure 5.35. In this region, the governing equation is (Polubarinova-Kochina 1962)

$$\frac{\partial^2 h^2}{\partial x^2} + \frac{\partial^2 h^2}{\partial y^2} = \frac{-w}{K\,(z_m + h)} \tag{9-3}$$

where z_m is the aquifer thickness below sea level. Both Equation 9–2 and Equation 9–3 can be solved for an infinite-strip coastline; that is, one with flow in only one direction. The x- and z-axes are shown on Figure 9.35.

The *Dupuit-Ghyben-Herzberg model* of one-dimensional flow in coastal aquifers yields the following expression for the x- and z-coordinates of the interface (Todd 1953):

$$z = \sqrt{\frac{2q'xG}{K}} \tag{9-4}$$

where q' is the discharge from the aquifer at the coastline, per unit width $[(L^3/T)/L]$.

One of the failings of the Dupuit-Ghyben-Herzberg model of coastal aquifers is that the salt-water interface intercepts the water table at the coastline. This does not allow for vertical components of flow and discharge of the fresh water into the sea floor. Therefore, a simple model has been developed in which the x- and z-coordinates of the interface are given by the following relation (Glover 1964):

$$z = \frac{Gq'}{K} + \sqrt{\frac{2Gq'x}{K}} \tag{9-5}$$

Note that Equation 9–5 is identical to 9–4, except that a constant, Gq'/K, has been added. Thus, when $x = 0$, z will still have a value. The interface is shown in Figure 9.37.

The width of the outflow face, x_0, may be found from the expression

$$x_0 = -\frac{Gq'}{2K} \tag{9-6}$$

and the height of the water table at any distance, x, from the coast is given by

$$h = \sqrt{\frac{2q'x}{GK}} \tag{9-7}$$

9.8.2 Oceanic Islands

Oceanic islands are underlain by salty ground water as well as being surrounded by sea water. The fresh water takes the form of a fresh-water lens floating on the

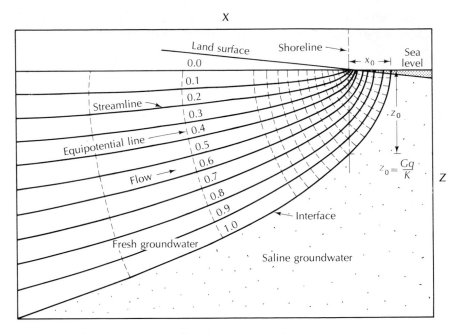

FIGURE 9.37   Flow pattern near a beach as computed using Equation 9–7. Source: R. E. Glover, U.S. Geological Survey Water-Supply Paper 1613-C, 1964.

more dense salty water. Equation 9–2 also describes the flow of water in an oceanic island. It can be solved for islands of regular shape, such as an infinite strip or circle.

For an infinite-strip island receiving recharge at a rate, w, with a width equal to $2a$, the head of the water table, h, at any distance, x, from the shoreline is given by (Fetter 1972b):

$$h^2 = \frac{w[a^2 - (a - x)^2]}{K(1 + G)} \tag{9–8}$$

A circular island with a radius of distance R can be evaluated using

$$h^2 = \frac{w(R^2 - r^2)}{2K(1 + G)} \tag{9–9}$$

where h is the head above sea level at some radial distance, r, from the center of the island.

EXAMPLE PROBLEM

Part A: An infinite-strip island has a width of 2 km. The permeability of the sediments is 10^{-2} cm/s and there is a daily accretion of 0.13 cm/day. The density of fresh water is 1.000 and the density of salty ground water is 1.025. Compute a water-table profile across the island using Equation 9–8. Then determine the interface depth using Equation 9–1.

$$G = \frac{1}{1.025 - 1} = 40$$

$$K = 10^{-2} \text{ cm/s} = 8.64 \text{ m/day}$$

$$w = 0.0013 \text{ m/day}$$

$$a = 1000 \text{ m}$$

$$h^2 = \frac{w[a^2 - (a - x)^2]}{K(1 + G)}$$

$$z = Gh$$

x (m)	h (m)	z (m)
1000	1.92	76.8
900	1.91	76.4
800	1.88	75.2
700	1.83	73.2
600	1.76	70.4
500	1.66	66.4
400	1.53	61.2
300	1.37	54.8
200	1.15	46.0
100	0.84	33.6
0	0	0

Part B: From the computed profile, it can be seen that the fresh-water lens thins very rapidly in the last 100 m as the shoreline is approached. As the Dupuit assumptions may not be valid near the coastline, the profile of the interface close to the coast can be computed by use of Equation 9–5. The outflow per unit width, q', is equal to the recharge rate times the half-width.

$$q' = 0.0013 \text{ m}^3\text{/day} \times 1000 \text{ m}$$

$$= 1.3 \text{ m}^3\text{/day/m}$$

$$z = \frac{Gq'}{K} + \sqrt{\frac{2Gq'x}{K}}$$

x (m)	z (m)
0	6.0
20	21.5
40	28.0
60	32.9
80	37.1
100	40.7

Find the width of the outflow face:

$$x_0 = -\frac{Gq'}{2K}$$

$$= -3.0 \text{ m}$$

Find the height of the water table at a distance from the coast of 100 m:

$$h = \sqrt{\frac{2q'x}{GK}}$$

$$= 0.87 \text{ m}$$

Comparison of the position of the salt-water interface in a coastal aquifer as computed by using Equation 9–4, which does not allow for an outflow face, with the position computed by Equation 9–5, which includes an outflow face, will show the results are similar (see Figure 9.36). However, the solution using Equation 9–5 will be more exact at the coastline and in that area will show a greater value of head and a greater depth to the interface. The coastal zone is about 1 to 5 percent of the total width of the island (Vacher 1987). The two equations will yield essentially the same results away from the coastal zone. Use of Equations 9–1, 9–4, 9–8, and 9–9 will result in a slight error near the coastline owing to the lack of allowance for an outflow face.

However, the Dupuit-Ghyben-Herzberg model for oceanic islands has been used very successfully in several field applications (for example, Fetter 1972b; Vacher 1988; Wallis, Vacher, & Stewart 1991; Vacher & Wallis 1992).

9.9   TIDAL EFFECTS

Aquifers located next to tidal bodies are subject to short-term fluctuations in the head, h, due to the tide. Water-level recorders located in coastal wells show a fluctuation in the hydraulic head that parallels the rise and fall of the tide. The amplitude of the fluctuation is greatest at the coast and diminishes as one goes inland.

If we have a confined aquifer, as shown in Figure 9.38, water can enter at the subsurface outcrop. The governing flow equation in one dimension can be used to describe the flow of water into and out of the aquifer as the tide changes.

The amplitude of the tidal change is H_0 and the tidal period, or time for the tide to go from one extreme to the other, is t_0. At any distance, x, inland from the coast, the amplitude of the tidal fluctuation, H_x, is given by (Jacob 1950)

$$H_x = H_0 \exp\left(-x\sqrt{\pi S/t_0 T}\right) \tag{9-10}$$

where S and T are the aquifer storativity and transmissivity.

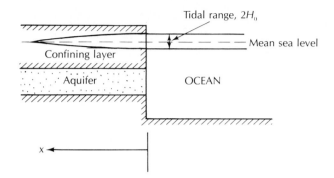

FIGURE 9.38 Coastal aquifer showing the tidal range, $2H_0$, and the effect of the tide on the potentiometric surface.

The time lag, t_τ, between the high tide and the peak of the water level (or low tide and the low point in the water level) is given by

$$t_\tau = x\sqrt{t_0 S/4\pi T} \qquad (9\text{--}11)$$

The preceding equations can also be applied to unconfined flow, as an approximation, if the range of tidal fluctuations is small compared with the saturated aquifer thickness (Erskine 1991).

Equations 9–10 and 9–11 show that the response of the hydraulic head in an aquifer to changes in the tides decreases exponentially with distance from the coast, whereas the time lag decreases linearly with distance.

9.10   GROUND-WATER REGIONS OF THE UNITED STATES

One useful generalization in the study of hydrogeology is the concept of ground-water regions. These are geographical areas of similar occurrence of ground water. If an area is subdivided into several smaller regions, useful comparisons can be made between areas of well-known hydrogeology and areas that are geologically similar but have not been as well studied.

The ground-water regions of the United States have been classified by several different authorities. In 1923, O. E. Meinzer, who could be considered to be the father of modern hydrogeology in the United States, proposed a classification system based on 21 different ground-water provinces (Meinzer, 1923a). Thomas proposed a system based on 10 ground-water regions in 1952 (Thomas 1952) and McGuinness revised Thomas' system in 1963 (McGuinness 1963). In 1984 Heath published yet another revision of Thomas' basic system (Heath 1984). Heath based his classification system on five features of ground-water systems:

1. The components of the system and their arrangement,
2. the nature of the water-bearing openings of the dominant aquifer or aquifers with respect to whether they are of primary or secondary origin,

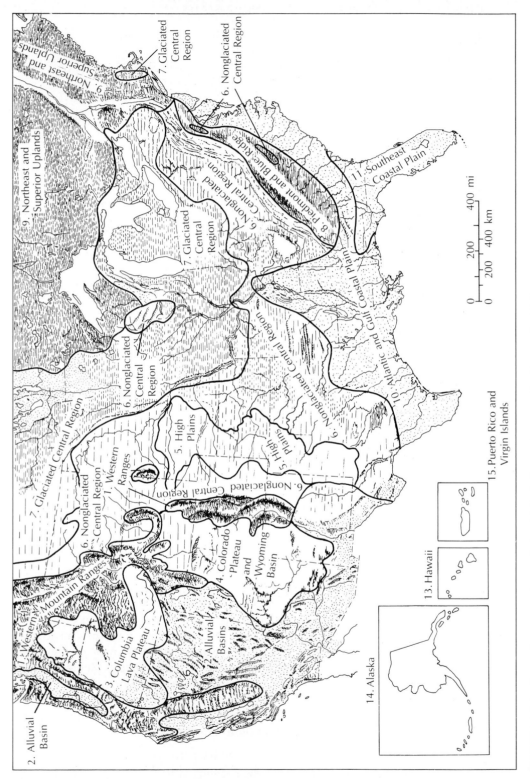

FIGURE 9.39 Ground-water regions of the United States. Source: R. C. Heath, U.S. Geological Survey Water-Supply Paper 2242, 1984.

1. Western Mountain Ranges
2. Alluvial Basin
3. Columbia Lava Plateau
4. Colorado Plateau and Wyoming Basin
5. High Plains
6. Nonglaciated Central Region
7. Glaciated Central Region
8. Piedmont and Blue Ridge
9. Northeast and Superior Uplands
10. Atlantic and Gulf Coastal Plain
11. Southeast Coastal Plain
13. Hawaii
14. Alaska
15. Puerto Rico and Virgin Islands

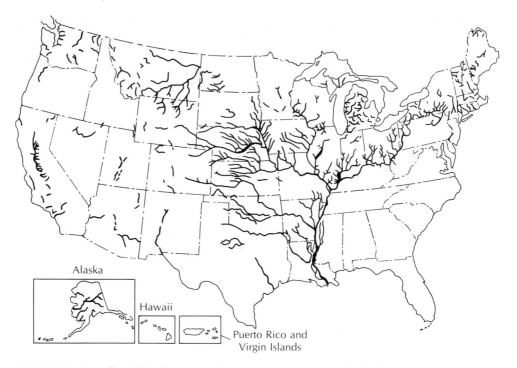

FIGURE 9.40 Alluvial Valleys ground-water region. Source: R. C. Heath, U.S. Geological Survey Water-Supply Paper 2242, 1984.

3. the mineral composition of the rock matrix of the dominant aquifers with respect to whether it is soluble or insoluble,

4. the water storage and transmission characteristics of the dominant aquifer or aquifers, and

5. the nature and location of recharge and discharge areas.

Eleven of the ground-water regions are based on physiography and are not necessarily contiguous. The locations of these areas are shown on Figure 9.39. Area 12 comprises alluvial valleys throughout the United States and is shown on Figure 9.40. Area 13 is Hawaii, area 14 is Alaska, and area 15 is Puerto Rico and the Virgin Islands.

In 1988 Heath defined the ground-water regions of North America as the organizing theme of the book: *Geology of North America, Vol. 0–2, Hydrogeology*. In addition to extending his coverage to include all Canada and Mexico, some of his 1984 regions in the United States were subdivided for editorial purposes. In this text we will retain the 1984 subdivisions, which are based on the character of the underlying rock units and recharge/discharge relationships.

1. Western Mountain Ranges

This region consists of mountains with thin soils over fractured rocks and alternating narrow alluvial valleys, some of which have been glaciated.

The mountains of the western United States are the headwater areas of many rivers, including the Columbia, Snake, San Joaquin, Sacramento, Missouri, Platte, Colorado, Arkansas, and Rio Grande. The mountains tend to be well watered, receiving higher levels of precipitation than the surrounding lowlands. The rocks in the western mountains are predominantly crystalline, although there are some sedimentary ranges. Intermontane valleys may contain aquifers of alluvial deposits of glacial outwash. Most consolidated rock aquifers are of local rather than regional extent. Mountain ranges included in this region are the northern Coast Ranges, the Sierra Nevada and Cascade Range, parts of the physiographic Basin and Range Province, and the Rocky Mountains.

2.   Alluvial Basins

This region consists of thick alluvial deposits in basins and valleys that are bordered by uplifted mountain ranges.

Lying between the ranges of the western mountains are tectonic troughs that have been partially filled with erosional products from the adjacent uplands and mountains. The sediment filling the valleys has been sorted into various size fractions by running water. Sand and gravel layers are found interspersed between silt and clay deposits. The valley-fill material is porous; hence, there are many permeable zones—especially, close to the mountains, where coarser debris was deposited in coalescing alluvial fans.

Runoff coming from the mountains can soak into the coarse deposits; however, rainfall over the basins themselves may be rather low. The deep sedimentary basins hold vast volumes of water in storage. Many of the aquifers can yield large amounts of water to individual wells. Demands for water are high, and the amount of ground-water recharge is relatively low. As a result, the water in storage is being withdrawn and ground-water levels are falling in many of the basins. Land subsidence due to the lowered ground-water levels in some of the basins is also a problem.

3.   Columbia Lava Plateau

This region consists of thin soils overlying a thick sequence of basalt flows with interbedded sediments.

The Columbia Lava Plateau is formed of intrusive volcanic rock with interbedded alluvium and lake sediments. The plateau is somewhat arid, but some of the higher areas and surrounding mountains are more humid.

The lava flows are nearly horizontal, with some layers and interflow surfaces highly permeable. Shrinkage cracks create vertical permeability. The amount of water in storage is high; it is replenished by local precipitation, runoff from nearby mountains, and excess irrigation water applied to the surface.

The plateau is deeply dissected by the Columbia and Snake rivers and their tributaries. There are many large springs flowing from the permeable basalt layers into incised rivers—e.g., the Thousand Springs area of the Snake River. Ground water is obtained from wells in the basalt, many of them 1000 ft (300 m) deep or more. Glacial outwash gravel deposits along many of the rivers also are productive aquifers.

4. Colorado Plateau and Wyoming Basin

This region consists of thin soils over consolidated sedimentary rocks.

The Colorado Plateau and Wyoming Basin are areas of extensive sedimentary rocks. The strata are flat-lying for the most part, but some areas have been tilted, folded, or faulted. At different levels, there are a number of plateaus that are deeply dissected by the Colorado River system or the Missouri River system. The sediments are chiefly shales and low-permeability sandstones.

The rate of ground-water recharge is very low, as the warm and arid climate causes most of the precipitation to be quickly evaporated. Some of the higher plateaus, such as the southern edge of the Colorado, do receive somewhat more precipitation. The most productive consolidated rock aquifer of the region, the Coconino Sandstone (Permian), crops out here. Other aquifers include the Dakota Sandstone (Cretaceous), the Entrada and Junction Creek sandstones (Jurassic), and the Wingate Sandstone (Triassic). Small amounts of water can be obtained from sandstone and limestone aquifers, which can receive local recharge, as well as from some alluvial deposits, such as the Uinta Basin at the edge of the Uinta Range. In general, the Colorado Plateau–Wyoming Basin is a water-poor region.

5. High Plains

This region consists of thick alluvial deposits over sedimentary rocks.

The High Plains ground-water region corresponds with the High Plains part of the physiographic Great Plains Province. This area is underlain by sedimentary rocks that dip gently to the east and are upturned at their contact with the Rocky Mountains and other places at which dome mountains, such as the Black Hills, have pushed through them. The entire surface is mantled with an alluvial apron representing material eroded from the Rocky Mountains. This alluvial cover, known as the Ogallala Formation, slopes to the east, but has been dissected in some areas.

The area of Nebraska between the Platte and Niobrara rivers is mantled with deposits of windblown sand—the Sand Hills. Both the Sand Hills and the Ogallala Formation (Pliocene) are highly porous and permeable. Rivers, which have cut all the way into bedrock, isolate these unconsolidated aquifers from recharge by streams draining from the Rocky Mountains. Recharge is by means of local precipitation, and the Sand Hills receive enough to be relatively well supplied compared with the rest of the High Plains. The Ogallala Formation is the major aquifer through most of the High Plains. It had a large volume of water in storage, but with heavy ground-water development for irrigation, water levels are rapidly falling—particularly in Texas and New Mexico.

Consolidated rock aquifers in Nebraska include units of the Arikaree Group (Miocene), Chadron Formation (Oligocene) and Dakota Group (Cretaceous). The Dakota Group is overlain by up to 3000 ft (1000 m) of impermeable Pierre Shale and is recharged in the Black Hills outcrop area. Water is also obtained from Quarternary-age river alluvium along modern rivers, which can furnish a local source for recharge.

6. Nonglaciated Central Region

This region consists of thin regolith over sedimentary rocks.

The Nonglaciated Central Region encompasses a large area of the central United States extending from the Rocky Mountains to the Appalachian Plateau and Valley and Ridge physiographic province and includes several small outlying sections such as the driftless area of Wisconsin and the Triassic sedimentary basins of the East. The bedrock underlying the region consists of level or gently tilted and folded sedimentary rocks ranging in age from Paleozoic to middle Tertiary. The land surface encompasses plains and plateaus and is gently rolling to sharply dissected. Climate ranges from semiarid to humid, but most of the area is subhumid to humid.

Consolidated rock aquifers include the Edwards Limestone (Cretaceous) of southwestern and central Texas, from which a flowing well in San Antonio yielded 24 million gallons per day (1050 L/s) when first drilled. Sands of the Trinity Group (early Cretaceous) and Woodbine Sand (late Cretaceous) are productive aquifers in north-central Texas and southern Oklahoma. These units include primarily clay, sand, and slightly consolidated sandstone. The Dakota Sandstone (late Cretaceous), Minnelusa Formation (Pennsylvanian and Permian), and Pahasapa Limestone (early Mississippian) are bedrock aquifers in South Dakota. In North Dakota, the Dakota Sandstone and Lakota Formation (early Cretaceous) are important. In eastern Montana, the Hell Creek Formation (late Cretaceous), Fox Hills Sandstone (late Cretaceous), and Fort Union Formation (Paleocene) are widespread. Other aquifers in Montana that are more restricted are the Judith River formation (late Cretaceous) and the Kootenai, Dakota, and Lakota formations (Cretaceous). Fresh water may be available at depth from the Madison Limestone (Mississippian). The Roswell Basin of New Mexico includes the San Andres Limestone (Permian), which is confined by the Artesia Group. The initially high artesian heads in the Roswell Basin have long since declined owing to the many flowing wells, as has the head in the Dakota Sandstone elsewhere. Limestone aquifers also occur in Arkansas, Missouri, Kentucky, and Tennessee. These units tend to be cavernous and many yield large springs, but generally they are not very productive because of the sporadic occurrence of water and the low storage capacity.

Carbonate aquifers, such as the Knox Dolomite of the Valley and Ridge physiographic province, provide ground water in many parts of the eastern sectors of this region.

7. Glaciated Central Region

This region consists of glacial deposits over sedimentary rocks.

The Glaciated Central Region extends from the Catskill Mountains in New York to the northern Great Plains in Montana and includes the Triassic

Valley of central Connecticut and Massachusetts. Only the eastern part of the region is underlain by uplands. The bedrock aquifers of the area are contiguous with those in the adjacent nonglaciated areas.

The principal feature of the Glaciated Central Region is the yield of ground water from glacial drift. Outwash plains and buried bedrock valleys are the principal sources. Most of the drift is fine grained, but there are numerous sand and gravel deposits. The region ranges from semiarid in North Dakota to humid in the Great Lakes states.

Major bedrock aquifers include limestones and sandstones of Paleozoic age. Some of these are widespread and include the following: Mt. Simon Sandstone, Galesville Sandstone, Ironton Sandstone, Franconia Sandstone, Jordan Sandstone, and Trempealeau Formation (all Cambrian); Prairie du Chien Formation and St. Peter Sandstone (both Ordovician). These are the bedrock aquifers of eastern Iowa, southern Minnesota, southern Wisconsin, and northern Illinois. The bedrock aquifers are collectively known locally as the deep aquifer, sandstone aquifer, or Cambrian-Ordovician aquifer. Elsewhere in this region the Paleozoic bedrock aquifers are not regional in extent, and saline water in them tends to be a problem. In areas underlain by Precambrian crystalline rocks, the yield to wells is generally small. Alluvium, especially glaciofluvial deposits, is a major source of ground water.

8. Piedmont–Blue Ridge Region

This region consists of a thick mantle of weathering residuum over fractured crystalline and metamorphic rock.

The Appalachian Mountains are a part of this region. Bedrock in the area includes metamorphic and crystalline rocks of the Piedmont and Blue Ridge. Precipitation ranges from 30 in. (76 cm) up to 75 in. (190 cm).

Weathered zones in the metamorphic rocks of the Piedmont yield small to moderate amounts of water almost anywhere, with larger-yield wells possible on fracture traces. More-productive wells are usually in valleys rather than on hilltops. Some sandstone units that lie in belts parallel to the structural trend are also moderately good aquifers. There is little ground water in the Blue Ridge; for the most part, the bedrock is impermeable.

9. Northeast and Superior Uplands

This region consists of glacial deposits over fractured crystalline rock.

There are two separate areas in this region. One is in northeast New York State and northern New England and the other in northern Wisconsin and Minnesota. The region extends into Canada as the Canadian shield. Glacial deposits of varying thickness mantle crystalline bedrock. Ground water is obtained from fractures in the bedrock or from permeable glacial deposits. Glacial outwash in the form of sand and gravel deposits forms the best aquifer systems in this region. The area is humid, with 30 to 45 in. (76 to 114 cm) of precipitation.

10. Atlantic and Gulf Coastal Plain

This region consists of complex sequences of interbedded sand, silt, clay, and limestone.

The Atlantic and Gulf coastal plain starts at Cape Cod and includes Long Island and southern New Jersey, most of each of the Atlantic coastal states, part of Alabama and Mississippi, parts of Missouri and Arkansas, all of Louisiana, and southeastern Texas. At the interior edge, the deposits thin to a featheredge and thicken toward the coast. They consist of unconsolidated to consolidated continental and marine sediments. Almost all of the coastal plain has a humid climate, with ample water available to recharge the aquifers.

Cape Cod and Long Island are covered with glacial deposits, including outwash, which is a good aquifer. Pleistocene sediments are also the most productive aquifer in Delaware. Elsewhere, the coastal plain deposits contain many excellent aquifers, including sands, sandstones, dolomites, and limestones. Saline water is present at the coasts and at depth in many of the areas. Otherwise, water quality is generally good.

Major aquifers of the Atlantic coastal plain include the Magothy Formation and Lloyd Sand Member of the Raritan Formation (both Cretaceous) of Long Island and northern New Jersey, Cheswold and Fredonia aquifers (Miocene) in Delaware, and Castle Hayne Limestone of North Carolina. A number of good aquifers exist in Texas and include the Wilcox Group and Carrizo Sand (Eocene); Catahoula Sandstone, Oakville Sandstone, and sand units of Lagarto Clay (Miocene); and Goliad Sand, Willis Sand, and sand units of the Lissie Formation (Pliocene to Pleistocene). Other states have equally productive aquifers, either unnamed or of local extent.

11. Southeast Coastal Plain

This region consists of thick layers of sand and clay over semiconsolidated and consolidated carbonate rock.

This area includes the Florida Peninsula and parts of coastal South Carolina, Georgia, and Alabama. The surface is underlain by unconsolidated deposits of clay, sand, gravel, and shell beds. Deeper layers consist of alternating layers of semiconsolidated and consolidated limestones and dolomites. The Floridan aquifer is located in the carbonate units and is one of the most prolific aquifers in the world. It is confined by the overlying Hawthorn Formation. Water may also be obtained from the surficial sand and gravel deposits. This region is described in greater detail as a case study in Chapter 8.

12. Alluvial Valleys

This region consists of thick sand and gravel deposits beneath floodplains and stream terrace deposits.

This region is not one of geographic continuity, but rather one of similar geologic origin. Many river systems have deposited thick sequences of sand and gravel, which are highly porous and permeable. Much of the sand and gravel is glacial outwash deposited by streams that carried water away from the melting ice front during the Pleistocene. These sediments were deposited far beyond the extent of the glaciers in some instances. They represent long, narrow aquifer systems and, as in most cases, are the foundations for modern river systems, with an ample supply of water to recharge that removed by pumping. The Miami River of Dayton, Ohio, is an example.

13. Hawaiian Islands

This region consists of lava flows interbedded with ash deposits and segmented by dikes.

The occurrence of ground water in the Hawaiian Islands has already been discussed as a case study in this chapter. Ground water occurs in several different types of aquifers, with some boundary conditions that are unique to the Hawaiian Islands. Different types of ground-water development schemes, such as Maui tunnels and inclined wells, are necessary to capture the ground water by skimming, thereby preventing saline-water encroachment.

14. Alaska

This region consists of glacial and alluvial deposits, occupied in part by permafrost and overlying bedrock of various types.

Alaska is placed in a single ground-water region for reasons of convenience. There are many different terranes in which ground water occurs in Alaska. The state is thinly populated and not a great deal of ground-water exploration has taken place. The occurrence of ground water is to a large extent controlled by the permafrost. In areas of continuous permafrost, ground-water supplies are limited to small, discontinuous thawed areas beneath large lakes and rivers. In the area of discontinuous permafrost, the water in the sand and gravel deposits is frequently not frozen and is available for development. Most inhabited areas of Alaska have glacial outwash or river alluvium that can serve as an aquifer.

15. Puerto Rico

This region consists of alluvium and limestones overlying and bordering fractured igneous rocks.

This region contains the islands of Puerto Rico and the U.S. Virgin Islands. These islands are generally hilly and underlain by both limestones and volcanic and intrusive igneous rocks. These islands receive high amounts of rainfall; ground-water recharge averages almost 6 ft (2 m) annually in Puerto Rico. Alluvium, which occurs in stream valleys and along the coast areas, is an effective aquifer where sand and gravel are present. The limestone areas are also aquifers; however, the volcanic rocks are metamorphosed and contain water mainly in fractures, as do other dense crystalline rocks.

NOTATION

a	Half-width of an infinite-strip oceanic island	**T**	Aquifer transmissivity
G	$\rho_w/(\rho_s - \rho_w)$	t_0	Tidal period from high to low tide
$h_{(x,y)}$	Elevation of the water table above sea level	**w**	Recharge to the aquifer
H_0	Tidal fluctuation	**x**	Distance from the coast
H_x	Fluctuation of water table due to tidal fluctuation at distance x from coast	x_0	Width of the outflow face
		$z_{(x,y)}$	Depth to the salt-water interface
K	Hydraulic conductivity	t_τ	Time lag between tidal high (or low) and ground water high (or low)
q'	Discharge per unit width		
R	Radius of circular oceanic island	ρ_s	Density of saline water
r	Radial distance from the center of a circular island	ρ_w	Density of fresh water
S	Aquifer storativity		

PROBLEMS

Answers to odd-numbered problems will appear at the end of the book.

1. At a tropical coastal aquifer the ground water is stagnant. The density of fresh water is 0.998 g/cm^3 and that of the underlying salt water is 1.024 g/cm^3. If the fresh-water head is 7.6 ft above mean sea level, what is the depth to the salt-water interface?

2. The fresh water at a coastal area has a density of 0.999 g/cm^3, and the underlying saline water has a density of 1.025 g/cm^3. If the fresh-water head is 2.14 m above mean sea level, what is the depth to the salt-water interface?

3. A coastal aquifer has a mean hydraulic conductivity of 1.25 m/day. The density of fresh water is 1.000 g/cm^3 and the density of underlying saline water is 1.024 g/cm^3. The ground-water discharge per unit width of the coastline is 0.00345 m^3/day.

   A. What is the depth to the salt-water interface at a point 100 m inland?

   B. What is the elevation of the water table above mean sea level at a point 100 m inland?

   C. What is the depth to the salt-water interface at the shoreline?

   D. What is the width of the outflow face?

4. A coastal aquifer has a mean hydraulic conductivity of 3.72 m/day. The density of fresh water is 1.000 g/cm^3 and the density of underlying saline water is 1.025 g/cm^3. The ground-water discharge per unit width of the coastline is 0.0127 m^3/day.

   A. What is the depth to the salt-water interface at a point 150 m inland?

   B. What is the elevation of the water table above mean sea level at a point 150 m inland?

   C. What is the depth to the salt-water interface at the shoreline?

   D. What is the width of the outflow face?

5. The aquifer beneath a circular oceanic island has a mean hydraulic conductivity of 122 ft/day. The amount of recharge is 0.00831 ft/day. The density of fresh water is 1.000 g/cm^3 and the density of underlying saline water is 1.024 g/cm^3. If the island is 5650 ft in diameter, what is the depth to the salt-water interface in the center of the island?

6. The aquifer beneath an infinite-strip oceanic island has a mean hydraulic conductivity of 122 ft/day. The amount of recharge is 0.00831 ft/day. The density of fresh water is 1.000 g/cm^3 and the density of underlying saline water is 1.024 g/cm^3. If the island is 5650 ft in width, what is the depth to the salt-water interface in the center of the island?

10 Water Chemistry

They [clouds] are often wafted about and borne by the winds from one region to another, where by their density they could become so heavy that they fall in thick rain; and if the heat of the sun is added to the power of the element of fire, the clouds are drawn up higher still and find a greater degree of cold, in which they form ice and fall in storms of hail. Now the same heat which holds up so great a weight of water as is seen to rain from the clouds, draws them from below upwards, from the foot of the mountains, and leads and holds them within the summits of the mountains and these, finding some fissure, issue continuously and cause rivers.

Leonardo da Vinci (1452–1519),
in *The Literary Works of Leonardo da Vinci*, J. P. Richter, ed., 1939

10.1   INTRODUCTION

For most of its uses, the chemical properties of water are as important as the physical properties and available quantity. In the next chapter, we will consider water quality, which involves the type and amount of substances dissolved in the water. In this chapter, our focus will be the chemical reactions that occur between water and the solids and gases it contacts.

Natural waters are never pure; they always contain at least small amounts of dissolved gases and solids. The composition of the aqueous solution is a function of a multiplicity of factors; for example, the initial composition of the water, the partial pressure of the gas phase, the type of mineral matter the water contacts, and the pH and oxidation potential of the solution. Water containing a biotic assemblage has an even more complex chemistry owing to the life processes of the biota.

The detailed study of water chemistry is far beyond the scope of this chapter. We will concentrate on the aspect of solubility of gases and liquids in dilute aqueous solutions. Further, we will assume that all reactions take place at a temperature of 25°C and a pressure of 1 atm. Small deviations from this assumption (a few atmospheres pressure and ±10° to 15°C) will not lead to significant error (Hem 1985). The systems considered will be presumed to be abiotic.

It is very difficult to collect a sample of ground water that is actually representative of the chemistry of the water as it exists in the ground. The process of drawing the sample up from the aquifer by means of a well and pump can change the pressure of the water. The sample may also be exposed to atmospheric oxygen during the sampling process. As a result, the Eh, pH, and equilibrium conditions of the water can change. Ground-water monitoring and methods of collecting representative ground-water samples are discussed in Chapter 11.

10.2   UNITS OF MEASUREMENT

Chemical analysis of an aqueous solution yields the amount of solute in a specified amount of water. There are several ways in which this can be reported.

Weight per weight units are dimensionless ratios of the weight of the solute divided by the weight of the solvent; for example, *parts per million* (ppm) and *parts per billion* (ppb). These units are no longer commonly used.

Weight per volume units are the more commonly used units today. They are expressed in terms of weight of solute per volume of water. Common units are **milligrams per liter** (mg/L) and **micrograms per liter** (μg/L). As a liter of pure water contains 1 million milligrams at 3.89°C, the temperature where it is most dense, it is commonly assumed that 1 ppm is equal to 1 mg/L. The density and hence weight of a liter of water will change with temperature and dissolved mineral matter. However, as a practical matter, the density corrections are necessary only if the dissolved solids of the water are in excess of 7000 mg/L (Hem 1985).

Equivalent weight units are very handy when the chemical behavior of the solute is being considered. The equivalent, or combining, weight of a dissolved ionic species is the formula weight divided by the electrical charge. By dividing a concentration in milligrams per liter by the equivalent weight of the ion, the result is a concentration expressed in **milliequivalents per liter** (meq/L). Dissolved species such as silica, which is not ionic, cannot be expressed in meq/L.

In chemical thermodynamics, units of **molality** are useful. One *mole* (mol) of a compound is its formula weight in grams. A solution of one mole per 1000 g of solvent is a one-*molal* solution; a solution of 1 mol of solute per liter of solvent is a one-*molar* solution. For dilute solutions, up to about 0.01 molal, the concentration expressed in either molality or molarity is equal.

For dilute solutions, it is not necessary to make density corrections; the following conversion factors may be used (Back & Hanshaw 1965):

$$\text{Molality} = \frac{\text{milligrams per liter} \times 10^{-3}}{\text{formula weight in grams}} \qquad \textbf{(10–1)}$$

$$\text{Molality} = \frac{\text{milliequivalents per liter} \times 10^{-3}}{\text{valence of ion}} \qquad \textbf{(10–2)}$$

EXAMPLE **Part A:** What is the weight of NaCl in a 0.01-molal solution?
PROBLEM

Atomic weight of sodium = 22.991 g

Atomic weight of chlorine = 35.457 g

One mole of NaCl = 58.448 g

0.01 mol = 0.01 × 58.448 g = 0.58448 g

0.01-molal solution = 0.58448 g NaCl in 1000 g H_2O

Part B: What is the concentration of NaCl in a 0.01-molal solution at 25°C?

At 25°C, the density of water is 0.99707 g/mL. The volume of 1000 g is 1000 g/0.99707 g/mL, or 1002.94 mL. The concentration is 0.58448 g in 1.00294 L, or 582.8 mg/L.

10.3 TYPES OF CHEMICAL REACTIONS IN WATER

Chemical reactions in an aqueous solution are either *reversible* or *irreversible*. Those that are reversible can reach equilibrium with their hydrochemical environment and are amenable to study by kinetic and thermodynamic methods.

The simplest aqueous reaction is the dissociation of an inorganic salt. If the salt is present in excess, it will tend to form a saturated solution:

$$NaCl \rightleftharpoons Na^+ + Cl^- \qquad \textbf{(10–3)}$$

Natural systems always tend toward equilibrium; thus, if the solution is undersaturated, more salt will dissolve. If it is supersaturated, salt will crystallize, although for kinetic reasons the solution may remain supersaturated. Notice that the water molecule does not actively participate in this reaction.

Water molecules can actively bond with either a gas or solid in a reversible reaction:

$$CaCO_3 + H_2O \rightleftharpoons Ca^{2+} + HCO_3^- + OH^- \qquad \textbf{(10–4)}$$

$$CO_2 + H_2O \rightleftharpoons HCO_3^- + H^+ \qquad \textbf{(10–5)}$$

In this type of reaction, the water molecule breaks into H^+ and OH^- radicals when combining with the species in solution.

Reversible oxidation-reduction reactions may also involve the transfer of electrons from one ion to another. When this happens, the species undergo a valence change:

$$4Fe^{2+} + 3O_2 + 8e^- \rightleftharpoons 2Fe_2O_3 \qquad \textbf{(10–6)}$$

In this example, the ferrous iron is oxidized to ferric iron by the transfer of an electron from the iron to the oxygen.

A gas or solid may also dissolve in an aqueous solution without dissociation:

$$O_{2\ (gas)} \rightleftharpoons O_{2\ (aqueous)} \qquad (10\text{-}7)$$

10.4   LAW OF MASS ACTION

If a reversible reaction can go in either of two directions, which way will it go? The answer to this basic question is found in the **law of mass action,** which suggests that the reaction will strive to reach equilibrium. In an aqueous mixture, both reactions are occurring simultaneously:

$$A + B \rightarrow C + D \qquad (10\text{-}8A)$$

and

$$C + D \rightarrow A + B \qquad (10\text{-}8B)$$

At chemical equilibrium, the two rates are equal; thus, if the mixture is not at chemical equilibrium, it will proceed in the direction that produces equilibrium. A chemical reaction may be expressed as

$$cC + dD \rightleftharpoons xX + yY \qquad (10\text{-}9)$$

where capital letters represent chemical constituents and lowercase letters represent coefficients. The *equilibrium concentration* of each chemical formula is [X], and the **equilibrium constant,** K, for the given reaction is

$$K = \frac{[X]^x[Y]^y}{[C]^c[D]^d} \qquad (10\text{-}10)$$

where [X] represents the molal concentration of the X ion. An equilibrium constant is valid only for a specific chemical reaction. It is either experimentally determined or calculated from thermodynamic properties. In equilibrium studies, the value of the concentration of a pure liquid or solid is defined as 1.

If AgCl is dissolved in water, it will eventually saturate the water and no more will dissolve. The reaction is

$$AgCl \rightleftharpoons Ag^+ + Cl^- \qquad (10\text{-}11)$$

The equilibrium reaction is given by

$$K_{sp} = \frac{[Ag^+][Cl^-]}{[AgCl]} \qquad (10-12)$$

The equilibrium constant for a slightly soluble salt is termed the **solubility product,** K_{sp}. The experimentally determined value of K_{sp} for the reaction is $10^{-9.8}$. Since [AgCl] is defined as 1,

$$K_{sp} = [Ag^+][Cl^-] = 10^{-9.8}$$

EXAMPLE PROBLEM

What is the solubility of Ag^+ at equilibrium?

The two ions have equal solubility:

$$[Ag^+] = [Cl^-] = \text{solubility}$$

The product of $[Ag^+][Cl^-]$ is the square of the solubility of either ion; thus, the solubility of either ion is the square root of the equilibrium constant:

$$\text{Solubility} = \sqrt{K_{sp}} = \sqrt{10^{-9.8}} = 1.26 \times 10^{-5} \text{ mol}$$

The solubility of Ag^+ is 1.26×10^{-5} mol/L.

The situation is more complex if it involves a salt, such as $PbCl_2$. The reaction is

$$PbCl_2 \rightleftharpoons Pb^{2+} + 2Cl^-$$

and the solubility product is given by

$$K_{sp} = \frac{[Pb^{2+}][Cl^-]^2}{[PbCl_2]} \qquad (10-13)$$

One mole of $PbCl_2$ yields one mole of Pb^{2+} and two moles of Cl^-. To solve the equation for the solubility, X, of $PbCl_2$, use the expression

$$K_{sp} = [X][2X]^2 \qquad (10-14)$$

The value of K_{sp} is $10^{-4.8}$ and, X, the solubility of $PbCl_2$, is found from

$$K_{sp} = 4X^3$$
$$X = \sqrt[3]{K_{sp}/4} = \sqrt[3]{0.25 \times 10^{-4.8}}$$
$$= 0.0158 \text{ mol}$$

EXAMPLE PROBLEM One thousand grams of water will dissolve 1.0×10^{-4} mol of $PbSO_4$. Calculate K_{sp}.

$$[Pb^{2+}] = [SO_4^{2-}] = 1 \times 10^{-4} \text{ mol}$$
$$K_{sp} = [Pb^{2+}][SO_4^{2-}] = 10^{-8}$$

10.5   COMMON-ION EFFECT

If the solvent contains another source for an ion also present in a salt, the **common-ion effect** will reduce the solubility of the salt. This applies to any salt in equilibrium with its saturated solution. If we dissolve AgCl in two solutions, one pure water and one containing 0.1 mol of NaCl, less of the AgCl will dissolve in the solution of NaCl. The solubility of NaCl is many orders of magnitude greater than that of the AgCl and is not affected. The total amount of the common ion in solution controls the amount of the less soluble salt that can dissolve. For example, consider the solution of AgCl in the 0.1-molal solution of NaCl. There will be X moles of AgCl and 0.1 mol of Cl^- from the NaCl. Thus, there are X moles of Ag^+ and $X + 0.1$ moles of Cl^-:

$$K_{sp} = [Ag^+][Cl^-] = [X][X + 0.1] = 10^{-9.8}$$

and

$$[0.1X] + [X^2] = 10^{-9.8}$$

Since $[X]$ is small, $[X^2]$ is very small and can be ignored; hence,

$$[X] = 10^{-8.8} = 1.58 \times 10^{-9}$$

The solubility of AgCl in pure water is 1.26×10^{-5} mol, whereas in a 0.1-molal solution of NaCl, it is only 1.58×10^{-9} mol. In general, ground and surface water contain ions from many sources, so that the common-ion effect must be considered.

10.6   CHEMICAL ACTIVITIES

In very dilute aqueous solutions, the molal concentrations can be used to determine equilibrium and solubility. For the general case, chemical activities must be computed from the concentration before the law of mass action can be applied. This is due to the fact that electrostatic forces cause the behavior of the solutes to be nonideal.

The **chemical activity** of an ion is equal to the molal concentrations times a factor known as the *activity coefficient:*

$$\alpha = \gamma m \qquad (10-15)$$

where

α is the chemical activity

m is the molal concentration

γ is the activity coefficient

In order to compute the activity coefficient of an individual ion, the *ionic strength* of the solution must be determined. For a mixture of electrolytes in solution, the ionic strength is given by

$$I = \frac{1}{2} \sum m_i z_i^2 \qquad (10-16)$$

where

I is the ionic strength

m_i is the molality of ith ion

z_i is the charge of ith ion

The ionic strength of 0.2-molal solution of $CaCl_2$ is

$$I = \frac{1}{2}(m_{Ca^{2+}} \times 2^2 + m_{Cl^-} \times 1^2)$$
$$= \frac{1}{2}(0.2 \times 2^2 + 0.4 \times 1^2) = 0.6$$

EXAMPLE PROBLEM
Compute the ionic strength of ground water from a Cambrian-age sandstone in Neenah, Wisconsin.

Chemical Analysis (mg/L)
(major ions only)

Ca^{2+}	Mg^{2+}	HCO_3^-	SO_4^{2-}
234	39	290	498

The concentrations must be converted to molality by Equation 10–1:

Chemical Analysis (molalities)

Ca^{2+}	Mg^{2+}	HCO_3^-	SO_4^{2-}
0.00584	0.0016	0.00475	0.00518

The ionic strength is then computed using Equation 10–16:

$$I = \tfrac{1}{2}(0.00584 \times 2^2 + 0.0016 \times 2^2 + 0.00475 \times 1^2 + 0.00518 \times 2^2)$$
$$= 0.0276$$

Once the ionic strength of a solution of electrolytes is known, the activity coefficient of the individual ion can be determined from the **Debye-Hückel equation:**

$$-\log \gamma_i = \frac{A z_i^2 \sqrt{I}}{1 + a_i B \sqrt{I}} \qquad \textbf{(10–17)}$$

where

γ_i is the activity coefficient of ionic species i

z_i is the charge of ionic species i

I is the ionic strength of the solution

A is a constant that is temperature-dependent (Table 10.1)

B is a constant that is temperature-dependent (Table 10.1)

a_i is the effective diameter of the ion (Table 10.2)

The Debye-Hückel equation is valid for solutions with an ionic strength of 0.1 or less (approximately 5000 mg/L).

Whereas Equation 10–10 is valid only for very dilute solutions where $\gamma \simeq 1$, the law of mass action, expressed in terms of chemical activities, is valid for solutions with any ionic strength:

TABLE 10.1 Values for constants A and B in the Debye-Hückel equation

Temperature, °C	A	B
0	0.4883	0.3241
5	0.4921	0.3249
10	0.4960	0.3258
15	0.5000	0.3262
20	0.5042	0.3273
25	0.5085	0.3281
30	0.5130	0.3290
35	0.5175	0.3297
40	0.5221	0.3305
45	0.5271	0.3314
50	0.5319	0.3321

Data from R. M. Garrels and C. L. Christ, *Solutions, Minerals and Equilibria.* San Francisco: Freeman Cooper, 1982.

TABLE 10.2 Values of the parameter a_i in the Debye-Hückel equation

a_i	Ion
11	Th^{4+}, Sn^{4+}
9	Al^{3+}, Fe^{3+}, Cr^{3+}, H^+
8	Mg^{2+}, Be^{2+}
6	Ca^{2+}, Cu^{2+}, Zn^{2+}, Sn^{2+}, Mn^{2+}, Fe^{2+}, Ni^{2+}, Co^{2+}, Li^+
5	$Fe(CN)_6^{4-}$, Sr^{2+}, Ba^{2+}, Cd^{2+}, Hg^{2+}, S^{2-}, Pb^{2+}, Co_3^{2-}, SO_3^{2-}, MoO_4^{2-}
4	PO_4^{3-}, $Fe(CN)_6^{3-}$, Hg_2^{2-}, SO_4^{2-}, SeO_4^{2-}, CrO_4^{3-}, HPO_4^{2-}, Na^+, HCO_3^-, $H_2PO_4^-$
3	OH^-, F^-, CNS^-, CNO^-, HS^-, ClO_4^-, K^+, Cl^-, Br^-, I^-, CN^-, NO_2^-, NO_3^-, Rb^+
	Cs^+, NH_4^+, Ag^+

Source: J. Kielland, "Individual Activity Coefficients of Ions in Aqueous Solutions," *American Chemical Society Journal* 59 (1937): 1676–78.

$$K = \frac{(\alpha_X)^x (\alpha_Y)^y}{(\alpha_C)^c (\alpha_D)^d} \qquad (10\text{–}18)$$

where $cC + dD \rightleftharpoons xX + yY$ and α_x is the activity of the X ion.

EXAMPLE PROBLEM Determine γ_i and α for Ca^{2+} in a solution where the molal concentration of Ca^{2+} is 0.00584 and $I = 0.0276$ at 25°C. The value of a_i for Ca^{2+} is 6.

The activity coefficient can be determined using Equation 10–17:

$$-\log \gamma_i = \frac{A z_i^2 \sqrt{I}}{1 + a_i B \sqrt{I}}$$

$$\log \gamma_i = -\frac{0.5085(2)^2 \sqrt{0.0276}}{1 + (6)(0.3281)\sqrt{0.0276}}$$

$$= -\frac{(0.5085)(4)(0.166)}{1 + (6)(0.3281)(0.166)}$$

$$= -0.255$$

$$\gamma_i = 0.556$$

The activity of calcium is then found from Equation 10–15:

$$\alpha = \gamma m$$

$$= (0.556)(0.00584)$$

$$= 0.00325$$

The *ion activity product* (K_{iap}), which is the product of the measured activities, can be calculated for any aqueous solution in order to test for saturation. The value of K_{iap} for a mineral equilibrium reaction in a natural water may be compared with the value of K_{sp}, the solubility product of the mineral. If the value of K_{iap} is equal to or greater than K_{sp}, the natural water is saturated or supersaturated with respect to the mineral. If K_{iap} is less than K_{sp}, the solution is undersaturated with respect to the mineral, and the mineral may be actively dissolving. For the case where the mineral C is being dissolved according to the reaction $cC \rightleftharpoons xX + yY$, K_{iap} is given by

$$K_{iap} = (\alpha_X)^x(\alpha_Y)^y \tag{10-19}$$

Solubility products for a number of compounds are given in Appendix 11.

10.7   IONIZATION CONSTANT OF WATER AND WEAK ACIDS

Water undergoes a dissociation into two ionic species:

$$H_2O \rightleftharpoons H^+ + OH^-$$

In reality, a hydrogen ion (H^+) cannot exist; it must be in the form H_3O^+, the hydronium ion, formed by the interaction of water with the hydrogen ion. For convenience, however, we will represent it as H^+. The equilibrium constant for water is

$$K_{eq} = \frac{\alpha_{H^+}\alpha_{OH^-}}{\alpha_{H_2O}} \tag{10-20}$$

For water that is neutral, there are exactly the same concentrations of H^+ and OH^- radicals, 10^{-7}. The negative logarithm of the concentration of H^+ ions in an aqueous solution is called the pH of the solution. For all aqueous solutions, either acidic or basic, the product $\alpha_{H^+}\alpha_{OH^-}$ is always 10^{-14} (at about 25°C). Since a neutral solution has equal amounts of H^+ and OH^- radicals, the pH is 7. If there are more H^+ ions than OH^- ions, the solution is acidic and the pH is less than 7.0. Basic solutions have more OH^- than H^+ ions and a pH between 7 and 14.

EXAMPLE PROBLEM   What is the $[H^+]$ and $[OH^-]$ of an aqueous solution of pH 3.2?

Since pH is the negative logarithm of $[H^+]$, the value of $[H^+]$ is $10^{-3.2}$. Since the product $[H^+][OH^-] = 10^{-14}$,

$$[OH^-] = 10^{-14}/[H^+] = 10^{-14}/10^{-3.2} = 10^{-10.8}$$

It is apparent that by measuring the pH of an aqueous solution, we can obtain the numerical value of both $[H^+]$ and $[OH^-]$. If the solution is nonideal, the pH meter measures the activity of H^+, since $\alpha_{H^+}\alpha_{OH^-} = 10^{-14}$.

An acid is a substance that can add H^+ (more properly, H_3O^+) ions to aqueous solutions. Strong acids will completely dissociate in water to release H^+ ions. Since the $[H^+][OH^-]$ product is constant at a given temperature, the concentration of OH^- ions decreases. A 1-molal solution of HCl will have a pH of 0 and a $[H^+]$ of 1. A 0.01-molal solution will have a pH of 2, and a $[H^+]$ of 10^{-2}. On the other hand, a 0.01-molal solution of H_2CO_3 will have a higher pH, as it is a weak acid. In dilute aqueous solution, the H_2CO_3 is only slightly broken down into ions. Weak acids with more than one H^+ per molecule ionize in steps; for example,

$$H_2CO_3 \rightleftharpoons H^+ + HCO_3^- \qquad \textbf{(10-21A)}$$
$$HCO_3^- \rightleftharpoons H^+ + CO_3^{2-} \qquad \textbf{(10-21B)}$$

The equilibrium constants at 25°C are

$$\frac{[H^+][HCO_3^-]}{[H_2CO_3]} = K_1 = 10^{-6.4} \qquad \text{(first ionization constant)}$$

$$\textbf{(10-22A)}$$

and

$$\frac{[H^+][CO_3^{2-}]}{[HCO_3^-]} = K_2 = 10^{-10.3} \qquad \text{(second ionization constant)}$$

$$\textbf{(10-22B)}$$

The value of K_{eq} for water varies significantly with temperature. Table 10.3 lists the equilibrium constants for the dissociation of water at temperatures between 0° and 60°C. At 0°C, a neutral solution has a pH of 7.5; at 60°C, neutrality occurs at pH 6.6.

TABLE 10.3 Equilibrium constants for dissociation of water

Temperature (°C)	K_{eq}	pH of a Neutral Solution
0	0.1139×10^{-14}	7.47
5	0.1846×10^{-14}	7.37
10	0.2920×10^{-14}	7.27
15	0.4505×10^{-14}	7.17
20	0.6809×10^{-14}	7.08
25	1.008×10^{-14}	7.00
30	1.469×10^{-14}	6.92
35	2.089×10^{-14}	6.84
40	2.919×10^{-14}	6.77
45	4.018×10^{-14}	6.70
50	5.474×10^{-14}	6.63
55	7.297×10^{-14}	6.57
60	9.614×10^{-14}	6.51

EXAMPLE PROBLEM

What is the pH of a 0.01-molal solution of H_2CO_3 at 25°C?

There are five ionic species present: H^+, OH^-, H_2CO_3, HCO_3^-, and CO_3^{2-}. The total of the carbonate species, H_2CO_3, HCO_3^-, and CO_3^{2-}, is 0.01 mol. The 0.01-molal solution is obtained by dissolving 0.01 mol of CO_2 in 1 L of water. For most geologic applications, some assumptions can be made to simplify the problem. From the values of K_1 and K_2, we see that K_2 is 10^4 smaller than K_1, so almost all of the H^+ ions come from $H_2CO_3 \rightleftharpoons H^+ + HCO_3^-$. There will also be a very small value of CO_3^{2-}, since K_2 is so small. Likewise, there will be relatively few OH^- ions since it is an acid solution.

From the dissociation reactions, we must balance the electrical charges:

$$[H^+] = [OH^-] + [HCO_3^-] + 2[CO_3^{2-}]$$

Since OH^- and CO_3^{2-} are relatively small,

$$[H^+] \simeq [HCO_3^-]$$

From the equilibrium equation,

$$\frac{[H^+][HCO_3^-]}{[H_2CO_3]} = K_1 = \frac{[H^+]^2}{[H_2CO_3]} = 10^{-6.4}$$

Since the solution has 0.01 mol CO_2, total,

$$[HCO_3^-] + [H_2CO_3] + [CO_3^{2-}] = 0.01 \text{ mol/1000 g}$$

With a small value for CO_3^{2-}, and HCO_3^- equal to H^+,

$$[H_2CO_3] + [H^+] \simeq 0.01$$

Since this is a weak acid, $[H^+]$ is very small compared with $[H_2CO_3]$, so that $[H_2CO_3] \simeq 0.01$. Putting these two results together,

$$\frac{[H^+]^2}{[H_2CO_3]} = 10^{-6.4} \text{ and } [H_2CO_3] = 0.01$$
$$[H^+]^2 = 0.01 \times 10^{-6.4} = 0.01 \times 3.98 \times 10^{-7} = 3.98 \times 10^{-9}$$
$$[H^+] = 6.31 \times 10^{-5} = 10^{-4.2}$$

Thus, pH = 4.2. Concentrations of other ions would be

$$[HCO_3^-] = [H^+] = 10^{-4.2}$$

$$[OH^-] = \frac{10^{-14}}{[H^+]} = 10^{-9.8}$$

$$[CO_3^{2-}] = \frac{10^{-10.3}[HCO_3^-]}{[H^+]} = 10^{-10.3}$$

These values are accurate to $\pm 1\%$ (Krauskopf 1967).

Although this type of problem can promote understanding of weak acids, in a real-world situation there may be many other ionic species present, increasing the ionic strength. This would necessitate the use of chemical activities and might also introduce the common ion effect.

10.8 CARBONATE EQUILIBRIUM

In hydrogeologic studies, the equilibrium of calcium carbonate in contact with natural water, either surface or ground water, is one of the most important geochemical reactions. Neutral water exposed to CO_2 in the atmosphere will dissolve CO_2 equal to the partial pressure. The CO_2 will react with H_2O to form H_2CO_3, a weak acid, and the resulting solution will have a pH of about 5.7. Soil CO_2 from organic decomposition is another source of even more importance in ground-water studies. As calcite and dolomite are soluble in acid solution, even rainwater will dissolve carbonate rocks. Likewise, a change in pH can result in a precipitation of $CaCO_3$ from a solution that was at equilibrium prior to the pH shift.

10.8.1 Carbonate Reactions

The following reactions must be considered in carbonate systems:

1. The solution of carbon dioxide in water to form carbonic acid:

$$H_2O + CO_2 \rightleftharpoons H_2CO_3 \qquad \text{(10–23A)}$$

$$K_{CO_2} = \frac{\alpha_{H_2CO_3}}{P_{CO_2}} \qquad \text{(10–23B)}$$

P_{CO_2} is the partial pressure of the carbon dioxide, which for most hydrogeological conditions is the gas activity.

2. The dissolution of carbonic acid in water to form bicarbonate:

$$H_2CO_3 \rightleftharpoons H^+ + HCO_3^- \qquad \text{(10–24A)}$$

$$K_{H_2CO_3} = \frac{\alpha_{H^+}\alpha_{HCO_3^-}}{\alpha_{H_2CO_3}} \qquad \text{(10–24B)}$$

3. The dissolution of bicarbonate in water to form carbonate:

$$HCO_3^- \rightleftharpoons H^+ + CO_3^{2-} \qquad \text{(10–25A)}$$

$$K_{HCO_3^-} = \frac{\alpha_{H^+}\alpha_{CO_3^{2-}}}{\alpha_{HCO_3^-}} \qquad \text{(10–25B)}$$

4. The dissolution of calcium carbonate in water to form calcium and carbonate: There are two common forms of calcium carbonate, calcite and aragonite. They each have different solubility products.

$$CaCO_3 \rightleftharpoons Ca^{2+} + CO_3^{2-} \qquad \text{(10–26A)}$$

$$K_{CaCO_3} = \frac{\alpha_{Ca^{2+}}\alpha_{CO_3^{2-}}}{\alpha_{CaCO_3}} \qquad \text{(10–26B)}$$

Table 10.4 contains the solubility products for each of these reactions for temperatures from 0° to 60°C.

Examination of Equations 10–24B and 10–25B reveals that the activity of the hydrogen ion plays an important role in determining the form in which carbonate is present. From Equation 10–24B, the relative proportions of H_2CO_3 and HCO_3^- vary with pH. At 20°C, the value of $K_{H_2CO_3}$ is $10^{-6.38}$. From Equation 10–24B:

$$K_{H_2CO_3} = \frac{\alpha_{H^+}\alpha_{HCO_3^-}}{\alpha_{H_2CO_3}} = 10^{-6.38}$$

This can be rearranged to yield

$$\frac{\alpha_{H_2CO_3}}{\alpha_{HCO_3^-}} = \frac{\alpha_{H^+}}{10^{-6.38}}$$

TABLE 10.4 Carbonate equilibria constants at 1 atm pressure

Temperature °C	K_{CO_2}	$K_{H_2CO_3}$	$K_{HCO_3^-}$	K_{CaCO_3} (cal.)	K_{CaCO_3} (arag.)
0	$10^{-1.11}$	$10^{-6.58}$	$10^{-10.63}$	$10^{-8.38}$	$10^{-8.22}$
5	$10^{-1.19}$	$10^{-6.52}$	$10^{-10.55}$	$10^{-8.39}$	$10^{-8.24}$
10	$10^{-1.27}$	$10^{-6.46}$	$10^{-10.49}$	$10^{-8.41}$	$10^{-8.26}$
15	$10^{-1.34}$	$10^{-6.42}$	$10^{-10.43}$	$10^{-8.43}$	$10^{-8.28}$
20	$10^{-1.41}$	$10^{-6.38}$	$10^{-10.38}$	$10^{-8.45}$	$10^{-8.31}$
25	$10^{-1.47}$	$10^{-6.35}$	$10^{-10.33}$	$10^{-8.48}$	$10^{-8.34}$
30	$10^{-1.52}$	$10^{-6.33}$	$10^{-10.29}$	$10^{-8.51}$	$10^{-8.37}$
45	$10^{-1.67}$	$10^{-6.29}$	$10^{-10.20}$	$10^{-8.62}$	$10^{-8.49}$
60	$10^{-1.78}$	$10^{-6.29}$	$10^{-10.14}$	$10^{-8.76}$	$10^{-8.64}$

Source: L. N. Plummer & E. Busenberg, *Geochemica et Cosmochemica Acta* 46 (1982):1011–1040.

TABLE 10.5 Distribution of carbonate species as a function of pH at 20°C

pH	Carbonic Acid	Bicarbonate Ion	Carbonate Ion
2.00	99.99%	0.01%	
3.00	99.96%	0.04%	
4.00	99.6%	0.4%	
5.00	96.0%	4.0%	
6.00	70.6%	29.4%	
6.38	50.0%	50.0%	
7.00	5.2%	94.8%	
8.00	2.3%	97.7%	
9.00		96.0%	4.0%
10.00		70.6%	29.4%
10.38		50.0%	50.0%
11.00		5.2%	94.8%
12.00		2.3%	97.7%
13.00		0.2%	99.8%

By substituting different values for α_{H^+} (expressed as pH), the ratio of bicarbonate ion to carbonic acid can be found. For example, at a pH of 6.38, α_{H^+} is 6.38 and the ratio of carbonic acid to bicarbonate ion is 1:1. This means that at a pH of 6.38, 50% of the solution is present as carbonic acid and 50%, as bicarbonate ion. At a pH of 6.00, the ratio is 2.4:1, which means that 70.59% of the carbonate is present as carbonic acid, whereas 29.41% is bicarbonate ion.

Equation 10–25B can be rearranged to find the ratio of bicarbonate ion to carbonate ion. At 20°C, the value of $K_{HCO_3^-}$ is $10^{-10.38}$. At a pH of 10.38, the ratio of bicarbonate to carbonate is 1:1. At a pH of 10, 70.59% is present as bicarbonate ion and 29.41%, as carbonate ion. At a pH of 11, only 5.17% is bicarbonate ion; 94.83% is carbonate ion. Table 10.5 shows the distribution of carbonic acid, bicarbonate ion, and carbonate ion as a function of pH. This relationship is also plotted on Figure 10.1. Table 10.5 and Figure 10.1 show that in the pH range of most natural waters, 4 to 9, carbonate is present as either carbonic acid or bicarbonate ion but not as carbonate ion. Carbonate ion is found in appreciable amounts only in very alkaline water. The ionization constants of the carbonate species vary slightly with temperature, so Table 10.5 would be slightly different at temperatures other than 20°C.

10.8.2 Carbonate Equilibrium in Water with Fixed Partial Pressure of CO_2

Water in streams and lakes is in contact with the atmosphere, in which CO_2 is present. The gas is dissolved in the water, adding to the carbonate content of the water. The system is described by

$$H_2O + CO_2 \rightleftharpoons H_2CO_3 \tag{10–27A}$$

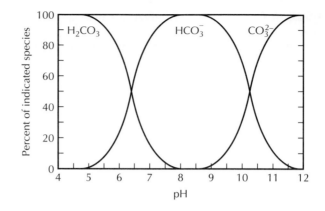

FIGURE 10.1 Distribution of major species of dissolved inorganic carbon at 20°C.

$$H_2CO_3 \rightleftharpoons H^+ + HCO_3^- \qquad \textbf{(10–27B)}$$
$$HCO_3^- \rightleftharpoons H^+ + CO_3^{2-} \qquad \textbf{(10–27C)}$$
$$CaCO_3 \rightleftharpoons Ca^{2+} + CO_3^{2-} \qquad \textbf{(10–27D)}$$
$$H_2O \rightleftharpoons H^+ + OH^- \qquad \textbf{(10–27E)}$$

The system is electrically neutral, so that

$$2m_{Ca^{2+}} + m_{H^+} = 2m_{CO_3^{2-}} + m_{HCO_3^-} + m_{OH^-} \qquad \textbf{(10–27F)}$$

The partial pressure of CO_2 in the atmosphere is $10^{-3.5}$. The value of the activity of H_2CO_3 can be computed from the activity product of CO_2 at 25°C:

$$K_{CO_2} = \frac{\alpha_{H_2CO_3}}{P_{CO_2}} = 10^{-1.5}$$

Therefore,

$$\alpha_{H_2CO_3} = P_{CO_2} \times 10^{-1.5} = 10^{-3.5} \times 10^{-1.5} = 10^{-5.0}$$

The activity values for the remaining ionic species can be determined in relationship to α_{H^+}. This enables computation of the pH of a solution of calcite that is in equilibrium with atmospheric CO_2. The H_2CO_3 will dissociate into H^+ and HCO_3^-:

$$K_{H_2CO_3} = \frac{\alpha_{H^+}\alpha_{HCO_3^-}}{\alpha_{H_2CO_3}} = 10^{-6.4}$$

As we know, $\alpha_{H_2CO_3} = 10^{-5.0}$:

$$\alpha_{HCO_3^-} = (10^{-6.4} \times 10^{-5.0})/\alpha_{H^+} = 10^{-11.4}/\alpha_{H^+}$$

The HCO_3^- will further dissociate into H^+ and CO_3^{2-}:

$$K_{HCO_3^-} = \frac{\alpha_{H^+}\alpha_{CO_3^{2-}}}{\alpha_{HCO_3^-}} = 10^{-10.3}$$

This can be rearranged to give $\alpha_{CO_3^{2-}}$:

$$\alpha_{CO_3^{2-}} = \frac{(\alpha_{HCO_3^-})(10^{-10.3})}{\alpha_{H^+}}$$

$$= \frac{10^{-11.4}}{\alpha_{H^+}} \times \frac{10^{-10.3}}{\alpha_{H^+}} = \frac{10^{-21.7}}{(\alpha_{H^+})^2}$$

The equilibrium constant for calcite is $10^{-8.3}$. For the solid phase of a substance at saturation, $\alpha = 1$; therefore, $\alpha_{CaCO_3} = 1$:

$$K_{CaCO_3} = \frac{\alpha_{Ca^{2+}}\alpha_{CO_3^{2-}}}{\alpha_{CaCO_3}} = 10^{-8.3}$$

The activity of Ca^{2+} can be found as

$$\alpha_{Ca^{2+}} = \frac{10^{-8.3}}{\alpha_{CO_3^{2-}}} = \frac{10^{-8.3}}{10^{-21.7}/(\alpha_{H^+})^2} = 10^{13.4}(\alpha_{H^+})^2$$

By definition, $\alpha_{OH^-} = 10^{-14}/\alpha_{H^+}$.

For very dilute solutions $\gamma_i \simeq 1$; therefore, $m_i = \alpha_i$, and the equation for electrical neutrality can be expressed as

$$2\alpha_{Ca^{2+}} + \alpha_{H^+} = 2\alpha_{CO_3^{2-}} + \alpha_{HCO_3^-} + \alpha_{OH^-}$$

Expressions for $\alpha_{Ca^{2+}}$, $\alpha_{CO_3^{2-}}$, $\alpha_{HCO_3^-}$, and α_{OH^-} have been determined with respect to α_{H^+}. The equation for electrical neutrality can be determined to be

$$2[10^{13.4}(\alpha_{H^+})^2] + \alpha_{H^+} = 2[10^{-21.7}/(\alpha_{H^+})^2] + 10^{-11.4}/\alpha_{H^+} + 10^{-14}/\alpha_{H^+}$$

Solution of the preceding yields $\alpha_{H^+} = 10^{-8.4}$.

Thus, the pH of a solution open to atmospheric CO_2 and in equilibrium with calcite is 8.4. This is lower than the pH of a solution of calcite with no external source of CO_2, which is about 9.9. This suggests that field measurements of pH should always be made, especially if the water is from a source with no external CO_2. Exposure of such a sample to the atmosphere for more than a few minutes would result in a lowering of the pH. Ground-water samples should always be tested for pH in the field as soon as the sample is collected. As a practical matter it is quite difficult to collect representative ground-water samples that don't undergo reactions during the process of collection.

10.8.3   Carbonate Equilibrium with External pH Control

In most ground-water and many surface-water bodies, there are ionic species other than H_2CO_3 that influence or control the pH. The hydrogeologist often has a set of chemical analyses and a measured pH for the total solution. If pH, total calcium, total carbonate, and ionic strength are known, the ion activity product, K_{iap}, can be calculated and compared with the solubility product, K_{sp}, to determine whether or not the water is in equilibrium with calcite. If K_{iap}/K_{sp} is greater than 1, the solution is supersaturated; if less than 1, it is undersaturated; if equal to 1, the solution is in equilibrium.

EXAMPLE PROBLEM   Determine whether the sample of ground water represented by the following analysis is saturated with respect to calcite. The field pH is 7.15. The total dissolved solids (TDS) is 371 mg/L.

	Ca^{2+}	Mg^{2+}	Na^+	K^+	HCO_3^-	SO_4^{2-}	Cl^-	NO_3^-
Concentration (mg/L)	82	9	25	7.6	252	17	40	38
Molality $\times 10^3$	2.046	0.37	1.087	0.194	4.13	0.177	1.128	0.613

1. Calculate the ionic strength.

$$I = \tfrac{1}{2} (0.002046 \times 2^2 + 0.00037 \times 2^2 + 0.001087 + 0.000194$$
$$+ 0.00413 + 0.000177 \times 2^2 + 0.001128 + 0.000613)$$
$$= 0.0088$$

2. Compute γ_i for Ca^{2+}, HCO_3^-, and CO_3^{2-}. For water at 25°C, the Debye-Hückel equation is

$$\log \gamma_i = - \frac{0.5085 z^2_i \sqrt{I}}{1 + 0.3281 a_i \sqrt{I}}$$

Values of a_i (from Table 10.2) are

$$Ca^{2+} = 6 \qquad HCO_3^- = 4 \qquad CO_3^{2-} = 5$$

$$\log \gamma_{Ca^{2+}} = - \frac{0.5085(2)^2 \sqrt{0.0088}}{1 + 0.3281(6)\sqrt{0.0088}}$$
$$= -0.161$$

$$\gamma_{Ca^{2+}} = 0.690$$

$$\alpha_{Ca^{2+}} = m_{Ca^{2+}}\gamma_{Ca^{2+}} = 0.002046 \times 0.690 = 0.0017 = 10^{-2.77}$$

$$\log \gamma_{HCO_3^-} = -\frac{0.5085(1)\sqrt{0.0088}}{1 + 0.3281(4)\sqrt{0.0088}}$$

$$= -0.0425$$

$$\gamma_{HCO_3^-} = 0.907$$

$$\alpha_{HCO_3^-} = m_{HCO_3^-}\gamma_{HCO_3^-} = 0.00413 \times 0.907 = 0.003746 = 10^{-2.43}$$

$$\log \gamma_{CO_3^{2-}} = -\frac{0.5085(2)^2\sqrt{0.0088}}{1 + 0.3281(5)\sqrt{0.0088}}$$

$$= -0.165$$

$$\gamma_{CO_3^{2-}} = 0.683$$

The molality of CO_3^{2-} is below detectable limits, but activity can be computed, since

$$\frac{\alpha_{H^+}\alpha_{CO_3^{2-}}}{\alpha_{HCO_3^-}} = 10^{-10.3}$$

From the pH, $\alpha_{H^+} = 10^{-7.15}$

$$\alpha_{CO_3^{2-}} = \frac{10^{-10.3} \times 10^{-2.43}}{10^{-7.15}} = 10^{-5.58}$$

3.   The calculated ion activity product is

$$K_{iap} = \alpha_{Ca^{2+}}\alpha_{CO_3^{2-}}$$

$$= 10^{-2.77} \times 10^{-5.58} = 10^{-8.35}$$

4.   The value of K_{sp} for calcite is $10^{-8.35}$. By comparing the ratio K_{iap}/K_{sp}, we determine whether or not the solution is saturated:

$$K_{iap}/K_{sp} = 10^{-8.35}/10^{-8.35} = 10^0 = 1.00$$

The water is saturated with respect to calcite.

10.9   FREE ENERGY

One of the functions in chemical thermodynamics is termed **free energy** (also known as *Gibbs free energy*). It is a measure of the driving energy of a reaction. At standard conditions, the standard Gibbs free energy of a reaction, ΔG_r^0 is the difference between the sum of the free energy of the products and the sum of the free energy of the reactants:

$$\Delta G_r^0 = \sum \Delta G_r^0 \text{ products} - \sum \Delta G_r^0 \text{ reactants} \qquad (10-28)$$

It is related to the equilibrium constant by the formula

$$\Delta G_r^0 = -RT \ln K_{sp} \qquad (10-29)$$

where

R is the gas constant, which is a conversion factor equal to 0.00199 kcal/(mol·K)

T is the temperature in kelvins

ΔG_r^0 is in kcal/mol

This is a useful relationship. Since the values of ΔG_r^0 at 25°C and 1 atm pressure have been measured for many reactions, the value of K_{sp} can be computed if ΔG_r^0 is known. At 1 atm pressure and 25°C, in base 10 logs (Hem 1985),

$$\log K_{sp} = -\frac{\Delta G_r^0}{1.364} \qquad (10-30)$$

10.10   OXIDATION POTENTIAL

For chemical reactions in which electrons are transferred from one ion to another (*oxidation-reduction reactions,* or redox reactions), the oxidation potential of an aqueous solution is called the Eh. A transfer of electrons is an electrical current; therefore, a redox equation has an electrical potential. At 25°C and 1 atm pressure, the standard potential, E^0 (in volts), has been measured for many reactions. The sign of the potential is positive if the reaction is oxidizing and negative if it is reducing. The absolute value of E^0 is a measure of the oxidizing or reducing tendency.

The oxidation potential of a reaction is given by the *Nernst equation:*

$$\text{Eh} = E^0 + \frac{RT}{nF} \ln K_{sp} \qquad (10-31)$$

where

R is the gas constant, 0.00199 kcal/(mol·K)

T is the temperature (kelvins)

F is the Faraday constant, 23.1 kcal/V

n is the number of electrons

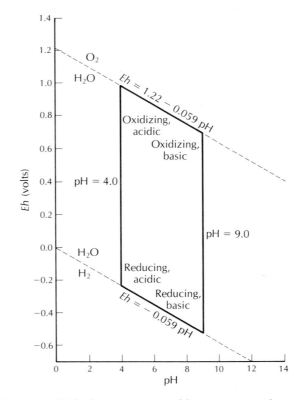

FIGURE 10.2 The Eh-pH field where water is a stable component. The usual limits of Eh and pH for near-surface environments are also indicated by solid lines. Source: K. Krauskopf, *Introduction to Geochemistry*. New York: McGraw-Hill Book Company, 1967. Used with permission.

Oxidation potential is measured with a specific ion electrode meter. A positive value indicates that the solution is oxidizing; a negative value indicates that it is chemically reducing.

If the pH and Eh of an aqueous solution are known, the stability of minerals in contact with the water may be determined. This stability relationship is best represented on an Eh-pH diagram. Water, itself, is stable only in a certain part of the Eh-pH field. Figure 10.2 shows the framework of aqueous Eh-pH fields. Water in nature at near-surface environments is usually between pH 4 and pH 9, although values that are more acidic or more basic can occur.

The Eh-pH diagram can be used to show the fields of stability for both solid and dissolved ionic species. It has been used very effectively for iron. The Eh-pH diagram depends upon the concentrations of all ionic species present. For the simple ions and hydroxides of iron, the fields depend upon the molality of the iron in solution. Figure 10.3 shows the stability-field diagram for a 10^{-7}-molal solution of iron. The iron may be either in the Fe^{3+} or Fe^{2+} valence state, depending upon its position in the stability field. As the iron concentration

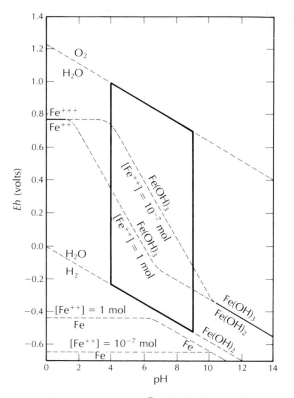

FIGURE 10.3   Stability-field diagram for a 10^{-7}-molal solution of iron. Source: K. Krauskopf, *Introduction to Geochemistry*. New York: McGraw-Hill Book Company, 1967. Used with permission.

increases, the line separating the ferrous and ferric state shifts to the left. This is demonstrated by a dashed line in Figure 10.3, which represents a 1-molal iron concentration. Ferrous iron can exist as Fe^{2+}, $Fe(OH)^+$, or $Fe(OH)_2$, depending upon the Eh and pH of the solution; ferric iron can be in the forms Fe^{3+}, $Fe(OH)_2^+$, and $Fe(OH)_3$. The pH at which these species change is also a function of the total amount of iron present. Procedures are available to compute an Eh-pH field for iron of any molality (Hem & Cropper 1959).

Of practical concern is the great difficulty in measuring the Eh of ground water under field conditions. Even for spring water, the measured Eh has been shown to be too great for the amount of ferrous iron in the sample (Hem & Cropper 1959). With very careful work, oxygen can be excluded from the sampling procedure and field Eh measured (Back & Barnes 1965).

High Eh is generally the direct result of dissolved oxygen in the water. For deep ground-water systems, the Eh is usually sufficiently low that, for a pH of less than about 8, iron is present as the soluble Fe^{2+} ion. Near a recharge zone, the ground water may have sufficient dissolved oxygen to elevate the Eh. As the water travels through the aquifer, the oxygen is chemically reduced by contact with reducing species, and the Eh is lowered. The oxygen can react with the small

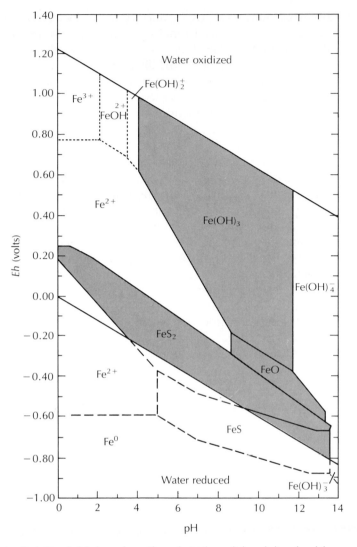

FIGURE 10.4 Stability fields based on Eh and pH for solid and dissolved forms of iron in an aqueous solution of 56 μ/L iron, 96 mg/L of sulfur as SO_4^{-2}, 61 mg/L carbon dioxide as HCO_3^- at 25°C, and 1 atm pressure. Source: J. D. Hem, U.S. Geological Survey Water-Supply Paper 2254, 1985.

amount of ferrous iron to form ferric hydroxide, $Fe(OH)_3$. Interestingly, the ferric hydroxide thus formed may be colloidal, and can move through the aquifer with the ground water. In the Eh-pH range where Fe^{2+} exists, large amounts of dissolved iron can be present.

Natural waters contain many ionic species. Again, using iron as an example, an Eh-pH diagram can be used to show the stable iron minerals in a mixed aqueous solution with iron, sulfur, and carbonate present. This is done in Figure 10.4; the given concentrations are iron 56 μg/L ($10^{-6.00}$ molar), 96 mg/L dissolved sulfur as SO_4^{2-}, and 61 mg/L dissolved carbon dioxide as HCO_3^-. Shaded areas indicate Eh-pH domains where solid species would be thermodynamically stable. The stable ionic species are also indicated on the figure. For a thorough discussion of Eh-pH diagrams, see Krauskopf (1967); Garrels and Christ (1965); Hem and Cropper (1959); Hem (1960); and Bass Becking (1960).

Because of the difficulty of measuring in situ Eh in ground water, there is not a great deal of information on the Eh-pH range of natural ground waters. Some data do suggest that a range of Eh from -0.2 to $+0.7$ V can occur (Back & Barnes 1965; Bass Becking, Kaplan, & Moore 1960). In one study, the measured Eh ranged from -0.04 to $+0.7$ V in ground water found in a single county in Maryland (Back & Barnes 1965). The ground water was in a coastal plain aquifer with a regional flow pattern toward the sea. The highest oxidation potentials were in shallow ground water of recharge areas. Eh was found to decrease with increasing length of flow from the aquifer recharge area. As might be expected, an inverse relationship was found between the oxidation potential and the amount of iron in solution. In the same study, field pH was found to range from 3.20 to 7.79, although, in general, higher and lower values are possible. For example, water draining from mineral deposits or mines can have a pH as low as 2 (Bass Becking, Kaplan, & Moore 1960).

Surface waters are usually oxidizing, although low Eh can occur in the anaerobic depths of some lakes. The pH of surface waters typically is in the range of 4 to 10 (Bass Becking, Kaplan, & Moore 1960).

10.11   ION EXCHANGE

Under certain conditions, the ions attracted to a solid surface may be exchanged for other ions in aqueous solution. This process is known as **ion exchange.** Both cation exchange and anion exchange can occur, but in some natural soils cation exchange is the dominant process. The presence of exchange sites is a function of the same general conditions affecting adsorption sites. The ion-exchange process can be conceptualized as the preferential absorption of selective ions with concomitant loss of other ions. Ion-exchange sites are found primarily on clays and soil organic materials (Mitchell 1932), although all soils and sediments have some ion-exchange capacity.

The ion-exchange reactions of different soils must be studied individually in the laboratory. Results are reported in terms of milliequivalents per 100 grams of soil. In one study of exchange capacities for stream sediments (Kennedy 1965), the results shown in Table 10.6 were reported.

A general ordering of cation exchangeability for common ions in ground water is

$$Na^+ > K^+ > Mg^{2+} > Ca^{2+}$$

The divalent ions are more strongly bonded and tend to replace monovalent ions. However, it is a reversible reaction and, at high activities, the monovalent ions

TABLE 10.6 Ion-exchange values for stream sediments

Size Fraction (μm)	Ion Exchange (meq/100 g soil)
4	14–65
4–61	4–30
61–1000	0.3–13

Source: V. C. Kennedy, U.S. Geological Survey Professional Paper 433-D, 1965.

can replace divalent ions. This is the concept behind the home water softener. The divalent Ca^{2+} and Mg^{2+} ions replace the monovalent Na^+ ions on the exchange media. The exchange medium is regenerated when a brine solution with very high Na^+ activity is forced through the softener. The Na^+ replaces the Ca^{2+} and Mg^{2+} at the exchange sites. Ion-exchange capacities of organic colloids and clays can also remove heavy metal cations and thus provide some protection to ground-water supplies (Wentink & Etzel 1972); but enough cases of ground-water pollution from heavy metals have been documented to demonstrate that such protection is limited to areas with clay in the soils.

One particularly well-studied ion-exchange reaction is the replacement of calcium in the soil with sodium. If water used for irrigation is high in sodium and low in calcium, the cation-exchange complex may become saturated with sodium. This can destroy the soil structure owing to dispersion of the clay particles. A simple method of evaluating the danger of high-sodium water is the *sodium-adsorption ratio,* or *SAR* (Richards 1954):

$$SAR = \frac{(Na^+)}{\left[\dfrac{(Ca^{2+}) + (Mg^{2+})}{2}\right]^{0.5}} \tag{10–32}$$

A low SAR (2 to 10) indicates little danger from sodium; medium hazards are between 7 and 18, high hazards between 11 and 26, and very high hazards above that. The lower the ionic strength of the solution, the greater the sodium hazard for a given SAR. Anions present in the water can affect calcium replacement (Pratt & Blair 1969; Bower, Ogata, & Tucker, 1968).

If the ion-exchange process is controlled by a reversible equilibrium process, the following equation applies:

$$b[\overline{A}] + a[B] \rightleftharpoons a[\overline{B}] + b[A] \tag{10–33}$$

where

A and B are chemically exchanging species, A with a valence of a and B with a valence of b

[A] is the concentration of solute A in terms of mass per unit volume of liquid

$[\overline{A}]$ is the amount of solute A adsorbed by ion exchange on a unit mass of sediment or soil.

When the exchanged ions are in equilibrium, the concentration of products and reactants at equilibrium is described by the ion-exchange selectivity coefficient, K_s:

$$K_s = \frac{[\overline{B}]^a\,[A]^b}{[\overline{A}]^b\,[B]^a} \qquad (10\text{--}34)$$

The **cation-exchange capacity** (CEC) is defined as $[\overline{A}] + [\overline{B}]$ and the total solute concentration, C_0, is equal to $[A] + [B]$. When the concentration of one of the exchanging ions is very low, the adsorbed phase of the other (dominant) ion is approximately equal to CEC, and the total solute concentration is almost entirely that of the dominant ion. Equation 10–34 can be rewritten under these conditions, if A is the major species, as

$$K_s = \frac{[\overline{B}]^a\,C_0^b}{CEC^b\,[B]^a} \qquad (10\text{--}35)$$

The *ion-exchange distribution coefficient, K_d,* is the ratio of the adsorbed species concentration to the concentration of the solute:

$$K_d = \frac{[\overline{B}]}{[B]} \qquad (10\text{--}36)$$

A standard laboratory test is available to determine the cation exchange capacity of soils. A 100-g sample of dry soil is mixed with a solution of ammonium acetate to saturate the exchange sites with NH_4^+ ions. The pH of the pore water is adjusted to a value of 7.0. The soil is leached with a strong NaCl solution to replace the NH_4^+ on the exchange sites with Na^+ ions. The sodium content of the leaching solution is then determined and the CEC computed as the difference between the sodium in the original solution and the sodium in the leaching solution at equilibrium. It is reported in milliequivalents per 100 g of soil. The CEC is frequently used as an indication of the potential of a soil to attenuate pollutants with exchangeable ions.

10.12   ISOTOPE HYDROLOGY

Isotopes of a particular element have the same atomic number but different atomic weights due to varying numbers of neutrons in the nucleus. *Stable* isotopes are not involved with any natural radioactive decay process. *Radioactive* isotopes undergo spontaneous radioactive decay to form new elements or isotopes. *Radiogenic* isotopes are the stable product of radioactive decay. Certain stable isotopes of hydrogen, oxygen, carbon, nitrogen, and sulfur can be used to study geologic processes that affect ground and surface water. Radioactive isotopes can be used to determine the age of ground water.

Environmental isotopes are those that are naturally occurring. Radioactive isotopes can also be introduced into the ground as part of a ground-water study, usually to determine the direction and/or velocity of ground-water flow.

10.12.1 Stable Isotopes

Stable-isotope studies are based on the tendency of some pairs of isotopes to *fractionate,* or separate into light and heavy fractions. This fractionation occurs during some geologic process, such as evaporation or heating. The five elements that are used in stable-isotope studies are able to fractionate readily, are fairly common, have a relatively large difference in mass between the two isotopes, and have one isotope that is much more abundant than the other. If R is the ratio of the heavy isotope to the light one, then the relative fractionation is expressed in del notation as:

$$\delta = \frac{R_{\text{sample}} - R_{\text{standard}}}{R_{\text{sample}}} \times 1000 \qquad \textbf{(10–37)}$$

Results are expressed as deviation in parts per thousand (‰). If the value of δ is positive, then the sample is enriched with the heavy isotope relative to the standard; a negative sample is isotopically light.

There are two stable isotopes of hydrogen, ^{1}H and ^{2}H (deuterium), as well as three stable isotopes of oxygen, ^{16}O, ^{17}O, and ^{18}O. There are nine different combinations of these isotopes that make stable water molecules with atomic masses ranging from 18 to 22. The most abundant water molecule, $^{1}\text{H}_2{}^{16}\text{O}$, which is the lightest, has a much higher vapor pressure than the heaviest form, $^{2}\text{H}_2{}^{18}\text{O}$. During phase changes of water between liquid and gas the heavier water molecules tend to concentrate in the liquid phase, which fractionates the hydrogen and oxygen isotopes. Water that evaporates from the ocean is isotopically lighter than the water remaining behind, and precipitation is isotopically heavier; that is, it contains more ^{2}H and ^{18}O than the vapor left behind in the atmosphere.

The use of mass spectrometry can determine the ratio of isotopes in a water sample. Important isotope ratios include $^{18}\text{O}/^{16}\text{O}$ and $^{2}\text{H}/^{1}\text{H}$. These isotopic ratios from an environmental water sample can be compared with the isotopic ratio of standard mean ocean water (SMOW). The comparison is made by means of the parameter δ, which is defined as

$$\delta^{18}\text{O} \ (‰) = \left[\frac{(^{18}\text{O}/^{16}\text{O})_{\text{sample}}}{(^{18}\text{O}/^{16}\text{O})_{\text{SMOW}}} - 1 \right] 10^3 \qquad \textbf{(10–38)}$$

$$\delta^2H\ (‰) = \left[\frac{(^2H/^1H)_{sample}}{(^2H/^1H)_{SMOW}} - 1 \right] 10^3 \qquad \textbf{(10–39)}$$

When δ^2H is plotted as a function of $\delta^{18}O$ for water found in continental precipitation, an experimental linear relationship is found that can be described by the equation (Mayo, Muller, & Ralston 1985)

$$\delta^2H = 8\delta^{18}O + 10 \qquad \textbf{(10–40)}$$

This is known as the *meteoric water line*. Continental precipitation samples will tend to group close to this line. Precipitation falling in areas with lower temperatures or at higher latitudes will tend to have lower δ^2H and $\delta^{18}O$ values. Naturally, oceanic water will fall below the meteoric water line as it is isotopically enriched. Deviations from the meteoric water line can be interpreted as being caused by precipitation that occurred during a warmer or colder climate than at present or by geochemical changes that occurred when the water was underground (Craig 1961). Geothermal water tends to be isotopically enriched with respect to $\delta^{18}O$ owing to equilibration of the oxygen in the ground water with respect to oxygen in the rocks (Mayo, Muller, & Ralston 1985). Figure 10.5 shows the meteoric water line and the results of stable isotope analyses for a number of spring water samples from the Meade thrust area of southeastern Idaho.

In some ground-water studies the plot of δ^2H as a function of $\delta^{18}O$ forms a straight line that is parallel to but below the meteoric water line (Figure 10.6). This has been interpreted as a "local" meteoric water line (White & Chuma, 1987; Mayo et al. 1992).

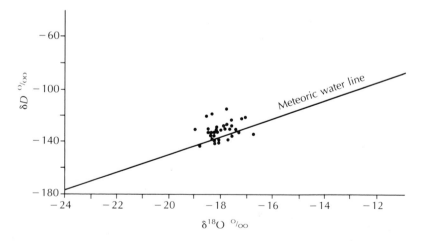

FIGURE 10.5 δD (δ^2H) and $\delta^{18}O$ values from spring and well waters in the Meade thrust area, southeastern Idaho. Source: A. L. Mayo. Ground-Water Flow Patterns in the Meade Thrust Allochthon, Idaho-Wyoming Thrust Belt, Southeastern Idaho. Ph.D. thesis, University of Idaho, 1982.

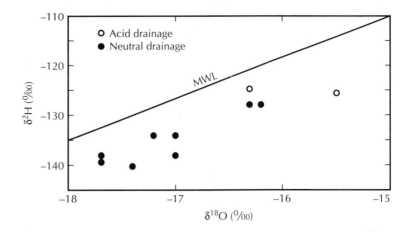

FIGURE 10.6   Plot of $\delta^2 H$ versus $\delta^{18}O$ for ground water in the central Wasatch Range, Utah, showing a "local" meteoric water line parallel to the global meteoric water line. Source: A. L. Mayo et al., *Ground Water* 30, no. 2 (1992): 243–249. Used with permission. © 1992 Ground Water Publishing Company.

Mayo and Klauk (1991) studied the isotopic composition of ground and surface water in the Great Salt Lake area. They found most samples were grouped on the meteoric water line, but that deviations occurred due to evaporation (for example, the brine in Great Salt Lake) and due to heating in thermal springs (Figure 10.7).

Stable carbon isotopes are ^{12}C and ^{13}C, with ^{13}C being relatively rare. The standard for carbon isotopes is a marine belemnite from the Pee Dee formation of South Carolina (PDB). Inorganic carbon in ground water can come from atmospheric carbon dioxide, carbon dioxide generated by biota in the soil zone, and dissolution of carbonate materials. The $\delta^{13}C$ of dissolved inorganic carbon in the ocean is about 0‰ PDB. Soil gas carbon dioxide is about -20‰, whereas atmospheric carbon dioxide is about -7‰ (Drever 1988). The contribution of atmospheric carbon dioxide is small due to the low partial pressure of atmospheric carbon dioxide (Mayo et al. 1992). The $\delta^{13}C$ of carbonate rocks is about 0‰, which is understandable because they precipitated from the ocean, which has a similar value. Mayo et al. (1992) found a mean $\delta^{13}C$ for carbonate rocks of the Wasatch Range of Utah of $+0.30$‰, and Muller and Mayo (1986) found a mean value of -1.85‰ $\pm$ 1.85 PDB for the Redwall Limestone of Arizona.

The $\delta^{13}C$ of ground water is estimated to come in roughly equal amounts from soil-zone gas and carbonate rocks in most carbonate rock aquifers (Mayo et al. 1992). If this is true, it should be about -10‰ PDB. If the actual value is between about 0‰ and -10‰, then carbonate rocks are contributing a greater amount. If the value is less than -10‰, then soil-zone carbon dioxide is more important than carbonate rock as a source.

Sulfur isotope fractionation studies can be used to distinguish the origin of dissolved sulfur in ground water. The sulfur isotope pair that is used is $^{34}S/^{32}S$.

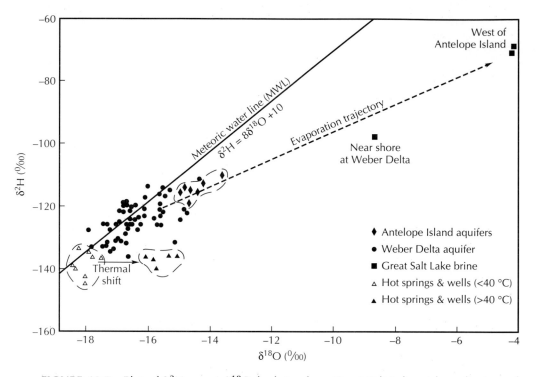

FIGURE 10.7   Plot of δ^2H versus $\delta^{18}O$ for brine from Great Salt Lake and nearby ground water showing deviations from the meteoric water line due to evaporation and heating. Source: A. L. Mayo and R. H. Klauk, *Journal of Hydrology* 127 (1991): 307–335. Used with permission.

The standard for $\delta^{34}S$ is the mineral troilite (FeS), which is found in the Canyon Diablo meteorite of Meteor Crater, Arizona. Sources of sulfur in ground water include atmospheric sulfur, dissolution of sulfate minerals such as gypsum, and oxidation of sulfide minerals, such as pyrite. The dissolution of gypsum will result in an enrichment of ^{34}S; that is, $\delta^{34}S$ becomes more positive. On the other hand, the oxidation of reduced sulfur species will result in an enrichment of ^{32}S, which means that $\delta^{34}S$ will decrease. Mayo and Klauk (1991) used sulfur isotopes to demonstrate that, although ground water from a sedimentary rock unit and a nearby crystalline rock unit had similar concentrations of dissolved sulfate, the ground waters had different sources because the $\delta^{34}S$ values of the two units were different.

The ratio of $^{15}N/^{14}N$ in inorganic nitrogen in ground water as compared with the ratio of atmospheric nitrogen, which is the standard, can be used to determine the source of the nitrogen. Kreitler and Jones (1975) used stable nitrogen isotopes to demonstrate that the source of nitrate contamination in a Texas aquifer was from natural soil nitrate. Kreitler, Ragone, and Katz (1978) demonstrated that in rural, eastern Long Island, New York, inorganic nitrogen in ground water was primarily from inorganic fertilizer, whereas in western, more urbanized, Long Island, the nitrogen was mostly from septic tank wastes.

10.12.2   Radioactive Isotopes Used for Age Dating

Tritium, ^3H, is an unstable isotope of hydrogen with a half-life of 12.3 y. Tritium in the atmosphere is typically in the form of the molecule H^3HO and enters the ground water as recharging precipitation. Prior to 1953, rainwater had less than 10 tritium units (TU). Starting in 1953, the manufacturing and testing of nuclear weapons has increased the amount of tritium in the atmosphere, with a resulting increase in tritium in the ground water. As a result, ^3H can be used in a qualitative manner to date ground water in the sense that ground water with less than 2 to 4 TU is dated prior to 1953; if the amount is significantly greater than 10 to 20 TU, it has been in contact with the atmosphere since 1953. Because of the great temporal and spatial variations in ^3H injected into the atmosphere since 1953, it cannot be used with more precision. Tritium has been used to trace the seepage of contaminated ground water from low-level nuclear-waste disposal areas (Foster 1982).

Radiocarbon dating methods can be applied to obtain the age of ground water. Carbon exists in several naturally occurring isotopes, ^{12}C, ^{13}C, and ^{14}C. Carbon 14 is formed in the atmosphere by the bombardment of ^{14}N by cosmic radiation (DeVries 1959). The ^{14}C forms CO_2, so that the atmospheric CO_2 has a constant radioactivity due to modern ^{14}C. If the CO_2 is incorporated into a form in which it is isolated from modern ^{14}C, age determinations can be made from the ^{14}C radioactivity as a percent of the original. The half-life of ^{14}C is 5730 y, so that if one-fourth of the original activity is present, two half-lives, or 11,460 y, have elapsed. When precipitation soaks into the ground, it is saturated with respect to CO_2, with a known ^{14}C activity. Once the water has entered the soil, additional carbon may come from soil CO_2 and the solution of carbonate minerals. The modern carbon is diluted by the inactive carbon from carbonate minerals. The raw dates obtained must be adjusted for this dilution.

If A_C is the measured ^{14}C radioactivity and A_0 is the activity at the time the sample was isolated, then the following equation may be used:

$$A_C = QA_0 2^{-t/T_C} \qquad\qquad \textbf{(10–41)}$$

where

> t is the age
>
> T_C is the half-life of ^{14}C
>
> Q is an adjustment factor to account for dilution by *dead carbon**
> (Wigley 1975)

---

*Dead carbon is carbon from a source old enough for any ^{14}C to have decayed below measurable limits.

The equation requires an estimation of initial value, A_0, and the adjustment factor, Q.

The value of A_0 will depend on the carbonate equilibria established under an open system in which the ground water was exposed to an infinite reservoir of CO_2. This occurs in nature in the soil zone and in shallow ground water. When the ground-water system becomes closed with respect to CO_2, then any added carbon would be only from carbonate rocks—i.e., dead carbon. The value of Q is generally in the range of 0.5 to 0.9. Carbon 14 dates of ground water thus tend to be somewhat less than raw dates as a result of the dilution by dead carbon from carbonate minerals. The determination of A_0 and Q is somewhat complex, and several different methods are available (Wigley 1975; Plines, Langmuir, & Harmon 1974). Radiocarbon dates of ground water of up to 50,000 to 80,000 y may be obtained, although the accuracy under the best of conditions is on the order of ±20% (Davis & Bentley 1982). Indeed, Muller and Mayo (1986) demonstrated that a variation of ±20% could be expected on the basis of variation of the ^{13}C content of the dead carbon in carbonate rocks.

A number of other isotopes have been used or have been proposed to be used to date ground water. Chlorine 36 is one of them (Bentley & Davis 1980). It has been proposed as a method to date ground water that is older than water that can be dated with carbon 14 as it has a half-life of 3.01×10^5 years. A ratio of ^{36}Cl to total Cl is determined; the higher the ratio, the younger the sample. Oceanic water is old enough that little, if any, ^{36}Cl is present. Young water near the coastline, which is likely to contain chloride produced as salt spray from the ocean, will appear to be much older owing to the large amount of "dead" chloride. Other isotopes with possible uses in age dating are ^{85}Kr, ^{81}Kr, ^{39}Ar, and ^{32}Si (Davis & Bentley 1982).

10.13   MAJOR ION CHEMISTRY

More than 90% of the dissolved solids in ground water can be attributed to eight ions: Na^+, Ca^{2+}, K^+, Mg^{2+}, SO_4^{2-}, Cl^-, HCO_3^-, and CO_3^{2-}. These ions are usually present at concentrations greater than 1 mg/L. Silica, SiO_2, a nonionic species, is also typically present at concentrations greater than 1 mg/L. Direct analysis can be done for the first six ions. Bicarbonate and carbonate concentrations are found by titrating the sample with acid to an endpoint with a pH of about 4.4. This is reported as total alkalinity. Based on the pH of the sample, the proportion of carbonate and bicarbonate can be calculated from Equation 10–25B. Field measurements of pH, temperature, and specific electrical conductance are usually made at the time the sample is collected. Without field pH, the concentrations of carbonate and bicarbonate cannot be determined from the alkalinity.

Other naturally occurring ions that may be present in amounts of 0.1 mg/L to 10 mg/L include iron, nitrate, fluoride, strontium, and boron. Iron and nitrate are typically included in water-chemistry studies, with fluoride, strontium, and boron being less commonly reported. Many other inorganic ions are important from a standpoint of water quality and are discussed in Chapter 11.

Total dissolved solids (TDS) can be determined by evaporating a known volume of the sample and weighing the residue. TDS can be estimated by summing the concentrations of the individual ions. This method does not account for any dissolved solids that might be present from unreported ions or other dissolved substances. For example, dissolved silica, SiO_2, may not be reported but contributes to TDS.

As a check on the chemical analysis, a cation-anion balance is usually performed. This is accomplished by converting all the ionic concentrations to units of equivalents per liter. The anions and cations are summed separately, and the results are compared. If the sum of the cations is not within a few percent of the sums of the anions, then either there is a problem with the chemical analysis or one or more ionic species that have not been identified are present in significant amounts.

10.14   PRESENTATION OF RESULTS OF CHEMICAL ANALYSES

Tables of data are the most common form in which the results of an analysis of water chemistry are reported. The data can be expressed in milligrams per liter (mg/L), milliequivalents per liter (meq/L), or millimoles per liter. For many purposes, the data may be also displayed in graphical form.

10.14.1   Piper Diagram

The major ionic species in most natural waters are Na^+, K^+, Ca^{2+}, Mg^{2+}, Cl^-, CO_3^{2-}, HCO_3^-, and SO_4^{2-}. A trilinear diagram can show the percentage composition of three ions. By grouping Na^+ and K^+ together, the major cations can be displayed on one trilinear diagram. Likewise, if CO_3^{2-} and HCO_3^- are grouped, there are also three groups of the major anions. Figure 10.8 shows the form of a trilinear diagram that is commonly used in water-chemistry studies (Piper 1944). Analyses are plotted on the basis of the percent of each cation (or anion).

Each apex of a triangle represents a 100% concentration of one of the three constituents. If a sample has two constituent groups present, then the point representing the percentage of each would be plotted on the line between the apexes for those two groups. If all three constituent groups are present, the analyses would fall in the interior of the field. The diamond-shaped field between the two triangles is used to represent the composition of water with respect to both cations and anions.

The cation point is projected onto the diamond-shaped field parallel to the side of the triangle labeled magnesium, and the anion point is similarly projected parallel to the side of the triangle labeled sulfate. The intersection of the two lines is plotted as a point on the diamond-shaped field.

As water flows through an aquifer it assumes a diagnostic chemical composition as a result of interaction with the lithologic framework. The term

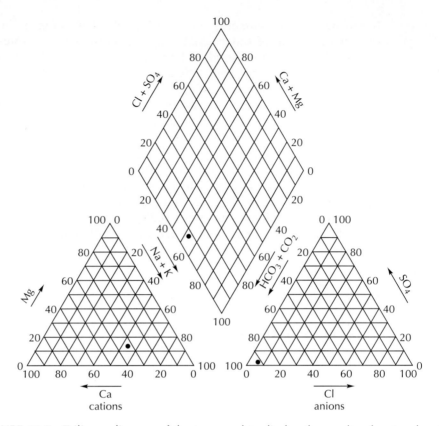

FIGURE 10.8   Trilinear diagram of the type used to display the results of water-chemistry studies. Results of the example problem from Section 10.14 are plotted on the diagram.

hydrochemical facies is used to describe the bodies of ground water, in an aquifer, that differ in their chemical composition. The facies are a function of the lithology, solution kinetics, and flow patterns of the aquifer (Back 1960, 1966). Hydrochemical facies can be classified on the basis of the dominant ions in the facies by means of the trilinear diagram (Figure 10.9).

EXAMPLE PROBLEM

Plot the results of the following analysis on a trilinear diagram:

	Ca^{2+}	Mg^{2+}	Na^+	K^+	HCO_3^-	CO_3^{2-}	SO_4^{2-}	Cl^-
mg/L	23	4.7	35	4.7	171	0	1.0	9.5
meq/L	1.15	0.39	1.52	0.12	2.80	0	0.02	0.27

The first step is to find the percent of each cation and anion group as a percentage of the total:

Cations	meq/L	% of Total	Anions	meq/L	% of Total
Ca^{2+}	1.15	36	Cl^-	0.27	9
Mg^{2+}	0.39	12	SO_4^{2-}	0.02	1
$Na^+ + K^+$	1.64	52	$CO_3^{2-} + HCO_3^-$	2.80	90
Total	3.18		Total	3.09	

Note: Due to analytical error and unreported minor constituents, the total equivalents of anions and cations do not exactly match. Theoretically, the total equivalent weight of the anions should be exactly that of the cations, as equivalent weights are based on the amount of the ion that would combine with O_2.

The points for both cations and anions are plotted on the appropriate triaxial diagrams in Figure 10.8. The positions of these points are projected onto the diamond-shaped field and the intersection of the projected lines is plotted.

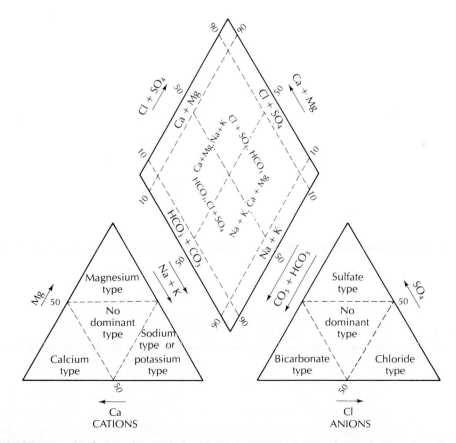

FIGURE 10.9 Hydrogeochemical classification system for natural waters using the trilinear diagram.

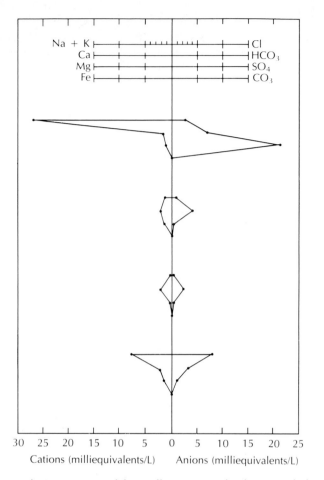

FIGURE 10.10 Analysis represented by Stiff patterns. The horizontal distance from the vertical axis is based on the number of milliequivalents per liter of each anion or cation. Use of the lower bar for iron and carbonate is optional. Source: J. D. Hem, U.S. Geological Survey Water-Supply Paper 2254, 1985.

10.14.2 Stiff Pattern

A second type of graphical presentation of chemical analyses is the **Stiff pattern** (Stiff 1951). A polygonal shape is created from four parallel horizontal axes extending on either side of a vertical zero axis. Cations are plotted in milliequivalents per liter on the left of the zero axis, one to each horizontal axis, and anions are plotted on the right. Figure 10.10 shows several Stiff patterns. The use of the lower horizontal bar with iron and carbonate is optional as in many waters they are close to zero. Stiff patterns are useful in making a rapid visual comparison between water from different sources. The larger the area of the polygonal shape, the greater the concentrations of the various ions. Figure 10.11 shows the use of Stiff patterns in an area where mineralized water exists in a portion of an aquifer

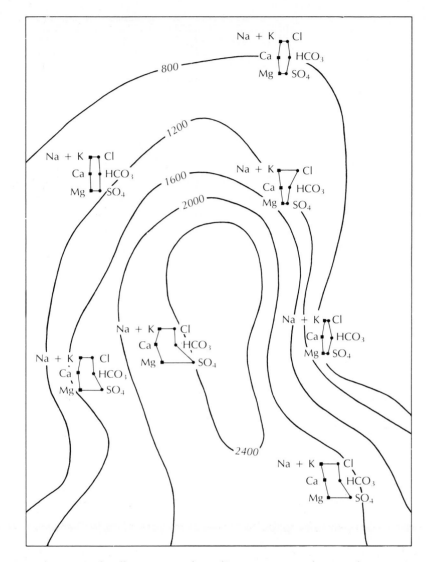

FIGURE 10.11 Use of Stiff patterns to show the varying ground-water chemistry in an area containing mineralized water from a waste-disposal operation. Solid lines represent isocons of total dissolved solids. Stiff patterns are centered over locations of wells.

system. The isocon lines represent lines of equal total dissolved solids and the Stiff patterns are centered over the location of a particular well. In these patterns the lower horizontal line for iron and carbonate was not used.

10.14.3 Schoeller Semilogarithmic Diagram

Schoeller (1955) proposed the use of semilogarithmic graph paper to plot the concentrations of the anions and cations. The concentrations are plotted in

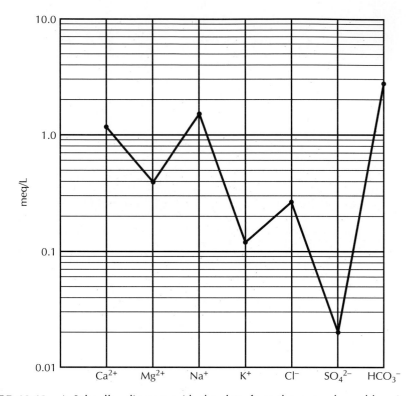

FIGURE 10.12   A Schoeller diagram with the data from the example problem in Section 10.14.1 plotted.

meq/L. This type of diagram allows us to make a visual comparison of the compositions of different waters. Figure 10.12 is an example of a Schoeller diagram, which uses the data from the example problem in Section 10.14.1. Sodium and potassium are often grouped, although since there can be more than three cations, this is not necessary.

CASE STUDY: CHEMICAL GEOHYDROLOGY OF THE FLORIDAN AQUIFER SYSTEM

The regional geohydrology of the Floridan aquifer was discussed in Section 8.6. The chemical geohydrology of this aquifer system is also well known. We will consider the chemical changes that take place as the water flows from the central recharge area at Polk City to the south. Data from five wells form the basis for the hydrochemical cross sections (Back, Cherry, & Hanshaw 1966; Back & Hanshaw 1970; Plummer 1977). Figure 10.13 indicates the locations of the wells on the potentiometric map of central Florida, with Polk City located at the southern edge of the recharge area. The five wells all tap the Floridan aquifer and lie approximately along the same flow path. Chemical analyses of water from the wells are given in Table 10.7. As water travels down the flow path, it increases in total dissolved solids, from 138 to 726 mg/L. All ions except bicarbonate show a progressive increase along the flow path (Figure 10.14). Computation of ion-activity products from the analyses show that both dolomite and calcite saturation increase along the flow path. For the most part, the K_{iap}/K_{sp} of the water is greater than 1, indicating supersaturation. Hydrochemical cross sections along the flow path are shown in Figure 10.14. A trilinear plot (Figure 10.15) of the well analyses indicates that the chemical composition of the water is shifting along the flow path. This is due to an increase in the Mg^{2+}/Ca^{2+} and the SO_4^{2-}/HCO_3^- ratios with increasing distance from the recharge area. The change in these ratios is due to the solution of gypsum ($CaSO_4 \cdot 2H_2O$) and dolomite ($CaMg(CO_3)_2$) along the flow paths. It is important to note that these reactions involve only the solution of minerals in fresh water. If this water mixes with sea water, water chemistry would rapidly change and be dominated by sodium and chloride, both of which are minor constituents in fresh water.

Calculation of the age of ground water from the Floridan aquifer on the basis of ^{14}C activity is complex owing to the solution of dead carbon from carbonate. Complicating the dating is the fact that the flow path from Polk City to Ft. Meade is partially open to soil CO_2, and then it is closed to CO_2 from Ft. Meade to Wauchula. However, from Wauchula to Arcadia, it is again open to CO_2, the source being the oxidation of lignite from sulfate reduction (Plummer 1977). Several authors have dealt with the ^{14}C dating of this water

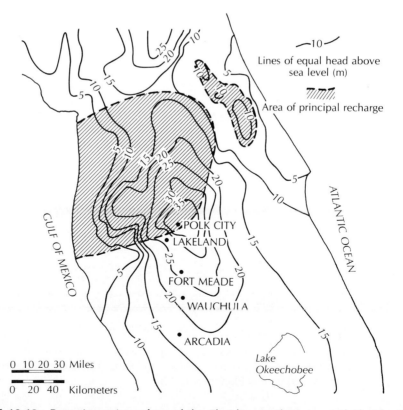

FIGURE 10.13 Potentiometric surface of the Floridan aquifer in central Florida. Source: Adapted from L. N. Plummer, *Water Resources Research* 13 (1977): 801–12.

TABLE 10.7   Chemical analysis of water from Floridan aquifer in central Florida

Well	Location	Temp % C	Field pH	SiO$_2$	Ca^{2+}	Mg^{2+}	Na$^+$	K$^+$	HCO$_3^-$	SO$_4^{2-}$	CL$^-$	TDS
									Milligrams per liter			
1	Polk City	23.8	8.0	12	34	5.6	3.2	0.5	124	2.4	4.5	138
2W	Lakeland	26.3	7.62	18	54	14	6.9	1.0	253	3.6	8.5	238
2S	Ft. Meade	26.6	7.75	16	58	17	6.1	0.7	163	71	9.0	272
3S	Wauchula	25.4	7.69	18	66	29	8.3	2.0	168	155	10	392
4S	Arcadia	26.3	7.44	31	106	60	21	3.7	206	344	28	726

Source: Data from W. Back and B. B. Hanshaw, *Journal of Hydrology* 10 (1970):330–68.

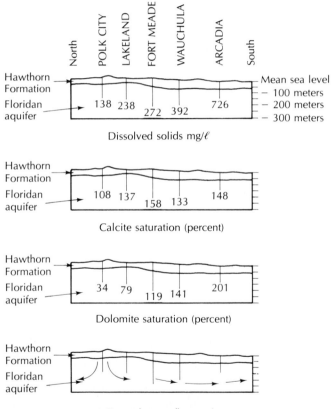

FIGURE 10.14   Hydrogeochemical cross sections from Polk City through Arcadia through the Floridan aquifer. Sources: Data from Back, Cherry, and Hanshaw (1966); Back and Hanshaw (1970); and Plummer (1977).

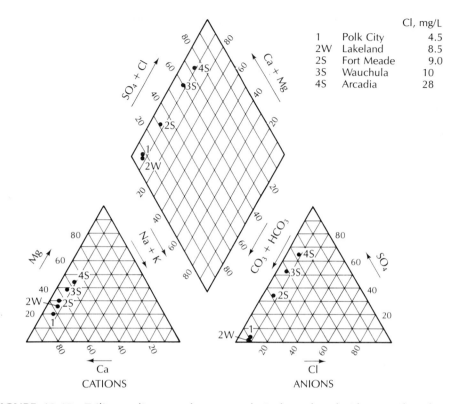

		Cl, mg/L
1	Polk City	4.5
2W	Lakeland	8.5
2S	Fort Meade	9.0
3S	Wauchula	10
4S	Arcadia	28

FIGURE 10.15 Trilinear diagram of water analysis from the Floridan aquifer of central Florida. Locations are shown on Figure 10.13. Source: W. Back and B. B. Hanshaw, *Journal of Hydrology* 10 (1970): 330–68 (Amsterdam: North-Holland Publishing Company).

(Wigley 1975; Back & Hanshaw 1970; Plummer 1977) and have cited dates ranging from 20,600 to 24,100 to 36,000 y B.P. Based on the isotopic dating and the reaction coefficients, apparent rates of solution for several mineral species have been calculated (Plummer 1977).

This type of study illustrates what can be accomplished in chemical geohydrology. It can be a companion to the more commonly performed hydrodynamic study of flow paths and rates. The flow velocity as determined from hydrodynamic considerations can be compared with radiocarbon estimates.

NOTATION

a	Effective diameter of an ion	K_{sp}	Solubility product
A_C	Measured ^{14}C radioactivity	**m**	Molal concentration
A_0	Radioactivity when a sample becomes geologically isolated	**n**	Number of electrons
		Q	Dead-carbon adjustment factor
A	Constant in the Debye-Hückel equation	**R**	Gas constant
B	Constant in the Debye-Hückel equation	**t**	Age
E^0	Standard potential	**T**	Temperature in kelvins
F	Faraday constant	T_C	Half-life of ^{14}C
I	Ionic strength	**z**	Charge of an ion
K_d	Ion-exchange distribution coefficient	α	Chemical activity
K_{eq}	Equilibrium constant	ΔG_r^0	Gibbs free energy
K_{iap}	Ion-exchange selectivity coefficient	γ	Activity coefficient
K_s	Ion-exchange selectivity coefficient		

PROBLEMS

1. How much KCl is in a 0.1-molar solution?

2. How much NaCl is in a 0.23-molar solution?

3. The solubility product for CuCl is $10^{-5.9}$. What is the solubility of Cu^+ at equilibrium with CuCl?

4. The solubility product of $CaSO_4$ is $10^{-4.5}$. What is the solubility of Ca^{2+} at equilibrium with $CaSO_4$?

5. The solubility product of fluorite, CaF_2, is $10^{-10.5}$.
   A. What is the solubility of Ca^{2+} at equilibrium with fluorite?
   B. If CaF_2 is dissolved in a solution of 0.001-molar F^-, how much will dissolve at equilibrium?

6. The solubility product of MgF_2 is $10^{-8.2}$.
   A. What is the solubility of Mg^{2+} at equilibrium with MgF_2?
   B. If MgF_2 is dissolved in a solution of 0.0072-molar F^-, how much will dissolve at equilibrium?

7. Given the following ground-water analysis:

Ca^{2+}	143 mg/L	SO_4^{2-}	254 mg/L	pH	= 8.0
Mg^{2+}	35 mg/L	HCO_3^-	317 mg/L	TDS	= 790 mg/L
Na^+	14 mg/L	Cl^-	4 mg/L		

   A. Convert all analyses into molal concentrations.
   B. Compute ionic strength.
   C. Compute activity coefficient for each ion.

    D. Find activity of each ion.

    E. Convert analyses to meq/L.

    F. Do a cation-anion balance.

    G. Find the K_{iap} of anhydrite ($CaSO_4$).

    H. Compare the K_{iap} of anhydrite with the K_{sp} of anhydrite ($10^{-4.5}$).

    I. Find the K_{iap} of calcite ($CaCO_3$).

    J. Compare the K_{iap} of calcite with the K_{sp} of calcite ($10^{-8.35}$).

8. Given the following ground-water analysis:

Ca^{2+}	76 mg/L	SO_4^{2-}	14 mg/L	pH = 7.8
Mg^{2+}	14 mg/L	HCO_3^-	267 mg/L	TDS = 460 mg/L
Na^+	22 mg/L	Cl^-	44 mg/L	
K^+	1.5 mg/L			

    A. Convert all analyses into molal concentrations.

    B. Compute ionic strength.

    C. Compute the activity coefficient for each ion.

    D. Find the activity of each ion.

    E. Convert analyses to meq/L.

    F. Do a cation-anion balance.

    G. Find the K_{iap} of anhydrite ($CaSO_4$).

    H. Compare the K_{iap} of anhydrite with the K_{sp} of anhydrite ($10^{-4.5}$).

    I. Find the K_{iap} of calcite ($CaCO_3$).

    J. Compare the K_{iap} of calcite with the K_{sp} of calcite ($10^{-8.35}$).

9. What are $[H^+]$ and $[OH^-]$ for an aqueous solution at pH 8.7?

10. What are $[H^+]$ and $[OH^-]$ for an aqueous solution at pH 4.53?

11. What is the pH of a 0.0027-molal solution of H_2CO_3?

12. What is the pH of a 0.0092-molal solution of H_2CO_3?

13. What is the pH of a 0.0027-molal solution of HCl?

14. Given the following analysis of ground water:

Ca^{2+}	76.8 mg/L	SO_4^{2-}	37 mg/L
Na^+	11.72 mg/L	Cl^-	13.25 mg/L
K^+	1.67 mg/L	HCO_3^-	305.4 mg/L
Mg^{2+}	20.54 mg/L	NO_3^-	2.03 mg/L
Fe^{2+}	6.38 mg/L	pH	8.3

    A. Convert all values to meq/L.

    B. Do an anion-cation balance.

    C. Plot the position on a trilinear diagram (Figure 10.16).

    D. Make a Stiff pattern of the analysis.

    E. Make a Schoeller diagram of the analysis.

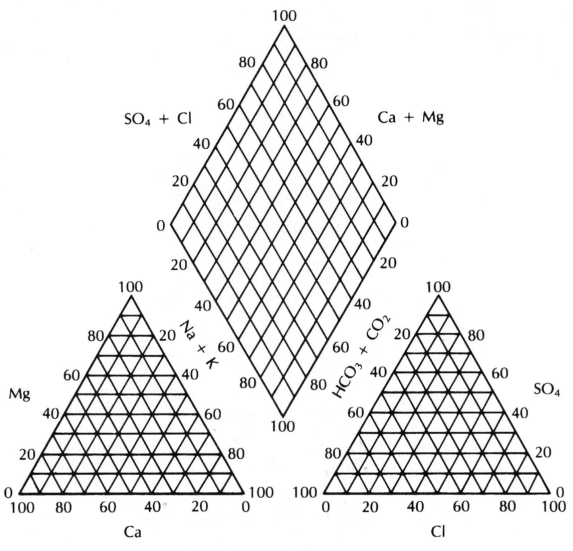

FIGURE 10.16 Trilinear diagram for Problem 14c.

11 Water Quality and Ground-Water Contamination

And these waters, falling on these mountains through the ground and cracks, always descend and do not stop until they find some region blocked by stones or rock very close-set and condensed. And then they rest on such a bottom and having found some channel or other opening, they flow out as fountains or brooks or rivers according to the size of the openings and receptacles; and since such a spring cannot throw itself (against nature) on the mountains, it descends into the valleys. And even though the beginnings of such springs coming from the mountains are not very large, they receive aid from all sides, to enlarge and augment them; and particularly from the lands and mountains to the right and left of these springs.

Discours admirables, Bernard Palissy (ca. 1510–1590)

11.1 INTRODUCTION

The quality of water that we ingest as well as the quality of water in our lakes, streams, rivers, and oceans is a critical parameter in determining the overall quality of our lives. Water quality is determined by the solutes and gases dissolved in the water, as well as the matter suspended in and floating on the water. Water quality is a consequence of the natural physical and chemical state of the water as well as any alterations that may have occurred as a consequence of human activity. The usefulness of water for a particular purpose is determined by the water quality. If human activity alters the natural water quality so that it is no longer fit for a use for which it had previously been suited, the water is said to be **polluted** or **contaminated.** It should be noted that in many areas water quality has been altered by human activity, but the water is still usable.

One basic measure of water quality is the *total dissolved solids (TDS),* which is the total amount of solids, in milligrams per liter, that remain when a water sample is evaporated to dryness. Table 11.1 gives a classification scheme for water based on the total dissolved solids.

Water naturally contains a number of different dissolved inorganic constituents. The major cations are calcium, magnesium, sodium, and potassium; the major anions are chloride, sulfate, carbonate, and bicarbonate. Although not in ionic form, silica can also be a major constituent. These *major constituents* constitute the bulk of the mineral matter contributing to total dissolved solids. In addition there may be *minor constituents* present, including iron, manganese,

TABLE 11.1   Classification of water based on total dissolved solids

Class	TDS (mg/L)
Fresh	0–1,000
Brackish	1,000–10,000
Saline	10,000–100,000
Brine	>100,000

fluoride, nitrate, strontium, and boron. *Trace elements* such as arsenic, lead, cadmium, and chromium may be present in amounts of only a few micrograms per liter, but they are very important from a water-quality standpoint.

Dissolved gases are present in both surface and ground water. The major gases of concern are oxygen and carbon dioxide. Nitrogen, which is more or less inert, is also present. Minor gases of concern include hydrogen sulfide and methane. Hydrogen sulfide is toxic and imparts a bad odor but is not present in water that contains dissolved oxygen.

Surface water may be adversely impacted by human activity. If organic matter, such as untreated human or animal waste, is placed into the surface-water body, *dissolved oxygen* levels diminish as microorganisms grow, using the organic matter as an energy source and consuming oxygen in the process. The total dissolved solids may increase owing to the disposal of wastewater, urban runoff, and increased erosion due to land-use changes in the drainage basin.

The concentration of dissolved solids in Lake Michigan increased by some 20 mg/L from 1895 to 1965 (Beeton 1965). This has been due to the discharge of water products into the lake as well as changes in the land uses of the basin that have altered the quality of water draining from the land. Air pollution has resulted in an increase in the dissolved solids of precipitation into the lake. A large-volume lake such as Lake Michigan can accept some increased amount of common dissolved salts and unreactive sediment without significant water-quality degradation. However, when a lake is lacking in a mineral critical to plant growth, the addition of only a small amount of the *limiting nutrient* can overly stimulate plant growth and result in a dramatic increase in the rooted vegetation and floating algae. This process is known as **eutrophication** (Beeton 1966). In Lake Michigan there is a greater concentration of phosphorus in the water near Milwaukee, Wisconsin, than there is at midlake. There is also a larger concentration of diatoms, a type of algae, near shore, where the phosphorus content is high (Hollard & Beeton 1972). The source of the increased phosphorus is agricultural and urban runoff, as well as sewage effluent carried into the lake by the Milwaukee River.

The natural quality of ground water varies substantially from place to place. It can range from total dissolved solids contents of 100 mg/L or less for some fresh ground water to more than 100,000 mg/L for some brines found in deep aquifers. The U.S. Environmental Protection Agency has developed a three-part classification system, taking this variability into account, for the ground waters of the United States (U.S. Environmental Protection Agency 1984):

Class I: *Special Ground Waters* are those that are highly vulnerable to contamination because of the hydrological characteristics of the areas under which they occur and that are also either an irreplaceable source of drinking water or ecologically vital in that they provide the baseflow for a particularly sensitive ecological system.

Class II: *Current and Potential Sources of Drinking Water and Waters Having Other Beneficial Uses* are all other ground waters except Class III.

Class III: *Ground Waters Not Considered Potential Sources of Drinking Water and of Limited Beneficial Use* because the salinity is greater than 10,000 mg/L or the ground water is otherwise contaminated beyond levels that can be removed using methods reasonably employed in public water-supply treatment.

The U.S. EPA uses this classification scheme in promulgating rules and regulations at the federal level. The highest degree of protection is given to Class I ground water.

Pollution of surface water frequently results in a situation where the contamination can be seen or smelled. However, contamination of ground water most often results in a situation that cannot be detected by human senses. Ground-water contamination can be due to bacteriological or toxic agents or simply to an increase in common chemical constituents to a concentration whereby the usefulness of the water is impaired. Figure 11.1 shows the extent of

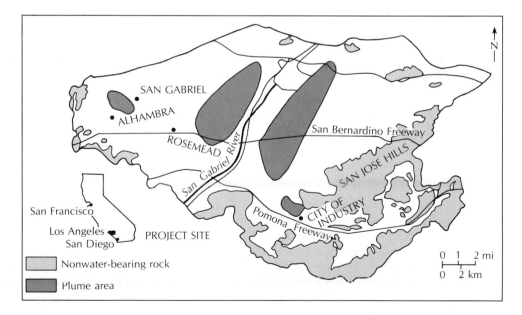

FIGURE 11.1 Plumes of contaminated ground water in the 170-mi² San Gabriel groundwater basin, southern California. Contaminants, which include trichloroethylene, perchloroethylene, carbon tetrachloride, and other suspected carcinogens, are found in more than 88 municipal supply wells to depths as great as 1000 ft (300 ft). Multiple sources of contamination are responsible for the common industrial solvents and degreasers found in the ground water. Source: J. J. Kosowatz & M. J. Sponseller, *Engineering News Record* 217, no. 21 (Nov. 21, 1986): 28–29.

four plumes of contaminated water that have developed in the San Gabriel ground-water basin of southern California.

In the past, water contamination was primarily due to microbiological agents. Although many advances in public health have been made, incidences of waterborne disease still occur in the United States and appear to be increasing. Of 672 cases of waterborne disease in the United States from 1946 to 1980, 52% were due to unknown causes, 22% were due to bacteria, 12% were viral in nature, 7% were due to parasites, 4% were caused by inorganic chemicals, and 3% were caused by organic chemicals (Lippy & Waltrip 1984). Use of untreated, contaminated ground water was responsible for 35% of the disease outbreaks in public water-supply systems during this period, whereas only 8% were due to untreated, contaminated surface water (Lippy & Waltrip 1984). In addition, many of the remaining outbreaks were caused by the failure of systems designed to treat contaminated ground water.

Reported incidences of waterborne disease in the United States peaked in 1980 at 553 and have shown a marked decrease since then. In 1987 there were 15 incidences; in 1988, 13; in 1989, 12; and in 1990, 14 (Herwaldt et al. 1992). For the 2-y period 1989–1990 half the cases were due to well water, 12% to spring water, and 38% to surface water. In more than half the cases, no disease agent was identified. The parasite *Giardia lamblia* was identified in 27% of the cases; viral agents, in 12%; the bacterium *E. coli* 0157:H7, in 4%; and cyanobacteria, in 4%. In the outbreak of E. coli 0157:H7, 243 people became ill, 32 were hospitalized, and 4 died (Herwaldt et al. 1992).

In March and April 1993, an estimated 200,000 people served by the Milwaukee, Wisconsin, public water supply became ill with severe intestinal distress and other flulike symptoms such as weakness, aches, and pains. At least eight people died, but for the most part they were already extremely ill from other causes and had weakened immune systems. The disease was eventually traced to a microscopic protozoan, *Cryptosporidium*. Improper filtration of the city water source, which is Lake Michigan, had failed to remove this parasite, which is fairly common in surface water.

Our understanding of the toxicology of carcinogenic compounds has increased along with the analytical capacity to detect low concentrations of organic compounds in aqueous samples. Recent regulations have greatly increased the amount of ground-water monitoring required. As a result, numerous instances of ground-water contamination have been revealed. Legal cases involving ground-water contamination have resulted in corporations paying millions of dollars to clean up contaminated ground water as well as paying for damages to families suffering from illness and death alleged to have been caused by organic chemicals in well water ingested by the plaintiffs (American Waterworks Association 1986).

The chemical and microbiological agents that are adversely impacting the quality of ground water are coming from a variety of sources, including land application of agricultural chemicals; animal wastes; septic-tank disposal systems for sewage; sewage-treatment lagoons; land application of organic wastes; municipal landfills; toxic- and hazardous-waste landfills; leaking underground storage tanks; faulty underground injection wells, pits, ponds, and lagoons used for storage; treatment and disposal of various liquid compounds; and chemical and petroleum product spills. In this chapter we will consider water-quality standards that have been developed in the United States, methods of monitoring ground-water quality, mass transport of contaminants in flowing ground water, sources of ground-water contamination, and methods of aquifer restoration.

11.2   WATER-QUALITY STANDARDS

Water-quality standards are regulations that set specific limitations on the quality of water that may be applied to a specific use. **Water-quality criteria** are values of dissolved substances in water and their toxicological and ecological meaning. These data can be used to set water-quality standards (U.S. Environmental Protection Agency 1976).

In Public Law 92-500, Section 302, the U.S. Congress directed each state to establish water-quality standards for surface-water bodies. These water-quality standards specify maximum concentrations of substances in surface water for the purpose of protecting aquatic life, users of surface water, and consumers of aquatic life. States are, of course, free to establish ground-water standards and other types of water-quality standards.

The U.S. Environmental Protection Agency (EPA) has been directed by Congress to establish *drinking water standards* under provisions of Public Law 93-523, the Safe Drinking Water Act, and its amendments. The goal of the Safe Drinking Water Act is to determine **maximum contaminant level goals (MCLGs)** and **maximum contaminant levels (MCLs)** for materials that may be found in drinking water. There are three criteria for selection of contaminants for regulation under this act: (1) the analytical ability to detect a contaminant in drinking water, (2) the potential health risk, and (3) the occurrence or potential for occurrence in drinking water. Maximum contaminant-level goals are nonenforceable health goals set at a level to prevent known or anticipated adverse effects with an adequate margin of safety. Maximum contaminant levels are enforceable standards that are to be set as close to the MCLGs as is feasible on the basis of the water-treatment technologies and cost. Compounds that are carcinogenic will have their MCLGs set at 0, whereas the MCLGs for chronically toxic compounds are based upon an acceptable daily intake, which takes into account total exposure from air, food, and drinking water. Risk from carcinogenic compounds is expressed in terms of additional cancer risk over a lifetime of exposure at a given level. A cancer risk of 10^{-6} means that there would be one additional cancer death out of a population of 1,000,000 people.

The drinking water standards set by the USEPA as of the date on the table are given in Table 11.2 (Pontius 1992a, 1992b, 1992c). The reader should be cognizant of the fact that these lists will be updated by the EPA from time to time and that the EPA should be consulted about any specific MCLG or MCL. Maximum contaminant levels will be enforced for all public water-supply systems by the various states. States are free to set MCLs that are more strict than the federal standards but not less stringent.

TABLE 11.2 USEPA drinking-water standards and health goals as of January 1993

Chemical	MCLG (μg/L)	MCL (μg/L)	SMCL (μg/L)
Synthetic organic chemicals			
Acrylamide (1)	0[d]	Treatment technique[d]	
Adipates (di(ethylhexyl)adipate)	500[k]	500[k]	
Alachlor	0[d]	2[d]	
Aldicarb	1[e]	3[e]	
Aldicarb sulfoxide	1[e]	4[e]	
Aldicarb sulfone	1[e]	2[e]	
Atrazine	3[d]	3[d]	
Benzene	0[a]	5[b]	
Benzo[a]anthracene (5)	0[f]	0.1[f]	
Benzo[a]pyrene	0[k]	0.2[k]	
Benzo[b]fluoranthene (5)	0[f]	0.2[f]	
Benzo[k]fluoranthene (5)	0[f]	0.2[f]	
Butylbenzyl phthalate (5)	100[f]	100[f]	
Carbofuran	40[d]	40[d]	
Carbontetrachloride	0[a]	5[b]	
Chlorodane	0[d]	2[d]	
Chrysene (5)	0[f]	0.2[f]	
Dalapon	200[k]	200[k]	
Dibenz[a,h]anthracene (5)	0[f]	0.3[f]	
Dibromochloropropane (DBCP)	0[d]	0.2[d]	
o-Dichlorobenzene (9)	600[d]	600[d]	10
p-Dichlorobenzene (9)	75[b]	75[b]	5
1,2-Dichloroethane	0[a]	5[b]	
1,1-Dichloroethylene	7[a]	7[b]	
cis-1,2-Dichloroethylene	70[a]	70[b]	
trans-1,2-Dichloroethylene	100[d]	100[d]	
1,2-Dichloropropane	0[d]	5[d]	
2,4-Dichlorophenoxyacetic acid (2,4-D)	70[d]	70[d]	
Di(ethylhexyl)phthalate	0[k]	6[k]	
Diguat	20[k]	20[k]	
Dinoseb	7[k]	7[k]	
Endothall	100[k]	100[k]	
Endrin	2[k]	2[k]	
Epichlorohydrin (1)	0[d]	Treatment technique[d]	
Ethylbenzene (9)	700[d]	700[d]	30
Ethylene dibromide (EDB)	0[d]	0.05[d]	
Glyphosate	700[k]	700[k]	
Heptachlor	0[d]	0.4[d]	
Heptachlor epoxide	0[d]	0.2[d]	
Hexachlorobenzene	0[k]	1[k]	
Hexachlorocyclopentadiene [HEX]	50[k]	50[k]	8[f]
Indenopyrene (5)	0[f]	0.4[f]	
Lindane	0.2[d]	0.2[d]	
Methoxychlor	40[d]	40[d]	

Chemical	MCLG (μg/L)	MCL (μg/L)	SMCL (μg/L)
Methylene chloride	0[k]	5[k]	
Monochlorobenzene	100[d]	100[d]	
Oxamyl (vydate)	200[k]	200[k]	
PCBs as decachlorobiphenol	0[d]	0.5[d]	
Pentachlorophenol	0[d]	1[d]	
Picloram	500[k]	500[k]	
Simaze	4[k]	4[k]	
Styrene (9)	100[d]	100[d]	10
2,3,7,8-TCDD (dioxin)	0[k]	3×10^{-8}[k]	
Tetrachloroethylene	0[d]	5[d]	
1,2,4-Trichlorobenzene	70[k]	70[k]	
1,1,2-Trichloroethane	3[k]	5[k]	
Trichloroethylene (TCE)	0[a]	5[b]	
1,1,1-Trichloroethane	200[a]	200[b]	
Toluene (9)	1000[d]	1000[d]	40
Toxaphene	0[d]	3[d]	
2-(2,4,5-Trichlorophenoxy)-propionic acid (2,4,5-TP, or Silvex)	50[d]	50[d]	
Vinyl chloride	0[a]	2[b]	
Xylenes (total) (9)	10,000[d]	10,000[d]	20
Inorganic chemicals			
Aluminum (2)			50–200[d]
Antimony	6[k]	6[k]	
Arsenic (8)	50[i]	50[i]	
Asbestos (fibers per liter)	7×10^6[d]	7×10^6[d]	
Barium	2000[e]	2000[e]	
Beryllium	4[k]	4[k]	
Cadmium	5[d]	5[d]	
Chromium	100[d]	100[d]	
Copper (7)	1,300[h]	1,300[h]	
Cyanide	200[k]	200[k]	
Fluoride (8)	4,000[a]	4,000[a]	2,000[a]
Lead (7)	0[h]	15[h]	
Mercury	2[d]	2[d]	
Nickel	100[k]	100[k]	
Nitrate (as N) (3)	10,000[d]	10,000[d]	
Nitrite (as N) (3)	1,000[d]	1,000[d]	
Selenium	50[d]	50[d]	
Silver			100[d]
Sulfate (4)	$4 \times 10^5 - 5 \times 10^5$[f]	$4 \times 10^5 - 5 \times 10^5$[f]	
Thallium	5[k]	5[k]	
Microbiological parameters			
Giardia lamblia	0 organisms[c]		
Legionella	0 organisms[c]		
Heterotrophic bacteria	0 organisms[c]		
Viruses	0 organisms[c]		

TABLE 11.2 *continued*

Chemical	MCLG (μg/L)	MCL (μg/L)	SMCL (μg/L)
Radionuclides			
Radium 226 (6)	0[g]	20 pCi/L[g]	
Radium 228 (6)	0[g]	20 pCi/L[g]	
Radon 222	0[g]	300 pCi/L[g]	
Uranium	0[g]	20 μg/L (30 pCi/L)[g]	
Beta and Photon emitters (excluding radium 228)	0[g]	4 mrem ede/yr[g]	
Adjusted gross alpha emitters (excluding radium 226, uranium, and radon 222)	0[g]	15 pCi/L[g]	

Note: A pCi (picocurie) is a measure of the rate of radioactive disintegrations. Mrem ede/yr is a measure of the dose of radiation received by either the whole body or a single organ.
1. This is a chemical used in treatment of drinking water supply. The USEPA specifies how much may be used in the treatment process.
2. Dual numbers were proposed for aluminum because it is a constituent of a chemical used in the treatment of drinking water and it might not be possible for all treatment systems to meet the lower limit.
3. The total of nitrate plus nitrite cannot exceed 10 mg/L.
4. The proposed rule has two levels being considered.
5. The establishment of MCLGs and MCLs is not required by the Safe Drinking Water Act for these compounds; however, MCLGs and MCLs for them are being considered at the indicated levels.
6. This MCL would replace the current MCL of 5 pCi/L for combined 226 Ra and 228 Ra.
7. There is no MCL for copper and lead. The indicated values are proposed action levels that, under a complicated set of rules, would require treatment of a water supply to reduce potential corrosion of the water mains and pipes. The usual source of these compounds in public water supplies is primarily from the corrosion of copper and lead pipe and solder containing lead.
8. Standard under review as of January 1992.
9. SMCL is a suggested value only. Concentrations above this level may cause adverse taste. See *Federal Register*, January 30, 1991.
[a]Final value. Published in *Federal Register*, April 2, 1986.
[b]Final value. Published in *Federal Register*, July 8, 1987.
[c]Final value. Published in *Federal Register*, June 28, 1989.
[d]Final value. Published in *Federal Register*, January 30, 1991.
[e]Final value. Published in *Federal Register*, July 1, 1991.
[f]Proposed value. Published in *Federal Register*, July 25, 1990.
[g]Proposed value. Published in *Federal Register*, July 18, 1991.
[h]Final value. Published in *Federal Register*, July 7, 1991.
[i]Proposed value. Published in *Federal Register*, Nov. 13, 1985.
[j]Proposed value. Published in *Federal Register*, February, 1978.
[k]Final value. Published in *Federal Register*, July 17, 1992.

Drinking-water standards are especially important for evaluating ground-water quality because many consumers utilize untreated ground water that is pumped directly from a well. Public water-supply systems that rely upon ground water are required to perform a complete analysis of the water for the drinking-water standards prior to the time a well is put into service and periodically thereafter. Private wells are often tested for bacteria and nitrate only when they are first drilled and then never tested again. It is important to maintain high quality in ground water in order to protect private well owners. State water-quality standards for ground water are sometimes based on the drinking-water standards.

The European Community has established water-quality standards for member nations that apply to all water used for drinking or food processing from both private and public supplies. There are standards set for 66 water-quality parameters including elements that are naturally occurring in water, toxic elements, organic compounds, and microbiological parameters. The European Community standards for organic compounds are not as specific as the USEPA drinking water standards. For example, there is a blanket guide (similar to a MCLG) of 1 µg/L for all chlorinated organic compounds, with the exception of pesticides and PCBs, which have a maximum admissible concentration of 0.1 µg/L individually and 0.5 µg/L in total (Carney 1991).

11.3   COLLECTION OF WATER SAMPLES

The practicing hydrogeologist rarely will perform the chemical analysis of water samples; this is typically done in a specialized analytical laboratory. However, the hydrogeologist will usually be involved with the collection of water samples in the field. This section will focus on methods of collecting representative water samples for chemical analysis.

The program to sample both ground and surface water must be carefully planned. There are four basic steps involved:

1.   Determination of the purpose of the sampling program. Is the objective to define the basic water chemistry, to determine if the water meets drinking-water standards, or to determine if there is contamination present? Will surface water, water in the vadose zone, or ground water be tested?

2.   Deciding how many sampling points will be tested. Will all possible points be tested, or will only selected sampling points be involved? Will new sampling points, such as ground-water monitoring wells, be needed?

3.   Determining which chemical constituents will be analyzed and the **quantification limits** that the lab will employ. Analytical instruments have a lower limit to the range in which the results can be quantified and below that a range where a compound can be detected but not quantified. Results can be expressed as detected, but not quantified, or as not detected. In some cases, the detection limit will be more sensitive for some instrumental methods than others for the same compound. The quantification limits selected should be based on the purpose of the sampling program.

4.   Development of a *quality assurance/quality control (QA/QC)* program. There are many aspects of QA/QC, the purpose of which is to ensure that the analytical results reported by the laboratory accurately express the actual concentrations of the solutes in the water as it existed in the field (Keith et al. 1983a, 1983b; Kirchmer 1983). This is not a trivial problem. It is beyond the scope of this book to discuss laboratory methods of QA/QC. The hydrogeologist has two basic methods of checking on the *accuracy* and *precision* of the laboratory. Accuracy—the ability of the laboratory to report what is in the sample—can be

measured by the use of **spiked samples,** where a set of samples with a known concentration of a solute is submitted to the lab. Precision—the ability of the laboratory to reproduce results—is determined by submitting *duplicate samples* from the same source. The duplicate sample should be thoroughly mixed before being split for shipment to the lab. A field sampling program should have duplicates submitted to the lab for 10 percent of the samples, and they should be submitted as *blind duplicates* so the lab doesn't know which samples are duplicates. **Field blanks** are used to assess the field sampling program. Highly purified water (HPLC grade) is taken into the field in a sealed container, run through the field sampling devices, placed into sample containers, and shipped to the laboratory for analysis. If trace amounts of solutes are reported in both the samples and the field blanks, then they can be assumed to be a result of the sampling or lab methodology and are not actually present in the water in the field.

Sampling protocols have been developed by the USEPA (Ford, Turina, & Seely 1983). They specify the type of sample that is needed (grab or composite), the type of container that is to be used for the sample, the method by which the sample container is cleaned and prepared, whether or not the sample is filtered, the type of preservative that is to be added to the sample in the field, and the maximum of time the sample can be held prior to analysis in the laboratory.

It is good field practice to clean thoroughly the sampling device prior to use. The method of cleaning should be such that no residue remains. For example, equipment rinsed in acetone should first be rinsed thoroughly with distilled water after the acetone has been used and then autoclaved to volatilize any acetone that may not have been rinsed away. Frequently the analytical methods employed will detect acetone in concentrations of a few parts per billion. The sampling devices and bottles should be rinsed with a sample of the water being sampled if they are not thoroughly dry. This will prevent the mixing of rinse water with the final sample.

11.4   GROUND-WATER MONITORING

11.4.1   Planning a Ground-Water Monitoring Program

The science of ground-water sampling has advanced greatly in recent years, not only in our understanding of the techniques to be used, but in the development of materials and equipment used in the sampling process (Fetter 1993). The first step in designing a ground-water monitoring program is to determine the purpose. There are at least four major reasons to monitor ground water: to determine the water quality and chemistry of a region, to determine the water quality and chemistry of a specific water-supply well or well field, to determine the extent of ground-water contamination from a known source, and to monitor a potential source of contamination to determine if the ground water becomes contaminated.

If the purpose of a study is to evaluate the existing water quality and chemistry of a region, then it is likely that only existing wells and springs would be sampled. In this event, it is necessary to know the construction details of the well and pump. One needs to know what aquifer the well is drawing water from in order to interpret the chemical analyses. Wells that tap more than one aquifer should not be used in these studies, which are usually designed to map the distribution of ions in a specific aquifer. A good geographical distribution of wells is needed. In studies of chemical hydrogeology, wells located in both recharge and discharge areas should be sampled. One must be careful that the water sample collected has not been altered by the well system. A common occurrence with existing private wells is that all of the water in the system passes through a water softener, which changes the water chemistry. Some older-model submersible pumps have capacitors that, if they leak, can introduce polychlorinated biphenols (PCBs) into the well water.

If the water quality of a potential well site is being investigated, nearby wells may be sampled to establish regional water quality. In most cases it is desirable to construct a test well before a permanent well is installed. As the test boring is being advanced, temporary test wells and screens can be installed at progressively deeper depths to sample multiple aquifers or potential water-yielding zones in the same aquifer. At each screen zone the drill column and bit are removed from the hole and a temporary casing and well screen are installed. The temporary well is pumped to develop it until the water is clear. Replicate water samples are then collected and analyzed. The hydraulic and water-quality results of the various tested zones are compared and the best one is selected for the permanent well construction. It has been shown that water samples from such temporary test wells are very similar to those from high-capacity municipal wells later constructed in the same aquifer (Fetter 1975). If there is only one potential aquifer zone, a small-capacity test well may be installed on a permanent basis. This well can later be used as an observation well for a pumping test on the permanent well and may be useful as a standby well for emergency use.

11.4.2 Installing Ground-Water Monitoring Wells

Most of the *ground-water monitoring wells* installed today are for the purpose of determining ground-water quality at localities such as waste-storage or waste-disposal facilities, underground storage tanks, mines, and areas of known or suspected ground-water contamination. Ground-water monitoring wells are installed for the specific purpose of determining the quality of the ground water in a specific aquifer and at a particular location.

The design of a typical ground-water monitoring well is given in Figure 11.2. There are a number of specific steps that need to be followed in order to properly install a ground-water monitoring well. One important consideration is to take care during the installation of the well that contaminating materials are not introduced into the ground (Fetter 1983; Hix 1992).

1.   A **boring** is advanced into the ground by means of a drilling rig. The drilling method needs to be selected on the basis of the local geology, size and depth of the monitoring well to be installed, and available expertise. Borings in unconsolidated materials are frequently installed by means of **hollow-stem augers,** which are rotated into the ground to bring up soil and create the boring (Figure 11.3). A plug at the bottom of the hollow stem keeps soil from going up into the interior of the stem (Hackett 1987, 1988; Keeley & Boateng 1987).

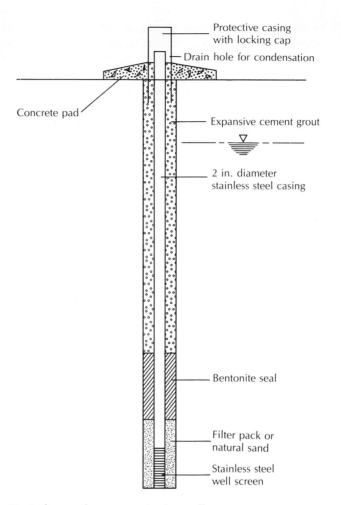

FIGURE 11.2   Typical ground-water monitoring well.

2.   Samples of the geologic materials are then collected so that a *geologic log* can be constructed. A **split-spoon sampler,** a hollow tube comprised of two halves, or a **Shelby tube,** a one-piece hollow tube, may be driven into the soil ahead of the *bit,* or cutting edge, at the bottom of the hollow-stem auger by temporarily removing the end plug. The sampling tube is retrieved and the soil sample is ejected and described by the geologist. It is also available for further testing, such as grain-size analysis.

3.   Once the auger is at the desired final depth of the bottom of the monitoring well, the plug at the end is removed. A *knockout plug* made of some noncontaminating material such as stainless steel may also be used if soil samples are not being collected as the auger is being advanced.

4.   The well consists of a **casing** and a **screen.** The casing is a piece of solid-wall pipe and the screen is a piece of pipe with holes, slots, gauze, or a continuous wire wrapped around it. The purpose of the screen is to allow water

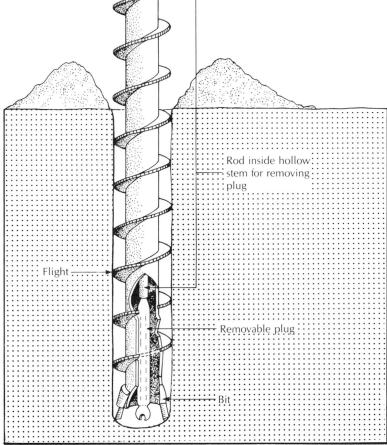

FIGURE 11.3 Hollow-stem auger drilling. The hollow-stem, continuous-flight auger bores into soft soils, carrying the cuttings upward along the flights. When the desired depth is reached, the plug is removed from the bit and withdrawn from inside the hollow stem. A well point on a casing (1¼-in. or 2-in.) can then be inserted to the bottom of the hollow stem and the auger pulled out, leaving the small-diameter monitoring well in place. Source: M. L. Scalf et al., *Manual of Ground Water Sampling Procedures*, National Water Well Association, 1981.

to enter the well but to keep out the soil. A commercially manufactured screen should be used rather than a pipe with slots cut into it. The screen is attached to the end of the casing by means of threaded joints. Casings and screens made from several different materials are commercially available. The hydrogeologist must select a material on the basis of cost, durability, and potential reactivity with the water in the aquifer. Teflon is the most costly, least durable, but most inert material. Stainless steel is the most durable material; it has moderate cost and is also essentially inert. Rigid polyvinyl chloride (PVC) pipe, because of low cost, is frequently used. Only PVC pipe with threaded joints is acceptable; PVC pipe joined with solvent-welded joints should never be used because solvent can add organic contaminants to the water. Because of its structural weakness and high cost for most uses Teflon is inferior to stainless steel as well-casing material. Stainless steel may react unfavorably with acidic or saline ground water; rigid PVC casing with threaded joints would be superior under such conditions.

In addition to monitoring wells, there are *multilevel sampling devices,* which can be installed in a single borehole to sample ground water. When monitoring wells are used to sample ground water from different depths, each monitoring well must be installed in a separate borehole. Drilling costs can be reduced substantially by the use of multilevel sampling devices. Figure 11.4 shows the design of a multilevel device used in sandy soil with a shallow water table and primarily horizontal flow (Pickens et al. 1981). The device consists of a rigid PVC pipe inside of which are multiple tubes, each of which ends at a sampling port at a different depth. Such a device can be used to collect ground-water samples at elevations in the aquifer as close as one to two feet. Only a small sample of water is withdrawn, so that each sample represents water from a very small portion of the aquifer. As a result, a very detailed picture of the vertical distribution of ground-water contamination can be developed. The multilevel device is installed by augering a boring to the desired depth with hollow-stem augers. The PVC pipe is lowered through the hollow stem and, as the augers are withdrawn, the sand heaves into the annular space around the pipe and seals it off. One disadvantage of the device is that water levels cannot usually be measured.

A second type of multilevel device can be used in a borehole in bedrock or cohesive material such as dense glacial till. Figure 11.5 shows the design of such a device (Cherry & Johnson 1982), which is similar to the one shown in Figure 11.4. However, inflatable packers are located above and below each sampling port. In this case the device is lowered into an open borehole and each zone to be sampled is isolated by inflating the packers above and below the sampling port. The device is intended to be permanently installed so that multiple samples over time can be taken. If needed, it could be removed from a well after a study is completed and reused.

Once a monitoring well or other device is installed, it is necessary to develop the well. The purpose of **well development** is to remove any fine material that may be blocking the well screen or port. A secondary purpose may be to remove any water from the aquifer that was introduced during well construction and that may not be representative of the local ground-water quality. Well development is usually accomplished by surging the well, that is, making the water in the well flow into and out of the well screen for a period of time. The well is then usually pumped for a while to remove the loose sediment from the well casing and screen. In order to maintain the integrity of the well for water-quality sampling, it is usually not good practice to add any water to the well during well development.

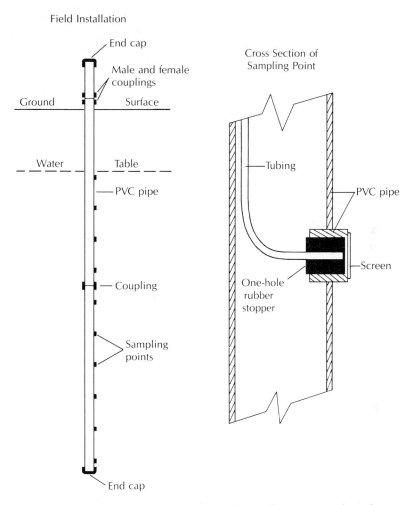

FIGURE 11.4 Multilevel ground-water sampling device for use in sandy soil. Source: J. F. Pickens et al., *Ground Water Monitoring Review* 1, no. 1 (1981): 48–51.

11.4.3 Withdrawing Water Samples from Monitoring Wells

Once the monitoring well is installed and developed, a method of removing the water from the well must be selected. There are a number of different ways to pump water from a monitoring well (Barcelona et al. 1984; Slawson, Kelly, & Everett 1982; Nielsen & Yeates, 1985). There are a large number of different devices commercially available for this task. The basic considerations in selecting a pumping device will be (1) does it collect a representative sample, (2) can it be easily cleaned and decontaminated if it is to be used in more than one well, (3) will it work in the application that is at hand, (4) can it lift the water from the water level in the well to the surface, (5) can it pump the well at a rate sufficient to purge it prior to sampling, (6) will the method of pumping or the materials from which

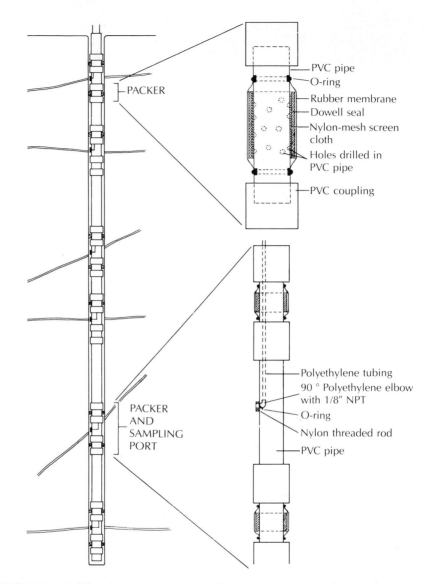

FIGURE 11.5 Multilevel ground-water sampling device for use in fractured rock borehole. Source: J. A. Cherry & P. E. Johnson, *Ground Water Monitoring Review* 2, no. 3 (1982): 41–44.

the pump is made change the water chemistry of the sample, (7) how easy is it to use the device, (8) how reliable is the device, and (9) how much does it cost to buy, maintain, and operate?

In selecting a sampling device it is important to pick one that will not alter the chemistry of the sample as it is brought to the surface. This could occur if the materials from which the pump is constructed would either leach compounds into the sample or absorb compounds from the sample. One of the first considerations will be the selection of the type of material used in the construction of the pumping device. Teflon and stainless steel are inert materials that could be used for the rigid parts of the pump; Teflon and polypropylene are inert materials that could be used for the flexible parts of the pump. Polyvinyl chloride may be acceptable for some uses but is not as inert as the other materials.

Changes in the pressure of the sample while it is being transported to the surface can cause loss of dissolved gases. This can result in a change in pH due to change in carbon dioxide, a change in Eh due to a change in oxygen, and a change in the dissolved volatile organic compounds due to a drop in pressure. The best sampling devices will not put the sample in contact with air or noninert gasses as it is brought to the surface and will maintain the sample under positive pressure. These considerations limit the usefulness of several sampling devices, such as suction pumps, peristaltic pumps, and air-lift pumps.

Bailers are tubes that can be made of any material and that have a check valve on the bottom. They are inexpensive and simple to operate as they can be lowered into the well on a wire or cord and yield a representative water sample if used carefully. **Bladder pumps** are positive-displacement devices that use a pulse of gas to push the sample to the surface. The gas does not come into contact with the sample and positive pressure is maintained at all times. The bladder pump yields a very representative water sample, provided the correct materials are used in its construction. The bladder pump is superior in performance to the bailer but costs more than an order of magnitude more.

One concern in sampling a monitoring well is to be assured that the sample does not contain any water that was standing in the well casing. One way to ensure this is to purge the well before pumping. Depending upon the design of the well and the type of pump, from one to five times the volume of water standing in the well casing and the screen should be removed prior to withdrawing the water sample. During the purging process it is good practice to monitor the pH and conductivity of the water until a stable condition is reached. It is also good practice to purge the well until any turbidity has cleared, although in some wells this may not be possible. If the well is slow to recharge, it is probably not a good idea to pump the well dry during the purging process. For such a well, one to two well volumes should be sufficient to purge any stagnant water.

Deep open-borehole wells in solid rock or mine shafts can present an opportunity to the hydrogeologist to obtain water-quality data at depths to which it might be uneconomical to construct dedicated monitoring wells. If the well is not pumped for some time and the borehole is not acting as a conduit for water flowing from one aquifer to another, the water quality at various levels in the borehole can reflect the ground-water quality at the same level. **Borehole geochemical probes** are water-quality monitoring devices that can be lowered into the well on a cable to measure such parameters as pH, Eh, temperature, and specific conductance. By lowering the probe and taking readings at discrete intervals, geochemical well logs can be constructed. In addition, the **Kemmerer sampler,** a sample collection device developed for drawing water samples from depths in lakes and oceans, can be used in deep wells. The Kemmerer sampler is lowered into the borehole in an open position until the desired depth is reached. At that point a weight is sent down on the cable, triggering a spring-loaded device on the sampler to close it. The closed Kemmerer sampler is then raised to the surface by the cable and the water sample is withdrawn. The Kemmerer sampler can be used to collect water samples at different depths so that a geochemical log can be developed.

A modification of the medical **syringe** can be used to draw a water sample from a specific depth in a monitoring well. The syringe is lowered into the well on the end of a length of tubing. A weight may be needed to sink the syringe and tubing. At the desired depth, a vacuum is applied to the tubing and the pressure of the water in the well will force the syringe plunger up, forcing a water sample into the syringe. The syringe can then be raised up by the tubing. The advantages of the syringe are that the sample does not undergo a pressure change or exposure to the atmosphere. The syringe can even be used as the sample storage container. However, the sample volume is small, and the syringe cannot be used to purge the well. If the syringe is used for sample storage, the chance of cross-contamination is very small if the tubing and other appurtenances are carefully cleaned between wells (Nielsen & Yeates 1985).

11.5   VADOSE-ZONE MONITORING

If a contaminant is detected in the ground water, it is an indication that a problem is at hand. Solutions to the problem may be difficult and expensive. It would be desirable to be able to determine if ground-water contamination is likely to occur at some time in the future so that corrective action could be taken early to prevent contamination from reaching the water table. For example, when a landfill or a lagoon is constructed, a ground-water monitoring system might be installed to detect leaks from the facilities. Storage facilities for toxic and hazardous materials can also be monitored for leaks. However, if a leak were to be detected prior to the time that the contaminant reaches the water table, corrective action, such as draining a lagoon and repairing the liner, might be possible before the ground water becomes contaminated. Leaks can potentially be detected by monitoring the water in the vadose (unsaturated) zone (Everett 1981; Johnson, Cartwright, & Schuller 1981; Everett et al. 1984; Robbins and Gemmell 1985; Wilson 1983, 1990; Robbins et al. 1990a, 1990b).

Contaminants in the unsaturated zone can move in both the liquid and vapor phases. Vapors can move in any direction depending upon their relative density and air-pressure gradients. Liquids can generally move downward only. Vadose-zone monitoring systems for liquids must be constructed beneath the facility to be monitored as that is the direction the liquid would be seeping. Therefore, such devices are usually installed prior to the time that the facility is constructed. It is difficult to install such devices after construction; for example, in order to put a collection device beneath an active landfill, either a vertical boring must be made through the landfill or an angular boring must be made from the side. Because of the fact that vapors can migrate laterally, gas-monitoring wells can be placed to the side of active facilities. Hence it is easier to retrofit a site with gas-monitoring devices than liquid-monitoring devices in the vadose zone.

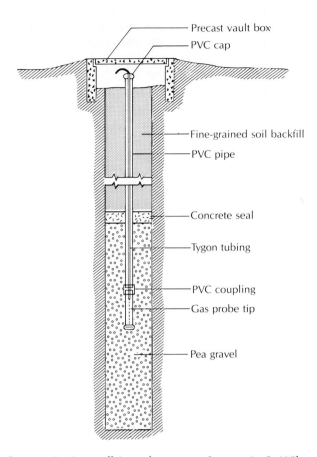

FIGURE 11.6 Gas monitoring well in vadose zone. Source: L. S. Wilson, *Ground Water Monitoring Review* 3, no. 1 (1983): 155–66.

Vapors can be detected by *gas-monitoring wells*, which are simply wells that terminate in the unsaturated zone (Figure 11.6). Gas-monitoring wells can be used to detect the movement of methane from municipal landfills. Methane has been implicated in explosions near landfills. Gas-monitoring wells can also detect the vapors of volatile organic chemicals in the vadose zone. These compounds move rapidly in ground water and many are known or suspected carcinogens. Gas-monitoring wells can be located beneath a landfill or lagoon if they are installed before construction. After construction, it is easier to place them to the side of an active facility.

The vadose zone can be monitored by indirect means. For example, a lagoon that holds aqueous liquids could be monitored for changes in soil moisture. Changes in soil moisture could mean a leak. If the lagoon holds a brine, changes in the electrical conductivity of resistivity blocks buried in the soil beneath the landfill could mean a leak. Temperature changes in the vadose zone could also indicate a leak.

The vadose zone can also be sampled by direct sampling of the soil. Soil samples can be removed by shovel, by a sampling tube being pushed into the ground, or by augering a borehole and then pushing a sampling tube ahead of the auger. Soil samples can be tested for moisture content, the moisture can be extracted and tested for chemical composition, or the soil sample can be leached with a standard leaching solution and the leachate tested. If the soil sample is placed in a vial for volatile organic compound testing within a few seconds of being exposed to the atmosphere, most of the volatile organics in the soil will be retained and testing of the volatile organics in the headspace in the vial will give a good indication of the organics in the soil.

There are a number of devices that can be placed in the vadose zone to collect a sample of the liquid in the pores. These fall into two categories: (1) **suction lysimeters,** which are devices that apply tension to a porous ceramic cup to induce the pore water that is under negative pressure to enter the porous cup, whereby it can be transported to the surface, and (2) **collection lysimeters,** which have a membrane liner buried in the vadose zone so that liquid collects on the liner until the soil above it becomes saturated, forcing the liquid to flow to a wet well, where it can be sampled.

The suction lysimeter has been used for many years in agricultural applications. It has been used in landfills to monitor for leachate with mixed success. The vacuum that can be applied is limited to 0.5 to 0.8 bar. Consequently, in some applications, the lysimeter fails to collect liquids, probably because it becomes clogged. It also is somewhat limited in that it samples liquids from only a very small portion of the aquifer. If it doesn't happen to be located under a leak, that leak may go undetected.

Collection lysimeters are less prone to failure and have the added advantage that they can be made any size; they can, therefore, be placed under spots where failure of a system component is more likely, such as a seam. Figure 11.7

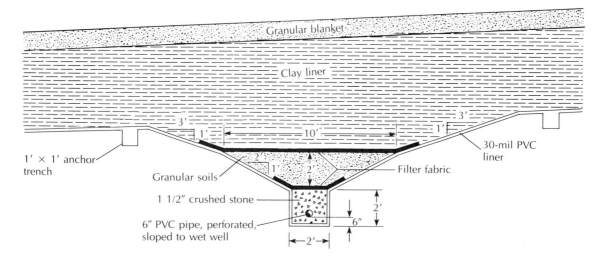

FIGURE 11.7   Cross section of leachate-collection lysimeter beneath clay liner of a landfill.

shows a collection lysimeter beneath a portion of a landfill. One landfill in Wisconsin was located over an aquifer where it appeared that ground-water monitoring would be very difficult to do reliably. The landfill was designed with a collection lysimeter beneath the entire landfill to serve as the primary means of detecting leaks in the clay liner of the landfill. Obviously, such a system must be installed as the facility is being constructed.

11.6   MASS TRANSPORT OF SOLUTES

11.6.1   Introduction

In studying ground-water contamination it is helpful to understand the basic theory behind the movement of solutes contained in ground water. In the study of water chemistry, the processes by which substances can become dissolved in water are examined. However, the processes by which these substances move through porous media are complex. They can be expressed mathematically, although in some instances we do not fully understand how to obtain the field data necessary to apply the theoretical equations.

There are two basic processes operating to transport solutes. **Diffusion** is the process by which both ionic and molecular species dissolved in water move from areas of higher concentration (i.e., chemical activity) to areas of lower concentration. **Advection** is the process by which moving ground water carries with it dissolved solutes. We will see how, as solutes are carried through porous media, the process of **dispersion** acts to dilute the solute and lower its concentration. Finally, there are chemical and physical processes that cause **retardation** of solute movement so that it may not move as fast as the advection rate would indicate.

11.6.2   Diffusion

The diffusion of a solute through water is described by Fick's laws. Fick's first law describes the flux of a solute under steady-state conditions:

$$F = -D \, dC/dx \qquad\qquad (11-1)$$

where

F = mass flux of solute per unit area per unit time
D = diffusion coefficient (area/time)
C = solute concentration (mass/volume)
dC/dx = concentration gradient (mass/volume/distance)

The negative sign indicates that the movement is from greater to lesser concentrations. Values for D are well known for electrolytes in water. For the major cations and anions in water, D ranges from 1×10^{-9} to 2×10^{-9} m^2/s.

For systems where the concentrations may be changing with time, Fick's second law may be applied:

$$\partial C/\partial t = D\ \partial^2 C/\partial x^2 \tag{11-2}$$

where $\partial C/\partial t$ = change in concentration with time.

Both Fick's first and second law as expressed above are for one-dimensional situations. For three-dimensional analysis, more general forms would be needed.

In porous media, diffusion cannot proceed as fast as it can in water because the ions must follow longer pathways as they travel around mineral grains. In addition, the diffusion can take place only through pore openings because mineral grains block many of the possible pathways. To take this into account, an effective diffusion coefficient must be used. This is termed D^*.

The value of D^* can be determined from the relationship

$$D^* = wD \tag{11-3}$$

where w is an empirical coefficient that is determined by laboratory experiments. For species that are not adsorbed onto the mineral surface it has been determined that w ranges from 0.5 to 0.01 (Freeze & Cherry 1979). Berner (1971) gave a nonempirical relationship between D^* and D that indicated that D^* was equal to D times the porosity divided by the square of the **tortuosity** of the flow path of the diffused species. Tortuosity is the actual length of the flow path, which is sinuous in form, divided by the straight-line distance between the ends of the flow path. Unfortunately, tortuosity cannot be determined in the field, and one is left with the experimental approach.

The process of diffusion is complicated by the fact that ions must maintain electrical neutrality as they diffuse. If we have a solution of NaCl, the Na^+ cannot diffuse faster than the Cl^-, unless there is some other negative ion in the region into which the Na^+ is diffusing.

It should also be mentioned at this point that if the solute is adsorbed onto the mineral surfaces of the porous medium, the net rate of diffusion will obviously be less than for a nonadsorbed species. This topic is addressed more fully in the section on retardation.

It is possible for solutes to move through a porous medium by diffusion, even though the ground water is not flowing. Thus, even if the hydraulic gradient is zero, a solute could still move. In rock and soil with very low permeability, the water may be moving very slowly. Under these conditions, diffusion might cause a solute to travel faster than the ground water is flowing. Under such conditions, diffusion is more important than advection.

11.6.3   Advection

The rate of flowing ground water can be determined from Darcy's law as

$$v_x = \frac{K}{n_e} \frac{dh}{dl} \qquad\qquad (11\text{--}4)$$

where

v_x = average linear velocity

K = hydraulic conductivity

n_e = effective porosity

$\dfrac{dh}{dl}$ = hydraulic gradient

Contaminants that are advecting are traveling at the same rate as the average linear velocity of the ground water.

11.6.4 Mechanical Dispersion

As a contaminated fluid flows through a porous medium, it will mix with noncontaminated water. The result will be a dilution of the contaminant by a process known as dispersion. The mixing that occurs along the streamline of fluid flow is called longitudinal dispersion. Dispersion that occurs normal to the pathway of fluid flow is lateral dispersion.

There are three basic causes of pore-scale longitudinal dispersion: (1) As fluid moves through pores, it will move faster through the center of the pore than along the edges. (2) Some of the fluid will travel in longer pathways than other fluid. (3) Fluid that travels through larger pores will travel faster than fluid moving in smaller pores. This is illustrated by Figure 11.8.

Lateral dispersion is caused by the fact that, as a fluid containing a contaminant flows through a porous medium, the flow paths can split and branch out to the side. This will occur even in the laminar flow conditions that are prevalent in ground-water flow (Figure 11.9).

The mechanical dispersion due to the preceding factors is equal to the product of the average linear velocity and a factor called the dynamic dispersivity (a_L).

$$\text{Mechanical dispersion} = a_L v_x \qquad\qquad (11\text{--}5)$$

11.6.5 Hydrodynamic Dispersion

The processes of molecular diffusion and mechanical dispersivity cannot be separated in flowing ground water. Instead, a factor termed the coefficient of hydrodynamic dispersion, D_L, is introduced. It takes into account both the mechanical mixing and diffusion. For one-dimensional flow it is represented by the following equation:

$$D_L = a_L v_x + D^* \qquad\qquad (11\text{--}6)$$

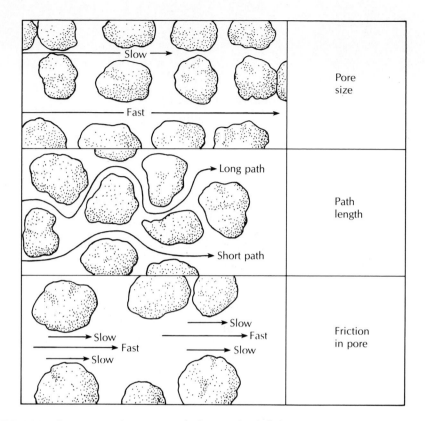

FIGURE 11.8   Factors causing pore-scale longitudinal dispersion.

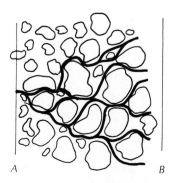

FIGURE 11.9   Flow paths in a porous medium that cause lateral hydrodynamic dispersion.

where

D_L = the longitudinal coefficient of hydrodynamic dispersion

a_L = the dynamic dispersivity

v_x = the average linear ground-water velocity

$D*$ = the molecular diffusion

The process of longitudinal hydrodynamic dispersion can be illustrated by the following simple experiment. A tube is filled with sand and set up so that distilled water is flowing through it at a constant rate. We then change the influent to a 1% saline solution and begin to monitor the effluent for chloride. The effluent has zero chloride initially, as distilled water is still flushing from the tube. Eventually we will begin to detect chloride in the effluent water. It arrives initially at a very low concentration. The "breakthrough" is not at the 1% concentration. This small amount gradually increases until the 1% saline concentration is reached. The first chloride ions to arrive traveled through the shortest flow paths. Diffusion in advance of the advecting water may have even caused some of the chloride to reach the outlet prior to water that was advecting it. The initial chloride was being diluted by the distilled water that was arriving at the same time. The amount of distilled water available for dilution continually decreased until the saline solution filled all the pores and the effluent water was at the influent concentration. Figure 11.10 illustrates this process.

The one-dimensional equation for hydrodynamic dispersion (Beruch & Street 1967; Hoopes & Harleman 1967) is given by

$$D_L \frac{\partial^2 C}{\partial x^2} - v_x \frac{\partial C}{\partial x} = \frac{\partial C}{\partial t} \qquad \textbf{(11–7)}$$

where

D_L is the coefficient of longitudinal hydrodynamic dispersion

C is the solute concentration

v_x is the average ground-water velocity in the x-direction

t is the time since start of solute invasion

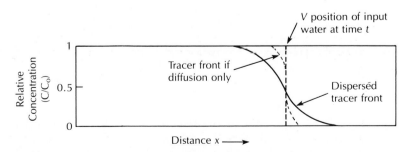

FIGURE 11.10   Influence of dispersion and diffusion on "breakthrough" of a solute.

The concentration, C, at some distance, L, from the source at concentration, C_0, at time, t, is given by the following expression (Ogata 1970), where erfc is the complementary error function:

$$C = \frac{C_0}{2}\left[\operatorname{erfc}\left(\frac{L - v_x t}{2\sqrt{D_L t}}\right) + \exp\left(\frac{v_x L}{D_L}\right)\operatorname{erfc}\left(\frac{L + v_x t}{2\sqrt{D_L t}}\right)\right]$$

(11–8)

where

C is the solute concentration (M/L^3, mg/L)

C_0 is the initial solute concentration (M/L^3, mg/L)

L is the flow path length (L; ft or m)

v_x is the average linear ground water velocity (L/T; ft/day or m/day)

t is the time since release of the solute (T; day)

D_L is the longitudinal dispersion coefficient (L; ft or m)

The advection-dispersion equation is based on the premise that the center of mass of the solute is moving at the same rate as the average linear ground-water velocity. Furthermore it is assumed that hydrodynamic dispersion causes the solute to spread out both ahead of and behind the center of mass in a pattern that follows a statistically normal distribution. This is the familiar bell-shaped curve. The normal distribution is also sometimes called a Gaussian distribution. Figure 11.11 shows the movement of a solute mass and the way that it spreads with time and distance. The solute front is moving at a rate that is greater than would be predicted by the average linear ground-water velocity.

Mechanical dispersion is also caused by the heterogeneities in the aquifer. As ground-water flow proceeds in an aquifer, regions of greater than average hydraulic conductivity and lesser than average conductivity will be encountered. The resulting variation in linear ground-water velocity results in much greater hydrodynamic dispersion than that caused by the pore-scale effects. The greater

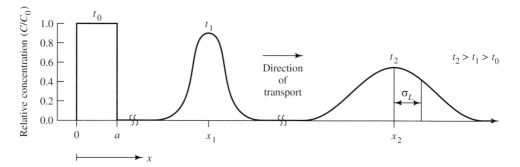

FIGURE 11.11 Transport and spreading of a solute slug with time due to advection and dispersion. A slug of solute was injected at $x = 0 + a$ at time t_0 with a resulting concentration of C_0. The ground-water flow is to the right. Source: C. W. Fetter, *Contaminant Hydrogeology*. New York: Macmillan Publishing Company, 1993.

the distance over which dispersivity is measured, the greater the value that is observed. This has been called the *scale effect* (Anderson 1979). Pore-scale dispersion measured in the laboratory is on the order of centimeters, whereas macrodispersion measured in the field is on the order of meters. Neuman (1990) made a study of the relationship between the apparent longitudinal dynamic dispersivity and the flow length. He found that for flow paths less than 3500 m long, these could be related by the equation:

$$a_L = 0.0175L^{1.46} \tag{11-9}$$

where

a_L is the apparent longitudinal dynamic dispersivity (L; ft or m)

L is the length of the flow path (L; ft or m)

Because of hydrodynamic dispersion, the concentration of a solute will decrease with distance from the source. The solute will spread in the direction of ground-water movement more than it will in the direction perpendicular to the flow. This is because the longitudinal dispersivity is greater than the lateral dispersivity. A continuous source will yield a plume, whereas a spill will yield a slug that grows with time as it moves down the ground-water flow path. This is illustrated by Figure 11.12.

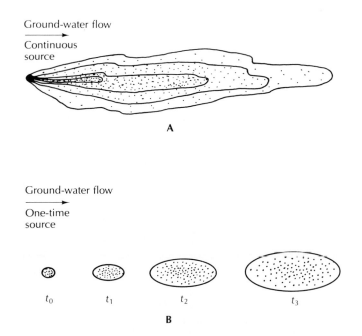

FIGURE 11.12 A. The development of a contamination plume from a continuous point source. **B.** The travel of a contaminant slug from a one-time point source. Density of dots indicates solute concentration.

Heterogeneities in the aquifer can cause the pattern of the solute movement to vary from what one might expect in homogeneous beds. Because flowing ground water always follows the most permeable pathways, those pathways will also have the most contaminant.

EXAMPLE PROBLEM   A landfill is leaking leachate with a chloride concentration of 725 mg/L, which enters an aquifer with the following properties:

$$\text{Hydraulic conductivity} = 3.0 \times 10^{-3} \text{ cm/s } (3.0 \times 10^{-5} \text{ m/s})$$
$$dh/dl = 0.0020$$
$$\text{Effective porosity} = 0.23$$
$$D^* = 1 \times 10^{-9} \text{ m}^2/s \text{ (estimated)}$$

Compute the concentration of chloride in 1 y at a distance 15 m from the point where the leachate entered the ground water.

1. Determine average linear velocity.

$$v_x = \frac{K(dh/dl)}{n_e} = (3.0 \times 10^{-5} \text{ m/s} \times 0.002)/(0.23)$$
$$= 2.6 \times 10^{-7} \text{ m/s}$$

2. Determine the coefficient of longitudinal hydrodynamic dispersion.
   a. Find the value of a_L from Equation 11–9.

$$a_L = 0.0175L^{1.46}$$
$$= 0.0175 (15 \text{ m})^{1.46}$$
$$= 0.91 \text{ m}$$

   b. Find the value of D_L.

$$D_L = a_L \times v_x + D^*$$
$$= (0.91 \text{ m} \times 2.6 \times 10^{-7} \text{ m/s}) + 1 \times 10^{-9} \text{ m}^2/s$$
$$= 2.4 \times 10^{-7} \text{ m}^2/s$$

3. Restate the 1-y time of travel in seconds.

$$t = 1 \text{ y} \times 60 \text{ s/min} \times 1440 \text{ min/day} \times 365 \text{ days/y}$$
$$= 3.15 \times 10^7 \text{ s}$$

4. Substitute values into Equation 11–8.

$$C_0 = 725 \text{ mg/L}$$
$$L = 15 \text{ m}$$
$$t = 3.15 \times 10^7 \text{ s}$$
$$D_L = 2.4 \times 10^{-7} \text{ m}^2/\text{s}$$
$$v_x = 2.6 \times 10^{-7} \text{ m/s}$$

$$
C = \frac{725}{2} \text{ erfc} \left(\frac{15 \text{ m} - (2.6 \times 10^{-7} \text{ m/s} \times 3.15 \times 10^7 \text{ s})}{2 \times [2.4 \times 10^{-7} \text{ m}^2/\text{s} \times 3.15 \times 10^7 \text{ s}]^{.5}} \right)
$$
$$
+ \left[\exp \left(\frac{2.6 \times 10^{-7} \text{ m/s} \times 15 \text{ m}}{2.4 \times 10^{-7} \text{ m}^2/\text{s}} \right) \right.
$$
$$
\left. \times \text{ erfc} \left(\frac{(15 \text{ m} + (2.6 \times 10^{-7} \text{ m/s} \times 3.15 \times 10^7 \text{ s}))}{2 \times [2.4 \times 10^{-7} \text{ m/s} \times 3.15 \times 10^7 \text{ s}]^{.5}} \right) \right] \text{ mg/L}
$$
$$
= 362.5 \left[\text{erfc} \left(\frac{15 \text{ m} - 8.19 \text{ m}}{5.5 \text{ m}} \right) + \exp(16.25) \times \text{erfc} \left(\frac{15 \text{ m} + 8.19 \text{ m}}{5.5 \text{ m}} \right) \right] \text{ mg/L}
$$
$$
= 362.5 \left[\text{erfc}(1.24) + \exp(16.25) \times \text{erfc}(4.22) \right] \text{ mg/L}
$$

The complementary error function can be determined from Appendix 13. Since the complementary error function of numbers greater than 3 is infinitesimally small, we may ignore the second term of the equation.

$$C = 362.5 \times 0.083 \text{ mg/L}$$
$$= 30 \text{ mg/L}$$

11.6.6 Retardation

We can consider solutes in two broad classes: conservative and reactive. Conservative solutes do not react with the soil and/or native ground water or undergo biological or radioactive decay. The chloride ion is a good example of a conservative solute.

The surfaces of solids, especially clays, have an electrical charge due to isomorphous replacement, broken bonds, or lattice defects. The electrical charge is imbalanced and may be satisfied by adsorbing a charged ion. The **adsorption** may be relatively weak, essentially a physical process caused by van der Waals forces. It may be stronger if chemical bonding occurs between the surface and the ion. Clays tend to be strong adsorbers, since they have both a high surface area per unit volume and significant electrical charges at the surface.

Most clay minerals have an excess of imbalanced negative charges in the crystal lattice. Adsorptive processes in soils thus favor the adsorption of cations. Divalent cations are usually more strongly adsorbed than monovalent ions. Some positively charged sites exist, but they are not as abundant as negative sites. In addition, some common negatively charged ions, such as HCO_3^-, SO_4^{2-}, and

NO_3^-, are too large to be effectively adsorbed. Chloride ions are larger than the common cations.

Laboratory studies of adsorption involve the use of a specific soil and given solutes. In one approach, batches of solutions of the ionic substances are prepared in differing concentrations. For each batch a preweighed mass of dry soil is mixed with measured portions of the solutions, each with a different initial concentration. In a second approach, different amounts of dry adsorbent are mixed with portions of the same solution, each with the same concentration. After equilibrium between the soil and the solution has been determined, the equilibrium concentration of the solute in each batch is determined. The amount of solute adsorbed by the soil will be in direct proportion to the chemical activity of the solute. The amount adsorbed onto the soil can be calculated as the difference in mass of the solute in solution before the test and in solution at equilibrium. As a result of the test for each batch, one knows the mass adsorbed per unit weight of soil, C^*, as a function of the equilibrium concentration of solute remaining in solution, C. A graphical plot of C as a function of C^* is known as an *adsorption isotherm*.

The adsorption characteristics are typically plotted on graph paper by showing the mass of solute adsorbed per unit mass of soil as a function of the concentration of the solute. Figure 11.13 shows an adsorption isotherm for lead. While either the Langmuir or Freundlich isotherm equation could be used to describe this relationship, the Freundlich equation is of special utility as it can easily be applied to retardation studies.

When an adsorption relationship can be plotted as a straight line on log-log paper, it is described by the Freundlich isotherm as

$$\log C^* = b \log C + \log K_f \qquad (11-10)$$

or

$$C^* = K_f C^j \qquad (11-11)$$

where

C^* = mass of solute sorbed per bulk unit dry mass of soil

C = solute concentration

K_f, j = coefficients

The slope of the curve on the log-log paper is represented by j. In a plot of C^* versus C, where the slope is a straight line, the relationship is linear, and b has a value of one. Under these conditions, the derivative of C^* with respect to C yields the relationship

$$dC^*/dC = K_d \qquad (11-12)$$

where K_d is known as a **distribution coefficient.**

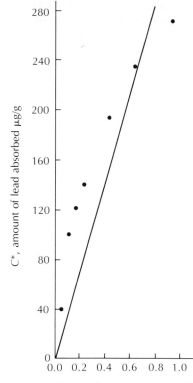

C, equilibrium lead concentration (mg/L)

FIGURE 11.13 Lead adsorption by Cecil clay loam at pH 4.5 and at 25°C described by a linear Freundlich equation through the origin. Source: W. R. Roy, I. G. Krapac, S. F. J. Chou, & R. A. Griffin, *Batch-Type Adsorption for Estimating Soil Attenuation of Chemicals.* Technical Resource Document, EPA/530-SW-87-006, 1987.

A second isotherm may be determined by plotting C/C^* versus C on arithmetic graph paper. If this falls on a straight line, it is the **Langmuir adsorption isotherm** (Olsen & Watanabe 1957) and is given by

$$\frac{C}{C^*} = \frac{1}{\beta_1 \beta_2} + \frac{C}{\beta_2} \qquad (11-13)$$

where

C is the equilibrium concentration of the ion in contact with the soil (mg/L)

C^* is the amount of the ion adsorbed per unit weight of soil (mg/g)

β_1 is an adsorption constant related to the binding energy

β_2 is the adsorption maximum for the soil

A plot of C/C^* as a function of C is made on rectilinear scales. The data points will fall on a straight line. Some experiments yield two straight-line segments—one at lower concentrations of the ion and one at higher concentrations with a lower slope (Syers et al. 1973). In some cases, the soil may have adsorbed some of the ion under natural conditions prior to the laboratory test. If this is the case, a correction must be made (Fitter & Sutton 1975).

The maximum ion adsorption, β_2, is the reciprocal of the slope of the straight line. The binding energy constant, β_1, is the slope of the line divided by the intercept. The experimental procedure usually tests for only a single ion at a time. Natural waters are more complex, and field reactions may differ from those determined by laboratory study. The Langmuir adsorption isotherm can be used for both anions and cations.

The K_d value can be used to compute the retardation of the solute front as it passes through the soil by the following equation:

$$\text{Retardation factor} = 1 + (\rho_b/\theta)(K_d) \qquad \textbf{(11–14)}$$

where

ρ_b is dry bulk mass density of the soil (M/L^3; gm/cm^3)

θ is volumetric moisture content of the soil (dimensionless)

K_d is distribution coefficient for the solute with the soil (L^3/M; mL/g)

If a solute is reactive, it will travel at a slower rate than the ground water owing to adsorption. The rate of solute movement can be determined by the retardation equation

$$v_c = v_x/[1 + (\rho_b/\theta)(K_d)] \qquad \textbf{(11–15)}$$

where

v_x is average linear velocity (L/T; ft/day or m/day)

v_c is velocity of the solute front where the solute concentration is one-half of the original value ($C/C_0 = 0.5$) (L/T; ft/day or m/day)

Figure 11.14 shows the impact of retardation on the movement of a solute front compared with a nonretarded species. It can be seen that the effect of retardation is to cause the solute front to advance more slowly.

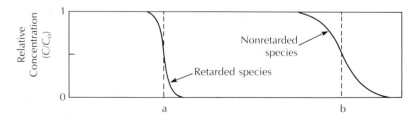

FIGURE 11.14   Influence of retardation on movement of a solute front.

EXAMPLE
PROBLEM

A calcareous glacial outwash sediment was present at the proposed site for an artificial recharge basin for wastewater (Fetter 1977a). The phosphorus-removal capacity of the soil was determined by laboratory adsorption studies. Equal weights of soil were shaken in various concentrations of disodium phosphate. The soil already had 0.016 mg of phosphorus per gram adsorbed prior to the test, and this was added to the value of C^* adsorbed during the test to determine total C^* in equilibrium with the solution. The equilibrium concentration of the solute is C.

A plot of C/C^* as a function of C is given in Figure 11.15. There are two straight-line segments, indicating that one type of adsorption is taking place at low activities of phosphorus and another type of adsorption, with a higher bonding energy, is occurring at greater concentrations. The slope of the straight line at lower concentrations is $\dfrac{20 \ (mg/L)/(mg/g)}{1 \ mg/L}$, so the reciprocal is 0.05 mg of phosphorus per gram of soil, which is the adsorption maximum (β_2). For higher concentrations, the adsorption maximum is 0.16 mg of phosphorus per gram of soil.

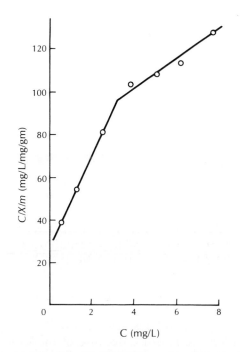

FIGURE 11.15 Langmuir adsorption isotherm for phosphorus adsorbed on calcareous glacial outwash. Source: C. W. Fetter, Jr., *Ground Water* 15 (1977): 365–71.

EXAMPLE PROBLEM

The following table shows the adsorption of varying amounts of lead by the Cecil clay. The aliquots of solution were each 200 mL and the amount of adsorbent in each case was 10.18 g. Find the value of K_d.

Initial Conc. (mg/L)	C Equilibrium Conc. (mg/L)	C* Amount Adsorbed (μg/g)
2.07	0.05	40
5.11	0.11	98
6.22	0.16	119
7.28	0.22	139
10.2	0.41	192
12.4	0.65	231
14.6	0.94	268

The value of C^*, the amount adsorbed, is found by taking the initial concentration, subtracting from that the equilibrium concentration, multiplying by the volume of the solution, and dividing the entire product by the weight of the adsorbent. In the first line of the above table, the computation is

$$C^* = \frac{(2.07 \text{ mg/L} - 0.05 \text{ mg/L}) \times 0.200 \, L}{10.18 \text{ g}} = 0.040 \text{ mg/g} = 40 \text{ μg/g}$$

Figure 11.13 shows the plot of the equilibrium concentration, C, versus the amount adsorbed, C^*, for the above data. The data can be reasonably approximated by a straight line drawn through the origin.

A statistical test called a linear regression can be used to find the straight-line equation that best fits the data set. Many scientific calculators have a built-in linear regression function. It can also be run on a personal computer with LOTUS 1-2-3, QUATTRO PRO, or EXCEL.

The straight-line equation, $y = mx + b$, where m is the slope of the line and b is the intercept with the y axis, simplifies to $y = mx$ when the intercept passes through the origin. In Figure 11.13 the data were forced through the origin; i.e., b was set to zero. The equation for the straight line in Figure 11.13 is $C^* = 232 \, C$ since the slope of the line, dC^*/dC, is 232. The value of K_d, the retardation coefficient, is thus 232 mL/g.

Data from W. R. Roy, I. G. Krapac, S. F. J. Chou, and R. A. Griffin, "Batch-type soil adsorption procedures for estimating soil attenuation of chemicals," U.S. Environmental Protection Agency, Technical Resource Document EPA/530-SW-87-006, 1987.

Potential solutes in ground water have a wide range of soil-solute specific distribution coefficients. Because of the large surface areas and numerous ion-exchange sites, clays will have the largest K_d values for specific inorganic solutes. Cations are more often strongly adsorbed than anions, and divalent cations will be adsorbed more readily than those of monovalent species. Substances such as chloride may be only very weakly adsorbed or even completely unattenuated by passage through a clay. Sodium is weakly attenuated; potassium, ammonia, magnesium, silicon, and iron are moderately attenuated; and lead, cadmium, mercury, and zinc can be strongly held.

Synthetic organic chemicals in solution can be adsorbed by the organic carbon in the soil. In determining the movement and retardation of organic chemicals when one does not have site-specific batch adsorption data, one must first consider the solubility of the organic compound in water. The relative tendency for an organic compound to remain dissolved in water rather than to be adsorbed onto soil organic carbon is related to the octanol-water partition coefficient of that chemical, which is the tendency for the chemical to be dissolved into either water or *n*-octanol when shaken in a solution of the two. The soil–water partition coefficient, K_{oc}, can be estimated from either the water solubility or the octanol-water partition coefficient. Table 11.3 gives the solubility and soil–water

TABLE 11.3 Solubility, K_{oc}, and mobility class for common organic pollutants

Compound	Solubility (ppm)	K_{oc} (mL/g)	Mobility Class
1,4-Dioxane	miscible	1	very high
4-hydroxy-4-methyl-2-pentanone	miscible	1	very high
acetone	miscible	1	very high
tetrahydrofuran	miscible	1	very high
N,N'-dimethylformamide		1	very high
N,N'-dimethylacetamide		2	very high
2-methyl-2-butanol	140000.	6	very high
2-butanol	125000.	6	very high
ethyl ether	84300.	8	very high
cyclohexanol	56700.	10	very high
3-methylbutanoic acid	42000.	12	very high
benzyl alcohol	40000.	12	very high
aniline	34000.	13	very high
2-hexanone (butylmethylketone)	35000.	14	very high
2-hydroxy-triethylamine		15	very high
2-methylphenol (o-cresol)	31000.	15	very high
2-methyl-2-propanol		16	very high
4-methylphenol (p-cresol)	24000.	17	very high
pentanoic acid	24000.	17	very high
cyclohexanone	23000.	18	very high
4-methyl-2-pentanone	19000.	20	very high
2,4-dimethyl phenol	17000.	21	very high
4-methyl-2-pentanol	17000.	21	very high
methylene chloride	13200.	25	very high
isophorone	12000.	26	very high
phenol	82000.	27	very high
2-chlorophenol	11087.	27	very high
hexanoic acid	11000.	28	very high
chloroform	7840	34	very high

TABLE 11.3 *continued*

Compound	Solubility (ppm)	K_{oc} (mL/g)	Mobility Class
1,2-dichloroethane	8450.	36	very high
1,2-trans-dichloroethene	6300.	39	very high
chloroethane	5700.	42	very high
5-methyl-2-hexanone	5400.	43	very high
chloromethane	5380.	43	very high
1,1-dichloroethane	5100.	45	very high
1,1,2-trichloroethane	4420.	49	very high
1,2-dichloropropane	3570.	51	high
benzoic acid	2900.	64	high
octanoic acid	2500.	70	high
heptanoic acid	2410.	71	high
1,1,2,2-tetrachloroethane	3230.	88	high
benzene	1780.	97	high
diethyl phthalate	1000.	123	high
2-nonanol	1000.	123	high
bromodichloromethane	900.	131	high
3-methylbenzoic acid	850.	136	high
trichloroethene	1100.	152	moderate
1,1,1-trichloroethane	700.	155	moderate
di-*n*-butyl phthalate	400.	217	moderate
1,1-dichloroethene	400.	217	moderate
carbon tetrachloride	800.	232	moderate
2-butanone (methylethylketone)	353.	235	moderate
4-methylbenzoic acid	340.	240	moderate
toluene	500.	242	moderate
tetrachloroethylene	200.	303	moderate
chlorobenzene	448.	318	moderate
1,2-dichlorobenzene	148.	343	moderate
o-xylene	170.	363	moderate
1,2,2-trifluoro-1,1,2-trichloroethane		372	moderate
styrene	162.	380	moderate
1,3-dichlorobenzene	118.	463	moderate
fluorotrichloromethane	110.	476	moderate
4,6-dinitro-2-methylphenol		477	moderate
p-xylene	156.	552	low
m-xylene	146.	588	low
1,4-dichlorobenzene	79.	594	low
ethyl benzene	150.	622	low
pentachlorophenol	14.	900	low
N-nitrosodiphenylamine	35.1	982	low
3,5-dimethylphenol		1038	low
BHC-delta	31.5	1052	low
2,6-dimethylphenol		1060	low
1,2,4-trichlorobenzene	30.	1080	low

TABLE 11.3 continued

Compound	Solubility (ppm)	K_{oc} (mL/g)	Mobility Class
naphthalene	31.7	1300	low
4-ethylphenol		1986	low
dibenzofuran	10.	2140	slight
hexachloroethane	8.	2450	slight
acenaphthene	7.4	2580	slight
tri-N-propylamine		2610	slight
BHC-alpha	8.5	2627	slight
BHC-beta	2.7	3619	slight
hexachlorobenzene	0.035	3910	slight
hexachlorobutadiene	3.2	4330	slight
di-n-octyl phthalate	3.	4510	slight
butyl benzyl phthalate	2.9	4606	slight
fluorene	1.98	5835	slight
2-methylnaphthalene	25.4	8500	slight
bis(2-ethylhexyl)phthalate	0.6	12200	slight
toxaphene	0.4	15700	slight
heptachlor epoxide	0.35	17087	slight
endosulfan II	0.28	19623	slight
fluoranthene	0.275	19800	slight
1,2-diphenylhydrazene (as azobenzene)	0.252	20947	immobile
endosulfan sulfate	0.22	22788	immobile
phenanthrene	1.29	23000	immobile
dieldrin	0.188	25120	immobile
anthracene	0.073	26000	immobile
BHC-gamma	0.15	28900	immobile
decanoic acid		39610	immobile
chlordane	0.056	53200	immobile
pyrene	0.135	63400	immobile
PCB-1254	0.042	63914	immobile
heptachlor	0.03	78400	immobile
endrin	0.024	90000	immobile
benzo(a)anthracene	0.014	125719	immobile
aldrin	0.013	132000	immobile
4,4'-DDE	0.01	155000	immobile
4,4'-DDT	0.0017	238000	immobile
4,4'-DDD	0.005	238000	immobile
benzo(a)pyrene	0.0038	282185	immobile
PCB-1260	0.0027	349462	immobile
chrysene	0.022	420108	immobile
benzo(b)fluoranthene		1148497	immobile
benzo(k)fluoranthene		2020971	immobile

Source: R. A. Griffin, 1985, personal communication, and W. R. Roy & R. A. Griffin, "Mobility of organic solvents in water-saturated soil materials," *Environmental Geology and Water Sciences*, 7 (1985):241–47.

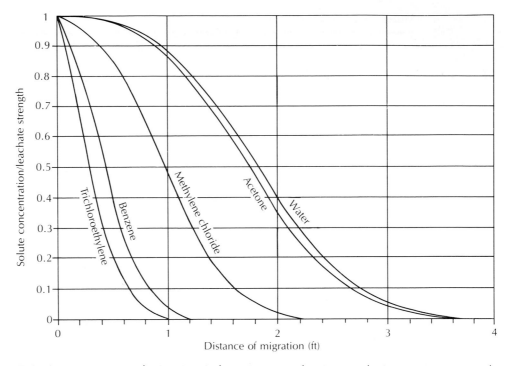

FIGURE 11.16   Vertical migration, in feet per 100 y, of various synthetic organic compounds through a soil with a hydraulic conductivity of 1.6×10^{-8} cm/s, hydraulic gradient of 0.222, bulk density of 2.00 g/cm³, particle density of 2.65, effective porosity of 0.22, and soil organic carbon content of 0.5%.

partition coefficient for a large number of organic chemicals that are potential ground-water contaminants. The K_d for an organic compound in a specific soil can easily be estimated as it is merely the K_{oc} for that compound times the weight fraction of organic carbon in the soil. If the soil is a pure silica sand, there will be very limited retardation.

   As a result of the processes of advection, dispersion, diffusion, and retardation, there is a pattern of solute distribution as one moves away from the source of the contamination. If the contaminant source contains multiple solutes, then each will have a retardation typified by the K_d for that solute and there will be a number of solute fronts. The resulting plume can be very complex. Figure 11.16 shows the relative rates of movement for water and several organic compounds of differing mobility classes. The movement is through a low-permeability glacial till with about 0.5% soil organic carbon. The curves were obtained from the one-dimensional solute transport equation and were solved on a personal computer using an algorithm written for the LOTUS 1-2-3 program.

EXAMPLE PROBLEM   Compute the relative velocity of the solute front of a solute-soil system with a distribution coefficient of 10 mL/g, a ρ_b value of 1.75 g/cm³ and θ of 0.20:

$$v_c = \frac{v_x}{1 + (\rho_b/\theta)(K_d)}$$

$$= \frac{v_x}{1 + \left(\dfrac{1.75 \text{ g/cm}^3}{0.20}\right)\left(10 \dfrac{\text{mL}}{\text{g}}\right)\left(\dfrac{1 \text{ cm}^3}{1 \text{ mL}}\right)}$$

$$= \frac{v_x}{88.5}$$

$$= 0.011 v_x$$

11.6.7 Degradation of Organic Compounds

It is well known that straight chain and aromatic hydrocarbons associated with petroleum products can undergo biological degradation (Barker, Patrick, & Major 1987; J. T. Wilson et al. 1986). Hydrogeologists have recently observed that halogenated organic solvents dissolved in ground water undergo transformations under natural conditions with the compounds undergoing progressive dehalogenation (Roberts, Schrainier, & Hopkins 1982; Parsons, Flood, & DeMarco 1984; Cline & Viste 1985).

The following classes of organic compounds have been found to undergo either biotic or abiotic degradation: chlorinated methanes, chlorinated ethanes, chlorinated propanes, chlorinated butanes, chlorinated ethenes, bromonated methanes, bromonated ethenes, alkylbenzenes, bromochloropropanes, halogenated acetates, and various aromatics such as benzene and toluene (for example, see Vogel, Criddle, & McCarty 1987; Vogel & McCarty 1987a, 1987b, 1985; Barrio-Lage, Parsons, & Nassar 1987; Barrio-Lage et al. 1987, 1986; Nelson et al. 1986, 1987; Vogel & Reinhard 1986; Grbic-Galic & Vogel 1987; Pignatello 1986).

The chlorinated ethanes and ethenes have been well studied and are a common ground-water contaminant. There are a number of reactions that can occur abiotically to break them into lower-molecular-weight compounds. *Substitution* is a reaction in which water interacts with the halogenated compound to substitute an OH^- for an X^-, creating an alcohol. Other groups can be substituted as well, such as the reaction with an HS^- radical under reducing conditions to release an X^- ion and form mercaptans. Substitution reactions proceed most rapidly for monohalogenated compounds. Monohalogenated compounds have reaction half-lives of about 1 mo. As the number of halogen ions increases, the half-life for reactions due to substitution increases rapidly into the range of years to hundreds of years. *Dehydrohalogenation* is a reaction where an alkane loses both an X^- and an H^+ ion to form a double bond between the carbon atoms, thus creating an alkene. The rate of dehydrohalogenation increases with increasing numbers of halogen ions; hence compounds that undergo substitution most slowly undergo dehydrohalogenation most rapidly. Bromine ions are more rapidly removed than chlorine ions in abiotic reactions.

Oxidations and *reductions* are typically biologically mediated. Oxidations include (1) the addition of an OH^- radical on an alkane in place of an H atom on a carbon that also contains a halogen ion, (2) oxidation of a halogen ion, (3) formation of an epoxy from the double bond of an alkene, and (4) addition of a halogen ion and a hydroxyl ion to the double bond of an alkene. Reductions start with the removal of an X^- ion by a reduced species, such as a reduced transition metal. The alkyl radical thus formed can react with an H^+ ion, which substitutes for the departed X^-. This process is called *hydrogenolysis*. If the adjacent carbon also contains a halogen ion, it too can be reduced. In this case the loss of halogens on adjacent carbon atoms creates an alkene by formation of a double bond between the carbon atoms.

Environmental conditions influencing the type and rate of the preceding reactions include pH, temperature, state of oxidation or reduction, microorganisms present, and types of other chemicals present. Reaction kinetics also play an important role in the determination of the abiotic and biotic fate of organic contaminants. Theoretically, the end products of the abiotic reactions are either ethane and ethene, both of which should be amenable to further biodegradation. However, under field conditions, such a favorable outcome might require many years to occur. Before it does, the contaminant might well flow from the point of origin to contaminate a large area of the aquifer system.

11.7   GROUND-WATER CONTAMINATION

11.7.1   Introduction

The specter of ground-water contamination looms over industrialized, suburban, and rural areas. The sources of ground-water contamination are many and the contaminants numerous. Common industrial solvents such as trichloroethylene, 1,1,1-trichloroethane, tetrachloroethane, benzene, and carbon tetrachloride have been found in widespread areas, with all indications being there are multiple sources (Fusillo, Hochreiter, & Lord 1985). Suburban areas have ground water with high levels of nitrate due to the use of lawn fertilizers as well as septic tank discharges (Flipse et al. 1984). Agricultural areas have not only high levels of fertilizers found in ground water (Pionke & Urban 1985), but specialized synthetic organic agricultural chemicals as well (Rothschild, Manser, & Anderson 1982). Landfills in urban and rural areas are known sources of contamination (Noss & Johnson 1984; McLeod 1984). Underground storage tanks holding petroleum products (Kramer 1982) and synthetic organic chemicals (Oliveira & Sitar 1985) have leaked and caused ground-water contamination.

It has been estimated that ground water has been contaminated in only 1% to 2% of the aquifers in the United States (Lehr 1981). However, these aquifers may well be in urban areas where the water is most needed. Contaminated ground water in most cases will not travel more than a few thousand feet from the source and in many cases not more than a few hundred feet. If there is a single source, then the contamination may be localized. If there are multiple sources, or if the contamination is a result of widespread land-use practices, then the contamination may cover a large area. Table 11.4 is a partial listing of contaminants reported to have been found in ground water.

Ground-water contamination is not an irreversible process. There are natural conditions that act to remove contaminants. Attenuation mechanisms include dilution, dispersion, mechanical filtration, volatilization, biological activity, ion exchange and adsorption on soil particle surfaces, chemical reactions, and radioactive decay. Even synthetic organic compounds can undergo biological decay (Cline & Viste 1984), although the decay products may also be toxic. In recent years a number of techniques have been developed for restoring the quality of ground water that has been contaminated. This is discussed in Section 11.3.

11.7.2 Septic Tanks and Cesspools

The disposal of domestic wastewater is accomplished in many areas through the use of septic tanks and drain tile fields. An estimated 800 billion gallons of water per year are discharged to the subsurface in the United States via septic tanks (U.S. Environmental Protection Agency 1977). Anaerobic decomposition of wastes takes place in the septic tank. The liquid waste is carried to a drain tile field, where it seeps through the vadose zone to the water table. An analysis of the typical septic tank effluent is given in Table 11.5.

Septic tank effluent contains bacteria and viruses. It is a major factor in the incidences of waterborne disease from private wells in the United States (Craun 1981) and presumably elsewhere as well. The most important factor that influences the development of ground-water contamination from septic tanks is the density of septic tank systems in the area (Yates 1985). Documented cases of widespread ground-water contamination from septic tank systems have been in areas where the lot sizes range from less than one-quarter of an acre to three acres (Yates 1985).

Several cases of infectious disease outbreaks due to septic tanks have been reported. In Polk County, Arkansas, in 1971, an outbreak of viral hepatitis was traced to a well that was contaminated by seepage from a septic tank located 95 ft away (Craun 1979). In 1972, typhoid in Yakima, Washington, was attributed to well water from driven well points. Waste water from the septic tank serving the home of a typhoid carrier was discharged into the ground 210 ft away from the contaminated well. A Norwalk-like virus was responsible for 400 cases of gastroenteritis at a resort camp in Colorado. A septic tank was located 50 ft above the spring supplying drinking water to the camp (Craun 1984).

Septic tanks are most likely to contribute to ground-water contamination in areas where (1) there is a high density of homes with septic tanks, (2) the soil layer over permeable bedrock is thin, (3) the soil is extremely permeable, such as gravel, or (4) the water table is within a couple of feet of the land surface. Areas with high population densities should not be served with septic tanks, and areas

TABLE 11.4 Chemicals and organisms known to have caused ground-water contamination (various sources)

Metals	Nonmetals	Organics	Extractable Organic Compounds	Volatile Organic Compounds	Organisms
aluminum	acids	aldrin	tri-n-propylamine	benzene	*Giardia lamblia*
arsenic	ammonia	BOD	3- and/or 4-methyl phenol	1,2-dichloro-ethane	*Salmonella* sp.
barium	boron	chlordane	4-methyl benzoic acid	1,1,1-trichloro-ethane	*Shigella* sp.
cadmium	chloride	DDT	1,4-dioxane	1,1-dichloroethane	typhoid
chromium	cyanide	detergents	4-methyl-2-pentanol	1,1,2-trichloro-ethane	*Yersinin entero-colitica*
copper	fluoride	ethyl acrylate	n,n-dimethyl-formamide	chloroethane	viral hepatitis
iron	nitrate	gasoline	2-hexanone	1,1-dichloro-ethene	
lead	phosphate	hydroquinone	4-methyl-2-pentanone	trans-1,2-di-chloroethene	
lithium	radium	lindane	1-methyl-2-pyrrolidinone	ethyl benzene	
manganese	selenium	paramethyl animo-phenol	2-hexanol	methylene chloride	
mercury	sulfate	PBB	3,5-dimethyl phenol and/or 4-ethyl phenol	tetrachloro-ethane	
molybdenum	various radio-active isotopes	PCB	benzoic acid	toluene	
nickel		DCPD (dicyclo-pentadiene)	hexanoic acid	trichloroethene	
silver		DIMP (diisopropyl-methyl-phospho-nate)	cyclohexanol	vinyl chloride	
uranium		DBCP (dibromo-chloropropane)	2-ethyl hexanoic acid	tetrahydrofuran	
zinc			octanoic acid	acetone	
			pentanoic acid	2-methyl-2-propanol	
			bis(2-ethylhexyl) phthalate	2-butanone	
			di-n-butyl phthalate	2-butanol	
			2,4-dimethyl phenol	2-propanol	
			isophorone		
			phenol		
			1,2-dichlorobenzene		

TABLE 11.5 Effluent quality from six septic tanks*

Site	Avg. Flow (gal/day)	BOD (mg/L)	COD (mg/L) (unfiltered)	COD (mg/L) (filtered)	TSS (mg/L)	Fecal Coliforms (no./mL)	Fecal Strep (no./mL)	Total N (mg/L)	Ammonia N (mg/L)	Nitrate-Nitrogen (mg/L)	Total P (mg/L)	Ortho P (mg/L)
A	75	131	325	249	69	2907	2.7	50.5	34.1	0.68	12.3	10.8
B	125	176	361	323	44	4127	39.7	57.8	42.5	0.46	14.1	13.6
C	245	272	542	386	68	27,931	1387	76.3	45.6	0.60	31.4	14.0
D	315	127	291	217	52	11,113	184	40.2	33.2	0.35	11.0	10.1
E	860†	120	294	245	51	2310	20.7	31.6	20.1	0.16	11.1	10.5
F	150	122	337	281	48	3246	25.3	56.7	38.3	0.83	11.6	10.5

Source: R. J. Otis, W. C. Boyle, & D. K. Sauer, Small-Scale Waste Management Program, University of Wisconsin-Madison, 1973.
*All values are means.
†Includes 340 gal/day sewer flow and 20 gal/day from foundation drain.

with thin soils, extremely permeable soils, and high water tables should be avoided.

11.7.3 Landfills

Burial in a landfill is the most common means of disposing of municipal refuse, ashes, garbage, leaves, demolition debris, and sludges from municipal and industrial wastewater treatment facilities. Radioactive, toxic, and hazardous wastes have also been subjected to land burial as a means of disposal. Precipitation that infiltrates the waste can mix with liquids already present in the waste and leach compounds from the solid waste. The result is a liquid known as **leachate.** Leachate can move downward from the landfill into the water table and cause ground-water contamination. If the waste is buried below the water table, moving ground water can leach compounds from the waste and become contaminated. Table 11.6 gives a chemical analysis of the leachate from a municipal waste landfill in Du Page, Illinois. Table 11.7 gives the overall range and range of median values for municipal solid-waste leachate in Wisconsin. Landfill leachates can contain very high concentrations of both inorganic and organic compounds.

When leachate from a landfill mixes with ground water, it forms a plume that spreads in the direction of the flowing ground water. As one goes away from the source, the concentration decreases owing to hydrodynamic dispersion and retardation. Figure 11.17 shows in cross section a plume of ground water with high sulfate from a fly ash* disposal pond located with the waste pile below the water table. In this particular case, the plume of high-sulfate water is moving directly toward the Mississippi River, where it is discharging into the river. The location

---

*Fly ash is the ash that is produced from the burning of coal at a power plant. Large amounts are produced from power generation, and this presents a major waste-disposal problem.

TABLE 11.6   Chemical analysis of landfill leachate at Du Page, Illinois

Na	748.	mg/L
K	501.	mg/L
Ca	46.8	mg/L
Mg	233.	mg/L
Cu	<0.1	mg/L
Zn	18.8	mg/L
Pb	4.46	mg/L
Cd	1.95	mg/L
Ni	0.3	mg/L
Hg	0.0008	mg/L
Cr	<0.10	mg/L
Fe	4.2	mg/L
Mn	<0.1	mg/L
Al	<0.1	mg/L
NH_4	862.	mg/L
As	0.11	mg/L
B	29.9	mg/L
Si	14.9	mg/L
Cl	3484.	mg/L
SO_4	<0.01	mg/L
PO_4	<0.1	mg/L
COD	1340.	mg/L
Organic acids	333.	mg/L
Carbonyls as acetophenone	57.6	mg/L
Carbohydrates as dextrose	12.	mg/L
pH	6.9	
Eh	+7 mv	
Conductivity	10.20 mmhos/cm	

Source: K. Cartwright, R. A. Griffin, & R. H. Gilkeson, *Ground Water* 15 (1977): 294–305.

of a waste-disposal area adjacent to a ground-water discharge zone limits the amount of ground water that can become contaminated, although surface-water contamination could occur.

The volume of leachate that is produced is a function of the amount of water percolating through the refuse. Land disposal of solid waste in humid areas is more likely to produce large volumes of leachate than land disposal in arid zones. The vadose zone in arid regions may receive little or no recharge. Under such conditions, solid-waste disposal is not likely to result in ground-water contamination.

Current state-of-the-art hydrogeologic and engineering practice for determining the locations of municipal waste landfills calls for careful geologic analysis

TABLE 11.7 Overall summary from the analysis of municipal solid-waste leachates in Wisconsin

Parameter	Overall Range*	Typical Range (range of site medians)*	Number of Analyses
TDS	584–50430	2180–25873	172
Specific conductance	480–72500	2840–15485	1167
Total susp. solids	2–140900	28–2835	2700
BOD	ND–195000	101–29200	2905
COD	6.6–97900	1120–50450	467
TOC	ND–30500	427–5890	52
pH	5–8.9	5.4–7.2	1900
Total alkalinity ($CaCO_3$)	ND–15050	960–6845	328
Hardness ($CaCO_3$)	52–225000	1050–9380	404
Chloride	2–11375	180–2651	303
Calcium	200–2500	200–2100	9
Sodium	12–6010	12–1630	192
Total Kjeldahl nitrogen	2–3320	47–1470	156
Iron	ND–1500	2.1–1400	416
Potassium	ND–2800	ND–1375	19
Magnesium	120–780	120–780	9
Ammonia-nitrogen	ND–1200	26–557	263
Sulfate	ND–1850	8.4–500	154
Aluminum	ND–85	ND–85	9
Zinc	ND–731	ND–54	158
Manganese	ND–31.1	0.03–25.9	67
Total phosphorus	ND–234	0.3–117	454
Boron	0.87–13	1.19–12.3	15
Barium	ND–12.5	ND–5	73
Nickel	ND–7.5	ND–1.65	133
Nitrate-nitrogen	ND–250	ND–1.4	88
Lead	ND–14.2	ND–1.11	142
Chromium	ND–5.6	ND–1.0	138
Antimony	ND–3.19	ND–0.56	76
Copper	ND–4.06	ND–0.32	138
Thallium	ND–0.78	ND–0.31	70
Cyanide	ND–6	ND–0.25	86
Arsenic	ND–70.2	ND–0.225	112
Molybdenum	0.01–1.43	0.034–0.193	7
Tin	ND–0.16	0.16	3
Nitrite-nitrogen	ND–1.46	ND–0.11	20
Selenium	ND–1.85	ND–0.09	121
Cadmium	ND–0.4	ND–0.07	158
Silver	ND–1.96	ND–0.024	106
Beryllium	ND–0.36	ND–0.008	76
Mercury	ND–0.01	ND–0.001	111

*All concentrations in mg/L except pH (std. units) and specific conductance (μmhos/cm). ND indicates no data.

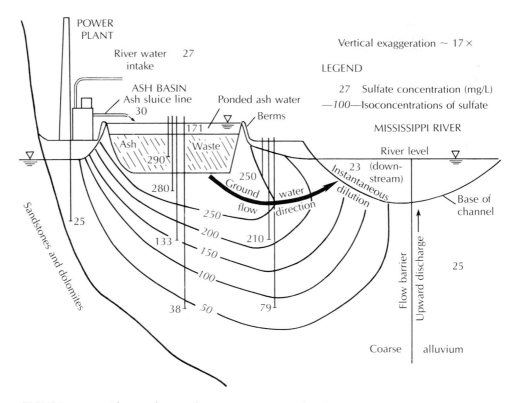

FIGURE 11.17 Plume of ground water contaminated with high-sulfate leaching from a fly ash landfill located below the water table. Source: Daniel R. Viste. *Proceedings, First Annual Conference of Applied Research and Practice on Municipal and Industrial Waste* (Madison: University of Wisconsin, 1978), pp. 327–40.

of several alternative sites in order to select the site least likely to result in ground-water contamination. Landfills may be designed to minimize the formation of leachate as well as to minimize the amount of leachate that escapes from the landfill. Leachate may also be collected and treated.

It is desirable in most cases to construct landfills above the water table. Some attenuation of leachate may occur as it passes through the unsaturated zone. A *natural-attenuation* landfill is one that relies totally on natural processes to attenuate any leachate formed. Such landfills should be well above the water table to promote maximum attenuation in the vadose zone. Soils with the greatest potential attenuation are clays because they have the most ion exchange and adsorption sites. Unfortunately, in humid areas, the water table in clay soils tends to be close to the land surface; this means that much of the landfill should be above grade. Leachate generation may be reduced by capping the landfill with two to three feet of compacted clay soil or a synthetic membrane. If large amounts of leachate are generated in a natural-attenuation landfill located in a low-permeability clay, there is a tendency for leachate to come to the land surface and form

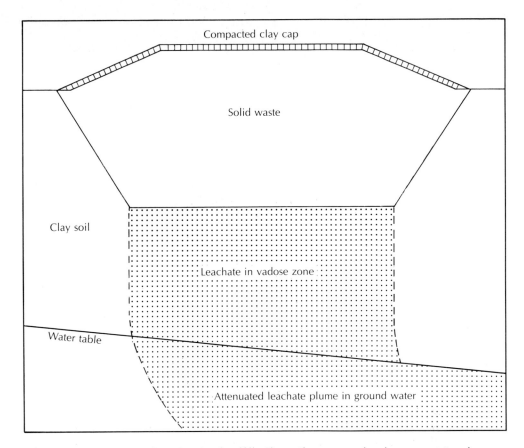

FIGURE 11.18 Natural-attenuation landfill. Clay soils attenuate leachate to varying degrees, depending upon soil type and leachate. Attenuated leachate passes through the unsaturated zone to the water table, where a plume is formed.

leachate springs. Figure 11.18 shows the design of a natural-attenuation landfill. Many natural-attenuation landfills have resulted in ground-water contamination from leachate.

A *lined* landfill is one designed to capture part of all of the leachate generated. Landfill liners are typically constructed of 3 to 10 ft (0.9 to 3 m) of compacted clay soils. The permeability of the liner should be no greater than 3 × 10^{-4} ft/day (1 × 10^{-7} cm/s). Alternatively, a synthetic membrane such as HDPE (high-density polyethylene) could be used as the liner. Because leachate will collect on the liner, a **leachate-collection system** is also needed. The leachate-collection system consists of a blanket of sand or gravel, with perforated drainage lines, lying on the liner. The base of the liner is sloped toward the drain tiles. Leachate drains through the leachate-collection system to a holding tank or sewer and is ultimately removed and treated. Clay-lined systems can be designed to collect about 70% to 90% of the leachate produced (Kmet, Quinn, & Slavik 1981).

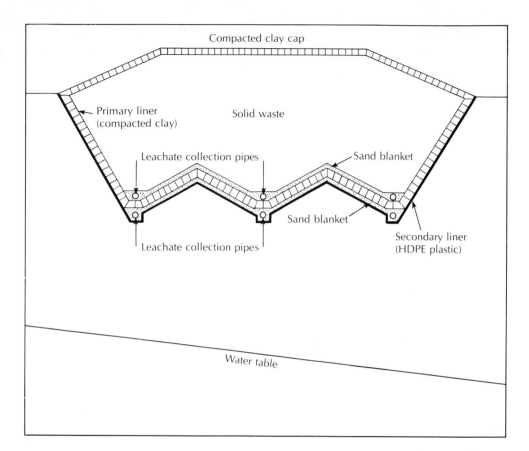

FIGURE 11.19   Double-lined landfill with leachate collection system. The primary liner consists of 5 ft of compacted clay soil with hydraulic conductivity of no more than 1×10^{-7} cm/s. The secondary liner is flexible membrane such as 40-mil HDPE plastic. The leachate collection system consists of 1-ft-thick sand layers with perforated pipes, which drain to a leachate collection tank.

The remainder of the leachate will seep through the liner. A double liner and secondary leachate-collection system installed beneath the primary liner can be constructed to capture the leakage through the primary liner. A membrane or clay liner could be used for the secondary liner. Figure 11.19 shows a double-lined system with leachate collection.

In areas with a high water table and low-permeability soils (3×10^{-3} ft/day, or 10^{-6} cm/s), a *zone-of-saturation* landfill could be constructed (Figure 11.20). An excavation below the water table is made. In clay soils this can easily be done because ground water seeps into the excavation at a very slow rate and evaporates; it does not accumulate. A recompacted clay liner with recompacted clay sidewalls is installed to further reduce the amount of seepage into the excavation. A leachate-collection system is installed. This collects not only the leachate that forms, but any ground water that seeps in as well. The zone-of-

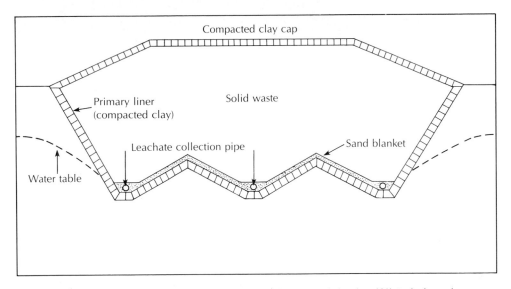

FIGURE 11.20 Zone-of-saturation landfill. The bottom of the landfill is below the water table. The leachate-collection system collects both leachate and ground water that seeps into the landfill.

saturation landfill is more efficient than the lined landfill because no leachate escapes. However, it will be necessary to collect the leachate long after the landfill is closed because the waste will be below the water table.

As an alternative to the zone-of-saturation landfill, a *hydraulic-gradient-control* landfill can be constructed in areas of a high water table (Figure 11.21). Underdrains are placed beneath the liner to lower the water table so that it is below the liner. A leachate-collection system is installed above the liner. The water discharging from the underdrains must be monitored to determine if it is being impacted by the portion of the leachate that drains through the liner. If it is, then it must be treated before being discharged. As with all of the systems, a well-designed, well-constructed cap is essential in reducing the amount of leachate formed, thereby reducing the amount of leachate that must be handled.

11.7.4 Chemical Spills and Leaking Underground Tanks

Ground-water contamination due to a variety of inorganic and organic compounds has occurred as a result of spills and leaks of toxic and hazardous chemicals. These discharges may be the result of a sudden action, such as a tank car accident, or may be the result of slow leakage. Typically more than one chemical may be released. As was seen in Section 11.6, different chemicals will travel through the ground at different rates owing to retardation effects. As a result, complex plumes of contaminated water may result.

If the contaminant dissolves in the water, it will flow along with the ground water. However, if a liquid discharged into the ground has a specific

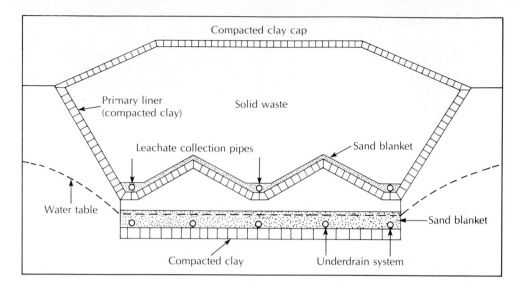

FIGURE 11.21 Lined landfill with water-table control system. Underdrains below the liner keep water table from saturating the liner and waste. Leachate that seeps through the liner is collected by the underdrain system.

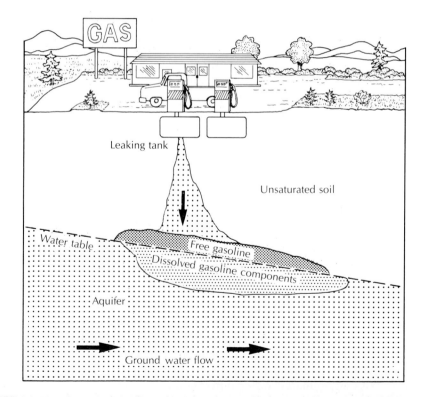

FIGURE 11.22 Organic liquids such as gasoline, which are only slightly soluble in water and are less dense than water, tend to float on the water table when a spill occurs.

gravity less than that of water, it can float on the water table. This is what happens when a petroleum product leaks into the ground. Figure 11.22 shows a gasoline plume moving along the surface of the water table. Note that the gasoline will spread upgradient a short way and that there will be some soluble components of the gasoline, such as benzene, dissolved in the water below the water table.

Dense liquids may sink to the bottom of the aquifer. In general, the chlorinated hydrocarbons are heavier than water. They have various solubilities in water. Table 11.8 gives the density and solubility of some organic compounds. For those that are denser than water, the pure product will sink to the bottom of the aquifer. Some of the product will go into the solution so that there will also be a plume of ground water with dissolved product. Figure 11.23 shows this.

The residual DNAPL in the vadose zone can partition into the air in the soil pores and fractures. The plume of DNAPL vapors can spread for many meters from the site of the spill (Mendoza & McAlary 1990; Mendoza & Friend 1990a, 1990b).

There are many sites in the United States where ground water has been contaminated with organic compounds. In some cases, contaminated water has

TABLE 11.8   Density and solubility in water of organic compounds

Compound	Specific Gravity[a]	Solubility[b] Milligrams compound/liter water (@ °C Temperature)
Acetone	0.79	Infinite
Benzene	0.88	1780 (20)
Carbon tetrachloride	1.59	800 (20), 1160 (25)
Chloroform	1.48	8000 (20), 9300 (25)
Methylene chloride	1.33	20,000 (20), 16,700 (25)
Chlorobenzene	1.11	500 (20), 488 (30)
Ethyl benzene	0.87	140 (15), 152 (20)
Hexachlorobenzene	1.60	0.11 (24)
Ethylene chloride	1.24	9200 (0), 8690 (20)
1, 1, 1-trichloroethane	1.34	4400 (20)
1, 1, 2-trichloroethane	1.44	4500 (20)
Trichloroethylene	1.46	1100 (25)
Tetrachloroethylene	1.62	150 (25)
Phenol	1.07	82,000 (15)
2-Chlorophenol	1.26	28,500 (20)
Pentachlorophenol	1.98	5 (0), 14 (20)
Toluene	0.87	470 (16), 515 (20)
Methyl ethyl ketone	0.81	353 (10)
Naphthalene	1.03	32 (25)
Vinyl chloride	0.91	1.1 (25)

[a]Source: R. Weast, "Handbook of Chemistry and Physics," 60th ed., CRC Press, Inc., 1979, 1980.
[b]From Verschueren, Karel. "Handbook of Environmental Data on Organic Chemicals." New York, Van Nostrand Reinhold, 1983. Numbers in parentheses = temperatures.

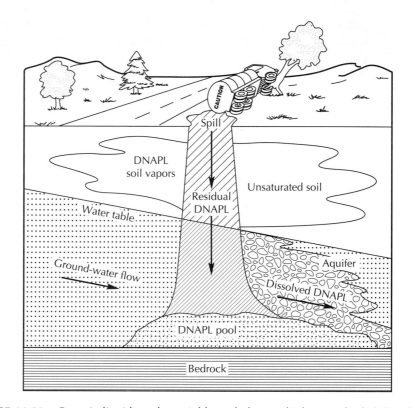

FIGURE 11.23   Organic liquids such as trichloroethylene, which are only slightly soluble in water and are more dense than water, may sink to the bottom of an aquifer when a spill occurs.

migrated from a spill area to a water-supply well. In a shallow aquifer in South Brunswick Township, New Jersey, 1,1,1-trichloroethane and tetrachloroethane migrated more than 3000 ft from a factory site to a water-supply well. At times the 1,1,1-trichloroethane was present in amounts in excess of 1000 μg/L and the tetrachloroethane in amounts ranging from 100 to 300 μg/L. The affected well was taken out of service (Roux & Althoff 1980). Figure 11.24 shows the plume of 1,1,1-trichloroethane. There are actually three plumes, each from a different source, although only one was affecting the supply well.

11.7.5   Mining

Extraction and processing of metallic ore and coal have been the source of both surface- and ground-water contamination. Ground water moving through mineralized rock zones may contain excessive amounts of heavy metals (Klusman & Edwards 1977). Mining and milling expose overburden and waste rock to oxidation. Oxidation of pyrite, a common mineral, can produce sulfuric acid. In the Appalachian region of the eastern United States, 6000 tons of sulfuric acid are

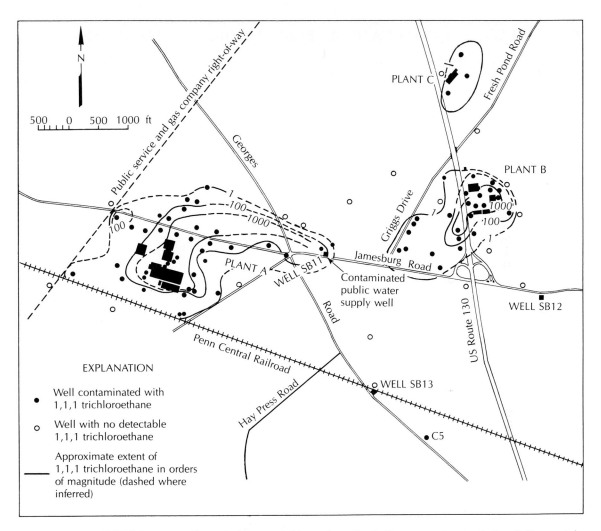

FIGURE 11.24 Plumes of 1,1,1-trichloroethane in shallow ground water in South Brunswick Township, New Jersey. The plume to the southwest has traveled to the east and has reached a public water-supply well. Source: P. H. Roux & W. F. Althoff, *Ground Water* 18 (1980): 464–72.

produced daily in this manner (Ahmad 1974). This results in highly acidic water draining from spoil piles and tailings deposits; hence, the shallow ground water and surface water of the region tend to have a low pH. The low-pH water draining through spoil piles and tailings can also leach heavy metals (Norbeck, Mink, & Williams 1974; Ralston & Morilla 1974), as well as soluble calcium, magnesium, sodium, and sulfate (McWhorter, Skogerboe, & Skogerboe 1974). Uranium and thorium mining and milling operations can release radioactive isotopes to the atmosphere, surface water, and ground water (Kaufman, Eadie, & Russell 1976).

CASE STUDY: CONTAMINATION FROM URANIUM TAILINGS PONDS

In one study of the ground water beneath two unlined uranium tailings ponds in Utah, a comparison was made of baseline natural uranium activity before the tailings ponds were put into service and the natural uranium activity after 11 y of operation (White & Gainer 1985). The tailings ponds, each of which covers about 40 ac, were constructed by placing earthen dams across a small valley. The surface geology is a fine-grained sandstone interfingered with siltstones and claystones. Ground-water flow is fracture-controlled. Figure 11.25A shows the locations of the tailings ponds and the fracture patterns affecting the ground water. Figure 11.25B shows the potentiometric surface. It was found that the base-

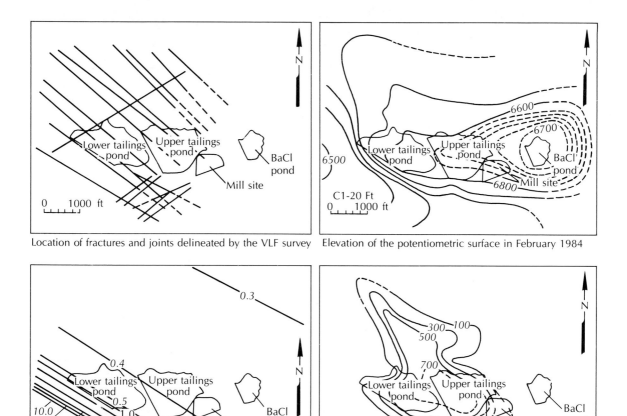

Location of fractures and joints delineated by the VLF survey

Elevation of the potentiometric surface in February 1984

Baseline natural uranium activity (μCi/mL $\times$ 10^7)

Elevated activity of natural uranium in the summer of 1983 (μCi/mL $\times$ 10^7)

FIGURE 11.25 Ground-water contamination by uranium beneath active uranium mill tailings ponds. Source: R. B. White & R. B. Gainer, *Ground Water Monitoring Review* 5, no. 2 (1985): 75–82.

line water quality showed somewhat elevated levels of natural uranium activity due to natural leaching of uranium-bearing rocks by water moving along a fracture system. Figure 11.25C shows the baseline natural uranium activity. It follows a linear pattern along the strike of a major fracture southwest of the lower tailings pond. After 11 y of operation, the natural uranium activity had increased significantly above the baseline measurements. Figure 11.25D shows the elevated activity, corrected for the baseline variation. The plume is spreading to the northwest, following the strike of the fractures. This illustrates one of the facets of ground-water contamination in fractured rock aquifers that are anisotropic. That is, the plume will tend to follow the fractures and not necessarily move normal to the potentiometric surface. If the aquifer were isotropic, from the potentiometric surface map one would expect that the plume would spread primarily to the southwest.

11.7.6 Other Sources of Ground-Water Contamination

In a 1977 report to Congress, the U.S. Environmental Protection Agency listed a number of waste-disposal practices and their potential impact on ground water (U.S. Environmental Protection Agency 1977). Waste-disposal practices mentioned in that report and not already covered in this section include liquid industrial waste disposal in lagoons and injection wells, oil-field brine disposal in lagoons and wells, land-spreading of sewage and industrial sludges, leakage from municipal wastewater sewers and lagoons, and land disposal of animal waste from feed lots. Other major causes of ground-water contamination listed in the report to Congress were spills and leaks; mine drainage; salt-water intrusion; poorly constructed or abandoned water, oil, and gas wells; infiltration of contaminated surface water; agricultural activities; highway deicing salts; and atmospheric contaminants.

11.8 GROUND-WATER RESTORATION

11.8.1 Introduction

Once ground water has been contaminated, it may take many years after the source of contamination has been eliminated for natural processes to remove the contaminants from an aquifer. During the 1980s, methods of restoring ground-water quality by implementing various types of remedial measures were developed. However, most of these methods are time-consuming and extremely expensive. There are two broad categories of remedial measures: one must either remove or isolate the source and/or pump and treat the ground water (JRB Associates, Inc. 1982; Canter & Knox 1985; Canter 1982).

11.8.2 Source-Control Measures

One extreme method of source control is to excavate and remove the source. This was done at a toxic chemical waste landfill in Wilsonville, Illinois, when the court

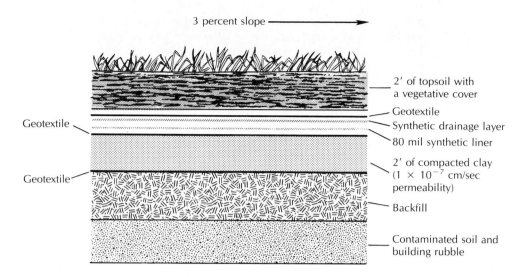

3 percent slope ⟶

2′ of topsoil with
a vegetative cover

Geotextile

Synthetic drainage layer

80 mil synthetic liner

2′ of compacted clay
$(1 \times 10^{-7}$ cm/sec
permeability)

Backfill

Contaminated soil and
building rubble

Geotextile

Geotextile

FIGURE 11.26   Design of a low-permeability multimedia cap to cover waste. Fill material is used above waste to create a 3% slope if the waste material or land surface over the waste material is not sloped.

ordered a landfill operator to exhume and remove a large number of drums of liquid waste that had been buried in a licensed landfill (T. J. Johnson et al. 1983). A less drastic approach is to isolate the waste in place. If the waste is entirely above the water table, this is much more easily done than if the waste extends below the water table.

Changing the surface drainage so that runoff from upland areas does not cross the land surface above the waste will reduce the amount of surface water that infiltrates into the waste and produces leachate. The construction of a low-permeability cap above the waste can also be very effective in reducing the amount of infiltration through the waste. Caps can be constructed of compacted clay, synthetic membranes, concrete, asphalt, and other types of materials. The most effective caps have several layers and include coarse granular layers between fine-grained layers to act as drains to divert infiltration away from the waste (Herzog et al. 1982). Figure 11.26 shows a multimedia cap design.

If the waste extends below the water table, then it is necessary to keep the ground water from flowing through it. This can be accomplished by installing a low-permeability vertical barrier around the waste body (Ayres, Lager, & Barvenik 1983; Brunsing & Cleary 1983; Fitzwater, Brassow, & Fetter 1983; Druback & Arlotta 1985; Lynch et al. 1984). Vertical barriers can be constructed by digging a trench and backfilling it with a slurry-type mixture of water, soil, and bentonitic clay. This is called a **slurry wall.** Slurry wall construction is limited to the depths that the trench can be constructed. A **grout curtain** can be installed by injecting any of a number of compounds into boreholes around the site. The materials fill the pore spaces in the rock or soil and harden. Grout curtains can be installed to great depths. Interlocking metal *sheet piling* can also be driven into soil to form a cutoff wall.

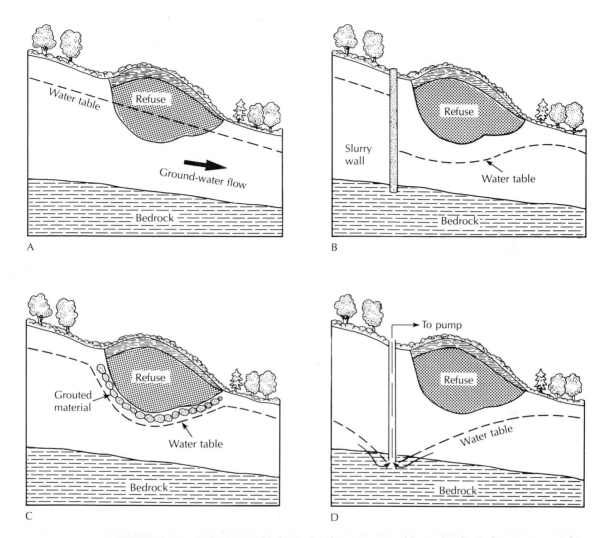

FIGURE 11.27 Isolation methods used to lower water table so that buried waste material is not in contact with ground water. **A.** No control measure. **B.** Up-gradient slurry wall to lower water table. **C.** Injection of grout to form a seal on sides and bottom. **D.** Gradient-control well to lower water table. Source: Modified from U.S. Environmental Protection Agency.

Figure 11.27A shows a waste material buried beneath the water table. Figure 11.27B indicates the installation of a low-permeability slurry wall to lower the water table and divert flowing ground water from the waste source. In this case, one wall installed upgradient of the waste is sufficient. In other cases, the waste could be surrounded with walls. Grout can even be injected through the waste to form a bottom seal. Used in conjunction with walls around the waste, the waste can be totally encapsulated. Figure 11.27C shows this.

Hydraulic gradient control measures can be used to lower the water table where it is in contact with the waste (Keely 1984; Schafer 1984; Campbell, Bost,

& Jacobsen 1984; Poulos & Laws 1985). Figure 11.27D shows a pumping well installed upgradient of buried waste material. The water table is lowered so that it no longer is in contact with the waste.

11.8.3 Plume Treatment

Once the contamination source is isolated, the task remains to restore the quality of the ground water that has become contaminated. One option is to take no action and let the contaminants be flushed from the aquifer by natural recharge. This process could be made more efficient in some hydrogeologic settings by artificially increasing the amount of water that enters the aquifer and hence accelerating the natural process. This approach is usually not desirable if the aquifer is used as a source of drinking water. Additionally, if the plume will eventually discharge to a surface-water body, it might cause contamination of the surface water. The time for natural restoration might be tens to hundreds of years. Even if the aquifer is not presently a drinking-water source, in the future it could be put to such a use. If the waste source were no longer present, future generations could unsuspectingly drill wells into the contaminated aquifer.

Plume treatment can occur in situ or by extraction of the water via wells (Lenzo 1984; Flathman, Quince, & Bottomley 1984; Brenoel & Brown 1985; Yaniga, Matson, & Demko 1985; Flathman et al. 1985). In situ treatment can be chemical or biological. Figure 11.28A shows a method of injecting nutrients and oxygen into a plume. Some compounds, such as hydrocarbons, can be biologically treated by natural soil bacteria with the proper mix of nutrients. If a chemical treatment scheme were proposed, the upgradient injection well of Figure 11.28A could be used to inject the proper chemicals. For example, an oxidizing agent could be added to the ground water. Figure 11.28B shows a permeable treatment bed installed where the water table is shallow. An acidic leachate could be neutralized by installing a permeable treatment bed with limestone gravel in it. Materials with ion-exchange properties could possibly be used to remove heavy metals.

Where the plume contains water contaminated with chlorinated solvents, many of which are believed to be carcinogenic at the low parts per billion range, the best option in most cases is to use *extraction wells* to remove the contaminated water. Figure 11.29 shows plume removal by means of shallow gradient-control wells. The spacing and pumping rate for the extraction wells is typically determined by computer modeling. The extraction wells are designed to capture the plume while at the same time removing as little of the uncontaminated water as possible. Extraction wells can also be planned as *plume-stabilization wells*. In this case they are located somewhere within the plume and sized to reverse the hydraulic gradient beyond the edge of the plume. They then prevent further movement of the plume. Locating an extraction well outside of the plume will tend to expand the plume boundaries.

Contaminated water that has been removed from the ground is usually treated before being discharged. Naturally, the treatment will depend upon the types and concentrations of the contaminants. Synthetic organic compounds are

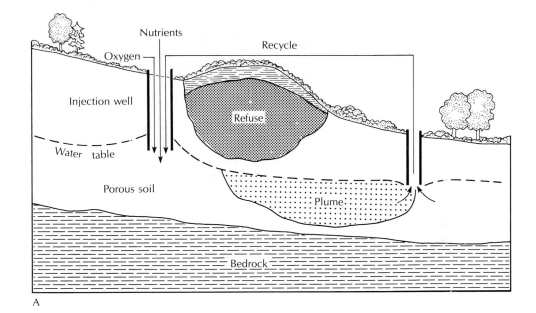

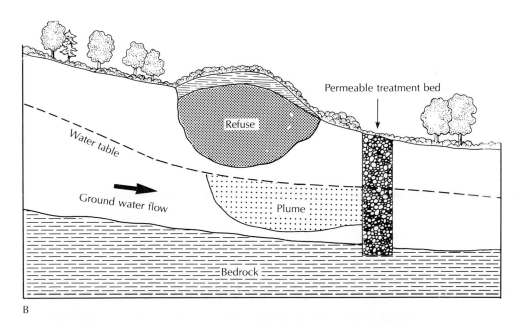

FIGURE 11.28 In situ treatment methods. **A.** Injecting nutrients into an aquifer to promote bioreclamation. **B.** Installation of a permeable treatment bed in path of plume to provide contact-type treatment such as ion exchange or neutralization. Source: Modified from U.S. Environmental Protection Agency.

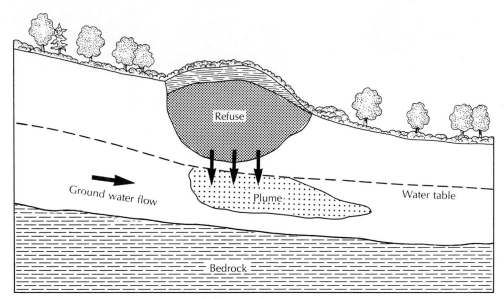

Before pumping

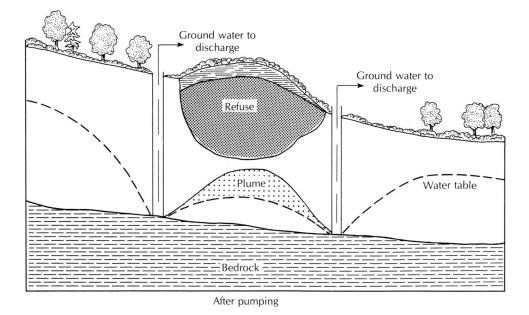

After pumping

FIGURE 11.29   Use of extraction wells to remove contaminated ground water. Source: U.S. Environmental Protection Agency.

usually removed from contaminated water by one or more of the following methods: volatilization to the atmosphere in an air-stripping column, adsorption on activated carbon, and biological treatment.

A large study of the efficiency of pump-and-treat methods of ground-water remediation has indicated that there would initially be a rapid decline in the concentration of dissolved constituents in the ground water but that eventually the concentration would show little change with additional pumping. If the initial concentrations were in excess of 1000 μg/L, then a reduction of 90% to 99% could be achieved before leveling occurred. If the initial concentration was less than 1000 μg/L, then less than 90% removal could be accomplished before no further reductions occurred (Doty & Travis 1991). Although a 90% removal is significant, the resulting ground water could still have an unacceptable concentration of a contaminant. Moreover, if residual nonaqueous phase organic liquids are present, the dissolved concentrations would soon rise if pumping is halted. In fact, it may be impossible to remediate some aquifers, especially fractured rock aquifers, that are contaminated with dense nonaqueous phase liquids (Mackay & Cherry 1989; Freeze & Cherry 1989).

11.9   CASE HISTORY: GROUND-WATER CONTAMINATION AT A SUPERFUND SITE

11.9.1   Background

From 1970 to 1979 a solvent recovery and recycling plant, Seymour Recycling Corporation, was operated in Seymour, Indiana. The company went bankrupt in 1979 and the owners abandoned 98 large tanks and approximately 50,000 drums, all filled with organic chemicals. Many of the drums were in a deteriorated condition, and prior to 1983, when they were all removed, an unknown amount of liquid synthetic organic compounds leaked into the soil. Recharging precipitation dissolved organic compounds from the soil and carried them into the ground water. Figure 11.30 shows the location.

From 1983 through 1985 a remedial investigation was performed for the U.S. Environmental Protection Agency (USEPA) to quantify the extent of contamination of soil and ground water. In 1986 a feasibility study was conducted to identify a practical method of remediating the contaminated soil and ground water. Following the completion of the feasibility study, the USEPA negotiated with the owners of the companies that had sent hazardous wastes to the site in order to force them to effect a site cleanup at their expense. Following lengthy negotiations, a consent decree was signed in 1988 by the USEPA and the responsible parties. The consent decree and the accompanying remedial action plan detailed the cleanup plan that was negotiated and that was to be implemented by the responsible parties. Actual construction of the remedial facilities was begun in 1988 and completed in late 1991.

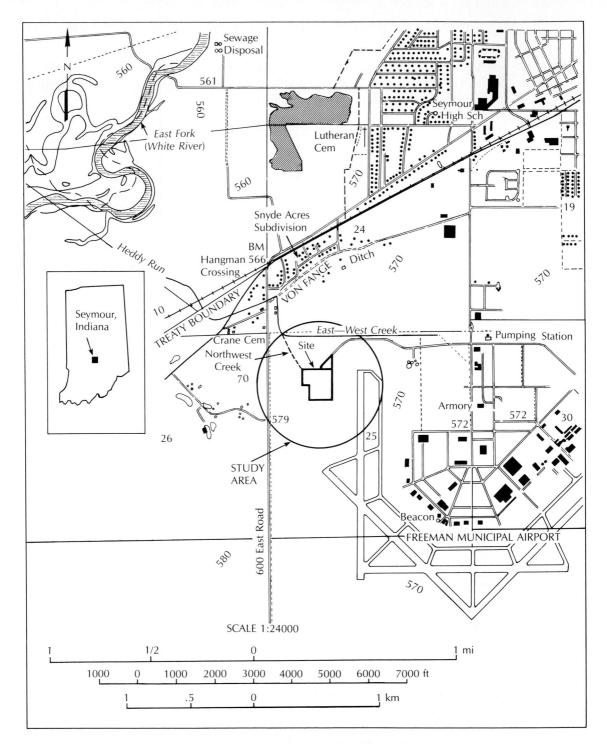

FIGURE 11.30 Location of the Seymour Recycling Corporation Superfund site. Source: CH2M-Hill, *Feasibility Study Report, Seymour Recycling Corporation Hazardous Waste Site,* U.S. Environmental Protection Agency, 1986.

11.9.2   Geology

The Seymour Recycling Corporation hazardous waste site is located on a nearly level plain at the edge of a glacial sluiceway now occupied by the East Fork of the White River. There are about 70 to 80 ft (21 to 24 m) of unconsolidated glacial fluvial sediments overlying shale bedrock at the site. The shale is reported to have a very low hydraulic conductivity and is not considered to be an aquifer. No bedrock wells are located in the area. There are both private and public water-supply wells screened in sand and gravel horizons of the glacial-fluvial deposits. A deposit of sand and gravel overlies the bedrock or a basal clay. In the immediate vicinity of the site, this deep aquifer is overlain by a confining layer of silty clay. The confining layer is as much as 50 ft (15 m) thick south of the hazardous waste site and is continuous beneath the site. However, it thins to the north and disappears about 2000 ft (600 m) northwest of the site. Figure 11.31 shows the thickness of the confining layer at a number of boring locations. Overlying the confining layer at the hazardous waste site is a sandy outwash layer called the shallow aquifer. This aquifer is relatively thin, 8 ft (2.4 m), at the southern end of the study area, but it thickens to the north. Eventually the shallow aquifer and the deep aquifer merge where the confining layer thins and disappears. There is a surface deposit of dune sand and loess, which is from 6 to 10 ft (2 to 3 m) thick. This acts as a semiconfining layer for the underlying layers of outwash sand. Figure 11.32 is a geologic cross section that goes roughly southeast to northwest; the location of the cross section is in Figure 11.33.

11.9.3   Hydrogeology

Ground water in the shallow aquifer is moving to the north beneath the site at a rate of about 400 ft (120 m) per year. Figure 11.34 is a water-table map of the site. There is a shallow stream, the East-West Creek, lying about 1000 ft (300 m) north

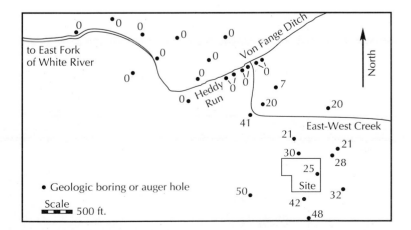

FIGURE 11.31   Thickness, in feet, of the confining layer as measured in borings in the vicinity of the Seymour Hazardous Waste Site.

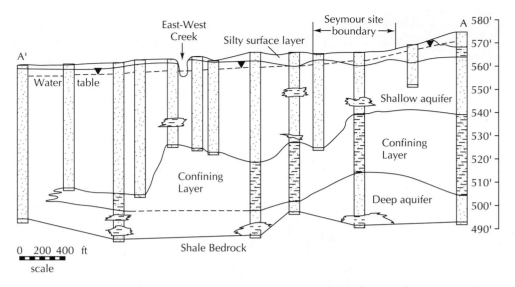

FIGURE 11.32 Southeast to northwest geological cross section through the Seymour Recycling Corporation Hazardous Waste Site. The elevation scale on the right is in feet above mean sea level.

of the site. During periods of high ground-water levels, the upper part of the shallow aquifer discharges in the East-West Creek. When ground water levels fall, the East-West Creek dries up, and all the ground water flows north beneath it. The East-West Creek discharges in the Von Fange Ditch, which drains to Heddy Run, a tributary of the East Fork of the White River. Ground-water flow north of the East-West Creek follows the course of the Von Fange Ditch and Heddy Run by turning west and draining toward the East Fork of the White River.

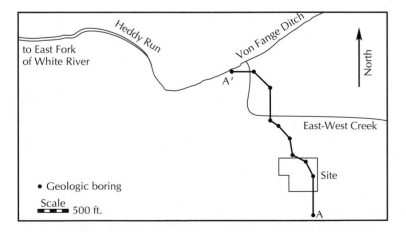

FIGURE 11.33 Location of the geological cross section shown on Figure 11.32.

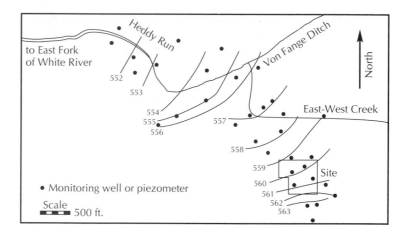

FIGURE 11.34 Elevation of the water table in the vicinity of the Seymour Recycling Corporation Hazardous Waste Site. Elevations are in feet above mean sea level.

In the part of the study area where the confining layer is present, the shallow aquifer and deep aquifer are hydraulically separated. The water table in the shallow aquifer is higher than the potentiometric surface of the deep aquifer, so there is downward leakage to the deep aquifer. Moreover, the deep aquifer beneath the site is flowing to the southeast, which is in the opposite direction of the flow in the shallow aquifer. Water levels in the deep aquifer are heavily influenced by pumping from an irrigation well located west of the site and high-capacity municipal supply wells located to the east.

11.9.4 Ground-Water Contamination

Three rounds of water-quality samples were collected from the monitoring wells in 1984 and 1985 as a part of the remedial investigation. A large number of hazardous organic compounds were found in the shallow aquifer beneath the site. Table 11.9 shows the maximum concentration of various organic compounds found in August 1984. The compounds that were quantified were on a list, the target organic compound list, that was used at most Superfund sites where the specific organic compounds that were present were unknown. A large number of tentatively identified compounds were not quantified.

Trans-1,2-dichloroethene had the highest concentration of any compound, as much as 240 mg/L. The water solubility of this compound is 600 mg/L at 20°C. With the concentration of a nonaqueous phase liquid at 40% of solubility, there is a possibility that pure phase product might be present in the aquifer. This is probably not the case at this site. First of all, no nonaqueous phase liquids were recovered from the monitoring wells. Moreover, in nested wells beneath the site the deeper wells within the shallow aquifer had lower concentrations than more shallow wells. If a dense nonaqueous phase liquid, such as trans-1,2-dichloroethene, were present in an aquifer, one would expect to find the product at the

TABLE 11.9 Maximum concentrations in µg/L of target organic compounds in August 1984 at the Seymour Recycling Corporation hazardous waste site

Semivolatile compounds	Maximum concentration
2,4-dimethylphenol	780 µg/L
Phenol	4,800
Benzoic acid	20,000
2-methylphenol	190
4-methylphenol	6,000
bis(2-ethylhexyl)phthalate	17
2-chlorophenol	10 K
Isophorone	120
Benzyl alcohol	200 K
Pentachlorophenol	10 K
Benzo(a)anthracene	10 K
Chrysene	20 K
Napthalene	100 K
1,2-dichlorobenzene	100 K
Volatile compounds	
Chloroethane	14,000
1,2-dichloroethane	3,200
1,1,1-trichloroethane	93,000
1,1,2-trichloroethane	2,300
Tetrachloroethene	500 K
1,1-dichloroethene	2,600
Trans-1,2-dichloroethene	240,000
Trichloroethene	500 K
Chloroform	3,300
Methylene chloride	32,000
Vinyl chloride	19,000
Acetone	37,000
Benzene	9,700
Ethylbenzene	500 K
Toluene	18,000
2-butanone	31,000
2-hexanone	9,800
4-methyl-2-pentanone	12,000
Total xylenes	500 K

K indicates a compound identified but at a concentration less than the indicated quantification limit

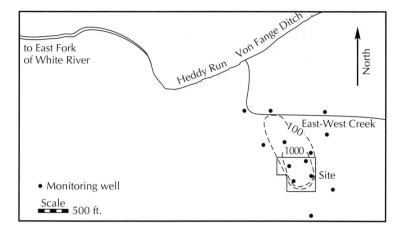

FIGURE 11.35 Extent of the 1,4-dioxane plume on December 1984. Concentrations in μg/L.

bottom of the aquifer, and hence the greatest dissolved concentrations at the bottom of the aquifer as well.

The extent of the plume of contamination in the shallow aquifer was based on the total of the hazardous organic compounds on the target organic list. As it turned out, one of the tentatively identified compounds was more mobile and less subject to natural biodegradation than any of the compounds on the target organic list. The reported extent of this compound, 1,4-dioxane, a cyclical ether, was contained within the target organic compound plume in December 1984 because there was a high analytical detection limit for the compound at that time. Figure 11.35 shows the extent of the 1,4-dioxane plume in December 1984.

Because of pending litigation and settlement negotiations, no water-quality samples were collected between 1985 and 1988. In 1989 it was discovered that 1,4-dioxane and another mobile compound, tetrahydrofuran, had spread much farther than expected. New analytical methods lowered the detection limit for 1,4-dioxane from 100 μg/L to 1 μg/L. Figure 11.36 shows the known 1,4-dioxane plume in 1990.

11.9.5 Site Remediation

Private water-supply wells were located to the north of the site in the shallow aquifer. As a part of the remedial effort at the site, public water supply was extended to the owners of these wells and the wells were sealed. Although this alleviated the immediate danger, the plume will eventually reach a point of surface-water discharge, and new wells could be installed in the future outside the area with public water. Therefore, an endangerment assessment recommended that the contaminated ground water be remediated.

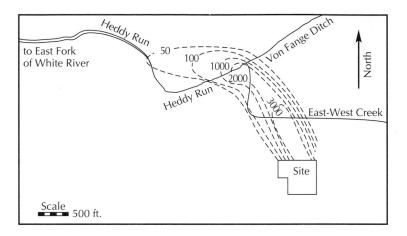

FIGURE 11.36   Extent of the 1,4-dioxane plume in 1990. Concentrations in μg/L.

The site-remediation plan had a number of elements:

1.   A plume-stabilization well was installed to prevent the further spread of the contamination. Unfortunately, the 1,4-dioxane plume was far past the area of influence of the plume-stabilization well by the time that it was built. However, the plume-stabilization well did capture the chlorinated organic plume, which was the basis for its design.

2.   Nutrients, including nitrogen, phosphorous, and potassium, were added to the soil at the site to promote the growth of soil microbes, which would biodegrade the residual organic compounds in the vadose zone.

3.   A horizontal soil-vapor-extraction system was built to flush volatile organic compounds from the soil at the site.

4.   Following the addition of the nutrients and the construction of the soil-vapor-extraction system, a multimedia cap was installed over the site to prevent the infiltration of rain water. This prevented any further release of organic compounds from the soil.

5.   A ground-water extraction system was installed. The first extraction well, the plume-stabilization well, is located some 600 ft (180 m) down-gradient of the site. As the ground water travels over this distance, substantial biodegradation occurs. The only chlorinated organic found in the closest extraction well is chloroethane. This is one of the end products of the biodegradation of chloronated organics. Two more extraction wells are located along the axis of the contamination plume. Figure 11.37 shows the three extraction wells and their calculated ground-water capture zones. The overlapping capture zones are greater than the extent of the 1,4-dioxane plume. The extracted water is pretreated at the site by air stripping and carbon extraction before being sent to the Seymour Waste Water Treatment Plant.

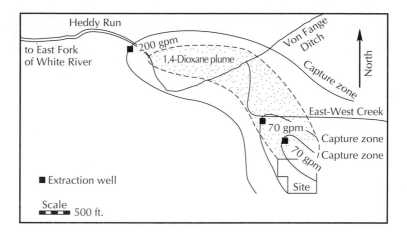

FIGURE 11.37 Location of extraction wells, with pumping rates in gallons per minute and corresponding capture zones.

6. The extraction well system will operate for a minimum of 12 y. It will continue until the water is cleaned to meet specific standards or until the pumping does not result in any further reduction in contamination. Air and ground-water monitoring will continue throughout the period of the remediation.

Total costs for remedial investigation, design, construction, and operation through 1992 were $25 million. An additional expenditure of $12 million is projected for operating expenses for the next 18 y.

References for this case study include: Nyer, Kramer, and Valkenburg (1991); Hauptman, Rumbaugh, and Valkenburg (1990); Kramer, Valkenburg, and Hauptman (1990); Fetter (1985, 1989, 1992); Miller and Valkenburg (1992); Walter et al. 1990; and Sghia-Hughes, Taddeo, and Fogel (1991).

11.10   CAPTURE ZONE ANALYSIS

In the design of pump-and-treat ground-water remediation systems, it is necessary to compute the *capture zone* of the extraction well(s). A capture zone consists of the up-gradient and down-gradient areas that will drain into a pumping well. If the water table is perfectly flat, the capture zone will be circular and will correspond to the cone of depression. However, in most cases the water table is sloping, so the capture zone and the cone of depression will not correspond. The capture zone will be an elongated area that extends slightly down-gradient of the pumping well and extends in an up-gradient direction. Capture zones are controlled by the time that it takes for water to flow from an up-gradient area to the pumping well. If sufficient pumping time elapses, the capture zone will eventually extend up-gradient to the closest ground-water divide.

In order to protect the quality of water draining toward municipal supply wells, *well-head protection zones* are being defined in many communities. Land uses that may result in ground-water contamination are being regulated in the well-head protection zones, which commonly correspond to the recharge areas for well fields completed in unconfined aquifers. The capture zone of an extraction well is equivalent to a well-head protection zone of a water-supply well.

Figure 11.38A shows a cross section of an unconfined aquifer with a sloping water table. The cone of depression of the pumping well will reverse the hydraulic gradient for a short distance down-gradient of the well. Thus, the capture zone extends for a short distance down-gradient of the well. The entire capture zone is surrounded by a ground-water divide. Outside the divide, flow will pass by the well, whereas inside the divide, the flow will be drawn into the well. This is illustrated in Figure 11.38B.

The equation to describe the edge of the capture zone for a confined aquifer when steady-state conditions have been reached is (Todd 1980; Grubb 1993):

$$x = \frac{-y}{\tan(2\pi Kbiy/Q)} \tag{11-16}$$

where

x and y are directions defined on Figure 11.38

Q is the pumping rate (L^3/T; ft^3/day or m^3/day)

K is the hydraulic conductivity (L/T; ft/day or m/day)

b is the initial saturated thickness of the aquifer (L; ft or m)

i is the hydraulic gradient of the flow field in the absence of the pumping well (dimensionless)

$\tan(y)$ is in radians

There are two bounding solutions to Equation 11–16 that do not follow directly from the preceding form of the equation.

1. The distance from the pumping well downstream to the stagnation point that marks the end of the capture zone is given by

$$x_0 = -Q/(2\pi Kbi) \tag{11-17}$$

where x_0 is the distance from the pumping well to the down-gradient edge of the capture zone (L; ft or m).

2. The maximum width of the capture zone as x approaches infinity is given by

$$y_{\max} = \pm Q/(2Kbi) \tag{11-18}$$

where $y_{\max}$ is the half-width of the capture zone as x approaches infinity.

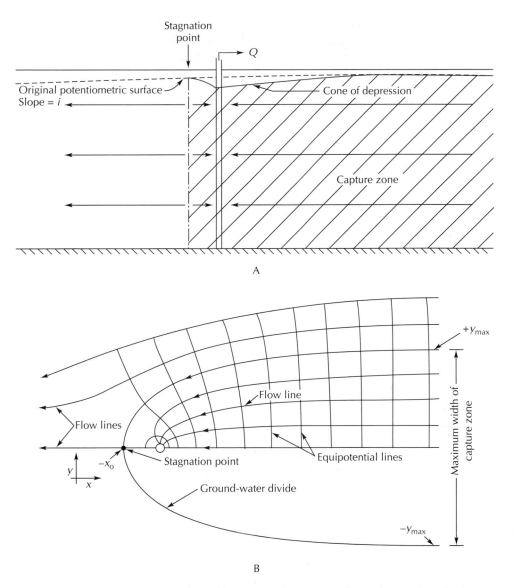

FIGURE 11.38 Capture zone of a well pumping from an aquifer with a uniform hydraulic gradient. **A.** Cross-sectional view. **B.** Top view of equipotential field.

Equations 11–16 through 11–18 are applicable to confined aquifers. In order to plot the shape of the capture zone, first use Equation 11–18 to find the maximum width of the capture zone. Once the value of y_{max} is known, substitute smaller values of y into Equation 11–16 to find the position of the capture zone. Finally, use Equation 11–17 to find the stagnation point where $y = 0$.

Grubb (1993) gave a solution for a capture zone in an unconfined aquifer. In this case one needs to know the head at two wells along the axis of the ground-water-flow direction. Let h_1 be the up-gradient head and h_2 be the down-gradient head, with L the distance between the two monitoring wells. In this case the shape of the capture zone is given by

$$x = \frac{-y}{\tan\left[\pi K(h_1^2 - h_2^2)y/QL\right]} \qquad (11-19)$$

where

x and y are directions defined in Figure 11.38
Q is the pumping rate (L^3/T; ft³/day or m³/day)
K is the hydraulic conductivity (L^2/T; ft²/day or m²/day)
tan(y)   is in radians

The maximum width of the capture zone as x approaches infinity is given by

$$y_{max} = \pm \frac{QL}{K(h_1^2 - h_2^2)} \qquad (11-20)$$

The position of the stagnation point where $y = 0$—that is, the down-gradient end of the capture zone—is given by

$$x_0 = \frac{-QL}{\pi K(h_1^2 - h_2^2)} \qquad (11-21)$$

EXAMPLE PROBLEM   Compute the capture zone of an interceptor well pumping at a rate of 190,000 ft³/day from a confined aquifer with a hydraulic conductivity of 1500 ft³/day, an original hydraulic gradient of 0.00300 and a saturated thickness of 75.0 ft.
Part A: Find the maximum width of the capture zone.

$$y_{max} = \pm \frac{Q}{2bKi}$$

$$= \pm \frac{190,000 \text{ ft}^3/\text{day}}{2 \times 75.0 \text{ ft} \times 1500 \text{ ft}^2/\text{day} \times 0.00300}$$

$$= \pm 281 \text{ ft}$$

Part B: Find the distance to the stagnation point.

$$x_0 = -\frac{Q}{2\pi Kbi}$$

$$= -\frac{190,000 \text{ ft}^3/\text{day}}{2 \times \pi \times 1500 \text{ ft}^2/\text{day} \times 75.0 \text{ ft} \times 0.00300}$$

$$= -89.6$$

Part C: Find the shape of the curve describing the capture zone. The following values are for one-half of the capture zone, which is symmetrical about the x-axis.

$$x = \frac{-y}{\tan\left(\dfrac{2\pi Kbiy}{Q}\right)}$$

$$= \frac{-y}{\tan\left(\dfrac{2 \times \pi \times 1500 \text{ ft}^2/\text{day} \times 75.0 \text{ ft} \times 0.00300 \times -y}{190{,}000 \text{ ft}^3/\text{day}}\right)}$$

$$= -\frac{y}{\tan(0.0116y)}$$

Remember that the tangent must be in radians.

y	x
250 ft	682 ft
200 ft	156 ft
175 ft	70.3 ft
150 ft	15.5 ft
140 ft	−1.2 ft
130 ft	−15.7 ft
100 ft	−48.9 ft
70 ft	−70.6 ft
30 ft	−86.2 ft
10 ft	−89.2 ft
1 ft	−89.6 ft

COMPUTER NOTES

QUICKFLOW is a ground-water model that—among other things—can be used to model capture zones for extraction wells. This model is based on analytical solutions to the two-dimensional ground-water-flow equation for both steady-state and transient conditions. The steady-state module can simulate wells, uniform recharge, recharge concentrated in circular areas (to simulate recharge ponds), and line sources and sinks. A uniform regional hydraulic gradient can also be simulated. The transient module simulates the effect of wells and a uniform regional gradient in a confined or leaky confined aquifer.

The program is interactive and features pull-down menus. Menus are accessed by moving the highlight bar with the Arrow keys and then pressing the Enter key. The model needs the following data for all modes: regional hydraulic gradient and direction of flow, hydraulic conductivity, aquifer top and bottom elevations, and a reference head, which is a point where the head is always known; it does not change during the simulation. The reference head should be located as far as possible from points where the head may change, such as recharge ponds and wells. For transient flow problems, the following additional data are needed: storage coefficient, Hantush leakage factor, and time to compute the solution. A digitized base map can be entered into the program to provide a graphical reference on the screen.

The main screen of the program has a menu bar across the top and a status window on the right that lists the current values of aquifer parameters and model settings. In order to start a modeling exercise, the aquifer parameters must first be entered. QUICKFLOW has default values for several of these parameters, but you may need to change them. Highlight **aqUifer** on the main menu bar and press the Enter key. A pull-down menu will appear for the following parameters, which are also listed in the status window on the right.

In order to specify a parameter, highlight the key word on the pull-down menu and press the Enter key. A window will then open; type in the parameter value and press the Enter key. Use consistent units, such as ft, ft/day, ft^3/day, and days or m, m/day, m^3/day, and days. For unconfined, steady-state solutions, set the aquifer top well above the reference elevation. An aquifer bottom of zero is typically used as the datum, with all elevations measured above that datum. The heads are used to compute transmissivity in the transient model. For the steady-state model, the aquifer is assumed to be unconfined if the head drops below the top.

In order to add wells, linesinks, ponds, streamlines, and/or particle traces to the steady-state model or wells and/or particle traces to the transient model, highlight **Add** on the main menu bar and press Return. Then move the highlight bar to the desired item and press the Enter key. Mark the location of the item on the screen by moving the cursor with the Arrow keys or a mouse. Once the cursor is moved to the correct location, press the Enter key to mark the location. Depending upon the feature, the program will ask for further information. For **wells,** you will need to enter the well radius (L) and the pumping rate (L^3/T). A positive pumping rate is used for a production well and a negative pumping rate for an injection well. For **head linesink** or **flux linesink,** use the Enter key to mark the beginning of the linesink; then move with the Cursor keys or a mouse to the end of the linesink and press the Enter key again. For a head linesink, enter a

Pull-Down Menu	Status Window	Parameter
K	K:	Hydraulic conductivity (L/T)
Bottom	Bot:	Aquifer bottom elevation (L)
Top	Top:	Aquifer top elevation (L)
Reference	Hr:	Reference head (L)
Gradient	I:	Hydraulic gradient (dimensionless)
rEcharge	R:	Uniform recharge rate (L/T)
Porosity	n:	Porosity (dimensionless)
Storage	S:	Storage coefficient (dimensionless)
Leakage	L:	Hantush leakage factor (L)
tIme	t:	Time to compute transient solution (T)

constant head value. For a flux linesink, divide the total flux from the linesink by its length to get a rate per unit length (L^2/T). A positive value is a drain and a negative value is a recharge source, such as an infiltration gallery. For **ponds,** select the center of the pond and press Enter. Then move the cursor to the edge of the pond and press Enter again. The pond rate is in units of $L/T,$ with a positive value being recharge and a negative value representing discharge via the pond. There is a pull-down menu for **Streamlines** that allows you to specify a single streamline or several streamlines starting in either a line or a circle. The point you select for a streamline is the beginning of the streamline, which will then be drawn down-gradient from that point. Arrowheads may be added to streamlines by selecting **Opts** on the main-menu bar and **Arrow** on the pull-down menu. **Particle Traces** are added in the same fashion as streamlines. However, they may be tracked up-gradient from a well or other sink as well as down-gradient. All these features can be edited using **Select** on the main-menu bar. The pull-down menu is the same as the **Add** menu selection. Move the highlight bar to the item and press the Enter key. Then mark the location of the item to be edited or deleted with the cursor. Press the **Edit** key and choose from **Edit selected item, Delete selected item,** or **delete All like items.** In order to move a well or linesink, it must first be deleted and then reentered. The default mode of the model is a steady-state solution. A transient solution can be selected using the **Model** option on the main-menu bar.

Once the aquifer parameters are entered and the options are selected, move the highlight bar to **Calc** on the main-menu bar and press the Enter key. Then select **reCalculate** on the pull-down menu and press the Enter key. The problem will be solved and the result will be displayed on the screen. To exit QUICKFLOW, highlight **Quit** and then type Y and O.

You will find a free student version of QUICKFLOW on the computer diskette accompanying this book. Instructions for running the program are found in Appendix 15. This student version will not support digitized maps, has a limited computational grid, and will allow a maximum of 5 wells, 5 linesinks, 5 ponds, 10 streamlines, and 10 particle traces. To start using the demonstration model, select **File** and then move the highlight bar to **Retrieve** and press Enter. The file demo.qfl will appear. Press the Up arrow to mark this, and use a mouse to click on the symbol ^ent or press the Control key and the Enter key simultaneously. A one-well model will be created and appear on the screen. You can add wells, linesinks, additional streamlines, etc., to the model with the **Add** menu and then recalculate to get a feel for the model.

If you wish to obtain additional information about this program or to purchase the fully configured model, contact:

Geraghty & Miller, Inc.
Modeling Group
10700 Parkridge Boulevard, Suite 600
Reston, VA 22091
(703) 758-1200

NOTATION

a_L	Dynamic dispersivity	K_f	Adsorption isotherm coefficient
b	Saturated thickness of an aquifer	L	Length of flow path
C	Solute concentration	n_e	Effective porosity
C^*	Amount of solute sorbed per unit dry mass of soil	Q	Pumping rate
		t	Time
C_0	Initial solute concentration	v_c	Solute front velocity, where $C/C_0 = 0.5$
D	Diffusion coefficient	v_x	Average linear ground-water velocity
D_L	Longitudinal coefficient of hydrodynamic dispersion	w	Diffusion coefficient
		x_0	Down-gradient distance from pumping well to edge of capture zone
D^*	Effective diffusion coefficient		
dC/dx	Concentration gradient	y_{max}	Maximum half-width of the capture zone
dh/dl	Hydraulic gradient	β_1	Adsorption isotherm coefficient
F	Mass solute flux	β_2	Adsorption isotherm coefficient
i	Hydraulic gradient	ρ_b	Dry bulk mass density
j	Adsorption isotherm coefficient	θ	Volumetric moisture content
K	Hydraulic conductivity		
K_d	Distribution coefficient		

PROBLEMS

1. A saline solution with a concentration of 370 mg/L is introduced into a 2-m-long sand column in which the pores are initially filled with distilled water. If the solution drains through the column at an average linear velocity of 0.79 m/day and the dynamic dispersivity of the sand column is 15 cm, what would the concentration of the effluent be 1.8 days after flow begins?

2. Given the flow situation in Problem 1, what would the effluent concentration be 2.1 days after flow begins?

3. A landfill is leaking an effluent with a concentration of sodium of 1250 mg/L. It seeps into an aquifer with a hydraulic conductivity of 7.3 m/day, a gradient of 0.0030, and an effective porosity of 0.25. A down-gradient monitoring well is located 25 m from the landfill. What would the sodium concentration be in this monitoring well 300 days after the leak begins? *Note:* In this problem you will need to find erfc($-x$), which is equal to $1 + $ erf(x).

4. What would the concentration of sodium be at the same time at a monitoring well located 32 m down-gradient of the leaking landfill?

5. What is the relative velocity of a solute front of a solute-soil system with a distribution coefficient of 23 mL/g, a dry bulk density of 2.12 gm/cm^3, and a volumetric water content of 0.20?

6. What is the relative velocity of a solute front of a solute-soil system with a distribution coefficient of 1.9 mL/g, a dry bulk density of 1.88 gm/cm^3, and a volumetric water content of 0.29?

7. A capture well is pumping at a rate of 50,000 ft^3/day from an unconfined aquifer with a hydraulic conductivity of 1250 ft/day, an initial hydraulic gradient of 0.0043, and an initial saturated thickness of 40 ft.

   A. What is the maximum width of the capture zone?

   B. What is the distance from the well to the stagnation point?

8. A capture well is pumping at a rate of 1500 m^3/day from an unconfined aquifer with a hydraulic conductivity of 425 m/day, an initial hydraulic gradient of 0.00098, and a saturated thickness of 22 m.

   A. What is the maximum width of the capture zone?

   B. What is the distance from the well to the stagnation point?

12 Ground-Water Development and Management

I approach now an account of the experiments that I carried out at Dijon together with Engineer Charles Ritter, to determine the laws of flow of water through sand. . . . Each experiment consisted of establishing a specified pressure in the upper chamber of the column by adjustment of the inflow tap; then, when it was established by means of two observations that the flow had become essentially uniform, the outflow from the filter during a certain time was noted, and the mean outflow per minute was calculated from it.

Les fontaines publiques de la ville de Dijon, Henry Darcy, 1856

12.1   INTRODUCTION

The area of water resources development and management is so broad that we will make no attempt to cover all aspects in this book. Instead we will focus on ground water, since it is this aspect of the general topic to which hydrogeology most closely relates. However, as ground water is not isolated from surface water, a study of ground-water development necessarily encompasses many aspects of surface-water flow.

Surface water occurs in readily discernible drainage basins. The boundaries are topographic and may be easily delineated on a topographic map. The water conveniently flows in the direction in which the land surface is sloping. Moreover, surface water does not cross topographic divides (except, perhaps, during floods) and the locations of the drainage divides are fixed. Ground water, on the other hand, occurs in aquifers that are hidden from view. The boundaries of an aquifer are physical: it can crop out, abut an impermeable rock unit, grade into a lower-permeability deposit, or thin and disappear. At a given location, the land surface may be underlain by several aquifers. Each aquifer may have different chemical makeup and different hydraulic potential; each may be recharged in a different location and flow in a different direction. Moreover, ground-water divides do not necessarily coincide with surface-water divides. Clearly, the development and management of ground water is more complicated than that of surface water, simply on the basis of the mode of occurrence.

12.2   DYNAMIC EQUILIBRIUM IN NATURAL AQUIFERS

Under natural conditions, an aquifer is usually in a state of *dynamic equilibrium* (Theis 1938). A volume of water recharges the aquifer and an equal volume is discharged. The potentiometric surface is steady and the amount of water in storage in the aquifer is a constant. The aquifer transmits the water from the recharge areas to the discharge zones. The maximum amount of water any section of the aquifer can transmit is a function of the transmissivity and the maximum gradient of the potentiometric surface. If the water table is close to the surface of an unconfined aquifer, the aquifer is full and is transmitting the maximum amount of water. If, however, the water table is far below the surface, the aquifer is not transmitting water at full capacity.

The amount of water that recharges an unconfined aquifer is determined by three factors: (1) the amount of precipitation that is not lost by evapotranspiration and runoff and is thus available for recharge; (2) the vertical hydraulic conductivity of surficial deposits and other strata in the recharge area of the aquifer, which determines the volume of recharged water capable of moving downward to the aquifer; and (3) the transmissivity of the aquifer and potentiometric gradient, which determine how much water can move away from the recharge area. Should an aquifer be transmitting the maximum volume of water, it is more than likely that some potential recharge is being rejected in the recharge area. This is often the case in humid areas. Should the water table be low, indicating that the aquifer is not flowing at full capacity, there is probably either a lack of potential recharge or low vertical hydraulic conductivity in the recharge area, retarding downward movement. Aquifers in arid regions typically have deep water tables in the recharge areas, indicating a deficiency in the amount of potential recharge.

Recharge to confined aquifers can occur in places in which the confining layer is absent. Under such conditions, the three factors affecting unconfined aquifer recharge are controlling. If there is a hydraulic gradient across a leaky confining layer in a direction that promotes flow into the aquifer, then recharge can occur across the confining layer. In this case, the vertical hydraulic conductivity of the confining layer, the thickness of the confining layer, and the head difference across it control the amount of recharge. Recharge to a confined aquifer may come from both downflow from a higher aquifer or upflow from a lower aquifer.

When a well begins to pump water from an aquifer, the water is withdrawn from storage around the well and from vertical leakage (Theis 1938). As the cone of depression grows, an increasingly larger portion of the aquifer will be contributing water from storage. The amount of water discharging naturally from the aquifer will remain at the predevelopment rate until the pumping cone reaches the recharge or discharge area. When the pumping cone reaches a discharge area, the potentiometric gradient toward the discharge area is lowered and the amount of natural discharge proportionally reduced. If the pumping cone reaches the recharge area of an aquifer, it may induce additional recharge of water that was previously rejected. It is even possible for a section of the aquifer to change from a discharge area to a recharge area. For example, drawdown near a river may eliminate ground-water discharge to the river and induce infiltration from the river into the aquifer, reversing the prior direction of flow.

In any event, the pumping cone will continue to grow until it has sufficiently reduced natural discharge or increased recharge to balance the volume of water removed by pumping. With this occurrence, a new condition of dynamic equilibrium is reached (Theis 1938). However, it is important to note that in order for this to happen a pumping cone must form. This means the potentiometric surface in parts of the aquifer must be lowered. The resultant cone of depression is usually a complex surface resulting from the withdrawals of hundreds of wells. Should the sum of the remaining natural discharge and the pumping withdrawals exceed the available recharge, the cone of depression will not stabilize and water levels will continue to fall.

The rate at which the cone of depression spreads is a function of the storativity of a confined aquifer or the specific yield of a water-table aquifer. As storativity values are 100 to 1000 times smaller than specific yields, the cone of depression will spread 100 to 1000 times faster in an artesian aquifer than in a water-table aquifer. It can take many years for the cone of depression to influence recharge or discharge areas sufficiently for an aquifer to regain dynamic equilibrium.

CASE STUDY: DEEP SANDSTONE AQUIFER OF NORTHEASTERN ILLINOIS

The deep sandstone aquifer beneath northeastern Illinois is confined by the Maquoketa Shale, a leaky layer. It has an outcrop area that receives direct recharge west of the limit of the shale. Under predevelopment conditions, the potentiometric surface stood at or above land level and had a very gentle slope toward Lake Michigan of about 3×10^{-4} foot per foot. Flow through the aquifer from the recharge area was about 0.9 million gallons (3400 m^3) per day (Suter et al. 1959). Natural discharge was by slow upward leakage through the shale into Silurian limestone beneath Lake Michigan. Most of the ground-water recharge in the recharge area of the regional aquifer was circulating in local flow systems. It could not move as a part of regional flow owing to the low carrying capacity of the aquifer caused by the gentle slope of the potentiometric surface. As ground-water development began, the well discharge soon exceeded the natural discharge and the pumping cone rapidly grew. As the hydraulic gradient steepened, increasingly greater amounts of water were drawn into the regional flow system. By 1958, an estimated 19 million gallons (72,000 m^3) per day were moving into the pumping cone (Suter et al. 1959). At that time, pumpage from the aquifer was 43 million gallons (163,000 m^3) per day. The difference between pumpage and recharge was coming from storage as the cone of depression grew. By 1975, ground-water pumpage was 150 million gallons (568,000 m^3) per day, and the hydraulic gradient was 4×10^{-3} ft/ft, with water levels falling several feet per year.

If pumpage from the deep sandstone aquifer in the Chicago region had been limited to 46 million gallons (174,000 m^3) per day, the cone of depression would have eventually stabilized. The period to reach a stable configuration would have been about 50 y from the time pumpage first reached 46 million gallons (174,000 m^3) per day (Fetter & Young 1978; Young 1976). However, so long as the rate of deep-well pumpage exceeds 46 million gallons per day, the cone of depression will continue to grow (Sasman et al. 1977).

12.3   GROUND-WATER BUDGETS

Some knowledge of the amount of natural recharge to an aquifer is mandatory in a ground-water development program. There are several methods available to make such an estimate. In Section 3.7, a method was described for estimating annual ground-water recharge from baseflow-recession curves. This method is useful for areas in which the ground-watershed and the river basin coincide.

A second method is to determine the flow in the aquifer across a vertical plane at the boundary of the recharge area and the discharge area (Walton 1960). If there is sufficient knowledge of the potential field of the aquifer, the area(s) of recharge and discharge can be determined. The rate of steady flow from recharge areas to discharge areas is determined using either Darcy's law for a confined aquifer or the Dupuit equation for an unconfined aquifer. The flow from the recharge areas is equal to the rate of recharge for aquifers in dynamic equilibrium.

A **water budget** for the recharge area of an aquifer is a very useful means of determining ground-water recharge. An advantage of the water-budget method is that the aquifer does not have to be in dynamic equilibrium in order to use it. Many of the parameters used for a hydrologic budget are measured directly: precipitation, streamflow, transported water, and reservoir evaporation. Ground-water inflow, outflow, and change in storage are computed from the hydraulic aquifer characteristics and measured potentiometric data. The amount of evapotranspiration could be measured in lysimeters, but it is more typically computed using an appropriate formula such as the Thornthwaite method (Section 2.3). Basinwide evapotranspiration may also be determined by a water-budget analysis as follows:

$$
\begin{aligned}
\text{Evapotranspiration} = {} & (\text{precipitation} + \text{surface-water inflow} \\
& + \text{imported water} + \text{ground-water inflow}) \\
& - (\text{surface-water outflow} + \text{ground-water} \\
& \quad \text{outflow} + \text{reservoir evaporation} \\
& \quad + \text{exported water}) \\
& \pm \text{changes in surface-water storage} \\
& \pm \text{changes in ground-water storage}
\end{aligned}
\tag{12–1}
$$

The natural recharge to an undeveloped aquifer may be determined by a water-budget analysis of the recharge area:

$$
\begin{aligned}
\text{Ground-water recharge} = {} & (\text{precipitation} + \text{surface-water inflow} \\
& + \text{imported water} + \text{ground-water inflow}) \\
& - (\text{evapotranspiration} + \text{reservoir} \\
& \quad \text{evaporation} + \text{surface-water outflow} \\
& \quad + \text{exported water} + \text{ground-water outflow}) \\
& \pm \text{changes in surface-water storage}
\end{aligned}
\tag{12–2A}
$$

Equation 12-2A can account for ground-water recharge not only from precipitation, but also from losing streams, irrigation water, unlined canals, and so forth. Its usefulness may be limited, however, if evapotranspiration cannot be determined. Detailed knowledge of all the factors is necessary if the computation of recharge is to be accurate.

If the land surface of the recharge area is developed for agriculture, industry, urban growth, and so forth, additional computations of recharge may be necessary. Water used for many purposes may be recharged to the ground-water reservoir. Excess irrigation water is one example. On Long Island, New York, all ground water pumped for cooling or air conditioning must be returned to the aquifer from which it was removed. Water used for domestic purposes is often recharged by septic tank drain fields or cesspools. The increasing emphases on the use of land systems* for municipal wastewater treatment means that treated sewage effluent that formerly flowed into rivers may now be recharging aquifer systems. The amount of recharge from such sources can be determined by a supplemental water-budget analysis:

$$
\begin{aligned}
\text{Additional ground-water recharge} = {}& (\text{industrial use} + \text{municipal use} \qquad \textbf{(12--2B)}\\
& + \text{domestic use} + \text{irrigation use})\\
& - (\text{cooling-water evaporation}\\
& + \text{irrigation-water evapotranspiration}\\
& + \text{water exported in products}\\
& + \text{sewage discharge into surface waters})
\end{aligned}
$$

The determination of the additional ground-water recharge is often more complicated than an analysis of the basic amount of recharge. Water-use records of dozens or hundreds of individual water users and sewage dischargers must be collected. These records may range from excellent to nonexistent. Many of the factors must be estimated. For example, the owners of private wells for home use will almost never know how much water they pump. An accurate accounting of this type involves long and tedious inventory analysis.

Should an attempt be made to balance the additions to the water supply of an area with the depletions, the result should be accurate to within the accuracy of measurement or estimation of the various parameters. Each parameter will have an accuracy of estimation dependent upon how precisely the measurement can be made. Measurement of streamflow might be accurate to $\pm 5\%$ for a good measurement. If total streamflow is 30 ft^3/s, then there is a variability of 30 ft^3/s $\times$ 0.05, or ± 1.5 ft^3/s. The overall accuracy of the estimate can be determined by taking a weighted average of the individual variability.

---

*Land systems of treatment include, among other techniques, spraying wastewater as irrigation water, recharging it through seepage basins, and allowing it to flow across the land surface. Biological, chemical, and physical processes act to remove pollutants and purify the water.

<table>
<tr><td>**EXAMPLE**
PROBLEM</td><td colspan="4">A water-budget analysis of a watershed indicates an estimated total outflow of 90 ft^3/s. The accuracy of estimation of each of the individual components is known. What is the accuracy of the estimate of the total discharge?</td></tr>
</table>

Factor	Estimated Flow (ft^3/s)	Accuracy of Estimate (%)	Variability of Factor (ft^3/s)
Evapotranspiration	60	±25	±15
Surface outflow	20	± 5	± 1
Ground-water outflow	5	±10	± 0.5
Exported water in canal	5	± 2	± 0.1
TOTAL	90		±16.6

The sum of the variability of each factor is ±16.6 ft^3/s; therefore, the accuracy of the total flow is 16.6/90 or ±18%.

The amount of water available for use from an aquifer is not the natural recharge. It is the increase in recharge or leakage from adjacent strata induced by development, along with the reduction in discharge. As water levels fall to accommodate the development, there will also be some water available from storage. A water-budget analysis is helpful in determining the amount of natural recharge and discharge. Further hydrologic analysis is then necessary to evaluate the effects that pumping will have on these figures.

12.4   MANAGEMENT POTENTIAL OF AQUIFERS

Aquifers can play many roles in the overall development of the water resources of an area. Some of the functions have been recognized for many years; others have been recognized only recently. The most obvious use of an aquifer is to supply water to wells—the *supply function*. One of the more vexing problems in ground-water management has been to determine how much water an aquifer can supply. This problem will be discussed in Section 12.5.

Aquifers also transmit water from one location to another. This has been called a *pipeline function* (Kazmann 1956). Many communities are dependent upon ground-water sources that are recharged elsewhere. In this case, the aquifer acts as an aqueduct. However, when the user community does not have zoning and land-use control, as, for example, when the aquifer is transmitting water some distance, there may be difficulty in protecting the recharge area of the aquifer from overdevelopment or contamination* (Hordon 1977). Aquifers are not as efficient in carrying large volumes of water as are surface canals. However, surface canals

---

*The 1986 Safe Drinking Water Act in the United States enables communities to establish ground-water protection zones around well fields.

can lose large amounts of water by evaporation; furthermore, they require capital for construction.

Ground water can be mined in the same manner as minerals. Whenever ground water is withdrawn at a rate greater than the rate of replenishment, mining is occurring. Under these conditions the water in storage in the aquifer must be considered a nonrenewable resource. This is the *mining function* of an aquifer.

In some aquifers, the rate of replenishment is so low that it is almost nonexistent. For example, the average annual recharge to the Ogallala aquifer of the southern High Plains of New Mexico and Texas is 1.5 in. (3.8 cm) per year (Brutsaert, Gross, & McGehee 1975). Based on the average rate of withdrawal for the 1951–1960 period, the ground-water mine will be exhausted in 100 y. That pumping rate is 28 times the natural recharge. Property owners of land in the High Plains are permitted an income-tax depletion allowance to compensate for the loss of property value due to a falling water table (Sellers 1973).

The unsaturated zone overlying an aquifer can act as a waste-treatment system. This has been called the *"filter-plant" function* of aquifers (Hordon 1977). However, the unsaturated zone can do much more than act as a physical filter to remove bacteria and viruses. It is also effective in removing phosphorus and heavy metals (Fetter 1977a). Passage of water through the saturated zone can also improve the quality. Degradation of native water quality can occur if the treatment potential of soil systems is exceeded.

Ground water can also have an *energy-source function*. The ground-water heat pump is a viable alternative to conventional heat pumps in some localities (Gass and Lehr 1977). As the thermal energy of the aquifer is removed by a heat pump in colder climates, it is another type of mining. In Wisconsin, the thermal impact is very small (Andrews 1978). In states farther south, where heating and air conditioning demands are more equal, the net impact on ground-water temperatures is even less.

Ground-water reservoirs sometimes also have a *storage function*. This is not true for an undeveloped aquifer in dynamic equilibrium if the recharge zone is rejecting potential recharge. However, if the ground-water reservoir has unused storage capacity, it can effectively store water from wet periods for use during time of drought (Ambroggi 1977; Greydanus 1978; Lehr 1978; Thomas 1978). Aquifers with available storage capacity may be either those that are not filled by the natural recharge to them or those in which the potentiometric surface has been lowered by pumping. Water put into storage could be from natural sources, especially extreme precipitation or flood events (Ambroggi 1977). Increasingly, treated wastewater effluent is being used to replenish aquifers (Fetter & Holzmacher 1974).

When ground water stored in aquifers can be used to replace surface-water reservoirs, a number of benefits accrue (Helwig 1978). The expense of surface-water reservoirs is circumvented, as there are no capital costs involved. There are no evaporative losses from ground-water storage, nor are there any infiltration losses. Surface-water reservoirs sometimes create great ecological disruption with the destruction of riverine and floodplain environments. Productive farmland may also be lost beneath surface reservoirs. Such disruption may be avoided with ground-water storage; furthermore, there is no worry over dam safety. On the other side of the coin, a person cannot waterski in a ground-water reservoir!

Aquifers are also used for the storage of natural gas. The aquifer must be confined, and a structural or stratigraphic feature, such as an anticline, is required to hold the gas in place. Wells are drilled through the structure and the gas is pumped into the aquifer under pressure. It displaces water and forms a bubble in the aquifer. Fresh water could also be stored in salt-water aquifers, as the former is less dense and would float as a bubble in the saline water.

12.5   PARADOX OF SAFE YIELD

It is a natural inclination of scientists to compare and classify phenomena in quantitative terms. Thus, it is to be expected that hydrogeologists have attempted to define the amount of water that could be developed from a ground-water reservoir. The term **safe yield** was apparently used in this regard as early as 1915 (Lee 1915). At that time, safe yield was regarded as the amount of water that could be pumped "regularly and permanently without dangerous depletion of the storage reserve." Later, other factors that need to be considered were added, such as economics of ground-water development (Meinzer 1923b), protection of the quality of the existing store of ground water (Conkling 1946), and protection of existing legal rights and potential environmental degradation (Banks 1953). Synonyms for safe yield appear in the literature, including "potential sustained yield" (Fetter 1972a), "permissive sustained yield" (American Society of Civil Engineers 1961), and "maximum basin yield" (Freeze 1971). A composite definition, based on the ideas of many authors, could be expressed as follows: Safe yield is the amount of naturally occurring ground water that can be withdrawn from an aquifer on a sustained basis, economically and legally, without impairing the native ground-water quality or creating an undesirable effect such as environmental damage.

The concept of ground-water withdrawals causing environmental damage warrants more than a mention. Many surface-water systems are dependent upon natural ground-water discharge. It has been shown by model studies that ground-water development may reduce streamflow and, as a consequence, lower lake levels and dry wetlands (Collins 1972). As these may be environmentally sensitive areas, the danger of environmental harm is real. Likewise, ground-water withdrawals have been linked to subsidence of the land surface (Bouwer 1977). This has resulted in land-surface cracking and damage to structures, highways, pipelines, dams, and tunnels. The gradients of irrigation canals have been changed—even reversed—and low areas have become flooded by sea water. In a broader sense, environmental impacts include ecological, economical, social, cultural, and political values (Fetter 1977b).

Many authorities are uneasy with the concept of safe yield. For some, the term is too vague (Thomas 1951). Obviously, the amount of ground water that can be produced will vary under varied patterns of pumping and development. In addition, the question of what would constitute an undesirable result to be avoided is open to debate (Anderson & Berkebile 1977). The abandonment of the term *safe yield* has been proposed on the grounds that it does not take into account the interrelationship of ground water and surface water and may preclude the development of the storage functions of an aquifer (Kazmann 1956).

However, in spite of the reservations of many hydrogeologists with regard to the concept of safe yield and its implications, the basic concept must be applied whenever the use of an aquifer is planned and managed. Ground-water management programs obviously imply that water must be pumped from the ground (Peters 1972). If there is no evaluation of the hydrologic and environmental impacts of various withdrawal programs, it is possible that uncontrolled withdrawals will exceed prudent levels.

A single value for the safe yield of an aquifer cannot be provided in the same sense as a quantity such as mean annual precipitation. Safe-yield values are based on a number of constraints; such values must be determined by a team of professionals, in the same manner that an environmental impact statement is prepared. Economists, engineers, engineering geologists, plant and wildlife ecologists, and lawyers might all participate with the hydrogeologist in preparing a safe-yield determination for an aquifer or a ground-water basin. The safe-yield evaluation should include a statement of the legal and economic constraints that were considered, as well as the limiting values of environmental damage that were considered. Indeed, such a study should provide a series of safe-yield values and the different factors that applied to each determination. This is obviously not a simple matter. Computer models of ground-water flow systems are ideal tools for estimating the series of values. All the hydraulic factors can be evaluated. R. A. Freeze has shown how a computer model can compute a "maximum basin yield" (Freeze 1971).

The safe yield of an aquifer system is only one facet of a ground-water management program. Artificial augmentation of precipitation or recharge could increase the amount of water that can be withdrawn on a sustained basis. The use of ground-water reservoirs for cyclic storage means that in drought years it is necessary, and desirable, to pump water on a temporary basis far in excess of the safe yield. Under these conditions, the ground-water supply would replace surface-water supplies that might be critically low or be used to irrigate crops normally watered by rainfall. In wet years, the ground-water reservoir would be replenished by above-average recharge and pumping at rates below the safe yield.

The underlying principle of ground-water development is that by withdrawing water from an aquifer, some of the natural discharge may be made available for use (Peters 1972).

12.6   WATER LAW

12.6.1   Legal Concepts

The development and management of water resources must take place within a framework of legal obligations, rights, and constraints. Naturally these factors differ from country to country; even in the United States, each individual state has its own body of water law. Because of this, it is not possible to fully discuss all water laws that might be applicable to a particular situation. However, we will look at some general principles.

There are two basic aspects of the legal framework: common law and legislative law (Tank 1983). *Common laws* are the traditional legal precepts laid down by court decisions. They are based primarily on precedent, but can be overturned by later courts if it is felt that societal needs have changed. In the United States, common law derives its legitimacy from the U.S. Constitution and the constitutions of the various states. The final arbitrator of common law is either the highest court in a state or the U.S. Supreme Court, depending upon whether an action is brought in state court or federal court.

Legislative law has two arms: *statutory law* and *administrative law*. Congress or state legislatures may pass laws that regulate water, thus creating statutory law when signed by the president or the governor. In addition, legislative bodies may enable their administrative bodies to write rules and regulations that have the power of law, thus creating administrative law.

Water law has traditionally been concerned with the quantity of water available. In this regard, a complex body of common law has risen. In recent years there has been a trend for legislation to be written to allocate water, especially in areas where it is scarce, rather than to rely upon common law. Society is acting to remedy what the majority sees as inadequacies and inequities in the common law.

The concept of protecting the quality of the water is relatively recent. Most common law has arisen with respect to water quantity. Water-quality issues that have been addressed under common law have generally been ones of some specific episode of pollution in which one or more of several common law theories (e.g., trespass, negligence, private nuisance, public nuisance, or strict liability) have been applied. Legislative law that addresses the issue of water quality has been passed at the federal, state, and local levels.

12.6.2 Laws Regulating Quantity of Surface Water

A **water right,** as defined by law, is not legal title to the water, but the legal right to use it in a manner dictated by state law.

In the eastern United States, the **riparian doctrine** of ownership of surface-water rights is recognized (Goldfarb 1988). This concept holds that the riparian landowner—the property owner adjacent to a surface-water body—has the first right to withdraw and use the water. This right is often controlled by the state in the sense that application to the state for a permit to withdraw is necessary. Ownership of the water right is held with ownership of the land, and all riparian owners have equal rights to the water. Water withdrawals are limited to "reasonable" use in comparison with other riparians. As all riparians have equal rights to use reasonable amounts of water—even new users—it is fortunate that there is a large amount of surface water available in the eastern United States. As eastern riparian owners generally return most of the water to the stream, albeit sometimes more polluted, there have been few crises due to lack of water except during droughts. Water problems in these areas typically involve quality rather than quantity of water.

Water in rivers was used in large quantities for mining during the 1849 California gold rush. Mining law recognized that the first to stake a claim to a property was the owner of the mineral therein. It was only natural that the same legal concept be applied to the water needed to extract, process, and transport the ore. Thus arose the **prior-appropriation doctrine** (Wilkinson 1986; Tarlock 1985), which is followed in 17 western states of the United States. This was first applied in *Irwin v. Phillips* (1855). The right to use water is separate from other property rights. The water-right holder does not necessarily need to be riparian, and riparian owners may have no water rights. The first person to divert water from a surface course has the primary water right, and it passes to successive owners. Junior users have lower rights and in time of drought may not receive any water, even though senior users always get their full share. However, some states have established a priority of use, with domestic use receiving top priority, irrigation receiving second priority, and industrial and commercial receiving lowest priority. Most states have some limitation on the transfer of appropriate rights, sometimes going so far as to link the water right to a specific use and piece of land (Hirshleifer, deHaven, & Milliman 1960). Some states provide for the forfeiture of a water right if it is not exercised for a given time period.

Courts have recently placed limitations on absolute ownership of surface water rights. In 1983 the California Supreme Court ruled in *National Audubon Society v. Superior Court of Alpine County* (1983) that the Los Angeles Department of Water and Power must limit its diversions from Mono Lake, even though the diversions were legal when initiated. The **public trust doctrine** was invoked. This basically held that the private right to use water was limited by the need to preserve the environmental aspects of a unique scenic, recreational, and scientific area, which benefited all (Goldfarb 1984). In 1984 the California State Water Control Board held that the Imperial Irrigation District of southeastern California could not permit Colorado River water to drain into the Salton Sea as return flow from irrigation (California Water Resources Control Board 1984). As this water was then not available for use by other parties, the Water Control Board ruled that the Irrigation District was wasting water, which is illegal under California law. (The Salton Sea is in a closed basin and has been formed by drainage of Colorado River water that had been diverted to California. Hence, there is no public trust to preserve the Salton Sea because it is not a natural feature.)

The **Winters Doctrine** also limits the doctrine of prior appropriations. In 1908 the U.S. Supreme Court held that when Congress created reservations for the Indian nations, the water rights to develop those reservations and make the land productive were reserved (*Winters v. United States* 1908). This included water on nonreservation land that had been opened to settlement. Native American water rights are senior to those granted by state law in most basins, since the reservations were established so long ago (Berry 1974; Collins 1986). Most Native American water rights have never been fully utilized, and it is estimated that in total they may amount to a quantity of water three times the annual flow of the

Colorado River (Wilkenson 1986). Resolution of water rights under the Winters Doctrine will be litigated for years to come (Foster 1978; Deloria 1985). In 1985 the State of Montana reached agreement on Native American water rights with the Assiniboine and Sioux tribes on the Fort Peck Reservation.* The Fort Peck–Montana Compact is one of the very few Native American water rights issues that have been settled.

Although the Winters Doctrine was first applied to surface water, in *Cappaert v. United States* (1976) the U.S. Supreme Court later extended it to ground water. The Tohono O'odham (Papago Nation) of Arizona have negotiated the rights to obtain ground water.[†] (Weatherford & Schupe 1986). Both the Tohono O'odham and the Fort Peck Assiniboine and Sioux are authorized to lease their water for off-reservation uses. This may be a way for others to maintain a source of water without infringing upon Native American water rights under Winters.

There are advantages for both the claimants of Native American water rights and others who hold water rights to reach settlements. The sooner the Native Americans receive water rights, the sooner they can begin to employ them for economic development to benefit the tribes. The value of irrigated farmland and the ability of nonindians to borrow money is depressed in areas where there are Native American claims to water rights (Tomsho 1992). In 1992 there were a total of seven settlements of tribal water rights including the Northern Cheyenne Tribe of Montana, Jicarilla Apache Tribe of New Mexico, Pueblo de Cochiti Tribe of New Mexico, Ak-Chin Indian Community of Arizona, San Carlos Apache Tribe of Arizona, Ute Indian Tribe of Utah, and the Southern Utes and Ute Mountain Utes of Colorado (*U.S. Water News* 1992a).

The Winters Doctrine also applies to other federally reserved land, such as national parks and monuments (Brookshire, Watts, & Merrill 1985). However, the water use there is much more limited than the potential for irrigation and other uses on Indian reservations.

Inasmuch as most major rivers cross state boundaries, it is not surprising that surface-water law has a strong federal component. The U.S. Congress, for example, apportioned the water from the Colorado River among various western states. The appropriation of surface water in the Colorado River Basin has been made on the basis of a concatenation of events that included the Colorado River Compact (1922), the Boulder Canyon Project Act (1928), a treaty with Mexico (1944), the Upper Colorado River Compact (1948), the Colorado River Storage Project Act (1956), a Supreme Court decision, *Arizona v. California* (1963), and the Colorado River Basin Project Act (1968) (Jacoby, Weatherford, & Wegner 1976).

In addition, Congress has also appropriated most of the money to develop the surface-water resources of the western United States. The Reclamation Act of

---

*State of Montana/Assiniboine and Sioux Tribes of Fort Peck Indian Reservation Compact, ratified in S.B. 467, 49th Leg., 1985 Montana Laws.

[†]P. L. 97-293, Title III, 96 Stat. 1274 (1982), San Xavier Papago Reservation of Arizona.

1902 is the linchpin of federal funding for western water projects. As a result, about 15 percent of all water in the west is supplied by the U.S. Bureau of Reclamation, all at a subsidized price (Wilkinson 1986). The Central Arizona Project is a massive canal system built mostly with federal funds to transfer the Colorado River water appropriated to Arizona to the users in Phoenix and Tucson (Weatherford & Schupe 1986).

Another example of the federal role in surface water is the case of *Wisconsin et al. v. Illinois et al.* (1967). Illinois began to divert surface water from the Lake Michigan Basin at Chicago in the early part of this century by reversing the flow of the Chicago River to divert sewage flows from the water system intakes in Lake Michigan. The diversion resulted in a lowering of lake levels; this impacted upon shipping and power production at Niagara Falls. The riparian states on the Great Lakes sued Illinois in federal court in a common law action starting in 1925. The U.S. Supreme Court ruled in 1967 that Illinois could not divert more than 3200 ft^3/s from the lake for all purposes.

The power of the federal government appears to be superior to that of the states in regulation of in-stream flows. In *California v. Federal Energy Regulatory Commission* (1990), the U.S. Supreme Court held that the states could not establish minimum streamflows for a hydroelectric project that were more restrictive than those established by the Federal Energy Regulatory Commission (Blain & Evans 1990). In this case California wanted to set minimum streamflows in order to protect native fish in a section of a river to be bypassed by a hydroelectric power project. The state's minimum flow was greater than those ordered by the Federal Energy Power Commission, and the power company argued that if it were to bypass the additional water, the project would not be economically feasible.

12.6.3 Laws Regulating Quantity of Ground Water

There are a number of different types of state law governing ground-water use, with a number of variations of the basic concepts. The right of absolute ownership of the water under a property holder's land is known as the **English rule.** In 1843 an English court held that a landowner could pump ground water at any rate, even if an adjoining property owner were harmed (*Acton v. Blundell* 1843). In *Acton* the court held that the land owner held proprietary interest in the ground water beneath his land. This rule was transported across the Atlantic and in 1861 was planted in the United States with an Ohio court decision, *Frazier v. Brown* (1861). In *Frazier* the court noted this about ground water: "Because the existence, origin, movement, and course of such waters, and the causes which govern and direct their movement, are so secret, occult and concealed, an attempt to administer any set of legal rules in respect to them would be therefore, practically impossible." In 1903 a Wisconsin court ruled in *Huber v. Merkel* (1903) that a property owner could pump unlimited ground water, even with malicious intent to harm a neighbor, the water being put to no good use. This was the ultimate application of the English rule. However, this ruling was overturned in 1974 by the Wisconsin Supreme Court. In writing the majority opinion Justice Landry held that: "the Huber case and its misconceived progeny can no longer be relied upon as conferring to the owner of the land an absolute right to use with impunity all the water that can be pumped from the subsoil underneath (*Wisconsin v. Michaels Pipeline Constructors, Inc.* 1974). Even in Ohio, *Frazier* was finally overturned in *Cline v. American Aggregates* (1984). In this case the dewatering necessary for a deep limestone quarry caused a general decline in the water table, which caused many wells to go dry, others to become intermittent producers, and still others to start to produce poor-quality water. Expert testimony from qualified hydrogeologists was used in court to relate these occurrences to the dewatering of the limestone quarry (Bair & Norris 1990). In issuing the ruling Justice Holmes wrote: "Scientific knowledge in the field of hydrology has advanced in the past decade to the point that water tables and sources are more readily discoverable. This knowledge can establish the cause and effect relationship of the tapping of underground water to the existing water level. Thus liability can now be fairly adjudicated with these advances which were sorely lacking when this court decided *Frazier* more than a century ago."

In contrast to *Frazier* in Ohio and *Huber* in Wisconsin, as early as 1862 a New Hampshire court ruled that a landowner had the right to use only a reasonable amount of ground water. The rights of adjacent landowners were also recognized (*Basset v. Salisbury Manufacturing Company, Inc.* 1862) This is known as the **American Rule**. In general American courts are moving from the English Rule to the American Rule (Bowman & Clark 1989). Ground-water law in states can be based entirely on common law, or state legislatures can pass specific statutes that control the pumping of ground water, typically through a permit system. Permit systems generally have a threshold value, so smaller users are not restricted. For example, the threshold is 100,000 gal/day in Wisconsin, 25,000 gal/day in Iowa, and 10,000 gal/day in Minnesota.

Some western states have ground-water law dictating that waters within the state boundaries belong to the public. Individuals can establish water rights to ground water by use; the first to draw from an aquifer has established a senior right. New Mexico is a state with this doctrine for both ground and surface water (Tarlock 1985). However, the state engineer of New Mexico can regulate the withdrawal of ground water from "declared" underground water basins.

One area of conflict arises when surface water is fed by discharge of ground water. Development of the ground-water reservoir typically depletes the baseflow of the stream. Holders of surface-water rights may find that the allocated surface water is diminishing in volume as ground-water withdrawals increase. On the South Platte River Basin of Colorado, the stream depletion due to ground-water withdrawals is about equal to the consumptive use of the ground-water pumped from the basin for irrigation (Danielson & Qazi 1972). In some western states, the surface-water right can be usurped from the holder by ground-water pumpage. In the case of Colorado, this conflict is being resolved by integrating the claims of both surface-water and ground-water users with regard to the same basin (Peak 1977). In some ground-water basins of New Mexico, would-be ground-water developers must acquire sufficient surface-water rights to compensate for the reduced flow from the stream (Tarlock 1985).

In California, a ground-water system based on correlative rights is followed. The right to use ground water belongs to the owner of the overlying land, so long as the use is reasonable. Water in excess of that used by the overlying owners can be allocated by appropriation. Thus, there are two systems of ownership of water rights. The determination of available water is based on the average recharge to the aquifer, termed the *safe yield* (although it is not safe yield as defined in Section 12.5). If withdrawals are less than the average annual recharge, the excess can be apportioned. If an *overdraft* (defined as "pumping in excess of the average annual recharge") exists for at least five years, the ground water is apportioned among all users in amounts proportional to their individual pumping rates, with the total pumping set at the average annual recharge. This is the **mutual-prescription doctrine,** which was first adjudicated in the case of the Raymond Basin (*Pasadena v. Alhambra* 1949).

The rights of states to regulate ground water was limited in 1982 by the federal courts in *Sporhase v. Nebraska* (*Sporhase v. Nebraska* 1982; Barnett 1984). Sporhase and Moss owned land in both Nebraska and Colorado. A well located in Nebraska pumped water from the underlying Ogallala aquifer, which was used to irrigate both the Nebraska and the Colorado land. Nebraska had a regulation that one who wished to pump ground water and export it to another state was required to first apply for and obtain a permit. Sporhase and Moss did not apply for the permit and were sued in Nebraska court. The Nebraska court ruled against them but was overruled by the Supreme Court. Justice Stevens wrote:

> Although water is indeed essential for human survival, studies indicate that over 80% of our water supplies is used for agricultural purposes. The agricultural markets supplied by irrigated farms are worldwide. They provide the archetypal example of commerce among the several states for which the framers of our Constitution intended to authorize federal regulation. The multi-state character of the Ogallala aquifer—underlying tracts of land in Colorado and Nebraska, as well as parts of Texas, New Mexico, Oklahoma and Kansas— confirms the view that there is a significant federal interest in conservation as well as in fair allocation of the diminishing resource.

The high court ruled that the states can regulate water within their boundaries but do not own it and cannot prohibit export except in times of severe water shortages within the state. This ruling was applied in 1983 in *City of El Paso v. Reynolds*. El Paso, Texas, wished to withdraw water from two large and scarcely used aquifers in New Mexico. The court ruled that New Mexico did not need the aquifer for human use and therefore New Mexico could not prohibit the transfer of water to El Paso (*El Paso v. Reynolds* 1983).

If one wishes to artificially recharge and store water in an aquifer, can someone else pump out that water? Can a public agency use the aquifer under a property holder's land for storage of imported water? These are questions central to the use of aquifers as storage reservoirs, and they must be answered if cyclic storage is to become an important procedure in water management. Two California cases dealing with aquifer storage have apparently settled these questions.

They also form a basis for resolution of these problems in other states (Gleason 1978).

The first California case involved the rights of the landowner to exploit an aquifer. Alameda County has a water district formed to protect the aquifer, especially from salt-water intrusion. As a conservation practice, local and imported water is injected into the aquifer to maintain the seaward hydraulic gradient. A sand and gravel pit was operating and, by 1969, the sand-mining operation extended 80 ft into the aquifer. The injected water was flooding the sand pit, requiring the mining company to pump large amounts of water from it. The mining company sued the water district to enjoin it from recharging the aquifer. The mining company lost the case, as the court held that public agencies can store water in an aquifer, up to the historic ground-water level, even if it decreases the usefulness of land to the property holders (*Niles Sand and Gravel Co., Inc. v. Alameda County Water District* 1974).

In the second case, a public entity injected water for underground storage with the intention of withdrawing it at a later time. This right to recapture injected waters was recognized by a court ruling in 1943 (*City of Los Angeles v. Glendale* 1943) and then reaffirmed in 1975 (*City of Los Angeles v. City of San Fernando* 1975). In both instances, the aquifers in question were in the Los Angeles area, where several cities were competing for use of ground water that comes from natural sources and artificial recharge.

CASE STUDY: ARIZONA'S GROUND-WATER CODE

In 1980 the state of Arizona passed a comprehensive new ground-water law designed to reduce the severe overdraft of ground water as well as to allocate the limited ground-water resources of the state (Ferris 1980; Tarlock 1985). A new state agency, the Department of Water Resources, was established to administer the code. Geographical areas known as Active Management Areas (AMAs) were established for those areas with severe overdraft problems. Within the designated AMAs, existing and future use of ground water is regulated. Ground-water users at the time the AMA was formed could claim a grandfather right. Persons can apply for a ground-water withdrawal permit for a new or expanded use for almost any purpose but irrigation. They must show that there is sufficient ground water available for the permit and that no water is available from other sources, such as the grandfather ground-water rights or the Central Arizona Project. Management plans must also be established for each AMA. For the three urban AMAs, the management plan has a goal of reducing ground-water withdrawals to the safe yield by 2025. In the agricultural AMA the goal is to preserve irrigated agriculture as long as possible while still preserving future water supplies for nonagricultural uses. Water conservation will be necessary to meet the goals of the management plans. The difference in urban-area water usage between Phoenix (267 gal, or 1011 L, per capita per day) and Tucson (160 gal, or 606 L, per capita per day) illustrates the possible scope of savings via conservation. However, the biggest potential savings is in agriculture; it accounts for 89% of the total water usage in the state. The real strength of the code is the provision that prohibits the establishment of new urban areas unless there is a 100-y supply of water available. This prevents new development of water-short areas and protects home buyers.

12.6.4 Laws Regulating the Quality of Water

Common law can be applied to a situation where damages to an individual have occurred because of contamination of ground or surface water. In such a case, a suit would be filed and one or more theories of common law advanced along with expert testimony about the technical facts surrounding the alleged contamination.

A number of laws have been written with the intent of preventing the contamination of ground and surface water. There are a number of relevant federal laws, which apply in all states, as well as specific state laws. As a general principle, state laws can be more strict, but cannot be less strict, than federal laws if they address the same topics.

National Environmental Policy Act of 1969 (P.L. 91-190)

Title I of the act establishes a national environmental policy and environmental goals. Environmental-impact statements are required for major federal projects. The Supreme Court has ruled that the policies and goals of Title I are not enforceable standards and the federal agencies are required only to consider them when making decisions. An environmental-impact statement is required to disclose the environmental impact of a proposed action, unavoidable adverse consequences, alternatives to the proposed action, the relationship of short-term uses of the environment to long-term productivity, and irretrievable commitments of resources.

Federal Water Pollution Control Act of 1972 (P.L. 92-500)
Clean Water Act Amendments of 1977 (P.L. 95-217)

The objective of this act and its amendments is "to restore and maintain the chemical, physical and biological integrity of the nation's water." National goals to achieve a degree of water quality that would make the waters of the United States "fishable and swimmable" by 1983 and to eliminate the discharge of pollutants into the waters of the United States by 1985 were established. The act sets water-quality standards, establishes minimum national effluent standards, requires pollution discharge permits, and has a construction grant program for public sewage-treatment plants. As a result of this program, surface-water quality has increased dramatically in many areas of the United States. The goal of eliminating discharges into surface waters has placed emphasis on land treatment and disposal of wastewater and sludges. In some cases this can pose a threat of ground-water contamination.

Safe Drinking Water Act of 1974 (P.L. 93-523) and Amendments

This law was passed to set standards for safe drinking water, protect "sole source" aquifers, and protect drinking-water aquifers from contamination resulting from underground injection of waste.

The Safe Drinking Water Act requires the U.S. Environmental Protection Agency (EPA) to set drinking-water standards to protect public health and welfare. The 1986 amendments to the act substantially enlarge the list of substances to be regulated. Maximum contaminant levels and maximum contaminant-level goals are to be established for toxic and carcinogenic compounds as well as secondary maximum contaminant levels for substances with no health risk but that do create aesthetic concerns.

The "Gonzales Amendment" authorized EPA to designate aquifers that are especially valuable because they are the only source of drinking water in an area. No federal financial assistance may be given to a project that might contaminate one of these "sole source" aquifers so as to create a significant hazard to public health. In the 1986 amendments, Congress directed the states to develop plans to protect the surface area around public water-supply wells from potential contamination from such sources as hazardous wastes, pesticides, and leaking underground storage tanks. These wellhead protection plans will need a very strong input from hydrogeologists if they are to be successful.

The underground injection of hazardous wastes and other materials is also regulated under the Safe Drinking Water Act. Current drinking-water aquifers as well as all other aquifers with total dissolved solids less than 10,000 mg/L are protected by regulating injection of liquid waste into deep boreholes. Class I wells dispose of hazardous waste into isolated strata below current or potential drinking-water aquifers. Class II wells are for the recirculation of oil-field brines. Class III wells are used in solution-mining of ores. Wastes are not to be injected into or above potable water aquifers.

Resource Conservation and Recovery Act of 1976 (RCRA) (P.L. 94-580)

The Resource Conservation and Recovery Act was the first major federal effort to deal with hazardous solid waste. It is a management system designed to regulate hazardous waste from the time it is created and continuing through its final disposition. The keystone of the system is a permit and manifest system used to keep track of hazardous waste. Waste generators and transporters as well as managers of hazardous waste storage, treatment, and disposal facilities are all regulated. Facilities may not be placed in the recharge zones of sole-source aquifers. New facilities are required to have ground-water monitoring systems, and old facilities must be retrofitted for ground-water monitoring. Hazardous-waste landfills and lagoons are required to monitor leachate and to have double-liner systems. Land disposal of liquid hazardous waste is prohibited, and land disposal of certain other hazardous wastes is to be phased out.

Comprehensive Environmental Response, Compensation and Liability Act of 1980 (CERCLA) (P.L. 96-510)

The Comprehensive Environmental Response, Compensation and Liability Act of 1980 is commonly referred to as Superfund. This act is targeted at the cleanup of releases of hazardous substances in the air, on land, or in the water. Parties who release contaminants from hazardous wastes are required to clean them up. If the responsible parties do not do so, the EPA can do the work and bill the cost to the responsible parties. Potentially responsible parties include the generators, transporters, and disposers of the waste. A provision known as "joint and several liability" permits the government to recover the full costs of the remedial action from any of the responsible parties, even if they were responsible for only part of the waste. The EPA is empowered to respond immediately to the release of a hazardous substance and then recover costs at some time in the future from the responsible parties—if any can be found and if they have any money!

CERCLA requires the EPA to establish a National Priorities List of sites to be targeted for remedial action. From time to time new sites are added to the National Priority List. Many of them are abandoned sites where the responsible owners are bankrupt or assetless. Other sites may be owned by Fortune 500 companies.

Superfund Amendments and Reauthorization Act of 1986 (SARA)

Superfund was reauthorized in 1986 with additional funds added to the program and additional mandates from Congress relating to the restoration of contaminated sites. Section 121 of SARA places emphasis on cleanup remedies that will reduce the volume, toxicity, and mobility of hazardous substances and contaminants to the maximum extent practicable.

Cleanup activities in the United States are governed by CERCLA and SARA. Of particular importance is the assignment of liability for the costs of environmental cleanup and restoration that have been established by these laws. Financial liability under Section 107 of SARA extends to operators as well as present and past site owners, whether or not the site owner had knowledge of the operations at the site. As some banks have found to their dismay, foreclosure on a contaminated property can bring with it financial liability for cleanup of the property.

There are several important concepts related to financial liability under CERCLA and SARA. *Strict liability* means that one did not have to contribute to the problem to be liable. The current owner of a site may be liable for cleanup costs even if the site was contaminated prior to purchase. Liability may be retroactive so that a past owner of a site may have current liability *even if they complied will all regulations in existence at the time of the activity*. The concept of *joint and several liability* holds that any responsible party could be held accountable for the entire cleanup cost, even if their contribution was a minor one. This might occur if all other potentially responsible parties either could not be located or had no money. A generator of hazardous waste can be held responsible for a share of the cleanup costs of a disposal site even if they did not know that a transporter had taken their waste to that site. If all the financial resources of all the potentially responsible parties are exhausted, then the federal government will spend funds from the Superfund account. The liability burdens under Superfund can be onerous, and many property transfers are now preceded by an environmental audit in order to help prospective buyers and lenders from unknowingly acquiring a contaminated site.

Surface Mining Control and Reclamation Act (SMCRA) (P.L. 95-87)

The Surface Mining Control and Reclamation Act was passed with the intent of preventing imminent danger to the health and safety of the public and significant, imminent environmental harm that might be caused by both underground- and surface-mining operations. One aspect of the law is that it requires that a hydrogeological study be performed prior to the covering or burial of acid-forming or toxic waste materials from mining or when any mine is to be filled with waste material. These waste materials are typically overburden spoils or mill tailings. The mine operator must prove that there will be only minimal disturbance of the hydrologic regime around the mine. If a mining activity contaminates, diminishes, or interrupts the ground-water or surface-water supply of an adjacent landowner, SMCRA requires the mine operator to replace the water supply.

Uranium Mill Tailings Radiation and Control Act of 1978 (UMTRCA) (P.L. 95-604 as amended by P.L. 95-106 and P.L. 97-415)

The Uranium Mill Tailings Radiation Control Act regulates the storage and disposal of mill tailings at both active and inactive uranium mill operations. The act provides that uranium mill tailings should be stabilized, controlled, and disposed of in an environmentally sound and safe manner. Title I of the act addresses remedial actions that must be taken at unsafe abandoned sites. Title II provides for the regulation of the handling and disposal of waste materials at active sites. Correction of adverse impacts on ground or surface water from uranium mill tailings is a feature of UMTRCA.

Toxic Substances Control Act (TOSCA) (P.L. 94-469 as amended by P.L. 97-129)

The purpose of this act is to protect human health and the environment by requiring testing and use restrictions for chemicals that may present an "unreasonable" risk to health and the environment. This is an umbrella act that regulates toxic compounds during research and development, manufacturing, processing, and distribution. It has been interpreted to permit controls on the use, disposal, or storage of toxic compounds where ground-water contamination from such compounds has occurred or may be expected to occur. Federal agencies and state governments come under the jurisdiction of this act.

Federal Insecticide, Fungicide and Rodenticide Act (FIFRA) (P.L. 92-516 as amended by P.L. 94-140, P.L. 95-396, P.L. 96-539, and P.L. 98-201)

The primary purpose of FIFRA is to regulate the manufacture, use, and disposal of pesticides. It requires pesticides to be labeled and restricts their use where appropriate. Use restrictions can be applied in areas where ground-water contamination by pesticides has occurred or could be expected to occur.

CASE STUDY: WISCONSIN'S GROUND-WATER PROTECTION LAW

State of Wisconsin Act 410 was enacted to protect the quality of the state's ground water. There are five areas that the act addresses: establishment of numerical standards for ground-water quality, a laboratory certification program, an expanded ground-water monitoring program, establishment of an environmental repair fund for remedial work at contaminated sites, and a compensation fund to pay part of the cost of drilling new wells for property owners whose wells have become contaminated with substances other than bacteria and nitrate. The most far-reaching provision is the numerical standards program. This is set out in administrative rules (NR 140). Ground-water quality standards are set out for substances that have been detected in or have a reasonable probability of entering ground water in Wisconsin. An enforcement standard is set for each substance on the basis of federal standards indicating a concentration above which there is a threat to public health or the environment. Enforcement standards are applicable to ground water that has been impacted by some regulated activity, such as a landfill or a hazardous-waste storage area. Enforcement standards apply at any point beyond the property boundary of a regulated facility, at any point of ground-water use, and at any point beyond what is known as a "design management zone" for a regulated activity. A design management zone is a three-dimensional portion of the earth beneath a regulated facility within which natural attenuation may be used to reduce the concentration of a contaminant. If the site is small, the property boundary is coincident with the design management zone boundary. If the site is larger, the design management zone is a designated horizontal distance from the edge of the facility. For example, for land treatment of wastewater, the design management zone is 250 ft (76 m), for wastewater treatment lagoons it is 100 ft (30 m), and for hazardous waste disposal facilities it is 0 ft. The law also specifies preventive action limits (PALS), which are set at some percentage of the enforcement standard. For substances of public health concerns, PALs are set at 20% of the enforcement standard, except for carcinogenic and mutagenic compounds, where the PAL is 10% of the enforcement standard. For public welfare–related substances the PAL is set at 50% of the enforcement standard. When a PAL is exceeded anywhere within the site, including the design management zone, a wide range of responses is available. These range from no action to requiring additional monitoring to revising operating procedures at the managed facility to remedial action to restore ground-water quality. If an enforcement standard is violated, some action must be taken. This includes requiring changes in operation of a facility, requiring redesign of a facility, requiring alternative methods of waste treatment and disposal, requiring closure of a facility, and requiring restoration of ground-water quality. Leeway has been given to the Wisconsin Department of Natural Resources in determining the appropriate response and the timeframe for it.

This water-quality protection law, in conjunction with other rules under which Wisconsin regulates waste treatment and disposal activities, creates a strong framework to protect the water quality of the state. A survey of water quality in both community and private wells in Wisconsin was conducted prior to the implementation of the ground-water protection law. It found that 65 of 1174 community wells tested and 82 of 617 private wells had detectable levels of volatile organic compounds. However, only 5 community wells and 14 private wells had volatile organic compounds above health advisory levels. In one area of Wisconsin, the central sand plains, about 900 wells were sampled for aldicarb, a pesticide. A substantial number, 201, had detectable aldicarb levels and 70 contained aldicarb in concentrations in excess of Wisconsin's recommended health advisory level of 10 μg/L (Krill & Sonzogni 1986). It would appear that, in general, Wisconsin has high-quality ground water, although some areas, such as the central sand plains, have substantial problems.

12.7 ARTIFICIAL RECHARGE

Resource management sometimes is interpreted as limiting the development and use of a resource in order to conserve it. Ground-water management has a somewhat broader scope, in that artificial recharge can be used to expand the amount of available water. *Water spreading* has been used in the western United States for decades to capture additional runoff. Much of the recharge of alluvial basin-fill aquifers comes from stream-bed infiltration during the wet season. In order to increase the amount of infiltration, stream water is diverted onto empty land below the mouths of the canyons carrying the streams (Conkling 1946).

Surface spreading is feasible given the following circumstances: The upper soil layers are permeable, the water table is not close to the surface, the land is relatively flat, and the aquifer to be recharged has a transmissivity great enough to carry the water away from the spreading area. It is extensively practiced simply by placing low dams across the ephemeral streams draining from the mountains. The dams act to flood the land along the stream channel.

Surface spreading tends to raise the water table over a rather extensive area. The same is true of other diffused types of artificial recharge. Irrigation is also a form of artificial recharge. Because of problems of salt accumulation in the soil, the amount of applied irrigation water is in excess of the needs of plants for evapotranspiration. The unevaporated water percolates through the unsaturated zone and recharges the water table. When spray irrigation is used as a means of wastewater treatment, the amount of water applied is far in excess of the plant requirement (Kardos 1967). The result may be a substantial increase in the elevation of the water table (Parizek & Meyers 1967). Similar diffused sources of artificially recharged water include septic tank drain fields. Wastewater from these sources percolates through the unsaturated zone and recharges the water table (Dudley & Stephenson 1973; Beatty & Bouma 1973).

Recharge basins are frequently used to recharge unconfined aquifers, especially where land costs are high. Basins are advantageous in that a substantial hydraulic head can be maintained in order to increase the infiltration rate. They are inexpensive to construct and operate. On Long Island, New York, basins are used to recharge storm-water runoff from roadways and parking lots (Seaburn 1970a, 1970b). They are also used in Peoria, Illinois (Smith 1967; Harmeson, Thomas, & Evans 1968), Fresno, California (Nightingale & Bianchi 1973; Salo, Harrison, & Archibald 1986), and Orange County, California (Matthews 1991), to recharge aquifers with river water. The infiltration capacity through coarse river gravels at Peoria is very high, with rates of 20 to 100 ft (6.1 to 30.5 m) per day being typical annual averages.

Recharge basins concentrate a large volume of infiltrating water in a small area. As a result, a ground-water mound forms beneath the basin. As the recharge starts, the mound begins to grow; when the recharge ceases, the mound decays as the water spreads through the aquifer. The growth and decay can be described mathematically (Hantush 1967; Singh 1976). Digital-computer models have been used to evaluate the impact of recharge basins on the water table (Bianchi & Haskell 1968; Hunt, 1971). It is important to maintain the unsaturated zone beneath the recharge basins in order to help maintain high infiltration rates, while still providing water-quality improvements (Fetter 1977a; Fetter & Holzmacher 1974).

The infiltration capacity of recharge basins is initially high and then declines as recharge progresses. This is due to surface clogging by fine sediments

(Behnke 1969) and biological growths in the uppermost few inches of the soil (Moravcoua, Masinova, & Bernatova 1968). It has been found that the operation of recharge basins with alternating flooding and drying-out periods will maintain the best infiltration rates (Fetter & Holzmacher 1974). Fine surface sediments may occasionally need to be removed mechanically (Sniegocki & Brown, 1970).

If the intent of a management program is to recharge a confined aquifer, then **recharge wells** must be used. The design of a well for artificial recharge is similar to that of a supply well. The principal difference is that water flows out of the recharge well and into the surrounding aquifer under either a gravity head or a head maintained by an injection pump.

Injection wells for artificial recharge are prone to clogging. A large amount of water is being pushed through a small volume of aquifer near the well face. Clogging can be due to a number of factors; for example,

1. Air entrainment caused by aeration of water falling into the well,
2. Filtration of suspended sediment and organic matter,
3. Development of bacterial growths in the aquifer,
4. Formation of precipitates due to geochemical reactions between the recharge water and native ground water,
5. Swelling of clay colloids in the aquifer,
6. Dispersal of clay particles due to ion exchange between recharging water and aquifer materials,
7. Precipitation of iron from native ground water due to recharge of water with a pH and Eh in the range of ferric iron,
8. Growth of iron bacteria, or
9. Mechanical compaction of aquifer materials due to high injection pressures.

Frequent maintenance of injection wells may be necessary. This typically consists of pumping water out of the wells to redevelop the flow capacities. Chlorination may also be used to reduce biological growth (Fetter & Holzmacher 1974). Wisconsin is one state that does not permit the use of injection wells for any purpose. The reason for this prohibition is to prevent well disposal of wastes that may result in ground-water contamination.

12.8   PROTECTION OF WATER QUALITY IN AQUIFERS

One of the most important aspects of ground-water management is the protection of the water quality in an aquifer. There are a number of artificial sources of potential ground-water contamination. Pollution from such sources as septic tanks, sanitary landfills, land-treatment systems for municipal wastewater, waste injection wells, toxic chemical disposal sites, cemeteries, mine tailings, acid mine drainage, water softener regeneration salt, highway deicing salt, oil-field brines, agricultural chemicals and fertilizers, and accidental oil, gasoline, and chemical spills have been well documented in scientific literature. An incident such as the burial in Michigan of slaughtered cattle that had been given feed contaminated by PBB, a chemical fire retardant, points out that the earth is typically thought of as the most convenient repository for material that society does not know how to handle.

Ground-water contamination can occur also when water of poor quality is drawn into a well field that originally has been developed in high-quality water. The best example of this is salt-water intrusion in coastal areas. Heavy pumping of coastal aquifers can cause a landward migration of the interface separating fresh and salty ground water.

One critical aspect of preventing ground-water pollution is the identification of the recharge areas of aquifers. In such areas, protection of the aquifer is vital. Recharge areas should be zoned as water-quality conservation areas, with close control of potential sources of contamination. The hydrogeology of sites for such facilities as sanitary landfills and land-treatment systems should be intensively studied to ensure that the condition of local soils and the rate of direction of local ground-water flow preclude possible aquifer contamination. Hazardous- and toxic-waste storage and disposal should be barred from aquifer recharge areas. Sanitary sewers for the collection of domestic wastewater are preferable to septic tanks in aquifer recharge areas. Under most conditions, land areas underlain by extensive confining layers or situated in the discharge areas of unconfined aquifers are preferable for uses that might contaminate the ground water. Agricultural practices in recharge areas should also be regulated. Overapplication of chemicals and fertilizers should be discouraged and controlled collection and disposal of animal waste encouraged.

One of the most difficult aspects of aquifer protection is the control of abandoned wells. Many states require that abandoned wells be backfilled. This regulation can be enforced for a newly drilled water, oil, or gas well that proves to be unproductive. However, in the case of wells that fall into disuse and casual abandonment, the owners might not be aware of the danger they can pose to ground-water quality. Likewise, improper well construction can cause shallow ground water or surface water to migrate downward into the aquifer. Abandoned wells may also be used for the disposal of liquid or solid waste, an obvious threat to water quality. Education is more efficient than regulation in preventing contamination of this type.

Salt-water encroachment can be prevented by regulating the spacing and withdrawal rates of wells. The objective is to avoid establishing a hydraulic gradient that slopes from the zone of contaminated water toward the well field. If the aquifer contains fresh water overlying salty water, pumping rates must be low to avoid drawing up salty water from below, a phenomenon known as *upconing* (Figure 12.1). Saline-water aquifers underlie up to two-thirds of the land areas of the conterminous United States, so the problems of water encroachment and upconing are not limited to coastal areas (Bruington 1972).

Control of sea-water intrusion in coastal areas has been practiced in a number of localities. In coastal areas of southern California, where there is a confined aquifer containing salt water and fresh water, a pressure-ridge system

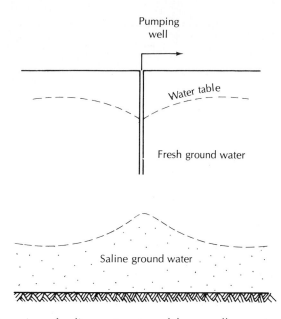

FIGURE 12.1 Upconing of saline water caused by a well pumping from an overlying fresh-water zone.

has been used (Bruington & Seares 1965). A line of injection wells parallel to the coast injects water into the aquifer. The result is a ridge in the potentiometric surface. Figure 12.2 shows this for an unconfined aquifer system. Water levels behind the barrier can be drawn down below sea level with no fear of salt-water encroachment. Similar barriers could be used in unconfined aquifers using artificial recharge from wells, pits, or trenches. Artificial recharge in the area of pumping-well fields could also be used to maintain the elevations of the potentio-

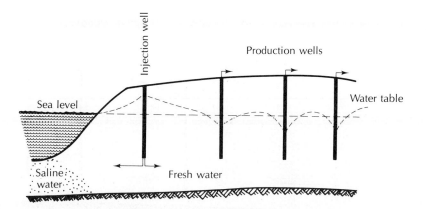

FIGURE 12.2 Use of injection wells to form a pressure ridge to prevent salt-water intrusion in an unconfined coastal aquifer.

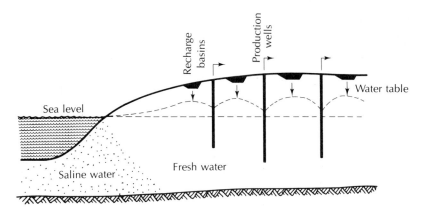

FIGURE 12.3   Use of artificial recharge in the area of production wells in an unconfined coastal aquifer. The artificially recharged water maintains the water table above sea level to prevent salt-water intrusion.

metric surface above sea level (Figure 12.3). A row of pumping wells could be installed parallel to the coast in either a confined or an unconfined aquifer. They could create a trough in the potentiometric surface lower than either sea level or the well-field areas behind the trough. The trough wells would pump salty water, which would not be suitable for most uses. However, wells behind the trough would pump fresh water (Figure 12.4).

12.9   GROUND-WATER MINING AND CYCLIC STORAGE

From the previous sections it is apparent that in order to develop the ground-water reservoir a cone of depression must be created. This means that water is

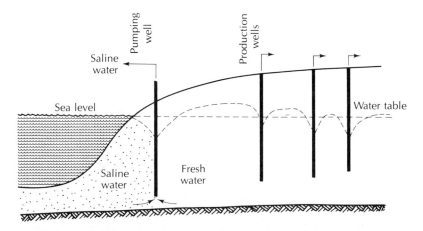

FIGURE 12.4   Use of pumping wells at the coastline to form a trench in the water table; the trench acts as a barrier to further salt-water encroachment.

removed from storage or is, in a sense, mined. Likewise, in order to utilize many aquifers for cyclic storage, the water levels must be drawn down in order to create storage space. Again, some water may have to be mined in order to prepare the aquifer to perform an essential service (Ralston 1973). Ground-water management programs with an objective to maintain natural equilibrium water levels may not be as efficient as other management plans.

The most dramatic need for limited ground-water mining is to provide supplemental water for use when precipitation and surface supplies are limited by drought. Normal rates of industrial and agricultural productivity can be maintained by the use of supplementary ground water (Lehr 1978). Water managers may find it difficult to convince themselves that the best time to deplete the ground-water reservoir artificially is when the amount of natural recharge is lowest. Ground-water levels will be falling at a seemingly alarming rate. However, the drawdown will create space for storage when nature provides an especially wet year (Ambroggi 1977). Water managers must be ready to capture this excess by artificial recharge to supplement the natural infiltration. It has been suggested that major recharge events that can result in a rapid rise in ground-water levels can be expected every 10 to 15 y (Ambroggi 1977). If water levels have not been depleted by "overpumping" during drought, much of the potential recharge may be lost owing to lack of aquifer storage space. This concept of *cyclic storage* is not applicable to aquifers to which the amount of possible recharge is limited by their capacity to transport and store ground water rather than by the amount of water available to them.

In some arid lands, considerable ground water is known to exist. Furthermore, there may be rather large hydraulic gradients, which could mean that this ground water is moving through the aquifers. This interpretation is at variance with the extremely low rates of recharge known to occur at present. Model studies have suggested that the gradients are "fossil" (Lloyd & Farag 1978); i.e., decayed remnants of higher ground-water gradients of 10,000 y ago, when pluvial periods during the late Pleistocene provided large amounts of recharge. Management programs, however, cannot be based on these extremely long cycles between recharge events. For practical purposes, such aquifers must be considered to be unreplenished, and all ground-water pumpage to be simply mining when recharge events are millennia apart.

The mining of ground water is occurring in many localities. The Ogallala aquifer of the southern High Plains of Texas and New Mexico is an example of a mined aquifer in arid regions, and the deep sandstone aquifer of northeastern Illinois is a humid-region example. Mining the deep sandstone aquifer appears to be an integral part of the areawide development of the ground-water and surface-water resources of northeastern Illinois (Schicht, Adams, & Stall 1976; Schicht & Adams 1977). Water levels will be drawn down by overdrafts until a "critical level" is reached. Ground-water pumping will then be curtailed to the extent that no further drawdown will occur.

Many areas of the world are underlain by aquifers containing saline ground water. Although desalinization is technically feasible, it is not economical except in some very limited applications. In water-short areas of the western United States, there is increasing competition between energy development and agricultural interests for the available water (Plotkin, Gold, & White, 1979). Some of this water is used for power-plant cooling, with subsequent evaporational losses. It has been suggested that saline ground-water resources could be mined to provide the necessary cooling water (Greydanus 1978). In this approach, it would be necessary to ensure that the available saline ground-water supply is sufficient for the design life of the power plant. Furthermore, the problem of the disposal of brine or salt would have to be dealt with. Saline ground water could also serve as a source of emergency cooling water for nuclear reactors.

12.10   CONJUNCTIVE USE OF GROUND AND SURFACE WATER

It is obvious that in stream-aquifer systems it is counterproductive to consider surface-water management and ground-water management as separate actions. When water can flow from the stream to the aquifer, or vice versa, it is tantamount to having two separate policies or plans for the same water. No one would open a joint checking account with a total stranger. However, the legal separation of ground water and surface water, or the administration of ground-water and surface-water management by separate agencies, could have a similar result: a rapid depletion of the total resources due to overuse.

As is the usual case, the need for management of ground water and surface water as elements of an interrelated system is most critical when demand exceeds supply. The greatest application of such management in the United States has come in the arid west. Overpumping of aquifers near a river could deplete the surface resources, while development of a losing river could reduce the normal recharge to an aquifer system. On a regional basis, the incremental drawdown caused by any one pumper is small. It is the cumulative effect of many pumpers that can cause a depletion of the total resource. The individual user has no economic incentive to reduce pumping (Bredehoeft & Young 1970). The situation is exacerbated when ground-water pumping results in a reduction in the flow of surface waters. Ground-water users may feel social pressure if a neighbor's well is adversely affected; however, this pressure is not experienced when reduction in availability of surface water affects an unknown user many miles away. Individual litigation among tens or hundreds of water users along a stream-aquifer system would present a legal nightmare. Some type of institutional control offers the most workable management approach.

One management plan might be a total ban on ground-water withdrawals during such times that the flow of the river falls below a specified value. This, however, would not prevent the streamflow from falling even more, as the response time of an aquifer is quite long (Young & Bredehoeft 1972). In order to fully protect the rights of senior surface-water users, ground-water pumping might be prohibited altogether. While this would serve one purpose, it would also eliminate the use of the ground-water reservoir for storage of excess water. In addition, ground-water development usually provides water supplies in excess of streamflow alone.

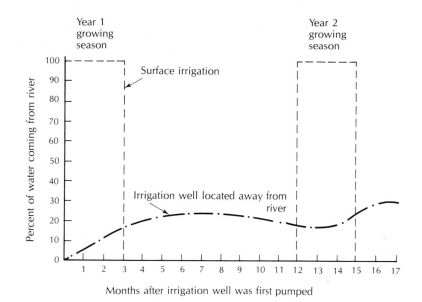

FIGURE 12.5 Comparison of withdrawal of irrigation water from a surface stream with induced infiltration to an irrigation well. The irrigation well delivers the same amount of water during the irrigation season as the surface withdrawal. Source: R. A. Young & J. D. Bredehoeft, *Water Resources Research* 8, no. 3 (1972): 533–56.

The most extreme case occurs when a river fed by mountain runoff flows across an aquifer system that receives no recharge from precipitation. Under such conditions, the river is the only source of water for both ground-water and surface-water users. Even in this extreme case, the river and aquifer system can be managed conjunctively to maximum benefit. If the water is used for irrigation, there is a demand only part of the year. The remainder of the year, the streamflow passes downstream. The maximum benefit would accrue if some of the water flowing in the nongrowing season could be stored for use when it is needed. This can be accomplished by taking advantage of the delayed response of the aquifer system (Young & Bredehoeft 1972). Wells located away from the river and pumped during the irrigation season would draw water from storage at that time, and then, when the cone of depression reaches the river, infiltration would be induced mostly during the nongrowing season (Figure 12.5). Wells located closer to the river would be pumped only near the end of the irrigation season.

While the benefits of conjunctive ground-water and surface-water use are obvious, the implementation of management is not easy. The legal framework for each state is different. In New Mexico, ground-water users in designated ground-water basins fall under the jurisdiction of the New Mexico state engineer. New appropriations of ground water are made only on the condition that the ground-water user acquire sufficient surface-water flow (Thomas 1972; Brutsaert & Gebhard 1975). In the state of Colorado, the Water Right Determination and Administration Act of 1969 is enforced by the state engineer. In part, this act reads

". . . it shall be the policy of this state to integrate the appropriation, use and administration of underground water tributary to a stream with the use of surface water in such a way as to maximize the beneficial use of all the waters of this state." However, the Colorado state engineer does not have unlimited license to maximize available water, as the act also states that "no reduction of any lawful diversion because of the operation of the priority system shall be permitted unless such reduction would increase the amount of water available to and required by water rights having senior priorities." A management plan that would increase the total amount of water but would result in junior users receiving more water at the expense of senior users is apparently not what the Colorado legislature intended.

It has been pointed out that, in most cases, a number of different management plans might be possible (Morel-Seytoux 1975). For example, an applicant for ground water might be allocated a large amount of water, but it would be available only during the nongrowing season. A total volume not nearly as great might be allocated if it were to be pumped during the irrigation season. Both plans could protect the senior water rights equally.

Conjunctive ground-water and surface-water management lends itself to the use of computer models (Young & Bredehoeft 1972; Brutsaert & Gebhard 1975; Morel-Seytoux 1975; Maddock 1974; Haimes & Dreizin 1977; Flores, Gutjahr, & Gelhar 1978). The hydrologic interaction in the stream-aquifer system can be described by partial differential equations, subsequently solved by numerical methods (Chapter 14). Techniques of operations theory can be used to maximize the amount of water available while minimizing the cost of development. The models can be established so that the holders of senior water rights are protected and water is allocated to the most beneficial use.

12.11   TRENDS IN WATER RESOURCES MANAGEMENT

Water is increasingly held to be a public rather than a private resource. In states where the right to use water is held to be a property right, it is protected by the U.S. Constitution. However, even in England, where the English Rule of absolute ownership of ground water originated, the Water Resources Act of 1963 changed concepts of ownership of water. Withdrawal of surface water and ground water in England and Wales is now subject to approval of a permit application (Trelease 1970).

In Texas, one of the last states that allowed unchecked pumping of ground water under the right of capture, a controversial catfish farm may cause a change in ground-water law. In 1991 a Bexar County catfish farm began to divert 33 million gallons a day of artesian ground water from the Edwards Aquifer. This amounted to 20% of the total water pumped from the entire county. There was an immediate drop in regional water levels. In 1992 the Texas attorney general issued an opinion that the Texas Water Commission has the right to regulate ground water withdrawals. The Texas Water Commission designated the Edwards Aquifer as the "Edwards Underground River" and began plans to regulate withdrawals. However, this was appealed by agricultural interests, and in September 1992 a district court judge declared that the Texas Water Commission was in over its head and under Texas law did not have the right to regulate ground water, no matter what they called it. The Texas Water Commission has appealed this ruling to a higher court. Stay tuned for the next edition of this text to see if Texas common law with regard to ground water will change!

In arid states of the United States West, competition for water between public and private users is increasing—especially as population shifts from rural to urban areas (Andersen & Wengert 1977). Municipalities are able to acquire water rights in the same manner as other entities: they may find an unallocated source or purchase a water right from its owner. However, they are also able to garner water rights by means not available to individuals or corporations: the exercise of the power of eminent domain. Through legal action, a municipal entity may acquire private property, including water rights (Radosevich & Sabey 1977). The use must generally be for a public purpose, and the owner must be compensated for the loss of property. As with any case of public condemnation of private property, the question of what constitutes "adequate compensation" is full of thorns. In Colorado, the compensation is set as "fair market value." This can be established on the basis of prior voluntary sales of water rights. Compensation can also include the loss of value to land that is separated from a water right necessary to put it to a given use (Radosevich & Sabey 1977).

Traditionally, water rights have been considered in connection with offstream purposes such as irrigation or municipal use. However, there are a number of in-stream uses that are important for public purposes. This includes the maintenance of water quality and fish and wildlife habitat. In Montana, there are increasing demands for water for energy development. In order to protect the ecological integrity of the Yellowstone River, the Montana Fish and Game Commission and the Montana State Water Quality Bureau have formally requested the appropriation of large amounts of water for in-stream flows (Thomas & Klarich 1979; Thorson 1986). This is a further example of the supersession of private needs by public in the allocation of a scarce resource.

The federal role in water use in the West underwent a major shift in emphasis with an omnibus water bill passed by Congress and signed by the President in October 1992. This bill will divert water that has traditionally been sold to farmers at below-market prices. About 1.2 million ac-ft a year of irrigation water that went to California's Central Valley will be used for environmental purposes, such as maintaining flows in rivers and into estuaries; this will result in significant improvements in fish and wildlife habitat (*U.S. Water News* 1992b).

Both the states and the federal government are moving to protect the quality of ground water. Federal laws such as the Resource Conservation and Recovery Act and the Comprehensive Environmental Response, Compensation and Liability Act have provisions for regulating activities that could degrade ground-water quality. CERCLA also has provisions to restore the quality of degraded aquifers. The Safe Drinking Water Act amendments of 1986 enable states to establish programs to regulate activities in the recharge areas of aquifers in order to protect ground-water quality. It is likely that more state laws, such as Wisconsin's ground-water protection law, will be enacted to protect the quality of ground water.

13 Field Methods

The rocks that form the crust of the earth are in few places, if anywhere, solid throughout. They contain numerous open spaces, called voids or interstices, and these spaces are the receptacles that hold the water that is found below the surface of the land and is recovered in part through springs and wells. There are many kinds of rocks, and they differ greatly in the number, size, shape and arrangement of their interstices and hence in their properties as containers of water. The occurrence of water in the rocks of any region is therefore determined by the character, distribution, and structure of the rocks it contains—that is by the geology of the region.

The Occurrence of Ground Water in the United States, Oscar Edward Meinzer, 1923

13.1 INTRODUCTION

The day is past when the only activity of the hydrogeologist was to locate and design a water well. Today, hydrogeologists are involved in many phases of resource management, including environmental impact analysis, as integral members of a multidisciplinary team. Hydrogeological studies are necessary and generally required by regulatory agencies for site studies prior to construction of such projects as sanitary landfills, land-treatment systems for wastewater, surface mines, power plants, artificial-recharge lagoons, nuclear-waste repositories, dams, and reservoirs.

In this chapter, we will introduce several techniques that can be applied both to the exploration for ground-water supplies and to various aspects of environmental hydrogeology. This includes the use of aerial photographs and other remote sensing data, as well as both surface and borehole geophysical methods. The methods of site evaluation will be examined.

13.2 FRACTURE-TRACE ANALYSIS

One technique that has been gaining acceptance among hydrogeologists is the use of *fracture-trace analysis.* As we discussed in Chapter 9, ground water is known to be concentrated in fracture zones found in many different rock types. The fracture traces are located by study of linear features on aerial or satellite photographs. On air photos, natural linear features consist of tonal variation in soils, alignment of vegetative patterns, straight stream segments or valleys, aligned surface depressions, gaps in ridges, or other features showing a linear orientation (Lattman 1958). Some linear features may be visible on the ground— for instance, surface sags or straight stream segments. Others, such as variation in soil tone, or alignment or height of vegetation of a certain type, may not be noticeable except on aerial photographs (Lattman 1958). Many of these natural linear features consist of interrupted segments, which may be of different types. For instance, a straight stream segment in a floodplain may align with a row of trees in a nearby woods. Natural linear features from 1000 ft (300 m) to around 4300 ft (1500 m) in length are fracture traces. Those greater than 4300 ft are termed **lineaments** (Lattman 1958). Some lineaments are up to 90 mi (150 km) long (Parizek 1976).

Fracture traces are surface expressions of joints, zones of joint concentration, or faults (Lattman & Matzke 1961). It is generally believed that the joint sets tend to be nearly perpendicular (Parizek 1976). They are known to extend to a depth of 3300 ft (1000 m) at one Arizona location (Lattman & Matzke 1961), although this is probably far deeper than is typical. Mapped fracture traces have been traced to cliffs where the fracture zone can be seen in cross section (Parizek 1976; Lattman & Matzke 1961). Under these observed conditions, the fracture zones dip at approximately 87° to 89°. Figure 13.1 shows an exposure of a fracture in a cliff in central Pennsylvania.

These fracture zones are less resistant to erosion than rock, which is less fractured. Hence, valley and stream segments tend to run along them. They may be zones of ground-water drainage, so that soils over them have a deeper water table or are not as moist as soils in surrounding areas. The soil color or vegetation

FIGURE 13.1   Cross section of a zone of fracture concentration revealed by a fracture trace near Spring Creek in Centre County, Pennsylvania. Photo courtesy of R. R. Parizek.

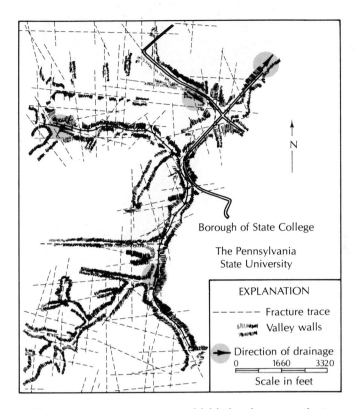

FIGURE 13.2   Valley development in an area of folded carbonate rocks in central Pennsylvania. The valleys tend to follow fracture traces. Source: R. R. Parizek, *Hydrogeologic Framework of Folded and Faulted Carbonates—Influence of Structure,* Mineral Conservation Series Circular 82, College of Earth and Mineral Sciences, Pennsylvania State University, 1971, pp. 28–65.

may appear to be different from that of surrounding soils. If they are zones of concentrated ground-water discharge, there may be a line of springs or seeps. Fracture traces in carbonate rocks are typically areas of solution. Aligned sinkholes or surface sags are typical surface expressions.

Fracture traces may be related to regional tectonic activity. They tend to be oriented at a constant angle to the regional structural trend; however, the orientation appears to be independent of local folds (Lattman & Matzke 1961). Lineaments are known to cut across rocks of many ages and cross folds and faults (Parizek 1976). They have been observed to be parallel to the major joint sets in flat-lying or gently dipping strata, but this is not the case if the strata are steeply dipping. If surface areas are separated by major faults, the individual fault blocks may have fracture traces of different orientation (Parizek 1976). The majority of fracture traces in an area appear to be grouped into two subparallel sets that are approximately perpendicular. Streams developed in rocks where fracture control is evident have been described as having a "stair-step" pattern (Setzer 1966). In the area of central Pennsylvania shown in Figure 13.2, the valley development

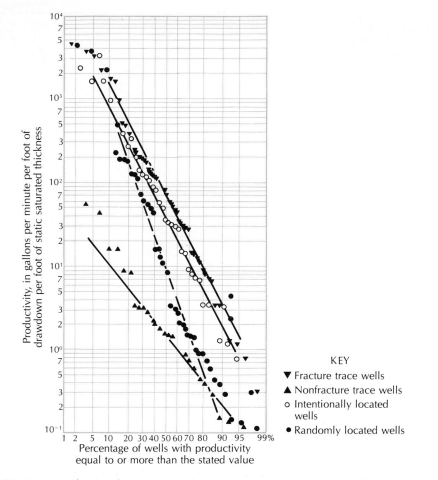

FIGURE 13.3 Production-frequency graph for water wells grouped according to whether or not they fall on a fracture trace. Source: S. H. Siddiqui and R. R. Parizek, *Water Resources Research*, 7 (1971): 1295–1312.

follows fracture traces. Most fractures are generally N-S or E-W, with lesser numbers running NW-SE and SW-NE.

Statistical studies of wells in carbonate terrane have shown that those located on fracture traces, either intentionally or accidentally, have a greater yield than those not on fracture traces (Siddiqui & Parizek 1971). Figure 13.3 illustrates that the productivity of fracture-trace wells is significantly above that of other wells not on fracture traces. The greatest yields come from wells located at the intersection of two fracture traces. Caliper logs of wells on fracture traces in carbonate-rock terrane showed many more cavernous openings and enlarged bedding planes than logs of those wells drilled in interfracture areas (Figure 13.4).

Many hydrogeologists are successfully using fracture-trace analysis to locate high-yield wells. The technique has been applied to carbonate-rock terrane

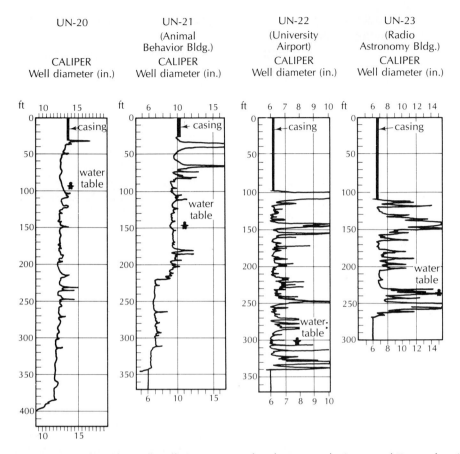

FIGURE 13.4 Caliper logs of wells in an area of carbonate rocks in central Pennsylvania. Wells UN-20 and UN-21 were drilled in interfracture areas; Wells UN-22 and UN-23 were located on fracture traces. Source: L. H. Lattman and R. R. Parizek, *Journal of Hydrology* (Elsevier Scientific Publishing Company) 2 (1964): 73–91. Used with permission.

(Lattman & Parizek 1964) but is also applicable to most other rock types (Parizek 1976). It is reportedly usable even if the bedrock is mantled by up to 170 ft (50 m) of glacial drift (Wobber 1967). Fracture-trace analysis is also widely used in selecting sites for sanitary landfills. Naturally, the landfill locations are most suitable if they fall in interfracture areas. Other uses include analysis of foundation and dam sites, evaluation of potential water problems in mines and tunnels, and control of mine drainage (Parizek 1976).

Fracture-trace analysis is also very useful in determining the locations of ground-water monitoring wells. Because ground-water flow preferentially follows the most permeable pathway, monitoring wells should be located on fracture traces. For example, if a hazardous-waste storage lagoon is located in an area of fractured bedrock, at least one of the down-gradient monitoring wells required under the Resource Conservation and Recovery Act should be located on a fracture trace.

In the identification of fracture traces on aerial photographs, a low-magnification stereoscope is generally used (Lattman 1958). Possible fracture traces are indicated by drawing directly on the photograph. One problem in identification is the confusion of linear features of human origin (fences, cowpaths, roads, power lines, plow and harvest patterns, etc.) with natural linear features. There is also a tendency to map fracture traces at oblique angles to regular grid systems on the photographs. As section lines almost always appear on air photos, especially in cultivated areas, there is a tendency to preferentially map NW-SE and NE-SW features as fracture traces. Following the stereoscopic mapping, the photos should be checked without use of the stereoscope to see if any other features are noticed. A typical air-photo scale for fracture-trace analysis is 1:20,000. Linear features that show up in more than one expression, and those crossing roads or fields, are more likely to represent fracture traces. In Figure 13.5, subtle fracture traces are indicated in an area of farmland in central Pennsylvania.

Following the mapping of linear features on air photos, it is necessary to make a field check. Some mapped features will usually turn out to be due to human activity. The more inexperienced the geologist, the more likely it is that this will occur. If a suspected fracture trace has a surface expression, it will be easier to locate in the field. Those without obvious surface expression must be

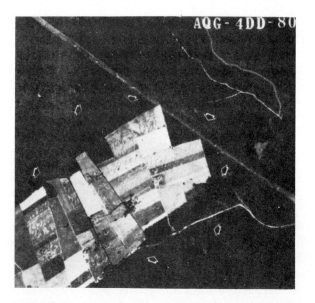

FIGURE 13.5 Aerial photograph of an area with 16 to 260 ft (5 to 80 m) of transported sediments overlying folded and faulted dolomite, limestone, and sandy dolomite, Centre County, Pennsylvania. The line of each fracture trace is indicated by arrows at both ends. Photo courtesy of R. R. Parizek.

located by virtue of their spatial relation to individual trees or buildings that are visible on the photographs and can be identified on the ground. In urbanizing areas, it may be possible to use older photographs taken before extensive development to map fracture traces. This makes the field location of the fracture traces even more difficult.

Yin and Brook (1992a, 1992b) point out that in crystalline rock areas, high-yield wells are generally associated with fracture traces, which may not necessarily correspond to topographic lows. They recommend that wells be located by fracture-trace methods rather than simply by drilling in a topographic low.

13.3   SURFICIAL METHODS OF GEOPHYSICAL INVESTIGATIONS

Geophysical surveys have been used by the mining and petroleum industries for many decades. Ground-water geologists soon discovered the usefulness of these surveys in exploring the shallow subsurface (within a few hundred feet), where ground-water supplies are usually found (McDonald & Wantland 1961; Heigold et al. 1979; Bays 1950). A number of different techniques are used, the most common of which are direct-current resistivity, seismic refraction, and gravity and magnetics methods. Seismic reflection is less widely used, although it is the preferred method in petroleum exploration.

Geophysical methods may be used to determine indirectly the extent and nature of the geologic materials beneath the surface. The thickness of unconsolidated surficial materials, the depth to the water table, the location of subsurface faults, and the depth of the basement rocks can all be determined. In some instances, the location, thickness, and extent of subsurface bodies, such as gravel deposits or clay layers, can also be evaluated. The correlation of geophysical data with well logs or test-boring data is generally more reliable than either type of information used by itself. As with all hydrogeological investigations, a careful definition of the problem and determination of the best type of information to solve the problem should be made before geophysical work is done. The geophysical survey should then be planned to yield the greatest amount of useful data for the budgeted cost.

13.3.1   Direct-Current Electrical Resistivity

Of the several electrical geophysical methods, *direct-current electrical resistivity* has found the greatest application to hydrogeology (Zohdy, Eaton, & Mabey 1974). A commutated direct current or a current of very low frequency (less than 1 cycle per second) is generated in the field or provided by storage batteries. It is introduced into the ground by means of two metal electrodes. If the soil is dry, water may be needed around the electrodes to establish a good connection. The voltage in the ground is measured between two other metal electrodes, also driven into the ground. By knowing that current flowing through the ground and the potential differences of voltage between two electrodes, it is possible to compute the resistivity of the earth materials between the electrodes. The resistivity of earth materials varies widely, from 10^{-6} $\Omega \cdot$m for graphite to 10^{12} $\Omega \cdot$m for quartzite. Dry materials have a higher resistivity than similar wet materials, as moisture increases the ability to conduct electricity. Gravel has a higher resistivity than silt or clay under similar moisture conditions, as the electrically charged surfaces of the fine particles are better conductors.

Electrical resistivity, R, is equal to the expression

$$R = \frac{A}{L} \frac{\Delta V}{I} \qquad (13\text{--}1)$$

where

   A is the cross-sectional area of current flow

   L is the length of the flow path

   ΔV is the voltage drop

   I is the electrical current

Electrical resistivity is measured in units of ohm-meters or ohm-feet. The four electrodes used can be designated as follows:

   A is the positive-current electrode

   B is the negative-current electrode

   $\left. \begin{array}{l} M \\ N \end{array} \right\}$ are the potential electrodes

If $\overline{XY}$ indicates the distance between electrode X and electrode Y, Equation 13–1 can be expressed as (Zohdy, Eaton, & Mabey 1974)

$$\overline{R} = \left(\frac{2\pi}{\dfrac{1}{\overline{AM}} - \dfrac{1}{\overline{BM}} - \dfrac{1}{\overline{AN}} + \dfrac{1}{\overline{BN}}} \right) \frac{\Delta V}{I} \qquad (13\text{--}2)$$

As earth materials are almost never homogeneous and electrically isotropic, the resistivity found by Equation 13–2 is an apparent resistivity, $\overline{R}$.

There are several electrode configurations in common usage. The **Wenner array** consists of the four electrodes spaced equal distances apart in a straight line: $\overline{AM} = \overline{MN} = \overline{NB} = a$. A current electrode is on each end (Figure 13.6A). In using the Wenner array, the apparent resistivity, $\overline{R}$, may be found from the expression

$$\overline{R} = 2\pi a \frac{\Delta V}{I} \qquad (13\text{--}3)$$

which is solved from Equation 13–2.

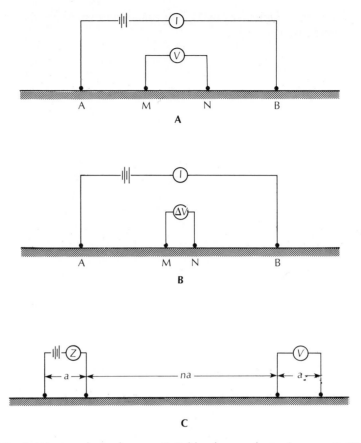

FIGURE 13.6   **A.** Wenner electrode array. **B.** Schlumberger electrode array. **C.** Dipole-dipole electrode array.

A second configuration is the **Schlumberger array.** It is a linear array, with potential electrodes placed close together (Figure 13.6B). Typically, $\overline{AB}$ is set equal to or greater than five times the value of $\overline{MN}$. The apparent resistivity is given by

$$\overline{R} = \pi \frac{(\overline{AB}/2)^2 - (\overline{MN}/2)^2}{\overline{MN}} \frac{\Delta V}{I} \qquad \textbf{(13-4)}$$

The **dipole-dipole** array is particularly convenient for making electrical *soundings*, which measure the changes in electrical properties with depth. The dipole-dipole configuration has a pair of current electrodes separated from a pair of potential electrodes. The same spacing, *a,* is used between the current electrodes as between the potential electrodes, and the distance between the electrode pairs, *na,* which is a multiple, *n,* of *a,* is much greater than the electrode spacing (Figure 13.6C). The apparent resistivity for the dipole-dipole array is given by

$$\overline{R} = n(n + 1)(n + 2)\, a\, \frac{\Delta V}{I} \qquad \textbf{(13–5)}$$

Geophysical instruments are available to measure the value of ΔV for a known I. The appropriate formula for the electrode array is used to compute the apparent resistivity.

Resistivity surveys are made in two fashions. An **electrical sounding** will reveal the variations of apparent resistivity with depth. **Horizontal profiling** is used to determine lateral variations in resistivity. When the electrode spacing is expanded in making an electrical sounding, the distance between the potential electrodes and the current electrodes increases. This means that the current will travel progressively deeper through the ground and will measure apparent resistivity to greater depths. Either the Wenner or the Schlumberger array may be used; however, the latter is more convenient for electrical sounding. This is because, for each incremental measurement, only the outer current electrodes must be moved every time. The inner electrodes are spread only occasionally. In the Wenner array, all four electrodes must be moved for each incremental measurement. The sounding is begun with the electrodes close together. After each reading, the electrodes are repositioned with a, or $\overline{AB}/2$, increased and a new measurement made. The apparent resistivity is plotted on logarithmic paper as a function of electrode spacing. For a number of reasons (Zohdy, Eaton, & Mabey, 1974), the Schlumberger array is superior to the Wenner array for electrical soundings. There is, however, a set of theoretical type curves of Wenner apparent resistivity for two-, three-, and four-layer earth models (Mooney & Wetzel 1956). This could be helpful in interpreting the results of a Wenner-array electrical sounding.

Radstake et al. (1991) developed a simple numerical model that can predict the expected response of a Wenner or Schlumberger resistivity sounding for a known earth profile. If one has some general knowledge of the geology of the area to be profiled, then this method can be used to select the optimum electrode spacing.

Goyal, Niwas, and Gupta (1991) evaluated a modified version of the Wenner array. The modified version starts with the inner potential electrodes very close to each other, as in the Schlumberger array. The outer current electrodes are kept in the same location, and the inner electrodes are gradually spread. They will pass through the standard Wenner array when the spacing between all four electrodes is the same and will eventually arrive at a configuration where the potential electrodes are close to the current electrodes. Goyal et al. found that this arrangement gave very good resolution for shallow resistivity soundings, especially in the vadose zone.

For a homogeneous earth, there is a definite relationship between the electrode spacing and the percent of the current that penetrates to a given depth

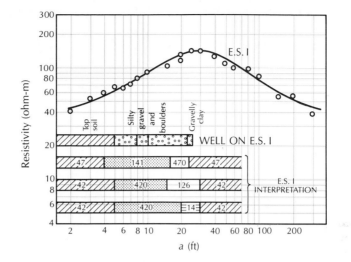

FIGURE 13.7 Wenner electrical-sounding curve of apparent resistivity as a function of electrode spacing. Three possible interpretations and a test boring are also given. Source: A. A. R. Zohdy, *Ground Water* 3, no. 3 (1965): 41–48.

(Zohdy 1965). For a nonhomogeneous and layered earth, the exact relationship cannot be easily determined. It is safe to assume that the greater the electrode spacing, the deeper the stratum influencing the apparent-resistivity curve. There are a number of possible earth models that could produce a given curve. In Figure 13.7, there are three possible theoretical interpretations expressed as resistivity and a test-boring log. The rise in apparent resistivity indicates a shallow zone of high resistivity. The test boring shows this to be a layer of silty gravel and boulders from 5 to 23 ft. It should be noted that the apparent-resistivity curve peaks at 30 ft. An interpretation that the layer of maximum resistivity lay at 30 ft would have been wrong.

In horizontal profiling, the electrode spacing is kept at a constant value. The electrodes are moved in a grid pattern over the land surface. The apparent resistivity of each point on the grid is marked on a map and isoresistivity contours are drawn. Figure 13.8 shows an apparent-resistivity map based on a large number of horizontal resistivity measurements. An area of buried stream-channel gravels is delineated where the apparent resistivity exceeds 80 $\Omega \cdot m$ (Zohdy 1965). A geologic cross section based on test borings and electrical resistivity soundings is shown in Figure 13.9. The location of the cross section is indicated in Figure 13.8 as line *AB*.

Some applications of electrical-resistivity surveys include identification of a buried fault in the Bunker Hill ground water basin in San Bernardino, California (Park, Lambert, & Lee 1990), areas of vertical fractures in crystalline rock terrain in Brazil (Medeiros & Lima 1990), and mapping the layered structure of a closed landfill in Dupage County, Illinois (Carpenter, Kaufmann, & Price 1990).

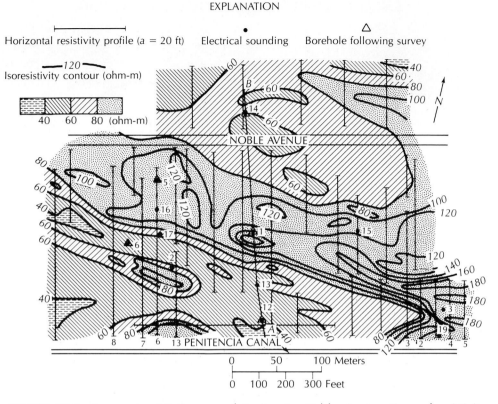

FIGURE 13.8 Apparent-resistivity map of Penitencia, California. Locations of resistivity profiles, soundings, and boreholes are shown. The location of Borehole and Electrical Sounding 1 is in the center of the high-resistivity zone. Source: A. A. R. Zohdy, *Ground Water*, 3, no. 3 (1965): 41–48.

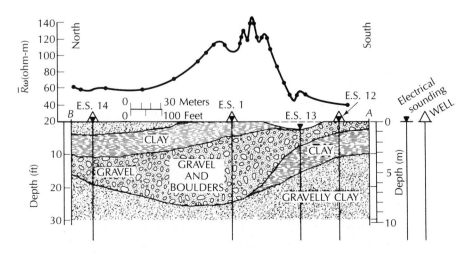

FIGURE 13.9 Geologic cross section based on test borings and electrical soundings. The position of the cross section is shown as line *AB* on Figure 13.8. Source: A. A. R. Zohdy, *Ground Water* 3, no. 3 (1965): 41–48.

Geoelectrical methods are useful in ground-water studies for such purposes as defining buried stream channels and areas of saline versus fresh ground water. Saline water has a much lower resistivity, as it is a better electrical conductor. Interpretation for such cases is relatively simple and may be done qualitatively. Layers of very low resistivity, such as clay, can also be discerned on sounding curves. It is often impossible to pick out the water table on an electrical sounding (Zohdy, Eaton, & Mabey 1974), although it is frequently attempted.

Electrical resistivity methods have been applied to many ground-water situations where the resistance of the fluid in the ground varies. For example, this occurs where there is a plume of saline ground water. Such a plume may be the result of salt-water intrusion, saline water seeping from a brine pit, or leachate from a landfill. The dissolved solids in the ground water can conduct electricity more readily and thus will have a lower apparent resistivity. A map of the apparent resistivity created by horizontal profiling can often show areas of contaminated ground water (Gilkeson & Cartwright 1983; Yazicigil & Sendlein 1982; Stewart, Layton, & Lizanec 1983; Urish 1983; Fretwell & Stewart 1981). Figure 13.10 shows the apparent resistivities around oil field brine ponds in Illinois. The data were obtained by horizontal profiling using a Wenner electrode array of 20 ft (Reed, Cartwright, & Osby 1981). The areas of low apparent resistivity indicate where the ground water contains higher levels of dissolved solids. These areas occur around both an active holding pond and abandoned ponds. There is a ground-water mound beneath the active brine pond, indicating that seepage from it is the cause of part of the ground-water contamination. However, the low apparent resistivity in the vicinity of the abandoned ponds indicates that there is also residual ground-water contamination.

13.3.2   Electromagnetic Conductivity

Electricity traveling through sedimentary units is transmitted more readily by the fluids in the pore spaces and by the fluid-grain interfaces than through the mineral grains themselves. As a result, the pore surface area and the pore fluid conductivity are important factors in determining overall bulk conductivity of earth units (Stewart 1982). Electromagnetic conductivity is the inverse of electrical resistivity. Field studies have shown similar results when both methods have been used at the same site (Sweeney 1984).

Electromagnetic methods use an electromagnetic field generated by a transmitter coil through which an alternating current is passed. This generates a magnetic field around the transmitter coil. When the transmitter coil is held near the earth, the magnetic field induces an electrical field in the earth. The electrical field will travel through the ground at different strengths depending upon the ground conductivity. The field strength is measured in a passive receiver coil. Changes in the phase, amplitude, and orientation of the primary field can be measured either with time or distance by using the receiver. These changes are related to the electrical properties of the earth.

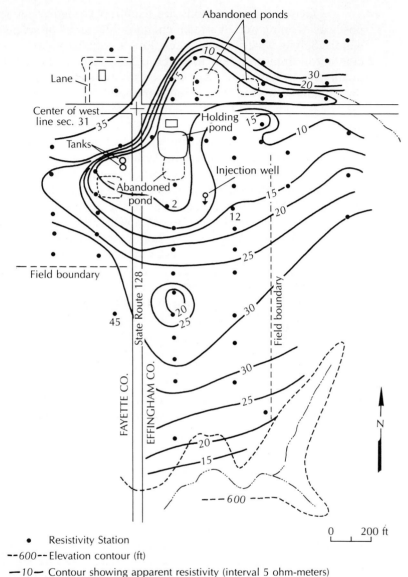

Abandoned ponds

Lane

Center of west
line sec. 31

Tanks

Abandoned
pond

Holding
pond

Injection well

Field boundary

State Route 128

FAYETTE CO.

EFFINGHAM CO.

Field boundary

N

0   200 ft

●   Resistivity Station

--600-- Elevation contour (ft)

—10— Contour showing apparent resistivity (interval 5 ohm-meters)

FIGURE 13.10 Apparent resistivity from horizontal profiling with a 20-ft spacing in the vicinity of oil-field brine holding ponds in Illinois. Source: Modified from P. C. Reed, K. Cartwright, and D. Osby, Illinois State Geological Survey Environmental Geology Note 95, 1981, p. 17.

There are several different electromagnetic methods available. They all have the advantage of being rapid, as none of them require the insertion of electrodes into the ground. Electromagnetic methods are not inherently more accurate than direct resistivity methods but are likely to be more cost-effective because field work can often be completed more rapidly. They can be used to detect changes in earth conductivity related to contaminant plumes, buried metallic wastes such as drums, or salt-water interfaces (Stewart 1982; Sweeney 1984; Stewart & Gay 1986; Greenhouse & Slaine, 1983).

Geonics, Ltd., of Mississisauga, Ontario, Canada, is the manufacturer of a line of electromagnetic equipment. This is not an endorsement of Geonics, Ltd. products; the brand name is given because their products are generally referenced by model number in the hydrogeological literature.

The Geonics EM-31 has the transmitter and receiver coils in the same unit. The coils are mounted on a long pole so that they have a fixed separation distance of 12 ft (3.66 m). The unit can be used by a single operator, who can walk along a line and note the meter readings at stations every 10 ft (3 m) or so. The output from the instruments is apparent conductivity in millimhos per meter. It can also be read continuously, allowing small-scale heterogeneities to be determined with greater precision than practical with electrical resistivity methods, where the electrodes must be moved for each separate reading. As the distance between the transmitter and the receiver cannot be varied, the depth of penetration of the electrical field is constant and relatively shallow, about 20 ft (6 m).

The Geonics EM-34-3 has separate units for the transmitter and the receiver coils. Two operators are needed, one for each coil. The coils can be held either horizontally or vertically. They are separated by a distance, L. With the coils oriented horizontally, the effective depth of penetration is about $0.75.L$. If the coils are oriented vertically, the effective depth of penetration is about $1.5L$ and the readings are less influenced by the near-surface layers. The EM-34-3 can be operated at three different intercoil spacings: 32.8, 65.6, and 131.1 ft (10, 20, and 40 m). The EM-34-3 unit can be used to study earth conductivity to much greater depths than the EM-31 as the spacing between transmitter and receiver can be much greater than the fixed 12 ft of the EM-31. Figure 13.11 illustrates the use of both the EM-31 and EM-34-3 in terrain conductivity mapping at a sanitary landfill site. The area was mapped by both electrical resistivity and electromagnetic conductivity using the Wenner array with a 20-m electrode spacing (Figure 13.11A), the EM-31 (Figure 13.11B), the EM-34-3 with a 15-m vertical coplanar coil spacing (Figure 13.11C), and the EM-34-3 with a 31-m vertical coplaner spacing (Figure 13.11D). The earth conductivity was converted to resistivity, in ohm-meters, by inverting and then dividing by 1000. The term *inductive resistivity* refers to conductivity converted to resistivity in this manner. Note that the terrain conductivity surveys took about one-sixth of the person days that were required for the resistivity survey.

Terrain conductivity surveys can be distorted by conductors such as buried pipelines and steel tanks. They can also be adversely impacted by high-tension power lines and electrical storms.

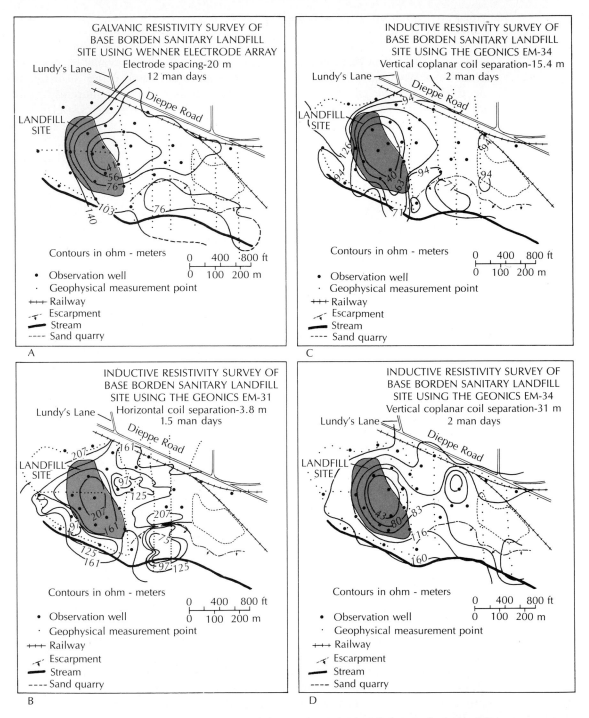

FIGURE 13.11 Comparison of electrical resistivity and electrical conductivity measurements at a sanitary landfill at Camp Bordon, Ontario. The low-resistivity area represents a plume of contaminated ground water with high total dissolved solids. The plume shows up more clearly on the surveys with greater depth of penetration (Wenner array and EM-34-3). Source: Geonics Ltd. Used with permission.

13.3.3 Seismic Methods

Seismic methods using artificially created seismic waves traveling through the ground are quite commonly employed in hydrogeology. These methods are useful in determining depth to bedrock, slope of the bedrock, depth to water table, and, in some cases, the general lithology. Applied seismology has been highly developed in the petroleum industry, where the *seismic reflection method* is used almost exclusively. Structural and formational boundaries can be indicated to great depth.

Hydrogeological studies often involve finding the thickness of unconsolidated material overlying bedrock. For this purpose, the **seismic refraction method** is superior. The loose material transmits seismic waves more slowly than consolidated bedrock. By studying the arrival times of seismic waves at various distances from the energy source, the depth to bedrock can be determined.

The energy source can be a small explosive charge set in a shallow drill hole. One or two sticks of dynamite is sufficient for depths to bedrock in excess of 100 to 200 ft (30 to 50 m). Of course, explosives should be handled only by persons trained and licensed to do so. A judgment of how large a charge to use must be made in each case. For shallower work, 15 to 50 ft (5 to 15 m), a sledgehammer struck on a steel plate lying on the ground may be a sufficient energy source. The seismic wave is detected by geophones placed in the earth in a line extending away from the energy source. A seismograph records the travel time for the wave to go from the energy source to the geophone. The more sophisticated seismographs are multichannel units with a number of geophones attached.

Figure 13.12 illustrates the travel paths of compressive seismic waves traveling through a two-layer earth. The seismic velocity in the lower layer is greater than that in the upper layer. As the energy travels faster in the lower layer, the wave passing through it gets ahead of the wave in the upper layer. At the boundary between the two layers, part of the energy is refracted back upward from the lower-layer boundary to the surface.

The angle of refraction of each wave front is called the *critical angle, i_c,* and is equal to the arc sin of the ratio of the velocities of the two layers:

$$i_c = \sin^{-1} \frac{V_1}{V_2} \tag{13-6}$$

EXAMPLE PROBLEM

Determine the critical angle, i_c, when $V_1 = 1000$ m/s and $V_2 = 4000$ m/s.

$$i_c = \sin^{-1} V_1/V_2 = \arcsin 0.25 = 14.5°$$

Figure 13.13 illustrates a wave front and the path of the refracted energy that travels along the lower-layer boundary. A direct wave in the upper layer is also shown.

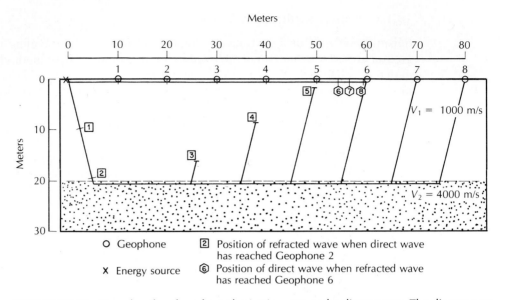

FIGURE 13.12 Travel paths of a refracted seismic wave and a direct wave. The direct wave will reach the first five geophones first, but for the more distant geophones the first arrival is from a refracted wave. Numbers inside symbols refer to distances traveled by wave paths going toward the indicated geophone.

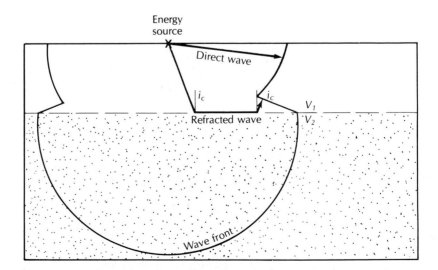

FIGURE 13.13 A seismic wave front at a given time after a charge is detonated.

If V_2 is less than V_1, the wave will be refracted downward and no energy will be directed upward. Thus, the refraction method will show higher-velocity layers but not lower-velocity layers that are overlain by a high-velocity layer.

Energy can travel directly through the upper layer from the source to the geophone. This is the shortest distance, but the waves do not travel as fast as those traveling along the top of the lower layer. The latter must go farther, but they do so with a higher velocity. In Figure 13.12, the positions of waves traveling to each geophone are indicated. Geophones 1 through 5 first receive waves that have traveled through only the upper layer. The sixth and succeeding geophones measure arrival times of refracted waves that have gone through the high-velocity layer as well. The position of the trailing wave front at each time the leading front reaches each geophone is indicated in the figure.

A graph is made of the arrival time of the first wave to reach the geophone versus the distance from the energy source to the geophone. This is known as a *travel-time* or *time-distance curve*. Figure 13.14 shows the time-distance curve for the shot in Figure 13.12. The reciprocal of the slope of each straight-line segment is the apparent velocity in the layer through which the first arriving wave passed. The slope of the first segment is 10 milliseconds per 10 meters, so that the reciprocal is 10 m per 10 ms, or 1000 m/s.

The projection of the second line segment backward to the time-axis ($X = 0$) yields a value known as the *intercept time, T_i.* This value can be determined graphically, as shown in Figure 13.14. T_i is 39 ms and X is 52 m. The depth to the lower layer, Z, is found from the equation (Dobrin 1976)

$$Z = \frac{T_i}{2} \frac{V_1 V_2}{\sqrt{V_2^2 - V_1^2}}$$

(13–7)

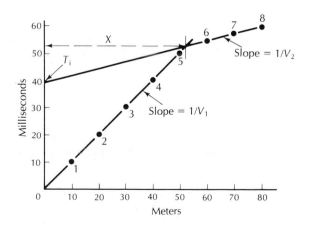

FIGURE 13.14 Arrival time–distance diagram for a two-layered seismic problem. Numbers refer to geophones in Figure 13.12.

The depth to the lower layer can also be found from the equation (Dobrin 1976)

$$Z = \frac{X}{2} \sqrt{\frac{V_2 - V_1}{V_1 + V_2}} \qquad (13\text{--}8)$$

where X is the distance from the shot to the point at which the direct wave and the refracted wave arrive simultaneously. This is shown on Figure 13.14 as the x-axis distance where the two line segments cross.

EXAMPLE PROBLEM

Find the value Z from Figure 13.14.

From the slope of each line segment, $V_1 = 1000$ m/s and $V_2 = 4000$ m/s. T_i is 39 ms and X is 52 m.

$$Z = \frac{T_i}{2} \frac{V_1 V_2}{\sqrt{V_2^2 - V_1^2}}$$

$$= \frac{0.039 \text{ s}}{2} \times \frac{1000 \text{ m/s} \times 4000 \text{ m/s}}{\sqrt{(4000 \text{ m/s})^2 - (1000 \text{ m/s})^2}}$$

$$= 20 \text{ m}$$

Also,

$$Z = \frac{X}{2} \sqrt{\frac{V_2 - V_1}{V_1 + V_2}}$$

$$= \frac{52 \text{ m}}{2} \sqrt{\frac{4000 \text{ m/s} - 1000 \text{ m/s}}{4000 \text{ m/s} + 1000 \text{ m/s}}}$$

$$= 20 \text{ m}$$

A more typical case in hydrogeology is a three-layer earth, the top layer being unsaturated, unconsolidated material. In the next layer, below the water table, the unconsolidated deposits are saturated, which yields a higher seismic velocity. The third layer is then bedrock. Under such conditions, the seismic method can be used to find the water table. However, similar velocities are possible from either saturated sand or unsaturated glacial till. Similar seismic refraction patterns could be obtained from a water table in a uniform sand deposit or a layer of unsaturated sand overlying unsaturated glacial till. This illustrates the point that geophysics is best interpreted in light of other data.

The three-layer seismic refraction case with $V_1 < V_2 < V_3$ is shown in Figure 13.15. The first arriving waves show three line segments. The reciprocal of the slope of each line is the seismic velocity of the respective layers. The intercept time for each of the two deeper layers is the projection of the line segment back

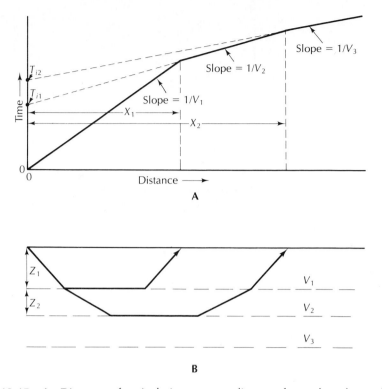

FIGURE 13.15 **A.** Diagram of arrival time versus distance for a three-layered seismic problem. **B.** Wave path for a three-layered seismic problem.

to the time-axis. Indicated on the figure is the distance, X_1, from the shot to the point at which waves from layers 1 and 2 arrive simultaneously and the distance, X_2, to the point at which waves from layers 2 and 3 arrive simultaneously. The thickness, Z_1, of layer 1 is found from the values of V_1 and V_2 and either T_{i1} or X_1 using Equation 13-7 or 13-8. The thickness of the second layer, Z_2, is found using (Dobrin 1976)

$$Z_2 = \frac{1}{2}\left(T_{i2} - 2Z_1 \frac{\sqrt{V_3^2 - V_1^2}}{V_3 V_1}\right)\left(\frac{V_2 V_3}{\sqrt{V_3^2 - V_2^2}}\right) \qquad \textbf{(13-9)}$$

The value of Z_1 must be computed before computing the value of Z_2.

The velocities computed from the reciprocals of the slope are called *apparent velocities*. If the lower layer is horizontal, they represent the actual velocity. However, if the lower layer is sloping, the arrival time for a shot measured downslope will be different from one measured upslope. Seismic lines are routinely run with a shot at either end, so that dipping beds can be determined. Time-distance curves for a dipping stratum are shown in Figure 13.16, with travel times measured from shots at either end of the line. The upper layer is unaffected

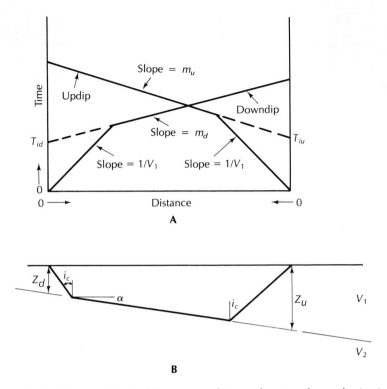

FIGURE 13.16 **A.** Diagram of arrival time versus distance for a two-layered seismic problem with a sloping lower layer. **B.** Wave path for the preceding problem.

by the dip of the lower bed, so that the reciprocal of the slope of the first line segment is V_1. In order to find the values of V_2 and the depth to the bedrock at the updip end of the line, Z_d, as well as at the downdip end, Z_u, a complex series of computations must be made (Dobrin 1976).

The slope of the second line segment of the downdip line is m_d, and the slope of the second line segment of the updip line is m_u. The value of the angle of refraction, i_c, is found from

$$i_c = \frac{1}{2}(\sin^{-1}V_1 m_d + \sin^{-1}V_1 m_u)$$ **(13–10)**

The value of V_2 is given by

$$V_2 = \frac{V_1}{\sin i_c}$$ **(13–11)**

The angle of slope of the dipping layer, α, is found from

$$\alpha = \frac{1}{2}(\sin^{-1}V_1 m_d - \sin^{-1}V_1 m_u)$$ **(13–12)**

Finally, the depths to the lower layer at either end of the shot line are found from the expressions

$$Z_u = \frac{V_1 T_{iu}}{(\cos \alpha)(2 \cos i_c)}$$ **(13–13)**

$$Z_d = \frac{V_1 T_{id}}{(\cos \alpha)(2 \cos i_c)}$$ **(13–14)**

If there are more than two dipping layers, then expressions of greater complexity must be used (Mota 1954).

EXAMPLE PROBLEM   A seismic survey yielded data for a dipping two-layer case in which the following values were obtained:

$$V_1 = 1570 \text{ m/s}$$
$$m_u = 1.67 \times 10^{-4} \text{ s/m}$$
$$m_d = 1.54 \times 10^{-4} \text{ s/m}$$
$$T_{id} = 0.046 \text{ s}$$
$$T_{iu} = 0.050 \text{ s}$$

Compute V_2, Z_u, and Z_d.

$$i_c = \tfrac{1}{2}(\sin^{-1}V_1 m_d + \sin^{-1}V_1 m_u)$$
$$= \tfrac{1}{2}(\sin^{-1}1570 \times 1.54 \times 10^{-4} + \sin^{-1}1570 \times 1.67 \times 10^{-4})$$
$$= \tfrac{1}{2}(13.99 + 15.20)$$
$$= 14.6°$$

$$V_2 = V_1/\sin i_c$$
$$= 1570/\sin 14.6$$
$$= 6230 \text{ m/s}$$

$$\alpha = \tfrac{1}{2}(\sin^{-1}V_1 m_d - \sin^{-1}V_1 m_u)$$
$$= \tfrac{1}{2}(\sin^{-1}1570 \times 1.54 \times 10^{-4} - \sin^{-1}1570 \times 1.67 \times 10^{-4})$$
$$= \tfrac{1}{2}(13.99 - 15.20)$$
$$= -0.6°$$

$$Z_u = \frac{V_1 T_{iu}}{(\cos \alpha)(2 \cos i_c)}$$
$$= \frac{1570 \times 0.050}{\cos -0.6 \times 2 \times \cos 14.6}$$
$$= 40.6 \text{ m}$$

$$Z_d = \frac{V_1 T_{id}}{(\cos \alpha)(2 \cos i_c)}$$

$$= \frac{1570 \times 0.046}{\cos -0.6 \times 2 \times \cos 14.6}$$

$$= 37.3 \text{ m}$$

The cases given in this section are but a few of the many possible cases that might be encountered. Figure 13.17 shows schematic travel-time curves for a number of nonhomogeneous earth models. Should the hydrogeologist suspect that the situation is more than a simple two- or three-layer model with homogeneous beds, an experienced geophysicist should interpret the field data.

The principal application of refraction seismology in hydrogeology has been in the identification of buried refractors such as the top of the water table—which, being saturated, has a greater seismic velocity than the unsaturated equivalent soil unit—and the bedrock surface (Sverdrup 1986). It has been very useful in the delineation of bedrock valleys buried in glacial drift (Denne et al. 1984).

13.3.4 Ground-Penetrating Radar and Magnetometer Surveys

There are thousands of abandoned waste-disposal sites in the United States alone. For many sites the records of trench locations and drum burial areas are sketchy or nonexistent. If a remedial action plan is to be developed for such a site, it is necessary to know the extent of the buried waste. From a safety standpoint, one needs to know the locations of buried drums of toxic material so that drilling operations for the installation of monitoring wells don't penetrate the drums. There are several methods that are used to locate areas of waste disposal. Two of the more common are **magnetometer surveys** and **ground-penetrating radar** (Koerner et al. 1982; Evans, Benson, & Rizzo 1982; Hitchcock & Harmon 1983; Horton et al. 1981; Koerner, Lord, & Bowders 1981; Gilkeson, Heigold, & Laymon 1986).

Magnetometer surveys measure the strength of the earth's magnetic field. A proton nuclear magnetic resonance magnetometer is frequently used. This is a hand-held instrument and one person can rapidly perform a survey over a site of a few acres in size. A grid system is set up and measurements are made of the magnetic field at each intersection of the grid. Areas with large amounts of buried metal, such as steel drums, will have magnetic anomalies associated with them. The strength of the anomaly will vary with the amount and depth of the buried metal.

Ground-penetrating radar (GPR) is based on the transmission into the ground of repetitive pulses of electromagnetic waves in the frequencies of 10 to 1000 MHz. The pulses are reflected back to the surface when the radiated energy encounters an interface between two materials of differing dielectric properties. The interfaces that cause the reflections may be due to such things as a change in

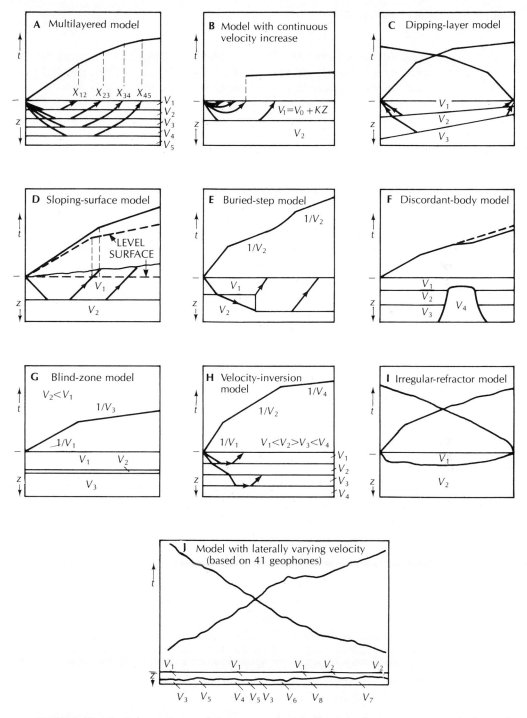

FIGURE 13.17 Schematic travel-time curves for idealized nonhomogeneous geologic models. Source: G. P. Eaton. "Seismology," in *Techniques of Water-Resources Investigations.* U.S. Geological Survey, 1974, Book 2, Chap. D1, pp. 67–84.

strata or buried objects. The ground-penetrating radar system has antennae to both send a signal and receive the reflected pulse. The reflected pulse may be recorded on a magnetic tape recorder and generally is also converted into voltage as a function of time and displayed on a graphic recorder. The graphic output from a GPR system looks like a sonar signal. Figure 13.18 shows the graphic output from a ground-penetrating radar survey line. A typical receiving unit will display the output in a number of shades of gray or colors that vary proportionally to the voltage level of the received wave.

The GPR unit is pulled along the ground and a continuous line profile is generated. Parallel lines are surveyed to cover an area being studied. The depth of penetration of the GPR is a function of the type of geological material and the frequency of the radar being used. The lower frequencies penetrate a given medium to the greatest depths; higher frequencies will not penetrate as deeply but give greater resolution. Greater resolution increases the ability of the unit to discriminate between interfaces and objects that are closely spaced. For waste-site studies, typical depths of study are 5 to 20 ft (1.5 to 6 m). GPR is capable of showing the location of a single 55-gal metal drum buried at depths of 6 to 9 ft (2 to 3 m) (Horton et al. 1981). It has been successfully used to show the limits of a buried crystalline mass below a concrete pavement. The great advantage of GPR is that it is able to give a continuous profile of the subsurface, an accomplishment not otherwise possible.

Beres and Haeni (1991) found that ground-penetrating radar records of the subsurface in glaciated terrain in Connecticut could be qualitatively interpreted with the aid of boreholes with lithologic logs. They distinguished fine-grained from coarse-grained sediments, bedrock, and boulders. The water table in coarse-grained sediments was easily identified. Very shallow water tables and water tables in fine-grained sediments where there was a significant capillary fringe were not easy to distinguish. Penetration depth of the GPR signal ranged from 20 ft (7 m) in fine-grained glaciolacustrine sediment to 70 ft (20 m) in very coarse-grained sand and gravel.

13.3.5 Gravity and Aeromagnetic Methods

Measurement of the gravimetric and magnetic fields of the earth are standard geophysical methods used to study the structure and composition of the earth. To the extent that the basic geology influences the hydrogeology, these methods are quite useful in ground-water studies. The collection of data can be relatively simple if ground stations are used. However, the reduction and correction of the data are fairly complex (Zohdy, Eaton, & Mabey 1974; Dobrin, 1976; Wilson, Peterson, & Ostrye 1983). Airborne geomagnetic surveys obviously involve specialized skills and equipment.

Both gravity and magnetics surveys can be used to delineate the area of unconsolidated basin fill or buried stream-channel aquifers. In Figure 13.19, a Cenozoic basin in Antelope Valley, California, is depicted in cross section. The presence of the basin and the area of the deepest part are shown on both an aeromagnetic profile and a gravity profile.

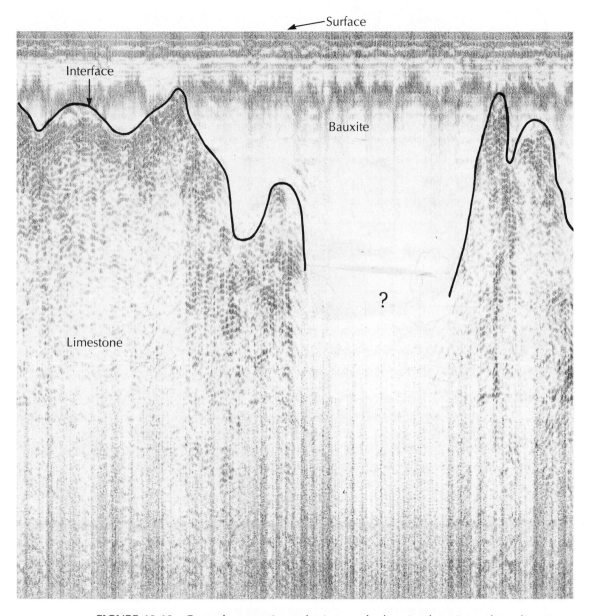

FIGURE 13.18 Ground-penetrating radar image of a bauxite deposit overlying limestone. No horizontal or vertical scale. Source: Courtesy of Joan Underwood, SEC Donohoe, Sheboygan, Wisconsin.

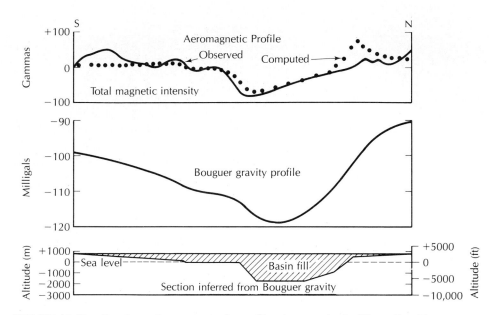

FIGURE 13.19 Gravity and aeromagnetic profiles across a basin-fill aquifer. The aeromagnetic profile was flown at 490 ft (150 m). Source: D. R. Mabey. "Magnetic Methods," in *Techniques of Water-Resources Investigations*, U.S. Geological Survey, 1972, Book 2, Chap. D1, pp. 85–115.

Magnetic anomalies are caused by distortions of the earth's magnetic field created by magnetic materials in the crust. Magnetic anomalies indicate the type of rocks in a very general way. In hydrogeologic studies, magnetic anomalies can be useful in indicating the depth to magnetic basement rocks. Sedimentary rocks are typically nonmagnetic; hence, they do not affect the magnetic field. Some magnetic rocks, such as basalt flows, can be important aquifers. Magnetic surveys might be useful in tracing basalt flows in areas of nonmagnetic rocks.

The mass of the rocks beneath a point on the earth will affect the local value of the acceleration of gravity. To be useful, the measured values must be referenced to a common datum—usually, mean sea level. A free-air correction is made to compensate for elevation differences. To correct for the gravitational attraction of the rock that lies between the gravity station and sea level, a *Bouguer correction* is made. Corrections must also be made for tidal effects, latitude, and terrain. After the measured gravity data are corrected, the result is a *Bouguer anomaly value,* which can be mapped with gravity contours drawn. Such a map could help to define the extent of a buried bedrock valley if there were a density difference between the sediments and the bedrock.

It should be emphasized that there are many possible earth models that would result in the same gravitational or magnetic anomaly. There is no unique solution to any set of geophysical data, and the person interpreting the data must keep this in mind.

13.4   GEOPHYSICAL WELL LOGGING

Direct access to the subsurface is gained wherever there is a well or test boring. When a well is drilled, a record may be made of the geologic formations encountered. The reliability of a **lithologic well log** depends on the method of drilling and sample recovery as well as the knowledge and skills of the person making the log. There are also many existing wells for which there is no available record of the subsurface geology.

Borehole geophysics offers a great deal in the way of practical application to hydrogeology. Borehole geophysical methods were developed primarily in the petroleum industry, and virtually all oil and gas wells are routinely logged when drilled. In the water-well industry, the use of geophysical logging is generally restricted to either research projects or high-capacity municipal and industrial wells. The cost of well logging is not justified by the marginal benefits gained for small-yield domestic wells.

Borehole geophysical data have a number of uses (see Table 13.1). The log of a well can indicate the areas of high porosity and permeability that would produce the most water. Zones of an aquifer with high-salinity water can be identified. If a number of wells are logged over an area, the logs can be used for stratigraphic correlation. The lithology of the rocks penetrated by the wells can be identified, especially if some core samples are available for baseline comparison. Regional ground-water flow patterns might be identified from such characteristics as fluid temperature. Nuclear well-logging techniques can be used in cased wells. This is the only way to get subsurface data under such conditions. Geophysical logs give a permanent record, based on repeatable measurements. Thus, data collected for one purpose are available for other, unanticipated uses in the future.

Because of the large number of borehole techniques that are applicable to water wells (Keys & MacCary 1971; Keys 1967; Baldwin & Miller 1979; Brown 1971; Crosby & Anderson 1971; Norris 1972; Keys & Brown 1978; MacCary 1983; Keys 1986; Kwader 1986), a discussion of only the more common methods will be included in this section, with emphasis on qualitative rather than quantitative interpretation of geophysical logs. Generally, a suite of geophysical logs is made, rather than only a single type. The methods tend to be complementary; one may confirm another. Likewise, certain interpretations are made on the basis of two or more logs. Figure 13.20 consists of six different geophysical logs made on the same borehole, along with a lithologic log. It can readily be seen that the logs deflect with the changes in lithology.

Continuously recorded well logs can be made as a pen-and-ink strip chart or as a digital record that can be displayed on a monitor and stored on magnetic media, such as a floppy disk. There also are more simple instruments available that give point readings at discrete depth intervals.

A probe is lowered into the borehole on a cable. The cable, which contains power lines from the surface to the probe, supports the weight of the probe and transmits signals from the probe to the recorder at the surface.

TABLE 13.1   Summary of log applications

Required Information on the Properties of Rocks, Fluid, Wells, or the Ground-Water System	Widely Available Logging Techniques That Might Be Utilized
Lithology and stratigraphic correlation of aquifers and associated rocks	Electric, sonic, or caliper logs made in open holes; nuclear logs made in open or cased holes
Total porosity or bulk density	Calibrated sonic logs in open holes, calibrated neutron or gamma-gamma logs in open or cased holes
Effective porosity or true resistivity	Calibrated long-normal resistivity logs
Clay or shale content	Gamma logs
Permeability	No direct measurement by logging. May be related to porosity, injectivity, sonic amplitude
Secondary permeability—fractures, solution openings	Caliper, sonic, or borehole televiewer or television logs
Specific yield of unconfined aquifers	Calibrated neutron logs
Grain size	Possible relation to formation factor derived from electric logs
Location of water level or saturated zones . .	Electric, temperature, or fluid conductivity in open hole or inside casing, neutron or gamma-gamma logs in open hole or outside casing
Moisture content	Calibrated neutron logs
Infiltration	Time-interval neutron logs under special circumstances or radioactive tracers
Direction, velocity, and path of ground water flow	Single-well tracer techniques—point dilution and single-well pulse; multiwell tracer techniques
Dispersion, dilution, and movement of waste	Fluid conductivity and temperature logs, gamma logs for some radioactive wastes, fluid sampler
Source and movement of water in a well . .	Injectivity profile; flowmeter or tracer logging during pumping or injection; temperature logs
Chemical and physical characteristics of water, including salinity, temperature, density, and viscosity.	Calibrated fluid conductivity and temperature in the well; neutron chloride logging outside casing; multi-electrode resistivity
Determining construction of existing wells, diameter and position of casing, perforations, screen	Gamma-gamma, caliper, collar, and perforation locator; borehole television
Guide to screen setting	All logs providing data on the lithology, water-bearing characteristics, and correlation and thickness of aquifers
Cementing.	Caliper, temperature, gamma-gamma; acoustic for cement bond
Casing corrosion	Under some conditions, caliper or collar locator
Casing leaks and/or plugged screen	Tracer and flowmeter

Source: W. S. Keys and L. M. MacCary, "Application of Borehole Geophysics to Water-Resources Investigations," in *Techniques of Water-Resources Investigations,* U.S. Geological Survey, 1971, Book 2, Chap. E1.

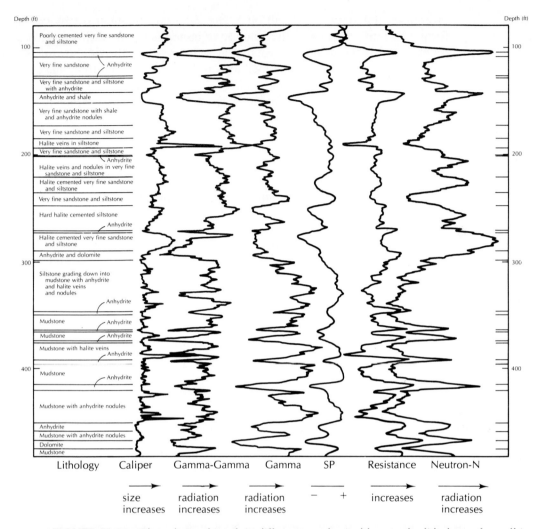

FIGURE 13.20 The relationship of six different geophysical logs to the lithology of a well in the upper Brazos River Basin, Texas. Source: W. S. Keys and L. M. MacCary, in *Techniques in Water-Resources Investigations*, U.S. Geological Survey, 1971, Book 2, Chap. E1.

The probe contains the necessary electronics, energy or nuclear sources, and detectors. Logging information can be obtained as the probe is either lowered or raised.

13.4.1 Caliper Logs

A **caliper log** is used to measure the diameter of an uncased borehole in bedrock units. It can also be used to find the casing depth. The nominal hole diameter is, of course, the size of the drill bit.* The hole may be enlarged by caving of the formations into the hole or by solution of minerals by the drill water. It also may be enlarged if the drill bit is rotated at a depth while the downward pressure is removed. Another use of caliper logs is to indicate solution-enlarged bedding planes and joints in carbonate aquifers. The hole diameter may be less than the drill-bit diameter if a mud cake has built up on the walls or if a plastic strata, such as a soft clay, was squeezed into the hole by the weight of the overlying, more competent rock or sediment.

13.4.2 Temperature Logs

A *temperature log* is a continuous vertical record of the temperature of the fluid in the borehole. This may or may not be indicative of the temperature of the fluid in the rocks opposite the borehole fluid. In a recently drilled well, the borehole fluid may be well mixed. After the well has had an opportunity to reach environmental equilibrium, the temperature log can reveal zones of differing temperature in the well. There will be a component of the geothermal gradient present; however, water in different aquifers may be at discrete temperatures, which may be detected on the log. Thermal logging has been used to trace the movement of water previously injected into an aquifer (Norris 1972). The recharged water had a daily thermal variation of up to 17°C, and the diurnal fluctuation was traced in a series of observation wells (Figure 13.21).

13.4.3 Single-Point Resistance

Electrical resistance can be measured in a borehole by a number of different methods. The simplest case is the **single-point resistance,** in which a single electrode is lowered into the borehole on an insulated cable. The other electrode is at the ground surface. As the electrode is lowered into the borehole, the resistance of the earth between the two electrodes is measured.

The single-point electrode is measuring the resistance of all of the rocks between the electrodes. Most of the variation in resistance is due to the changes in the conductivity of the borehole fluid and to a small volume of rock around the borehole near the downhole electrode. If the borehole fluid is homogeneous, the variation in resistance will be due to the lithologic variations near the borehole.

Lithologies with a high resistance include sand, sand and gravel, sandstone, and lignite. Clay and shale have the lowest resistance. Increasing salinity will cause a decrease in resistance. If the borehole is enlarged (for instance, by a fracture), the resistance will also decrease. If the caliper log indicates a cavity, and the resistance log decreases, the decrease is due to the hole enlargement. Should the borehole be straight, a decrease in resistance might be due to either a

---

*When a hole is drilled, the drilling rig turns a pipe (drill stem). The rock at the bottom of the hole is broken up by the rotating drill bit, which cuts it into pieces. Drilling fluids circulating through the hole and drill stem bring the broken rock to the surface.

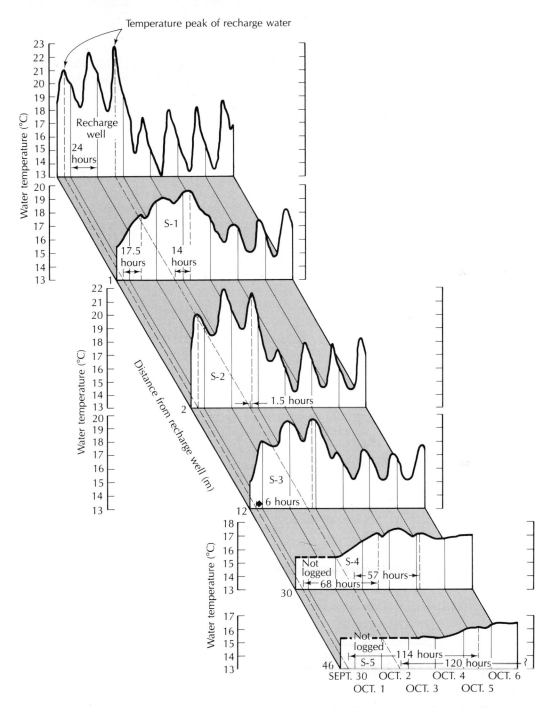

FIGURE 13.21 Fluctuations in the temperature of ground water in a number of wells near an injection well for artificial recharge. All temperatures are measured at a depth of 161 ft (49 m). The diurnal variation in the temperature of the recharge water dies out by 98 ft (30 m). Source: W. S. Keys and R. F. Brown, *Ground Water* 16 (1978): 32–48.

shale layer or, perhaps, a sandstone containing a brine. However, other logs can distinguish brine from fresh water and sandstone from shale. Figure 13.22 shows a single-point resistance log.

13.4.4   Resistivity

Earth **resistivity** may be measured in a borehole by lowering two current electrodes and measuring the resistivity between two additional electrodes. Resistivity is measured in ohm-meters and is different from resistance, the latter being measured in ohms. Single-point resistance logging measures the total resistance of the earth materials, while resistivity measures a specific property of the rock and the contained pore waters. The trace of a resistivity log is similar to that of a single-point resistance log. However, resistivity logs can be calibrated and used quantitatively.

A number of different electrode configurations are used in resistivity logging. These include the *short-normal, long-normal,* and *lateral configurations,* as shown in Figure 13.23. The three logs have similar traces (Figure 13.24). The short-normal curve indicates the resistivity of the zone close to the borehole. It is in this area that the drilling fluid may have invaded the formations. The long-normal spacing has more spacing between electrodes and thus measures the resistivity farther away from the borehole—presumably, beyond the influence of the drilling fluid. Both the short-normal and long-normal resistivity measure a greater radius of influence than the single-point resistance.

Lateral devices have very widely spaced electrodes for measuring zones that are far from the borehole. Because of the wide spacing, lateral devices will not pick out thin beds of different resistivity. For example, the 18-ft by 8-in. lateral log will be best for beds at least 40 ft (12 m) thick.

13.4.5   Spontaneous Potential

Along with resistance and resistivity, **spontaneous potential** is another form of electrical logging. It is a measure of the natural electrical potential that develops between the formation and the borehole fluids. It is run only on an open hole filled with fluid, as are resistance and resistivity. It can be used below a casing in a partially cased well. The spontaneous-potential (SP) curve can be used for determination of bed thickness, geologic correlation, and also for delineation of permeable rocks. An SP device consists of a surface electrode and a borehole electrode with a voltmeter to measure potential.

One use of the SP curve is to distinguish shale from sandstone lithology. Shale has a positive SP response and sandstone a negative one if the salinity of the formation fluid is greater than that of the borehole fluid.

13.4.6   Nuclear Logging

Some of the most useful logging methods involve the measurement of either natural radioactivity of the rock and fluids or their attenuation of induced radia-

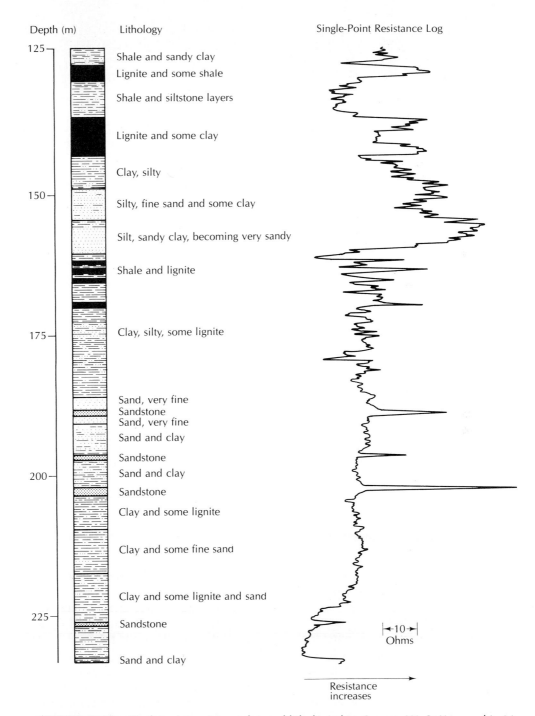

FIGURE 13.22 Single-point resistance log and lithologic log. Source: W. S. Keys and L. M. MacCary, in *Techniques of Water-Resources Investigations,* U.S. Geological Survey, 1971, Book 2, Chap. E1.

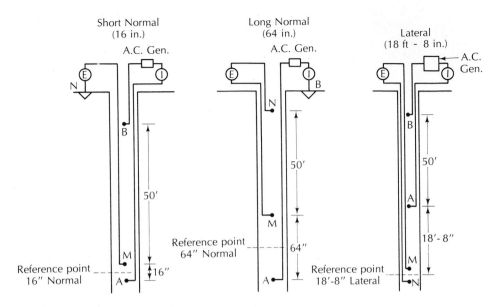

FIGURE 13.23 Electrode configuration for various resistivity logging devices.

tion. *Nuclear logging* can be done in either a cased or an uncased hole, and the logs are not affected by the type of drilling mud. The use of radioactive isotopes involves special safety precautions.

Radioactive decay is a process with a random component; hence, the instantaneous rate of decay fluctuates. Over a long time period, the rate of decay per time period is constant. However, as the time period decreases, the variation in the number of decay events per time period will increase. Nuclear logging records the number of disintegrations over a fixed time period, called the *time constant*. The longer the time constant, the less the likelihood that a variation in radiation intensity is due to random decay and, hence, the more likely the variation reflects different lithology.

Consideration must also be given to the speed at which the probe is moved up or down the hole. If the speed is too great, the probe may pass a thin bed before the time constant has elapsed. Consequently, the selection of the proper time constant and logging speed is very important and depends upon the equipment, logging technique, and lithology (Keys & MacCary 1971).

Nuclear logs do not have exact reproducibility, owing to the statistical nature of the decay process. Repeat logging runs are necessary to determine if an observed variation represents a lithologic change or a statistical fluctuation in the decay rate. In Figure 13.25, the first two neutron-gamma logs were made first going up and then going down the hole. The same peaks are present, but there is variability in the exact radiation count. To the right is a third log of the same hole

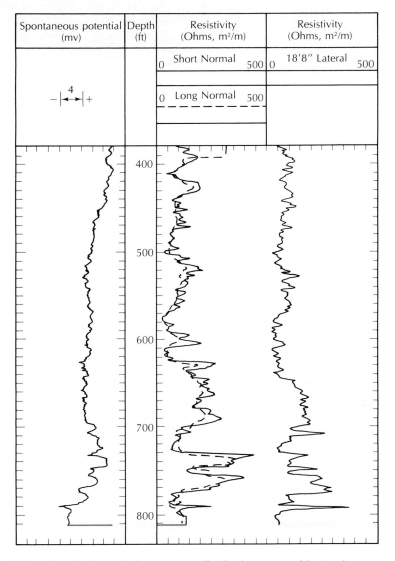

FIGURE 13.24 Electric logs of a limestone well. The long-normal log is shown as a dashed line. Source: W. S. Keys and L. M. MacCary, in *Techniques of Water-Resources Investigations*, U.S. Geological Survey, 1971, Book 2, Chap. E1.

made with a different radiation source having a longer time constant—10 s versus 3 s. The right-hand log has a poor ratio of time constant and logging speed. It does not distinguish thin beds, and positions of the lithologic contacts are incorrect.

The thickness of individual strata can be determined from nuclear logs if there is a change in lithology or porosity from one unit to the next. This is assumed

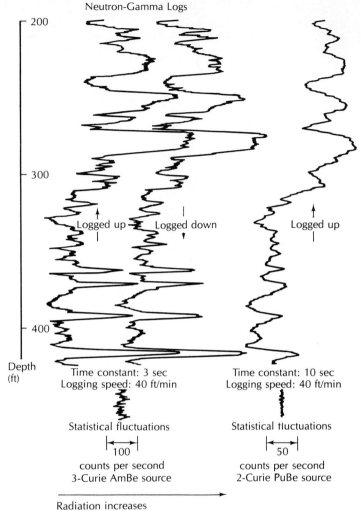

FIGURE 13.25 Statistical variation in neutron-gamma logs made of the same hole. The log on the right had a poor ratio of time constant and logging speed. Source: W. S. Keys and L. M. MacCary, in *Techniques of Water Resources Investigations,* U. S. Geological Survey, 1971, Book 2, Chap. E1.

to be equal to the thickness of the anomaly at one-half the maximum amplitude. This method will overestimate the thickness of thin layers. By convention, radiation increases to the right in nuclear logging. In a reversed log, it increases to the left.

There are three nuclear methods that can be used for composite identification (Keys & MacCary 1971). The neutron count rate increases with decreasing porosity, while the gamma-gamma count rate decreases. The natural gamma radiation increases with an increasing clay or shale content as well as with increased phosphate and K-feldspars, but has no direct relationship to porosity. A stratum with low natural gamma radiation and a low neutron count (or high gamma-gamma count) could be interpreted as a porous sandstone. A low natural gamma radiation and high neutron count could be a dense quartz sandstone or quartzite (see Figure 13.20).

Natural Gamma Radiation

This is the nuclear-logging method most commonly employed in hydrogeology. It is a measure of the natural radiation of rocks as determined by the emission of gamma activity by potassium 40, the uranium 238 decay series, and the thorium 232 decay series. These are constituent materials for some shales and clays with high gamma activity. Certain feldspars and micas are high in ^{40}K. A natural gamma log shows increasing radiation opposite sedimentary beds that contain potassium-rich shale, or clay or phosphate rock. Thus, a shaly sandstone could be distinguished from a clean quartz sandstone. The natural gamma log can be used for lithologic determination—especially of detrital sediments—on the basis of differences in radiation intensity. No calibration of the unit is necessary in this nuclear-logging method. Another advantage is that no radiation sources need be used.

The **natural gamma radiation log** has a tremendous advantage in that it can be performed on an existing cased well. Part of the radiation penetrates the well casing, and the amount absorbed by the casing is constant. Therefore, the variations in radiation due to lithology show up on the log. The method works equally well in both plastic and steel casing. It may not work inside hollow-stem augers if the thickness of the steel wall of the augers varies at the joints where the augers are joined. In this case the varying steel wall thickness will create a situation where the amount of radiation absorbed by the augers is not constant over the length of the augers and the resulting log is distorted.

Neutron Logging

A neutron probe contains a radioactive element, such as PbBe, which is a source of neutrons, and a detector. The emitted neutrons are slowed and scattered by collisions with nuclei of hydrogen atoms. Detectors are available to measure gamma radiation produced by the neutron-hydrogen atom collision, or the number of neutrons present at different energy levels. Thus a **neutron log** will be identified as a *neutron-thermal neutron,* a *neutron-epithermal neutron,* or a *neutron-gamma log* on the basis of the method of detection.

Hydrogen is present in the ground primarily in the form of water or hydrocarbons. In almost all rocks of interest to hydrogeologists there are no naturally occurring hydrocarbons. Therefore, apart from hydrated minerals, water is present as moisture in the pore spaces of the rock. An increase in the amount of water will result in an increase in the number of neutrons that are captured or moderated. As a result, saturated rocks with a high porosity will have a lower neutron count than low-porosity rocks. Above the water table, the neutron-logging equipment can be used to measure the moisture content but not the porosity. Neutron logging can be used to determine the specific yield of unconfined aquifers (Meyer 1967). It also can distinguish gypsum, with a high proportion of hydrated water, from anhydrite. Both have a very low natural gamma radiation; however, anhydrite has a high neutron count and gypsum has a low count.

Gamma-Gamma Radiation

In this type of logging, a source of gamma radiation, such as cobalt 60, is lowered into the borehole. Gamma photons are absorbed or scattered by all material with which the cobalt 60 comes in contact. This includes fluid, casing, and rock. The absorption is proportional to the bulk density of the earth material. *Bulk density* is defined as the weight of the rock divided by the total volume, including the porosity. Thus, gamma-gamma radiation increases with decreasing bulk density (increasing porosity). Bulk density can be determined from a calibrated gamma-gamma log. The formation porosity can be determined from the equation

$$\text{Porosity} = \frac{\text{grain density} - \text{bulk density}}{\text{grain density} - \text{fluid density}} \qquad (13\text{--}15)$$

Grain density may be determined from drill cuttings or assumed to be 2.65 g/cm^3 for quartz sandstone. Fluid density is 1.000 g/cm^3 for mud-free fresh water. The drilling fluid may contain additives to increase the fluid density.

13.5   HYDROGEOLOGIC SITE EVALUATIONS

Many types of construction and similar projects require evaluation of the hydrogeology of a site. This may be a part of a comprehensive study of the engineering geology of a proposed dam or tunnel. The siting of sanitary landfills or areas for toxic- or hazardous-waste disposal or storage almost always requires a hydrogeologic site evaluation. Environmental impact studies often include sections on geology and hydrogeology. Hydrogeological analysis can help ensure the integrity and safety of structures located in areas of high water table, spring discharge, or quicksand potential. Likewise, ground-water pollution generally can be avoided if sanitary landfill sites are selected and engineered on the basis of a thorough hydrogeological investigation. Well-planned and -executed field investigations of the hydrogeology of areas of ground-water contamination are also necessary in order to locate the source of the contamination, define the extent of the plume, and plan for corrective action.

While the detailed design of a hydrogeological site study will be dictated by the specific purpose, there are some elements common to all studies. The first is that they are all relatively expensive. A great expense is in the collection of the basic data. This includes the cost of geophysical surveys, test borings, test wells, pumping tests, permeability lab tests, and so forth. It is frequently the case that data collection is more costly than the evaluation of the data and the formulation of conclusions and recommendations.

The results of a hydrogeological study will generally need to be presented in part on maps. At a minimum, a site-location map will be needed. Standard U.S. Geological Survey topographic maps can be used for site-location maps and perhaps as base maps. A base map should show the important surface features and topography, as well as the locations of all test borings, wells, sampling points, geophysical station locations, and so forth. If a custom base map is needed of an area, it is generally ordered from an aerometric engineering company, which will prepare it from aerial photographs with some surveyed ground elevations. Contour intervals of 1 or 2 ft are desirable for most base maps unless the site has great relief.

A map of the surficial geology can be prepared using standard geologic field techniques. Attention should be given to the nature of bedding planes, fractures, and porosity of rock outcrops. A fracture-trace analysis on air photos can enhance the surficial geology map in areas of fractured bedrock. Detailed mapping of surficial geology can be facilitated if there are soil maps available. Interpretation sheets for the soil-map units indicate the parent material on which the soil formed. They also give a general range of infiltration rate for each soil type.

Surface geophysical surveys may be made as a part of the preliminary site evaluation. Seismic refraction, geoelectric soundings, and geoelectric surveys would be useful for those areas in which the unconsolidated deposits are important. The data from the geophysical surveys can guide the selection of the location of test borings and provide data to correlate between test borings. These borings, or wells, are needed to determine the geologic units present. *Borings* refer to uncased holes drilled in unconsolidated overburden, whereas **test wells** either have a casing or extend into bedrock. A geologist should examine the drill samples or cuttings, prepare a detailed sedimentological and petrographic description, and make a well log showing the depth and thickness of each stratum.

Slightly disturbed samples of unconsolidated deposits can be collected by driving a thin-walled steel tube (either a Shelby tube or split-spoon sampler) into the bottom of the drilled hole. In sandy soils, this must be done through hollow-stem drill pipe or auger-flights. The slightly disturbed sample can be used for laboratory testing, including permeameter tests for hydraulic conductivity and grain-size analysis. Test borings and wells can also be used for slug and baildown tests for hydraulic conductivity.

Piezometers should be installed to determine the configuration of the potentiometric surface. These are small-diameter cased wells with a short well screen. They may be steel or plastic pipe. A minimum of three piezometers is necessary to determine the horizontal direction of ground-water flow if the potentiometric surface is a sloping plane. If it has a more complex shape, then more piezometers are necessary. In order to determine vertical head differences, closely spaced piezometers set at different depths are needed. A set of piezometers at different depths is called a **piezometer nest.** There should be at least one piezometer nest with separate shallow and deep piezometers for a small site, with more for larger sites.

The elevation of the ground surface and the top of the casing of each piezometer should be accurately surveyed. Measurements of the depth to water of each piezometer can then be used to construct a potentiometric map. Successive water-level measurements are necessary, as the elevation and configuration of the potentiometric surface may change with time. Likewise, the amount and direction of vertical flow components may vary.

In addition to the potentiometric maps, the water table may be shown on geologic cross section. These are prepared on the basis of the test borings and wells. Piezometers are generally installed in a test-boring hole so that both geologic and hydraulic data are collected.

A great deal of hydrogeological judgment is needed in drawing ground-water maps. The number of data points may be limited and several interpretations possible; under such conditions draw the map that makes the most sense hydrogeologically.

13.6   RESPONSIBILITIES OF THE FIELD HYDROGEOLOGIST

A ground-water study can be of no better quality than the data that are collected in the field by or under the direction of the field hydrogeologist. Excellent field hydrogeology is the foundation upon which a competent hydrogeologic study must be based. This places great responsibility on the shoulders of the field hydrogeologist.

The mark of an excellent field hydrogeologist is the creation of a first-class field notebook. A field notebook should be a contemporaneous record of the day's events and data. Data should be recorded as they are collected, not later in the day or in the evening. A field notebook is not a personal diary, but rather it is an official business document of the firm that employs the hydrogeologist. It is not a place for personal observations or comments that the author would not like others to read. The author of the field notebook may not be available at some future time when it is necessary for someone to obtain information from it; therefore, it should be well organized and written in a neat hand so that others can read and understand it. Carefully drawn sketches are a good way to record many types of observations.

A field notebook should be a bound book with prenumbered, weatherproof pages. In the front of the book should be recorded the names of the author and his or her company, the company address, and the name and location of the project. Writing should be in waterproof ink. Basic information that should be included on a field notebook includes the following:

1. Date.
2. Time the author arrived at the site.
3. Names and affiliations of all persons at the site that day.
4. Weather conditions that day.
5. Basic description of all of the work that is being done that day.
6. Specific description of the work the author is doing that day.
7. Specific data being collected.
8. Any observations of occurrences that were out of the ordinary.
9. Any times that the author was not present at the site during the day.
10. Time that the author left the site at the end of the day.

Detailed information that is to be collected depends upon the task. If, for example, a well is being sampled, the field notebook should record

1. Well number and its general location.
2. Time of day that the well was opened.
3. Any observations that the condition of the well was out of the ordinary, such as a broken lock.
4. Depth to water in the well, how it was measured, and how the measuring device was cleaned prior to use.
5. Total depth of the well and how it was measured.
6. Computation of the volume of water standing in the well.
7. How the well was purged, how the purging equipment was cleaned prior to use, and how many well volumes were removed.
8. What was done with the purged water.
9. Calibration of the pH meter, the model and serial number of the pH meter, and the pH of the well both prior to and after purging.

10. How the water sample was withdrawn from the well, including the model and serial number of any pump that was used.

11. Description of how the sampling equipment was cleaned prior to being used for sampling.

12. How many and what types of sample containers were used and what preservatives were added.

13. Sample numbers of the containers that were used for that specific well.

14. What was done with the collected samples.

15. What time the well was finally capped.

A major portion of the field work for site assessments and hazardous-waste site work involves the installation of test borings and the construction of monitoring wells. This work is usually done by a contractor under the direction of the field hydrogeologist. The hydrogeological or engineering consultant will have prepared a detailed set of specifications for the work to be performed. These specifications should conform to all state codes and requirements and spell out the exact work to be done, the materials that are to be used, and the methods that are to be used in performing the work. The bidding procedure will also establish the manner in which the amount of money that the contractor will be paid is determined.

The field hydrogeologist is responsible for confirming that the work done by the drilling contractor is in exact conformity with the specifications. He or she is also responsible for keeping track of the materials that the contractor uses, which helps to determine if the specifications are followed and may also be a factor in determining the amount that the contractor is paid.

In order to monitor effectively the performance of the drilling contractor, there should be a hydrogeologist on the job on a full-time basis during drilling of borings, collection of core samples, and installation of monitoring wells. Once a well is installed in the ground, it is very difficult to inspect it to see if it fully conforms with the specifications. For example, the specifications will list the type of well casing and screen to be used, the diameter, and the wall thickness. These dimensions should be verified in the field by the hydrogeologist prior to installation. Once the well screen is in the ground, it is next to impossible to determine the wall thickness.

Most well contractors are honest businesspeople. However, in order to get a job, a contractor is usually the low bidder. This might lead a contractor to try to cut corners if possible. Sometimes the drilling contractor may not see the need for a step in the specifications that the hydrogeologist feels is critical, such as steam cleaning all the new well casing. Although the field hydrogeologist doesn't need to watch all the casing being steam cleaned, he or she should monitor the activities of the driller to see that it is done. The field hydrogeologist should listen to the driller if the latter has a suggestion, but the hydrogeologist must be ready to stand up to the driller to say that the work will be done according to the best judgment of the field hydrogeologist and in accordance with the specifications.

On rare occasions a drilling contractor might turn out to be less than honest. One site owner was quite surprised to find that the state inspector required that a backhoe be used to excavate the outside of a completed monitoring well. The site owner was even more surprised to find the contractor had filled the annular space outside the casing with empty paper bags and had placed a short plug of concrete at the surface to conceal them. This design was not in accordance with the specifications! Full-time oversight by a qualified field hydrogeologist could have prevented this fraud. The field hydrogeologist should always inspect any tools that are being placed in the borehole and any material used to construct the well and then observe the placement of all materials in the borehole.

In overseeing the installation of a monitoring well, the field hydrogeologist should personally make the following types of measurements, using his or her own tape and ruler (the driller should also be making these same measurements):

1. Outside diameter of the augers.
2. Inside diameter of the augers.
3. Final depth of the augered hole.
4. Length of the casing and screen to be placed in the hole.
5. Depth of the casing and screen after it is installed.
6. Number of bags of sand used in a sand pack.
7. Depth to the top of the sand pack prior to the emplacement of the bentonite seal.
8. Number of buckets of bentonite used in the bentonite seal.
9. Depth to the top of the bentonite seal prior to emplacement of the grout.
10. Number of bags of materials, such as cement and bentonite, used to make the grout.
11. Ratio of water to solids used in the grout (extra water will result in a loose grout that is easy for the contractor to install but might not make a good seal).
12. Depth to the top of the grout after placement.
13. Diameter and length of the protective surface casing.
14. Final height the casing extends above the ground.
15. Final height the protective surface casing extends above the ground.
16. Volume of water that is removed during well development.

The field hydrogeologist should keep track of the time that the drilling contractor spends on each task at each well. Finally, the field hydrogeologist should be in charge of checking the invoice submitted by the drilling contractor against his or her field record of the time spent and materials used in the construction of the wells.

13.7   PROJECT REPORTS

The culmination of a hydrogeological study is usually a report. Some reports, such as a thesis or dissertation, will be written for a highly technical audience. Most project reports will have a readership that will include both technical and non-technical persons. In these cases, the writer must take care to utilize an approach that can serve the needs of both. This is frequently done by making sure that there are some basic maps and illustrations that convey the general principles of the study. These may be accompanied by an ''executive summary,'' which is written in nontechnical language. This will normally serve the needs of the nontechnical reader (who may well be the client who sponsored the project). In those sections meant for the nontechnical reader, it may be necessary to define terms and explain basic concepts in nontechnical language.

The report will generally need to have sections that describe the technical aspects of the study in some detail. If the report is going to a reviewing agency, such as a permit-granting body, then sufficient background information, basic data, and calculations should be included so that the work can be independently checked and verified. Usually the data and calculations are best put into an appendix so that they don't impede the reader who is not making an independent verification of the work.

According to Moore (1991) the first step in report preparation is to determine its purpose and the intended audience. A topical outline is then prepared, with detailed headings and subheadings. The topical outline should be sent to the client for review and comment in order to ensure that it meets his or her needs. An annotated outline is then prepared, in which a sentence or two is used to explain each heading in the topical outline. It is first peer-reviewed by someone else in the organization and then sent to the client for further review and comment. A first draft of the introduction to the report is then written. This includes a background statement of the problem and a description of the purpose and scope of the study. Prior studies that have been done in the area are reviewed and summarized. At this stage of project planning, the methods of study to be used and the approach that will be taken are selected and described in the draft introduction.

The preceding steps should be done during project planning and before data collection begins. The data needs of the planned report will drive the design of the data-collection effort. If no thought is given to the report until the field work is complete, there is a strong possibility that critical data will not be available when it comes time to write the final report.

Once the data are collected, it is time to begin to write the final report. The annotated topical outline is the template upon which the report is prepared. The already-written draft introduction is rewritten to reflect any changes in methods and approaches that were necessitated by field conditions and results. A draft of the body of the report, summary, and conclusion is then prepared. The entire draft report is subjected to *peer review,* which is a process where another professional reviews the entire report and prepares a written critique. The purpose of peer review is to help avoid errors in the preparation of the report. Following peer review, any necessary changes in the report are made, and it is submitted to the client for review to determine if the client's needs are met. After client review and any additional revisions, the report is finalized and submitted where needed. It should always be kept in mind that the final report is the most important outcome of the project. It may be the only work product that the client and the reviewing agencies see and may be the basis of a permit or license application.

The following outline covers the material that would go into a final project report.

1. A title page describing the report, who prepared it, for whom it was prepared, and a date.
2. An executive summary.
3. An introduction stating the purpose of the study, why it was made, and the general conditions under which it was made.
4. The conclusions that can be drawn about the site on the basis of the study.
5. Recommendations to the client or agency regarding the use of the site or the need for additional studies.
6. The body of the report, consisting of
   a. a review of previous work that was done on the site,
   b. a description of the procedures and methods used in the study,
   c. the general results of the field study and laboratory analyses,
   d. an interpretation of the findings, and
   e. appropriate maps and cross sections.
7. An appendix, consisting of
   a. acknowledgments of help given by others during the study,
   b. a bibliography, and
   c. technical data, computations, and other supporting evidence.

NOTATION

a	Wenner array spacing	V_1	Seismic velocity in top layer
A	Cross-sectional area of electrical flow	V_2	Seismic velocity in middle layer
L	Length of flow path	V_3	Seismic velocity in bottom layer
ΔV	Voltage drop	Z	Depth to lower layer
I	Electrical current	Z_1	Thickness of top layer
i_c	Critical angle	Z_2	Thickness of middle layer

T_i Intercept time

X Distance from shot to place where direct wave and refracted wave arrive simultaneously

m_u Slope of updip line

m_d Slope of downdip line

α Angle of slope of dipping layer

PROBLEMS

1. This is a two-layer, seismic-refraction problem. A seismic line was run with the following first-arrival-time results:

Distance (ft)	Time (ms)
50	10
100	20
150	30
200	32.5
250	36
300	38.5
350	41.5
400	44
450	47
500	49.5
550	53
600	55

A. Find V_1.
B. Find V_2.
C. Find i_c.
D. Find Z.

2. This is a three-layer seismic problem. The following first-arrival times were measured in both the forward and reverse directions. Find the seismic velocities for each layer; then average them and use the average values to find Z_1 and Z_2 as well as the critical angle between layer 1 and layer 2.

Distance (ft)	Forward Time (ms)	Reversed Time (ms)	Distance (ft)
0	—	91.0	960
20	11.5	91.5	940
40	15.5	92.5	920
60	20.0	92.0	900
80	22.0	92.0	880
100	26.5	92.0	860
120	30.5	90.5	840
140	32.0	88.0	820
160	36.5	86.0	800
180	39.0	85.5	780
200	43.0	85.0	760
220	46.0	85.0	740
240	49.0	86.0	720
260	52.0	84.0	700
280	56.5	82.0	680
300	59.0	81.0	660
320	61.0	80.0	640
340	62.5	77.5	620
360	63.0	75.5	600
380	65.0	75.0	580
400	65.5	73.0	560
420	66.5	72.0	540
440	67.5	72.0	520
460	68.5	71.0	500
480	68.5	71.5	480
500	71.5	71.0	460
520	72.0	69.5	440
540	72.5	68.5	420
560	74.0	68.0	400
580	76.0	67.5	380
600	77.0	66.5	360
620	78.0	66.0	340
640	80.0	62.0	320
660	80.5	57.5	300
680	81.0	52.0	280
700	81.0	50.0	260
720	81.0	—	240
740	83.5	—	220
760	84.0	—	200
780	85.0	40.5	180
800	86.0	36.5	160
820	86.5	33.0	140
840	87.5	29.0	120
860	87.5	26.0	100
880	88.5	22.0	80
900	89.0	18.0	60
920	90.5	14.0	40
940	91.0	9.5	20
960	91.5	—	0

14 Ground-Water Models

In nature the hydraulic system in an aquifer is in balance; the discharge is equal to the recharge and the water table or other piezometric surface is more or less fixed in position. Discharge by wells is a new discharge superimposed on the previous system. Before a new equilibrium can be established water levels must fall throughout the aquifer to an extent sufficient to reduce the natural discharge or increase the recharge by an amount equal to the amount discharged by the well. Until this new equilibrium is established water must be withdrawn from storage in the aquifer and conversely the new equilibrium cannot be established until an amount of water is withdrawn from storage by the well sufficient to depress the piezometric surface enough to change the recharge or natural discharge the proper amount. The depression of the piezometric surface is called the cone of depression.
The Significance and Nature of the Cone of Depression in Ground-water Bodies,
Charles V. Theis, 1938

14.1   INTRODUCTION

There are two areas of hydrogeology where we need to rely upon models of a real hydrogeologic system: to understand why a flow system is behaving in a particular observed manner and to predict how a flow system will behave in the future. In addition, models can be used to analyze hypothetical flow situations in order to gain generic understanding of that type of flow system (Anderson & Woessner 1992). The term *model* refers to any representation of a *real system*. In studying a ground-water flow system we develop a *conceptual model*. For example, we might describe a ground-water system as being contained in deposits of glacial sand and gravel overlying a nearly level bedrock surface consisting of Precambrian-age granite. The sand and gravel unit contains a water table near the land surface, and ground water flows from a nearby moraine into a stream. Such a conceptual model is less complex than a real system. No matter how much field work is performed in describing the above system, our conceptual model could never fully describe all of the minute details of the real system. Nevertheless, conceptual models are necessary for us to be able to understand flow-system behavior.

Conceptual models are static. They describe the present condition of a system. In order to make predictions of future behavior, it is necessary to have some sort of dynamic model that is capable of manipulation. There are many types of dynamic models of ground-water flow. They include *physical scale models, analog models,* and *mathematical models* (Prickett 1975, 1979).

A *scale model* is made from the same materials as those of the natural system. For instance, a plastic container may be manufactured to scale and filled with sand or glass beads with a hydraulic conductivity scaled to the actual aquifer material. Dye added to the water helps the observer trace the flow. Pressure is measured in piezometers inserted through the walls. Water can be added to the model aquifer to simulate recharge and can also be pumped from scale-model wells. Sand models have been used for a variety of studies (e.g., Kimbler 1970; Peter 1970; James & Rubin 1972). Scale models using sand tanks have been used to study the movement of dense, nonaqueous phase liquids, both above and below the water table (Schwille 1988).

The flow of water through porous media is governed by equations similar to those governing the flow of electricity through a conductor. This also is true for the flow of a viscous fluid between two very closely spaced parallel plates. Models can be constructed using electrical circuits or viscous-fluid flow to simulate real or ideal aquifers. These are called *analog models,* since the model is analogous to the actual aquifer. Analog models are typically constructed to model two-dimensional flow. The models can be either horizontal or vertical. If areal flow patterns are being studied, a horizontal model is indicated. In order to study vertical flow, a cross-sectional or vertical model is the one of choice.

Electrical models (e.g., Anderson 1972; Spieker 1968) are made with a network of resistors scaled to represent the framework of the aquifer, with capacitors to provide for storage. The flow of electrical current in amperes through the model is representative of fluid flow. The voltage in the model corresponds to hydraulic potential, and the volume of water in storage is analogous to coulombs of electricity. Resistivity is inversely proportional to the hydraulic conductivity of the aquifer, while the capacitance network is scaled to aquifer storativity. Measurements of current and voltage made at various points in the model represent the flow of water and hydraulic head in the aquifer. Electrical analog models can be made to represent three-dimensional flow by linking a series of horizontal models together. While electrical analogs can be used to study the flow of water, they are not amenable to studying the flow in unconfined aquifers, where aquifer transmissivity decreases with pumpage as the water table declines. They are also unable to simulate mass transport, dispersion, and diffusion (Prickett 1975).

The *viscous-fluid model* is also known as a **Hele-Shaw model** after H. S. Hele-Shaw, who first used it. It has been applied to many problems in hydrogeology. Hele-Shaw models are especially adapted to the study of immiscible fluids with different densities; for example, as encountered in salt-water intrusion (Collins, Gelhar, & Wilson 1972). Two liquids with different densities are used to simulate the fresh and salt water. The injection of wastewater into flow fields can be visually studied using Hele-Shaw models (Peterson, Williams, & Wheatcraft 1978). Both horizontal and vertical models have been built. Because of physical constraints, it has not been possible to model three-dimensional flow with a viscous-fluid model.

A modification of the Hele-Shaw apparatus, in which a single layer of glass spheres of equal diameter was sandwiched between two parallel glass plates, was used to study the movement of dense, nonaqueous phase liquids (Schwille 1988).

Conrad et al. (1992) have built etched glass micromodels in which pore-scale movement of residual organic liquids trapped in an aquifer can be studied under microscopic magnification.

All analog models—and scale models, as well—have some definite disadvantages. Not the least of these is that the models must be constructed by someone handy at carpentry, plumbing, and wiring. The time and materials cost for large models is substantial. There must be room to construct and house the models, and possibly to store them when not in use. Finally, the models are not very flexible. It is difficult to change the aquifer geometry and hydraulic characteristics built into a model.

Scale models and Hele-Shaw models are advantageous in that the fluid movement is visible through the use of dyes. This is particularly helpful in presentations to those not familiar with subsurface flow. Under some conditions, such as transient flow with closely spaced wells and nonlinear boundaries, the resistance-capacitance electrical analog may be more accurate than digital-computer models (Prickett 1975).

Mathematical models rely upon the solution of the basic equations of ground-water flow, heat flow, and mass transport. The most simple mathematical model of ground-water flow is Darcy's law. To apply Darcy's law we need to have a conceptual model of the aquifer and to develop data on the physical properties of the aquifer system, the potential field, and the fluid properties. Darcy's law is an example of an *analytical model*. In order to solve an analytical model we need to know the initial and boundary conditions of the flow problem. These conditions need to be simple enough so that the flow equation can be solved directly by using calculus. Analytical flow models have been developed to simulate the flow of water to wells and streams (Walton 1984) as well as for heat and mass transport (Javendel, Doughty, & Tsang 1984).

Analytical models can be solved rapidly, accurately, and inexpensively with a programmable calculator or small personal computer. A minimal amount of data is needed since all sectors of a parameter are assigned the same value. For example, an average value for hydraulic conductivity would be used for an aquifer. If the problem involved a boundary or multiple wells, image well theory can be applied to the solution (Walton 1979).

Numerical models must be used when there are complex boundary conditions or where the value of parameters varies within the model area (Van der Heijde et al. 1985). Numerical solutions to the flow, heat, and mass transport equations require that they be recast in an algebraic form. These recast equations are numerical approximations and the answers obtained are also approximations. The equations are most commonly in matrix form and they are solved on a digital computer. Numerical models are one of the most important developments in hydrogeology of the last 15 y.

Stochastic models of ground-water flow are based on statistical theories (Dagan 1986; Gelhar 1986). They have been applied to determination of head and velocity fields (e.g., Rubin & Dagan 1992), as well as solute transport problems (e.g., Dagan, Cvetkovic, & Shapiro 1992; Welty & Gelhar 1992). Since the early 1980s a very large number of papers have been published, in which many different types of stochastic models of ground-water flow have been described. In general these papers are difficult for people not trained in the specific mathematics utilized to read and understand. Despite this massive research effort, most ground-water practitioners who use ground-water models rely upon numerical rather than stochastic models. In part, this is probably due to the evolving state of stochastic models as well as the difficulty of the mathematics employed in stochastic modeling.

14.2   APPLICATIONS OF GROUND-WATER MODELS

Ground-water models have been applied to four general types of problems: ground-water flow, solute transport, heat flow, and aquifer deformation (Mercer & Faust 1981; Wang & Anderson 1982; Trescott, Pinder, & Larson 1976; Mc-Donald & Harbaugh 1984; Pinder & Gray 1977; Konikow & Bredehoeft 1978; Prickett & Lonquist 1971). Numerical models start with the basic equation of ground-water flow (Equation 5–42). This is solved for the head distribution in the aquifer. Solute-transport models add an equation for the changes in chemical concentration in the ground water. Heat-transport equations utilize an equation for transfer of heat in the aquifer. Aquifer deformation models combine the flow equation with other equations that describe the changes in the physical structure of the aquifer with changes in the aquifer head.

Two broad classes of models have appeared: those that deal with flow through porous media and those that deal with flow through fractured media. A sand and gravel or sandstone aquifer is an example of an aquifer best described by use of a porous-media model. A fractured crystalline rock with widely spaced fractures would be an example of an aquifer best described by use of a fractured-rock model.

Models have been written for the flow of two or more immiscible fluids in an aquifer and for both saturated and unsaturated flow.

Ground-water flow models have been applied to the study of regional steady-state flow in aquifer systems; regional changes in hydraulic head caused by changes in discharge or recharge; changes in head near a well field, dewatering well system, injection well, or infiltration basin; and surface-water–ground-water interactions.

Solute-transport models have been used in studies of salt-water intrusion, leachate movement from landfills and other waste-disposal sites, contamination plumes from seepage ponds, radionuclide movement from radioactive-waste sites, movement of pesticides from agricultural fields, and other types of ground-water pollution.

Heat-transport models have been used in the analysis of thermal impact from high-level radioactive-waste storage, storage of thermal energy in aquifers, and ground-water heat-pump impacts. Land subsidence due to ground-water withdrawals has been studied with the use of deformation models.

14.3 DATA REQUIREMENTS FOR MODELS

In order to successfully transform a conceptual model into a mathematical, analog, or scale model, it is necessary to have a data base that provides adequate information to apply the requisite equations. All the models start with a ground-water flow model. For this, one needs to know the physical configuration of the aquifer. This includes the location, areal extent, and thickness of all the aquifers and confining layers; the locations of the surface water bodies and streams; and the boundary conditions of all aquifers. Hydraulic properties that need to be known include the variation of transmissivity or permeability and storage coefficient of the aquifers, the variation of permeability and specific storage of the confining layers, and the hydraulic connection between the aquifers and surface-water bodies. Hydraulic energy, as indicated by water-table or potentiometric-surface maps, and the amounts of natural-aquifer recharge and natural streamflow are also needed.

In order to model stresses on the natural ground-water flow system, the modeler must know the locations, types, and amounts through time of any artificial recharge, such as results from recharge basins and wells or return flow from irrigation, as well as the amounts and locations through time of ground-water withdrawals from wells. Changes in the amounts of water flowing in the streams and changes in the water levels of surface-water bodies should also be known.

Solute-transport models need, in addition to the flow-model data, the following information: distribution of effective porosity, aquifer dispersivity factors, fluid-density variations, and natural concentrations of solutes distributed throughout the ground-water reservoir. The locations and strengths of the sources of contamination must be known as well as retardation factors for the specific solutes with the specific rocks and soils of the area. The flow model is used to compute direction and rate of fluid movement, and the solute-transport equations are "piggybacked" onto the flow model to derive movement and retardation values of contaminants.

A model is initially calibrated by taking the initial estimates of the model parameters and solving the model to see how well it reproduces some known condition of the aquifer. Most models are initially calibrated against the steady-state ground-water heads. A water-table or potentiometric-surface map is required for this type of calibration. As this is almost always known with better accuracy than the distribution of aquifer parameters and/or amount of recharge, the values for the aquifer parameters and recharge are varied until the model closely reproduces the known water-table or potentiometric-surface condition.

This process is known as **model calibration.** As the same result can often be obtained by changing two variables simultaneously, it is highly desirable to *verify* a model **(model verification)** once it has been calibrated. This is usually done by history matching. A transient response of the model is obtained and compared with a known transient condition in the aquifer. For example, the water levels in the aquifer may have declined over time owing to withdrawal from wells. If the water levels through time and the locations and pumping rates of the wells are known, the model is verified if it reproduces the known water-level changes by inputting the known ground-water withdrawal history. If the known history is not reproduced with a desired degree of accuracy, the model parameters can be changed to recalibrate the model with a new set of model parameters and the verification run repeated. This process will eventually result in a calibrated and verified model.

In verifying against a transient condition, the aquifer storage coefficient is likely to be the parameter that must be adjusted as it will not have been utilized in the calibration against steady-state conditions. Once the model has been verified against a transient event, it should be checked against the steady-state condition to ensure that it still is calibrated. **Model field verification** can then be performed by actually stressing the aquifer to see if the model correctly predicts the response of the aquifer as it is stressed; for example, by performing a pumping test (Anderson 1986). Most computer models are not subjected to field verification as this is a very time-consuming and expensive step. Transient events for history-matching can include pumping tests in addition to long-term water-level declines. In some cases it may be necessary to use a calibrated model that has not been verified. This can occur if there is no history of water-level changes against which to verify the model. Such a model can certainly be useful, but it should be recognized that any predictions are less likely to be valid than those made with a verified model. The more accurate the data that initially go into a model, and the more detailed the data against which it is verified, the more confidence the modeler can have in the model results.

The process of model calibration and verification frequently requires many changes in the data parameters that make up the model. Scale and analog models might require rebuilding each time a change is made in data values. Numerical models can be easily recomputed with the new data. This is the primary reason that numerical models have apparently almost totally replaced other types of models. The rapid growth in computing power and fall in price of personal computers has put the ability to cheaply and easily solve numerical models on the desks of many, if not most, hydrogeologists. The remainder of this chapter will look at computer models of ground-water flow.

14.4   FINITE-DIFFERENCE MODELS

14.4.1   Finite-Difference Grids

In a conceptual model that is continuous, we presume to know the properties of the aquifer at every point within the boundaries. We can replace the continuous

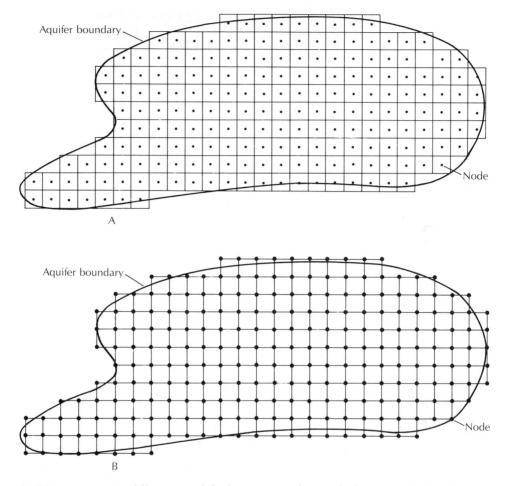

FIGURE 14.1 Finite-difference grids laid over an aquifer. **A.** Block-centered finite-difference grid. **B.** Mesh-centered finite-difference grid.

model with a set of discrete points arranged in a grid pattern. Figures 14.1A and 14.1B show two variations of a *finite-difference grid*. Associated with the grid are *node points,* where the equations are solved to obtain unknown values. Also associated with each node are the known values of such parameters as transmissivity and storativity. A *block-centered grid* (Figure 14.1A) is one where the node points fall in the center of the grid; a *mesh-centered grid* (Figure 14.1B) has the node points at the intersections of grid lines. The choice of whether to use a block-centered or a mesh-centered grid depends upon the boundary conditions. A block-centered grid is most useful when a flux is specified across a boundary, and a mesh-centered grid is most convenient for situations where the head is specified at the boundary.

The basic grid is regular, with the rows and columns being normal to each other and the distance in the x direction, Δx, being equal to the distance in the y direction, Δy. Often it is convenient to vary the size of the rows and columns so

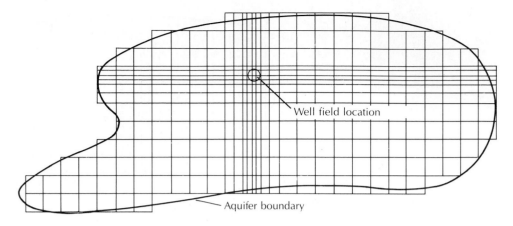

FIGURE 14.2   Variable-spacing finite-difference grid.

that there are more node points in certain parts of the aquifer than others. This might be desirable in the area around a well field, for example, where there are greater changes in head expected than in other parts of the aquifer (Figure 14.2). The size of the difference in Δx or Δy from one column or row to an adjacent one should not change by more than 50% in a variable grid spacing.

14.4.2   Finite-Difference Notation

Special notation is used to describe the positions of the nodes in the finite-difference grids. Figure 14.3A shows a finite-difference grid centered on (x, y). Adjacent points on the grid are located at a distance Δx away to the right or left and Δy away up or down; x is positive to the right and y is positive downward.

In the computer codes the locations of the nodes are designated with reference to node i, j, where i represents the column and j represents the row. The notation for i is positive to the right and for j it is positive downward. Thus the row above the j row is row $j - 1$ and the row below the j row is $j + 1$. The column to the left of column i is column $i - 1$ and the column to the right of column i is $i + 1$.

14.4.3   Boundary Conditions

In order to solve the ground-water flow equation we must be able to specify the boundary conditions. There are two basic types of boundary conditions (Wang & Anderson 1982). If the head is known at the boundary of the flow region, this is known as a **Dirichlet condition.** If the flux across a boundary to the flow region is known, this is a **Neumann condition.** In some cases the boundary conditions will be mixed, with some portions having known head and some portions having known flux.

As an example of boundary conditions, examine Figure 14.4. This is a sand and gravel aquifer overlying an impermeable basement rock. There are

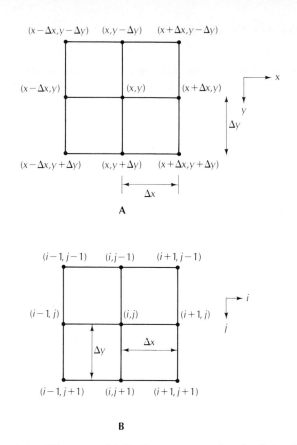

A

B

FIGURE 14.3 **A.** Finite-difference grid. **B.** Computer notation for finite-difference grid.

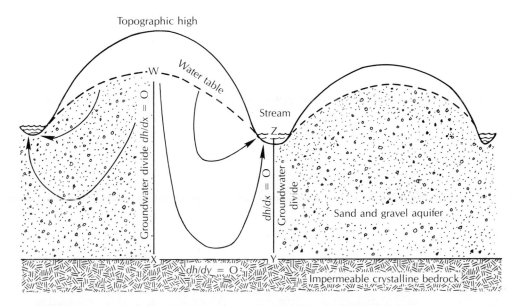

FIGURE 14.4 Boundary conditions for a cross section of a regional aquifer.

several flow regions present, each leading away from a ground-water divide toward a stream. We shall look at the regional flow in the region bounded by the letters W, X, Y, and Z.

The plane of W to X is a ground-water divide. There is no flow across the divide and the boundary condition along the divide is $dh/dx = 0$. The plane of Z to Y, which underlies the center of the stream, is also a ground-water divide, across which no flow takes place. The boundary condition along this divide is $dh/dx = 0$. No flow occurs from the sand deposit into the impermeable bedrock, so along the X to Y plane there is no flow and the boundary condition is $dh/dy = 0$. These first three boundaries are Neumann boundaries, where we have specified the flow condition.

There are two ways in which we could treat the upper boundary—the W to Z surface—which is the water table. We could specify that the position of the water table is fixed. This would be a Dirichlet boundary. For steady-state conditions—that is, no change of head with time—we could solve this model to give us the amount of recharge needed to maintain the water table in the observed position. The second approach would be to specify the flux across the water table—that is, the amount of recharge. This then becomes a Neumann boundary. This model could be solved to determine the position of the water table under various amounts of recharge. Both of the above models require that the aquifer permeability is known. If we wish to simulate the movement of the water table with time under Neumann conditions, we would also need to know the aquifer storativity.

14.4.4 Methods of Solution for Steady-State Case for Square Grid Spacing

In the absence of recharge to the aquifer, the finite-difference equation for the steady-state case (Laplace equation) is (Wang & Anderson 1982)

$$h_{i-1,j} + h_{i+1,j} + h_{i,j-1} + h_{i,j+1} - 4h_{i,j} = 0 \qquad (14\text{-}1)$$

When solved for $h_{i,j}$, Equation 14-1 becomes (Wang & Anderson 1982)

$$h_{i,j} = (1/4)(h_{i-1,j} + h_{i+1,j} + h_{i,j-1} + h_{i,j+1}) \qquad (14\text{-}2)$$

The value of $h_{i,j}$ is the average of the heads at the four closest nodes in the nodal mesh.

If there is recharge to the aquifer, then the finite-difference equation for the steady-state case (Poisson's equation) is (Wang & Anderson 1982)

$$(h_{i-1,j} - 2h_{i,j} + h_{i+1,j})/(\Delta x)^2 + (h_{i,j-1} - 2h_{i,j} + h_{i,j+1})/(\Delta y)^2 = -R/T \qquad (14\text{-}3)$$

where

Δx and Δy are the distances between nodes in the x- and y-directions

R is the recharge

T is the aquifer transmissivity

In a finite-difference mesh there are from tens to hundreds of nodes. Each node requires the solution of a form of Equation 14–2 or 14–3. The method of solution is to make an initial guess of the value of head for each of the nodes in the mesh. For Dirichlet boundary nodes the head will be fixed. For all other boundary nodes and interior nodes the value of the head will not be fixed.

The finite-difference equation is solved by what are known as *iterative methods*. On the basis of the fixed head values, plus the initial guesses, Equation 14–1 is solved for each node on the basis of the values at the surrounding four nodes. A computer code sweeps the solution path through the finite-difference grid so that for each node, other than the first one and the last one, the head values at some of the adjacent nodes will be based on the initial guess, while at the remainder of the adjacent nodes the head value will already have been recomputed by Equation 14–1. Once the head at each node has been recomputed, the difference between the initial guess and the recomputed head is determined. The process is repeated until the maximum difference in head values from one iteration to the next is less than some preset value known as the *convergence criterion*. When the solution has converged, the equation has been solved. The solution is approximate as there is some finite value to the convergence criterion. The smaller that value is, the more iterations, and hence the longer period of time, it takes to reach the solution. There is some practical trade-off between accuracy of the solution and the amount of computer time expended to reach it.

The **Gauss-Seidel** iteration method computes the value of $h_{i,j}$ during an iteration on the basis of heads at two adjacent nodes, which have been computed during the current iteration, and heads at two adjacent nodes, which were computed during the prior iteration.

The Gauss-Seidel equation for the Laplace equation is

$$h_{i,j}^{m+1} = (1/4)(h_{i-1,j}^{m+1} + h_{i,j-1}^{m+1} + h_{i+1,j}^{m} + h_{i,j+1}^{m}) \qquad \textbf{(14–4)}$$

where the superscript m indicates a value computed during a prior iteration and $m + 1$ indicates a value computed during the current iteration.

The Gauss-Seidel equation for Poisson's equation in a grid where $\Delta x = \Delta y$ is (Wang & Anderson 1982)

$$h_{i,j}^{m+1} = (1/4)(h_{i-1,j}^{m+1} + h_{i,j-1}^{m+1} + h_{i+1,j}^{m} + h_{i,j+1}^{m} + \Delta x^2 R/T) \qquad \textbf{(14–5)}$$

The method of **successive overrelaxation** (SOR) is a variation of the Gauss-Seidel method. The difference in head value computed at a node by two Gauss-Seidel iterations is known as the residual. During each Gauss-Seidel iteration, the residual shrinks until the convergence criterion is reached. In successive overrelaxation, a factor is added to increase the rate of convergence. The overrelaxation factor is a number between 1 and 2, and an acceptable value is normally determined by trial and error. If we designate the value of $h_{i,j}$ as determined by a Gauss-Seidel iteration as $(h_{i,j}^{m+1})$ and the overrelaxation factor as

f, the value of $h_{i,j}^{m+1}$ as determined by the SOR method is (Wang & Anderson 1982)

$$h_{i,j}^{m+1} = h_{i,j}^m + f(h_{i,j}^{m+1} - h_{i,j}^m) \qquad (14\text{--}6)$$

where $(h_{i,j}^{m+1} - h_{i,j}^m)$ is the residual.

14.4.5 Methods of Solution for the Transient Case

In a transient problem the head is a function of time. This is the type of solution that would be applied to problems such as determining the change in head around a pumping well or the growth of a ground-water mound beneath a recharge basin. Equation 5–45 is the transient-flow equation in two dimensions with a factor for recharge. This equation can be adapted to unsaturated aquifers if the transmissivity varies with time as the saturated thickness of the aquifer changes with changes in head.

The *alternating direction implicit method* of solution is used in a number of published computer codes (Trescott, Pinder, & Larson 1976; McDonald & Harbaugh 1984). Equation 5–43 is the two-dimensional flow equation without recharge. The fully implicit finite-difference approximation is (Wang & Anderson 1982)

$$h_{i+1,j}^{n+1} + h_{i-1,j}^{n+1} + h_{i,j+1}^{n+1} + h_{i,j-1}^{n+1} - 4h_{i,j}^{n+1}$$
$$= [(1/T)(Sa^2)][(1/\Delta t)(h_{i,j}^{n+1} - h_{i,j}^n)] \qquad (14\text{--}7)$$

where

S is storativity

T is transmissivity

Δt is the length of the time step

$a = \Delta x = \Delta y$ = the dimensions of the finite-difference grid

n represents the nth time step

The fully implicit finite-difference approximation can be combined with an iterative solution, such as the Gauss-Seidel. In this method the head is first determined along columns and then along rows. For each time step, a solution is obtained for the head at every point on the finite-difference grid by convergence to a residual value less than the convergence criterion. The convergence criterion is checked after each two iterations as the program must go through all the points along columns and then rows for a complete iteration cycle. The program then steps to the next time increment and the process is repeated. In the following equation, the time step counter, n, is now shown as a subscript while the iteration step counter, m, is shown as a superscript. The heads in column i at time step $n + 1$ may be found from the following equation (Wang & Anderson 1982):

$$h_{i,j-1,n+1}^{m+1} + [-4 - (Sa^2)/(T\Delta t)]h_{i,j,n+1}^{m+1} + h_{i,j+1,n+1}^{m+1}$$
$$= [(-Sa^2)/(T\Delta t)]h_{i,j,n} - h_{i+1,j,n+1}^m - h_{i-1,j,n+1}^{m+1} \qquad (14\text{--}8)$$

For the next iteration $(m + 2)$ the solution is oriented along rows. The heads in row j at time step $n + 1$ may be found from the equation (Wang & Anderson 1982)

$$h_{i-1,j,n+1}^{m+2} + [-4 - (Sa^2)/(T\Delta t)]h_{i,j,n+1}^{m+2} + h_{i+1,j,n+1}^{m+2}$$
$$= [(-Sa^2/(T\Delta t)]h_{i,j,n} - h_{i,j+1,n+1}^{m+1} - h_{i,j-1,n+1}^{m+2} \quad \textbf{(14–9)}$$

The solution of a transient problem takes much more time than the solution of a steady-state problem as the method is to solve a series of steady-state problems, each separated by a time step. The amount of central processor time in the computer used by the transient type of problem is thus greatly increased.

There are several other methods of solution for transient finite-difference models. These include line-successive overrelaxation and the strongly implicit procedure. Several of the ground-water model codes, including the Trescott, Pinder, and Larson model (1976) and the McDonald and Harbaugh model (1984), offer the user the choice of more than one method of solution. For a particular problem it may be that the user would find one method faster than another. This should be determined during the calibration procedure for each area modeled.

14.5   FINITE-ELEMENT MODELS

Finite-element models offer an alternative approach to the numerical modeling of ground-water flow. Rather than use the rectangular network of nodes that is used in the finite-difference method, the aquifer is divided into polygonal cells, typically triangular, but not necessarily so. Figure 14.5 shows the finite-element cells

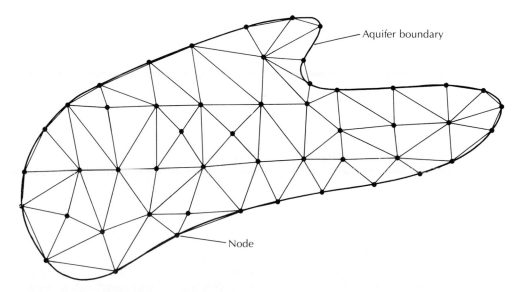

FIGURE 14.5 Finite-element grid for an aquifer.

for an aquifer to be modeled. The triangles intersect at nodes that represent the points at which the unknown values, such as heads, will be computed. The value of the head in the interior of each cell is determined by interpolation between the nodal points.

The mathematical basis for the finite-element method is much more complex than that of the finite-difference method. Most finite-element solutions are based on Galerkin's method. This method is described elsewhere (Wang & Anderson 1982; Pinder & Gray 1977). The mathematics are not intuitively obvious, unlike those of the finite-difference method, and will not be pursued further here.

Finite-element models are reported to be somewhat superior to finite-difference models for problems that have a moving boundary, such as a cross-sectional model with a water table that is transient, as well as coupled problems, such as contaminant transport (Wang & Anderson 1982). The finite-element model also has the advantage of being much more flexible in terms of mimicking the geometry of an aquifer system than the finite-difference method and requires fewer nodes.

14.6   METHOD OF CHARACTERISTICS

This is a useful method for simulating mass transport in ground water. It is usually used in conjunction with a finite-difference model that simulates the flow of the ground water. The finite-difference grid then becomes the coordinate system for the movement of the fluid and the contained solute. The method of characteristics follows the fluid and computes the changing concentrations of the solute. In the two-dimensional problem there are three equations: one each for the x-velocity, the y-velocity, and the concentration. The solution of each yields what is known as a characteristics curve (Mercer & Faust 1981). Points, representing solutes, are placed in various cells of the finite-difference grid. The computer code then moves each of them a distance proportional to the length of the time increment and the ground-water velocity at that point. This simulates convective transport and the remaining factors of the mass-transport equation are solved with finite-difference techniques.

14.7   USE OF PUBLISHED MODELS

Most hydrogeologists will never devise a new method of solution for the flow equations or write a new computer code for ground-water modeling. A large number of computer models have already been developed and more will be developed in research institutions. Published models have certain advantages over custom-designed ones. It is much less expensive to apply an already developed model, if an available model will accomplish the desired goal. The model will already have been "debugged" and most often applied to a field situation so the user has a good chance that the model will actually work in the intended application. Finally, if the published model has been widely used and accepted by practicing hydrogeologists, the results of another application will be more readily accepted than if an entirely new model code has been developed. This may be very important if the model results are to be used in a regulatory process or in litigation.

The U.S. Geological Survey has published several computer models. These are well documented and well tested. They are in the public domain and versions are readily available at moderate cost for both mainframe and personal computers.

The *McDonald and Harbaugh model* (MODFLOW) (McDonald & Harbaugh 1984) is an extremely versatile finite-difference ground-water flow model. Documentation is available from the U.S. Geological Survey. It is three-dimensional and the layers can be either water table, confined, or a combination of both. The model simulates recharge, evapotranspiration, areal recharge, flow to wells, flow to drains, and flow through river beds. It is set up as a series of separate modules, which are independent; the user selects only the modules needed for the particular system that is under study. The user may choose between different methods of solution. The code is written in FORTRAN and some knowledge of that language is helpful in formatting the data for input into the model. The model is simple to use and is available on floppy disks for use in an IBM-compatible personal computer. At least 256K of random access memory is needed, but 640K would be better. The solution time is speeded up if the computer has a math coprocessor (an optional computer chip that is plugged into the motherboard of the computer).

The *Konikow and Bredehoeft model* (MOC) (Konikow & Bredehoeft 1978) is a widely used solute-transport model. Documentation is available from the U.S. Geological Survey. It is based on a finite-difference grid and uses the method of characteristics for solution. It is two-dimensional and computes changes of concentration with time caused by advection, hydrodynamic dispersion, and mixing. The model assumes that the solute is nonreactive and that gradients in fluid density, viscosity, and temperature do not affect the flow velocity. This program is also available on floppy disks for use on an IBM-compatible personal computer. A full 640K of memory is required. A preprocessor is available to aid in setting up the program and entering the data.

The *Prickett Lonquist Aquifer Simulation Model* (PLASM) (Prickett & Lonquist 1971) is an earlier finite-difference, ground-water–flow model, but it is still very useful. Documentation is available from the Illinois State Geological Survey. In essence this is a series of related programs that can be used for distinct applications. Much less memory is required than for the U.S. Geological Survey programs as the user inputs only the portion of the model that is needed. (In the U.S. Geological Survey program, the entire program code is in memory and the user then selects the portions that are needed.) Both two- and three-dimensional as well as water-table and confined-aquifer models are included.

SUTRA (saturated-unsaturated transport) (Voss 1984) is a finite-element simulation model for fluid movement and the transport of either dissolved solutes

TABLE 14.1 Use of Ground-Water Models by
Practicing Hydrogeologists

Model	Percentage of Respondents Using the Model
MODFLOW	73
MOC	30
PLASM	13
SUTRA	13
RNDWALK	13
MODPATH	10
FLOWPATH	10

or energy under either saturated or unsaturated conditions. Under the solute-transport mode, the solute may be subjected to adsorption on the porous medium as well as decay or production. In addition, the flow of variable-density fluids, such as saline water or leachate, may be simulated. The heat-transport mode can simulate such things as subsurface heat conduction, geothermal reservoirs, and thermal storage in aquifers. The model may be employed in either a cross-sectional or horizontal configuration.

Another public-domain solute transport model is RNDWALK (Prickett, Naymik, & Lonquist 1981). This is a particle-tracking model in which the computer code moves particles representing contamination by advection and then disperses them using a normal distribution to account for dispersion.

In 1992 a survey was conducted to determine the ground-water models that were most frequently used by practicing ground-water modelers. There were 876 respondents to the survey. Table 14.1 shows the seven most popular ground-water-modeling packages and the percentage of respondents that used them (Geraghty & Miller 1992).

There are a number of different versions of many of these models. Software publishers have created preprocessors to create the data sets needed to run the models and postprocessors to present the results of the model run.

There are many different vendors of ground-water models. Training in the use of different modeling packages is available from the International Ground Water Modeling Center,* the National Ground Water Association,[†] and Environmental Education Enterprises.[‡]

Anderson and Woessner (1992) list the following steps in the development of a ground-water model:

1. Determine if and why a model is needed (*purpose* of the model).

2. Gather available geologic and hydrologic data and create a *conceptual model* of the system.

---

*Colorado School of Mines, Golden, CO 80401-1887; 303-273-3103.

[†]6375 Riverside Drive, Dublin, OH 43017; 800-551-7379.

[‡]2764 Sawbury Boulevard, Columbus, OH 43235; 614-792-0005.

3. Select the *computer code* that will be used. The computer code is what is normally termed the ground-water model—e.g., MODFLOW, MOC, etc. The computer code must be capable of fulfilling the purpose of the model and must be compatible with the available data. The public-domain codes listed in this section have undergone a process of *code verification* to determine if they accurately solve the governing flow equations.

4. The next step is *model design*. The model grid, boundary, and initial conditions are selected. These must conform to the conceptual model.

5. The model must undergo *calibration*. Different runs of the model are made, and the model parameters adjusted to determine if the model output can reproduce field data for head and flow. At this point a model has the lowest level of confidence.

6. In order to increase confidence in the model, a *sensitivity analysis* is performed on the calibrated model to see how changes in each parameter affect the model results.

7. If a second set of field data exists—for example, head changes during a pumping test—a second stage of calibration, called *verification,* can be accomplished. If the calibrated model can be verified, that increases the level of confidence in the model.

8. The calibrated and verified model can be used for *prediction*. If a model that has been calibrated but not verified is used for prediction, it must be recognized that there is a lower level of confidence in the result.

9. Recognizing that there is uncertainty even with a calibrated and verified model, *prediction sensitivity analysis* is performed. This demonstrates that there is a range of uncertainty in the predicted results.

10. The modeler is finally ready to *present results* of the study in a well-designed and written report.

Tsang (1991) points out that at each step in the process, it is necessary to determine if sufficient data exist for the chosen model to be effective. It may be necessary to stop the modeling process and go back to the field or laboratory to collect additional data before modeling resumes. If time and budget do not permit this, then the modeling process should be abandoned rather than continued to a faulty conclusion. In fact, preliminary modeling after an initial data collection program is often a good idea to help the hydrogeologist plan a more extensive data collection effort.

According to Anderson (1983), ''All models require the talents of a skilled model user, a tailor, to design hydrogeologically valid boundary conditions and initial conditions and select meaningful values for model parameters.''

The computer codes available today are capable of producing professional-looking graphical output that is based on little actual data or data of questionable validity. It is like a movie set, an impressive façade from afar, but if one looks closely, there is nothing behind it. If the model does not have a strong foundation of quality field and laboratory data, then the user should beware of the results.

CASE STUDY: GROUND-WATER MODELING FOR A PLANNED UNDERGROUND MINE

A large massive sulfide ore body was discovered in northeastern Wisconsin (D'Appolonia Consulting Engineers, Inc. 1980; Exxon Minerals Company 1982; Wisconsin Department of Natural Resources 1986). The ore body has recoverable reserves of 67.4 million tons of zinc, copper, and lead ore. It is tabular in shape, with a length of 4200 ft (1280 m) and a width ranging from 0 to 200 ft (61 m), and extends from the bedrock surface to a depth of 1800 ft (550 m) below the bedrock surface. The bedrock surface is overlaid by some 200 ft (60 m) of glacial drift, primarily permeable outwash.

The bedrock in which the ore body is found is Precambrian-age metasedimentary and metavolcanic rock. The top of the ore body is weathered and more permeable than the surrounding bedrock. The planned mine is to be constructed as an underground working. Because of the large amount of ground water in the ore body and the overlying glacial sediments, mine dewatering will be necessary. The dewatering system is expected to lower the water table in the glacial drift aquifer above the mine. In order to be able to predict how much water will be pumped from the mine, the surface extent and depth of the resulting cone of depression, and the environmental impact of the lowered water table on lakes, streams, and wetlands of the region, a ground-water model of the project was developed.

Extensive test borings in both the glacial drift materials and the bedrock materials were made in order to establish the geologic framework for the model. Numerous piezometers were installed to determine the hydraulic head in the various aquifers. Several lakes are in the area of the expected cone of depression, and the water budgets for the lakes and their hydraulic relationship to the water table were studied. Over a period of several years, streamflows were gaged, the heads in piezometers measured, and the stages in lakes measured to obtain baseline hydrologic data. Pumping tests in both glacial-drift aquifers and bedrock wells were performed to measure the hydraulic parameters of the aquifer materials. Figure 14.6 shows the water-table elevations of the glacial-drift aquifer. Some of the small lakes over the mine area are seepage lakes, which are in contact with the water table, and some are perched above the water table on low-permeability lake-bed sediments. Also shown on this map are the locations of the ore body and the mine/mill facility.

Once the physical boundaries and hydrologic parameters of the area were established, a model grid system was set up for a two-layer, horizontal finite-element model. Figure 14.7 shows the grid system. The boundaries of the model were set at major streams and lakes and were all constant-head boundaries. The model used is proprietary (Anderson 1986).

The model was calibrated against the steady-state water-table position. Steady-state models are calibrated against the ratio of the rate of ground-water recharge to the hydraulic conductivity of the aquifer. Figure 14.8 shows the modeled water-table elevation based on the calibrated model. The amount of recharge was reduced in some places to reflect wetland areas where evapotranspiration is greater than in the upland areas so that there is less recharge. Recharge from the seepage lakes to the water table was also included. The aquifer transmissivity reflected the saturated thickness and the local hydraulic conductivity at each node. The recharge rate and the saturated hydraulic conductivity of the model were varied as the model was calibrated. Figure 14.8 shows the computer-generated water table.

The model was not verified as there were no events that could be used for history matching. The value of the storage coefficient was obtained from a short-term pumping test. However, a mass-balance analysis of the model indicated that the model-computed

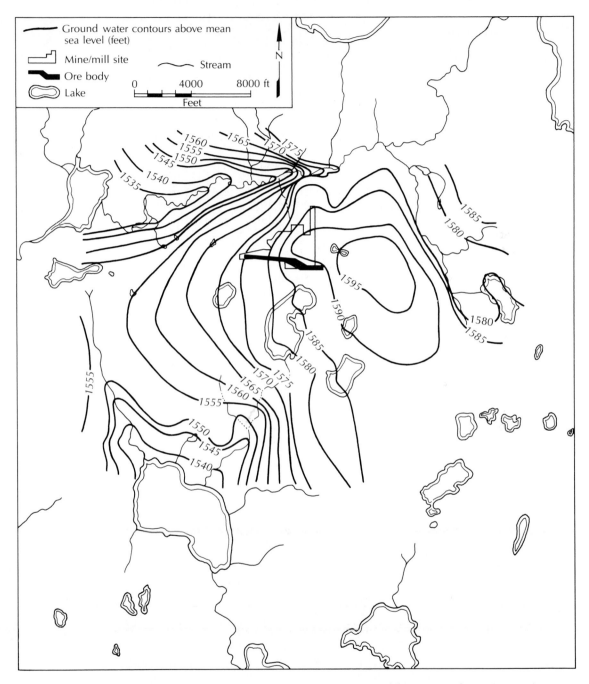

FIGURE 14.6 Existing ground-water table at the site of the proposed mine in northeastern Wisconsin. Source: Wisconsin Department of Natural Resources, Final Environmental Impact Statement, Exxon Coal and Minerals Co., Zinc-Copper Mine, Crandon, Wisconsin, 1986.

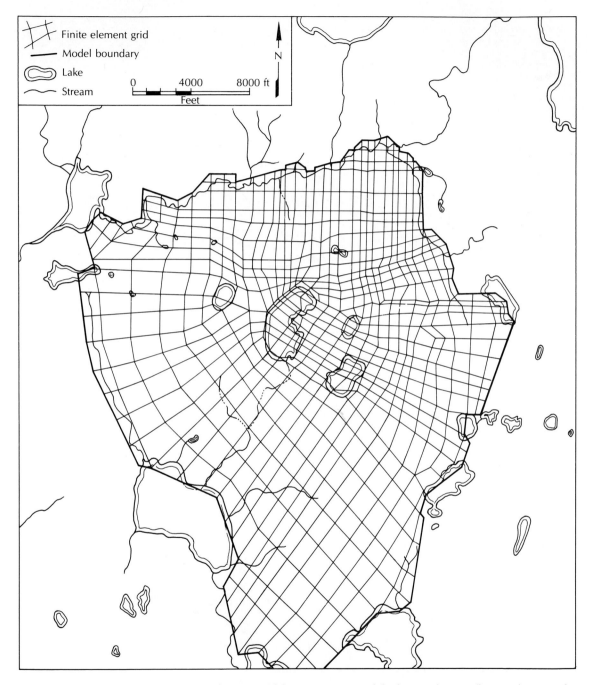

FIGURE 14.7   Finite-element grid for computer model of ground-water flow at the site of a proposed underground mine in northeastern Wisconsin. Source: Exxon Minerals Company, Preliminary Environmental Report, Crandon Project, 1984.

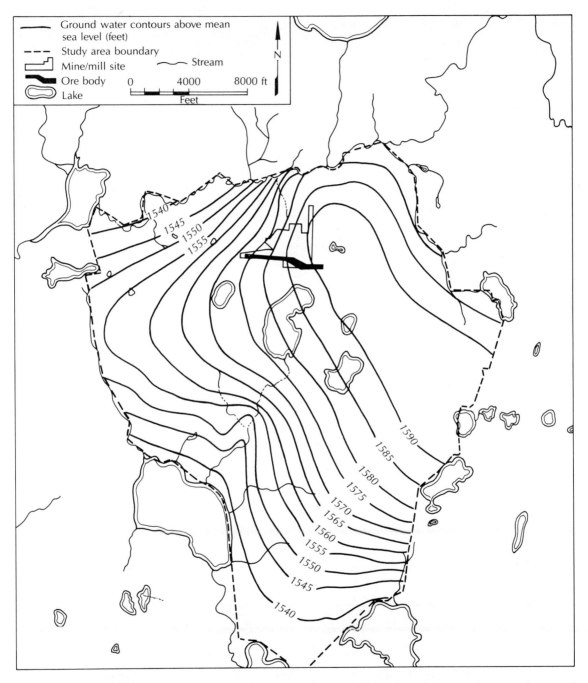

FIGURE 14.8 Computer model of ground-water table under existing conditions at the site of a proposed underground mine in northeastern Wisconsin. Source: Wisconsin Department of Natural Resources, Final Environmental Impact Statement, Exxon Coal and Minerals Co., Zinc-Copper Mine, Crandon, Wisconsin, 1986.

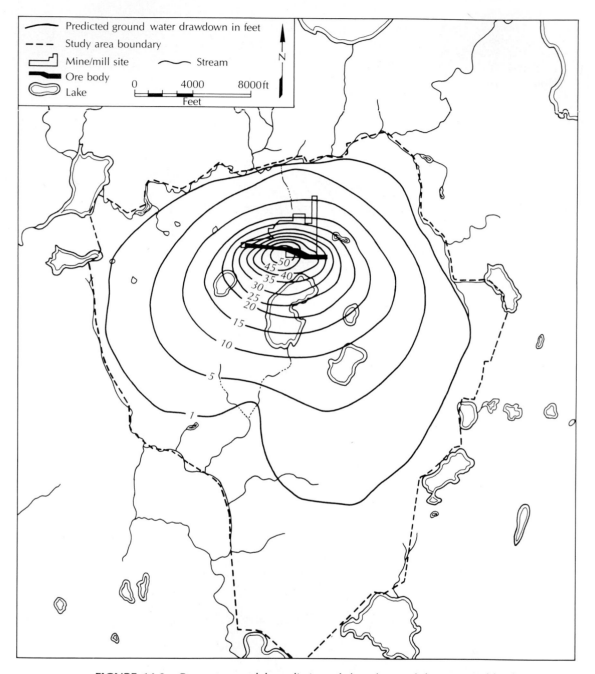

FIGURE 14.9 Computer model prediction of drawdown of the water table due to mine dewatering after 28 y of mining at the site of a proposed underground mine in northeastern Wisconsin. Source: Wisconsin Department of Natural Resources, Final Environmental Impact Statement, Exxon Coal and Minerals Co., Zinc-Copper Mine, Crandon, Wisconsin, 1986.

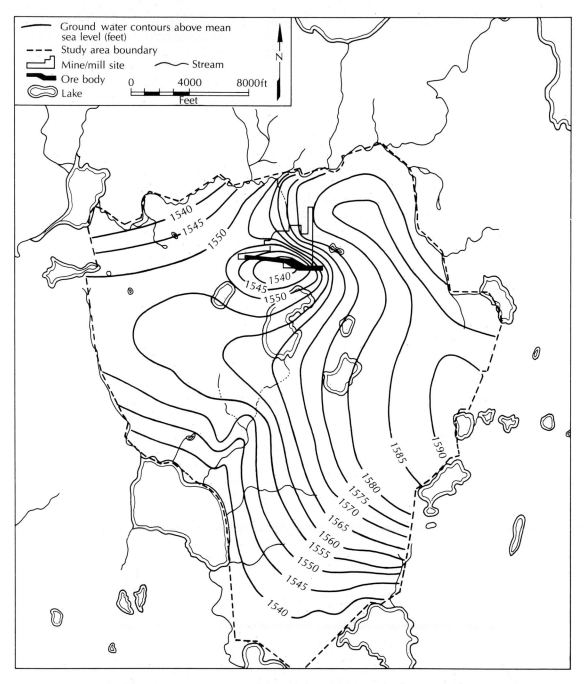

FIGURE 14.10 Computer model prediction of the configuration of the water table after 28 y of mining at the site of a proposed underground mine in northeastern Wisconsin. Source: Wisconsin Department of Natural Resources, Final Environmental Impact Statement, Exxon Coal and Minerals Co., Zinc-Copper Mine, Crandon, Wisconsin, 1986.

values of baseflow to the streams in the area were close to the measured baseflows. It was felt that this was a good indication of the reliability of the model.

A transient form of the model was stressed to simulate the drawdown conditions that would form when the mine becomes fully operational. Mine-dewatering activity was simulated by making the heads in the aquifer above the mine zero. In addition, provisions were made for a potable water-supply well pumping 50 gal (190 m) per minute and seepage from a soil absorption field for wastewater at 50 gal/min.

Figure 14.9 shows the drawdown from the original water levels after 28 y of mining. This is the end of the planned mining period and represents maximum drawdown. If the cone of depression is superimposed on the initial water table, Figure 14.8, the water table after 28 y of mining can be obtained. This is represented by Figure 14.10. Comparison of Figure 14.10 with Figure 14.8 shows a large cone of depression over the mine as well as many changes in the directions of ground-water flow that are a result of the predicted changes in the configuration of the water table. The model also predicted that reductions in baseflow would occur in many of the streams and that the stage of the seepage lakes would decline.

As neither the recharge rate nor the hydraulic conductivity of the aquifer was known with exact precision, three calibrated versions of the model were developed, each with a different recharge rate. The three recharge rates were 6 in. (15.2 cm) per year, 8.5 in. (21.6 cm) per year, and 11 in. (27.9 cm) per year. The model based on the middle recharge rate of 8.5 in./y is considered the most credible as it gave the best match of computed baseflows to streams with measured baseflows. The three versions of the model were not necessary in order to determine the steady-state position of the water table, against which they all were calibrated. Obviously, different hydraulic conductivity values were used for each recharge rate. As the recharge rate increases, so does the hydraulic conductivity necessary for calibration.

Greater volumes of water move through the aquifer for the models with higher recharge and greater permeability. This becomes important in predicting the amount of water that must be pumped in order to effect the desired mine dewatering. The maximum pumping rate for mine dewatering was predicted to be 933 gal (3531 L) per minute for the low-recharge model, 1270 gal (4807 L) per minute for the middle-recharge model, and 1590 gal (6018 L) per minute for the high-recharge model.

It was proposed that, to obviate the impact of the lower regional water table on the seepage lakes, ground water be pumped into the lakes to raise their water levels. This would result in additional seepage into the area of the cone of depression of the mine-dewatering system and add about 200 gal (757 L) to the amount of water that must be pumped from the mine. Knowledge of the approximate volume of mine-dewatering pumpage is needed to design the pumping system as well as the treatment system for the mine drainage water.

Although extensive environmental and economic geology studies were completed, the site owner shelved plans for construction of the mine, and as of early 1993 there do not appear to be any plans to develop it.

NOTATION

h Head

R Aquifer recharge

S Aquifer storativity

T Aquifer transmissivity

Δt Length of time step

Δx and Δy Distances between nodes

Appendices

APPENDIX 1
Values of the function $W(u)$ for various values of u

u	$W(u)$	u	$W(u)$	u	$W(u)$	u	$W(u)$
1×10^{-10}	22.45	7×10^{-8}	15.90	4×10^{-5}	9.55	1×10^{-2}	4.04
2	21.76	8	15.76	5	9.33	2	3.35
3	21.35	9	15.65	6	9.14	3	2.96
4	21.06	1×10^{-7}	15.54	7	8.99	4	2.68
5	20.84	2	14.85	8	8.86	5	2.47
6	20.66	3	14.44	9	8.74	6	2.30
7	20.50	4	14.15	1×10^{-4}	8.63	7	2.15
8	20.37	5	13.93	2	7.94	8	2.03
9	20.25	6	13.75	3	7.53	9	1.92
1×10^{-9}	20.15	7	13.60	4	7.25	1×10^{-1}	1.823
2	19.45	8	13.46	5	7.02	2	1.223
3	19.05	9	13.34	6	6.84	3	0.906
4	18.76	1×10^{-6}	13.24	7	6.69	4	0.702
5	18.54	2	12.55	8	6.55	5	0.560
6	18.35	3	12.14	9	6.44	6	0.454
7	18.20	4	11.85	1×10^{-3}	6.33	7	0.374
8	18.07	5	11.63	2	5.64	8	0.311
9	17.95	6	11.45	3	5.23	9	0.260
1×10^{-8}	17.84	7	11.29	4	4.95	1×10^{0}	0.219
2	17.15	8	11.16	5	4.73	2	0.049
3	16.74	9	11.04	6	4.54	3	0.013
4	16.46	1×10^{-5}	10.94	7	4.39	4	0.004
5	16.23	2	10.24	8	4.26	5	0.001
6	16.05	3	9.84	9	4.14		

Source: Adapted from L. K. Wenzel, *Methods for Determining Permeability of Water-Bearing Materials with Special Reference to Discharging Well Methods*. U.S. Geological Survey Water-Supply Paper 887, 1942.

APPENDIX 2

Values of the function $F(\eta, \mu)$ for various values of η and μ

$\eta = Tt/r_c^2$	$\mu = 10^{-6}$	$\mu = 10^{-7}$	$\mu = 10^{-8}$	$\mu = 10^{-9}$	$\mu = 10^{-10}$
0.001	0.9994	0.9996	0.9996	0.9997	0.9997
0.002	0.9989	0.9992	0.9993	0.9994	0.9995
0.004	0.9980	0.9985	0.9987	0.9989	0.9991
0.006	0.9972	0.9978	0.9982	0.9984	0.9986
0.008	0.9964	0.9971	0.9976	0.9980	0.9982
0.01	0.9956	0.9965	0.9971	0.9975	0.9978
0.02	0.9919	0.9934	0.9944	0.9952	0.9958
0.04	0.9848	0.9875	0.9894	0.9908	0.9919
0.06	0.9782	0.9819	0.9846	0.9866	0.9881
0.08	0.9718	0.9765	0.9799	0.9824	0.9844
0.1	0.9655	0.9712	0.9753	0.9784	0.9807
0.2	0.9361	0.9459	0.9532	0.9587	0.9631
0.4	0.8828	0.8995	0.9122	0.9220	0.9298
0.6	0.8345	0.8569	0.8741	0.8875	0.8984
0.8	0.7901	0.8173	0.8383	0.8550	0.8686
1.0	0.7489	0.7801	0.8045	0.8240	0.8401
2.0	0.5800	0.6235	0.6591	0.6889	0.7139
3.0	0.4554	0.5033	0.5442	0.5792	0.6096
4.0	0.3613	0.4093	0.4517	0.4891	0.5222
5.0	0.2893	0.3351	0.3768	0.4146	0.4487
6.0	0.2337	0.2759	0.3157	0.3525	0.3865
7.0	0.1903	0.2285	0.2655	0.3007	0.3337
8.0	0.1562	0.1903	0.2243	0.2573	0.2888
9.0	0.1292	0.1594	0.1902	0.2208	0.2505
10.0	0.1078	0.1343	0.1620	0.1900	0.2178
20.0	0.02720	0.03343	0.04129	0.05071	0.06149
30.0	0.01286	0.01448	0.01667	0.01956	0.02320
40.0	0.008337	0.008898	0.009637	0.01062	0.01190
50.0	0.006209	0.006470	0.006789	0.007192	0.007709
60.0	0.004961	0.005111	0.005283	0.005487	0.005735
80.0	0.003547	0.003617	0.003691	0.003773	0.003863
100.0	0.002763	0.002803	0.002845	0.002890	0.002938
200.0	0.001313	0.001322	0.001330	0.001339	0.001348

Source: After I. S. Papadopulos, J. D. Bredehoeft, and H. H. Cooper, Jr., "On the Analysis of 'Slug Test' Data," *Water Resources Research*, 9 (1973):1087–89.

Note: For slug tests on wells of finite diameter, $H/H_0 = F(\eta, \mu)$.

APPENDIX 3

Values of the functions $W(u, r/B)$ for various values of u

u \ r/B	0.002	0.004	0.006	0.008	0.01	0.02	0.04	0.06	0.08	0.1	0.2	0.4	0.6	0.8	1	2	4	6	8
0	12.7	11.3	10.5	9.89	9.44	8.06	6.67	5.87	5.29	4.85	3.51	2.23	1.55	1.13	0.842	0.228	0.0223	0.0025	0.0003
0.000002	12.1	11.2	10.5	9.89	9.44														
0.000004	11.6	11.1	10.4	9.88	9.44														
0.000006	11.3	10.9	10.4	9.87	9.44														
0.000008	11.0	10.7	10.3	9.84	9.43														
0.00001	10.8	10.6	10.2	9.80	9.42														
0.00002	10.2	10.1	9.84	9.58	9.30	8.06													
0.00004	9.52	9.45	9.34	9.19	9.01	8.03													
0.00006	9.13	9.08	9.00	8.89	8.77	7.98	6.67												
0.00008	8.84	8.81	8.75	8.67	8.57	7.91	6.67												
0.0001	8.62	8.59	8.55	8.48	8.40	7.84	6.67	5.87											
0.0002	7.94	7.92	7.90	7.86	7.82	7.50	6.62	5.86	5.29										
0.0004	7.24	7.24	7.22	7.21	7.19	7.01	6.45	5.83	5.29	4.85									
0.0006	6.84	6.84	6.83	6.82	6.80	6.68	6.27	5.77	5.27	4.85									
0.0008	6.55	6.55	6.54	6.53	6.52	6.43	6.11	5.69	5.25	4.84									
0.001	6.33	6.33	6.32	6.32	6.31	6.23	5.97	5.61	5.21	4.83	3.51								
0.002	5.64	5.64	5.63	5.63	5.63	5.59	5.45	5.24	4.98	4.71	3.50								
0.004	4.95	4.95	4.95	4.94	4.94	4.92	4.85	4.74	4.59	4.42	3.48	2.23							
0.006	4.54				4.54	4.53	4.48	4.41	4.30	4.18	3.43	2.23							
0.008	4.26				4.26	4.25	4.21	4.15	4.08	3.98	3.36	2.23							
0.01	4.04				4.04	4.03	4.00	3.95	3.89	3.81	3.29	2.23	1.55	1.13					
0.02	3.35				3.35	3.35	3.34	3.31	3.28	3.24	2.95	2.18	1.55	1.13					
0.04	2.68				2.68	2.68	2.67	2.66	2.65	2.63	2.48	2.02	1.52	1.13	0.842				
0.06	2.30				2.30	2.29	2.29	2.28	2.27	2.26	2.17	1.85	1.46	1.11	0.839				
0.08	2.03					2.03	2.02	2.02	2.01	2.00	1.94	1.69	1.39	1.08	0.832				
0.1	1.82						1.82	1.82	1.81	1.80	1.75	1.56	1.31	1.05	0.819	0.228			
0.2	1.22						1.22	1.22	1.22	1.22	1.19	1.11	0.996	0.857	0.715	0.227			
0.4	0.702						0.702	0.702	0.701	0.700	0.693	0.665	0.621	0.565	0.502	0.210			
0.6	0.454						0.454	0.454	0.454	0.453	0.450	0.436	0.415	0.387	0.354	0.177	0.0222		
0.8	0.311						0.311	0.310	0.310	0.310	0.308	0.301	0.289	0.273	0.254	0.144	0.0218		
1	0.219									0.219	0.218	0.213	0.206	0.197	0.185	0.114	0.0207		
2	0.049										0.049	0.048	0.047	0.046	0.044	0.034	0.011	0.0021	0.0003
4	0.0038										0.0038	0.0038	0.0037	0.0037	0.0036	0.0031	0.0016	0.0006	0.0002
6	0.0004														0.0004	0.0003	0.0002	0.0001	0
8	0																		

Source: After M. S. Hantush, "Analysis of Data from Pumping Test in Leaky Aquifers," *Transactions, American Geophysical Union*, 37 (1956):702–14.

APPENDIX 4
Values of the function $H(\mu, \beta)$

μ \ β	0.001	0.005	0.01	0.05	0.10	0.20	0.50	1.0	2.0	5.0	10.0	20.0
0.000001	11.9842	10.5908	9.9259	8.3395	7.6497	6.9590	6.0463	5.3575	4.6721	3.7756	3.1110	2.4671
0.000005	10.8958	9.7174	9.0866	7.5284	6.8427	6.1548	5.2459	4.5617	3.8836	3.0055	2.3661	1.7633
0.00001	10.3739	9.3203	8.7142	7.1771	6.4944	5.8085	4.9024	4.2212	3.5481	2.6822	2.0590	1.4816
0.00005	9.0422	8.3171	7.8031	6.3523	5.6821	5.0045	4.1090	3.4394	2.7848	1.9622	1.3943	0.8994
0.0001	8.4258	7.8386	7.3803	5.9906	5.3297	4.6581	3.7700	3.1082	2.4658	1.6704	1.1359	0.6878
0.0005	6.9273	6.6024	6.2934	5.1223	4.4996	3.8527	2.9933	2.3601	1.7604	1.0564	0.6252	0.3089
0.001	6.2624	6.0193	5.7727	4.7290	4.1337	3.5045	2.6650	2.0506	1.4776	0.8271	0.4513	0.1976
0.005	4.6951	4.5786	4.4474	3.7415	3.2483	2.6891	1.9250	1.3767	0.8915	0.4001	0.1677	0.0493
0.01	4.0163	3.9334	3.8374	3.2752	2.8443	2.3325	1.6193	1.1122	0.6775	0.2670	0.0955	0.0221
0.05	2.4590	2.4243	2.3826	2.1007	1.8401	1.4872	0.9540	0.5812	0.2923	0.0755	0.0160	0.00164
0.1	1.8172	1.7949	1.7677	1.5768	1.3893	1.1207	0.6947	0.3970	0.1789	0.0359	0.00552	0.00034
0.5	0.5584	0.5530	0.5463	0.4969	0.4436	0.3591	0.2083	0.1006	0.0325	0.00288	0.00015	
1.0	0.2189	0.2169	0.2144	0.1961	0.1758	0.1427	0.0812	0.0365	0.00993	0.00055	0.00002	
5.0	0.00115	0.00114	0.00112	0.00104	0.00093	0.00076	0.00042	0.00017	0.00003			

Source: Condensed from M. S. Hantush, "Modification of the Theory of Leaky Aquifers," *Journal of Geophysical Research*, 65 (1960): 3713–25.

APPENDIX 5
Values of the functions $K_0(x)$ and $\exp(x)K_0(x)$

x	$K_0(x)$	$\exp(x)K_0(x)$	x	$K_0(x)$	$\exp(x)K_0(x)$
0.001	7.02	7.03	0.25	1.54	1.98
0.005	5.41	5.44	0.30	1.37	1.85
0.01	4.72	4.77	0.35	1.23	1.75
0.015	4.32	4.38	0.40	1.11	1.66
0.02	4.03	4.11	0.45	1.01	1.59
0.025	3.81	3.91	0.50	0.92	1.52
0.03	3.62	3.73	0.55	0.85	1.47
0.035	3.47	3.59	0.60	0.78	1.42
0.04	3.34	3.47	0.65	0.72	1.37
0.045	3.22	3.37	0.70	0.66	1.33
0.05	3.11	3.27	0.75	0.61	1.29
0.055	3.02	3.19	0.80	0.57	1.26
0.06	2.93	3.11	0.85	0.52	1.23
0.065	2.85	3.05	0.90	0.49	1.20
0.07	2.78	2.98	0.95	0.45	1.17
0.075	2.71	2.92	1.0	0.42	1.14
0.08	2.65	2.87	1.5	0.21	0.96
0.085	2.59	2.82	2.0	0.11	0.84
0.09	2.53	2.77	2.5	0.062	0.760
0.095	2.48	2.72	3.0	0.035	0.698
0.10	2.43	2.68	3.5	0.020	0.649
0.15	2.03	2.36	4.0	0.011	0.609
0.20	1.75	2.14	4.5	0.006	0.576
			5.0	0.004	0.548

Source: Adapted from M. S. Hantush, "Analysis of Data from Pumping Tests in Leaky Aquifers," *Transactions, American Geophysical Union,* 37 (1956):702–14.

APPENDIX 6A

Values of the function $W(u_A, \Gamma)$ for water-table aquifers

$1/u_A$	$\Gamma = 0.001$	$\Gamma = 0.01$	$\Gamma = 0.06$	$\Gamma = 0.2$	$\Gamma = 0.6$	$\Gamma = 1.0$	$\Gamma = 2.0$	$\Gamma = 4.0$	$\Gamma = 6.0$
4.0×10^{-1}	2.48×10^{-2}	2.41×10^{-2}	2.30×10^{-2}	2.14×10^{-2}	1.88×10^{-2}	1.70×10^{-2}	1.38×10^{-2}	9.33×10^{-3}	6.39×10^{-3}
8.0×10^{-1}	1.45×10^{-1}	1.40×10^{-1}	1.31×10^{-1}	1.19×10^{-1}	9.88×10^{-2}	8.49×10^{-2}	6.03×10^{-2}	3.17×10^{-2}	1.74×10^{-2}
1.4×10^{0}	3.58×10^{-1}	3.45×10^{-1}	3.18×10^{-1}	2.79×10^{-1}	2.17×10^{-1}	1.75×10^{-1}	1.07×10^{-1}	4.45×10^{-2}	2.10×10^{-2}
2.4×10^{0}	6.62×10^{-1}	6.33×10^{-1}	5.70×10^{-1}	4.83×10^{-1}	3.43×10^{-1}	2.56×10^{-1}	1.33×10^{-1}	4.76×10^{-2}	2.14×10^{-2}
4.0×10^{0}	1.02×10^{0}	9.63×10^{-1}	8.49×10^{-1}	6.88×10^{-1}	4.38×10^{-1}	3.00×10^{-1}	1.40×10^{-1}	4.78×10^{-2}	2.15×10^{-2}
8.0×10^{0}	1.57×10^{0}	1.46×10^{0}	1.23×10^{0}	9.18×10^{-1}	4.97×10^{-1}	3.17×10^{-1}	1.41×10^{-1}		
1.4×10^{1}	2.05×10^{0}	1.88×10^{0}	1.51×10^{0}	1.03×10^{0}	5.07×10^{-1}				
2.4×10^{1}	2.52×10^{0}	2.27×10^{0}	1.73×10^{0}	1.07×10^{0}					
4.0×10^{1}	2.97×10^{0}	2.61×10^{0}	1.85×10^{0}	1.08×10^{0}					
8.0×10^{1}	3.56×10^{0}	3.00×10^{0}	1.92×10^{0}						
1.4×10^{2}	4.01×10^{0}	3.23×10^{0}	1.93×10^{0}						
2.4×10^{2}	4.42×10^{0}	3.37×10^{0}	1.94×10^{0}						
4.0×10^{2}	4.77×10^{0}	3.43×10^{0}							
8.0×10^{2}	5.16×10^{0}	3.45×10^{0}							
1.4×10^{3}	5.40×10^{0}	3.46×10^{0}							
2.4×10^{3}	5.54×10^{0}								
4.0×10^{3}	5.59×10^{0}								
8.0×10^{3}	5.62×10^{0}								
1.4×10^{4}	5.62×10^{0}	3.46×10^{0}	1.94×10^{0}	1.08×10^{0}	5.07×10^{-1}	3.17×10^{-1}	1.41×10^{-1}	4.78×10^{-2}	2.15×10^{-2}

APPENDIX 6B

Values of the function $W(u_B, \Gamma)$ for water-table aquifers

$1/u_B$	$\Gamma = 0.001$	$\Gamma = 0.01$	$\Gamma = 0.06$	$\Gamma = 0.2$	$\Gamma = 0.6$	$\Gamma = 1.0$	$\Gamma = 2.0$	$\Gamma = 4.0$	$\Gamma = 6.0$
4.0×10^{-4}	5.62×10^{0}	3.46×10^{0}	1.94×10^{0}	1.09×10^{0}	5.08×10^{-1}	3.18×10^{-1}	1.42×10^{-1}	4.79×10^{-2}	2.15×10^{-2}
8.0×10^{-4}								4.80×10^{-2}	2.16×10^{-2}
1.4×10^{-3}								4.81×10^{-2}	2.17×10^{-2}
2.4×10^{-3}								4.84×10^{-2}	2.19×10^{-2}
4.0×10^{-3}					5.08×10^{-1}	3.18×10^{-1}	1.42×10^{-1}	4.88×10^{-2}	2.21×10^{-2}
8.0×10^{-3}					5.09×10^{-1}	3.19×10^{-1}	1.43×10^{-1}	4.96×10^{-2}	2.28×10^{-2}
1.4×10^{-2}					5.10×10^{-1}	3.21×10^{-1}	1.45×10^{-1}	5.09×10^{-2}	2.39×10^{-2}
2.4×10^{-2}					5.12×10^{-1}	3.23×10^{-1}	1.47×10^{-1}	5.32×10^{-2}	2.57×10^{-2}
4.0×10^{-2}					5.16×10^{-1}	3.27×10^{-1}	1.52×10^{-1}	5.68×10^{-2}	2.86×10^{-2}
8.0×10^{-2}				1.09×10^{0}	5.24×10^{-1}	3.37×10^{-1}	1.62×10^{-1}	6.61×10^{-2}	3.62×10^{-2}
1.4×10^{-1}			1.94×10^{0}	1.10×10^{0}	5.37×10^{-1}	3.50×10^{-1}	1.78×10^{-1}	8.06×10^{-2}	4.86×10^{-2}
2.4×10^{-1}			1.95×10^{0}	1.11×10^{0}	5.57×10^{-1}	3.74×10^{-1}	2.05×10^{-1}	1.06×10^{-1}	7.14×10^{-2}
4.0×10^{-1}			1.96×10^{0}	1.13×10^{0}	5.89×10^{-1}	4.12×10^{-1}	2.48×10^{-1}	1.49×10^{-1}	1.13×10^{-1}
8.0×10^{-1}	5.62×10^{0}	3.46×10^{0}	1.98×10^{0}	1.18×10^{0}	6.67×10^{-1}	5.06×10^{-1}	3.57×10^{-1}	2.66×10^{-1}	2.31×10^{-1}
1.4×10^{0}	5.63×10^{0}	3.47×10^{0}	2.01×10^{0}	1.24×10^{0}	7.80×10^{-1}	6.42×10^{-1}	5.17×10^{-1}	4.45×10^{-1}	4.19×10^{-1}
2.4×10^{0}	5.63×10^{0}	3.49×10^{0}	2.06×10^{0}	1.35×10^{0}	9.54×10^{-1}	8.50×10^{-1}	7.63×10^{-1}	7.18×10^{-1}	7.03×10^{-1}
4.0×10^{0}	5.63×10^{0}	3.51×10^{0}	2.13×10^{0}	1.50×10^{0}	1.20×10^{0}	1.13×10^{0}	1.08×10^{0}	1.06×10^{0}	1.05×10^{0}
8.0×10^{0}	5.64×10^{0}	3.56×10^{0}	2.31×10^{0}	1.85×10^{0}	1.68×10^{0}	1.65×10^{0}	1.63×10^{0}	1.63×10^{0}	1.63×10^{0}
1.4×10^{1}	5.65×10^{0}	3.63×10^{0}	2.55×10^{0}	2.23×10^{0}	2.15×10^{0}	2.14×10^{0}	2.14×10^{0}	2.14×10^{0}	2.14×10^{0}
2.4×10^{1}	5.67×10^{0}	3.74×10^{0}	2.86×10^{0}	2.68×10^{0}	2.65×10^{0}	2.65×10^{0}	2.64×10^{0}	2.64×10^{0}	2.64×10^{0}
4.0×10^{1}	5.70×10^{0}	3.90×10^{0}	3.24×10^{0}	3.15×10^{0}	3.14×10^{0}	3.14×10^{0}	3.14×10^{0}	3.14×10^{0}	3.14×10^{0}
8.0×10^{1}	5.76×10^{0}	4.22×10^{0}	3.85×10^{0}	3.82×10^{0}	3.82×10^{0}	3.82×10^{0}	3.82×10^{0}	3.82×10^{0}	3.82×10^{0}
1.4×10^{2}	5.85×10^{0}	4.58×10^{0}	4.38×10^{0}	4.37×10^{0}	4.37×10^{0}	4.37×10^{0}	4.37×10^{0}	4.37×10^{0}	4.37×10^{0}
2.4×10^{2}	5.99×10^{0}	5.00×10^{0}	4.91×10^{0}	4.91×10^{0}	4.91×10^{0}	4.91×10^{0}	4.91×10^{0}	4.91×10^{0}	4.91×10^{0}
4.0×10^{2}	6.16×10^{0}	5.46×10^{0}	5.42×10^{0}	5.42×10^{0}	5.42×10^{0}	5.42×10^{0}	5.42×10^{0}	5.42×10^{0}	5.42×10^{0}
8.0×10^{2}	6.47×10^{0}	6.11×10^{0}	6.11×10^{0}	6.11×10^{0}	6.11×10^{0}	6.11×10^{0}	6.11×10^{0}	6.11×10^{0}	6.11×10^{0}
1.4×10^{3}	6.67×10^{0}	6.67×10^{0}	6.67×10^{0}	6.67×10^{0}	6.67×10^{0}	6.67×10^{0}	6.67×10^{0}	6.67×10^{0}	6.67×10^{0}
2.4×10^{3}	7.21×10^{0}	7.21×10^{0}	7.21×10^{0}	7.21×10^{0}	7.21×10^{0}	7.21×10^{0}	7.21×10^{0}	7.21×10^{0}	7.21×10^{0}
4.0×10^{3}	7.72×10^{0}	7.72×10^{0}	7.72×10^{0}	7.72×10^{0}	7.72×10^{0}	7.72×10^{0}	7.72×10^{0}	7.72×10^{0}	7.72×10^{0}
8.0×10^{3}	8.41×10^{0}	8.41×10^{0}	8.41×10^{0}	8.41×10^{0}	8.41×10^{0}	8.41×10^{0}	8.41×10^{0}	8.41×10^{0}	8.41×10^{0}
1.4×10^{4}	8.97×10^{0}	8.97×10^{0}	8.97×10^{0}	8.97×10^{0}	8.97×10^{0}	8.97×10^{0}	8.97×10^{0}	8.97×10^{0}	8.97×10^{0}
2.4×10^{4}	9.51×10^{0}	9.51×10^{0}	9.51×10^{0}	9.51×10^{0}	9.51×10^{0}	9.51×10^{0}	9.51×10^{0}	9.51×10^{0}	9.51×10^{0}
4.0×10^{4}	1.94×10^{1}	1.94×10^{1}	1.94×10^{1}	1.94×10^{1}	1.94×10^{1}	1.94×10^{1}	1.94×10^{1}	1.94×10^{1}	1.94×10^{1}

Source: Adapted from S. P. Neuman, *Water Resources Research*, 11 (1975):329–42.

APPENDIX 7
Table for length conversion

Unit	mm	cm	m	km	in	ft	yd	mi
1 millimeter	1	0.1	0.001	10^{-6}	0.0397	0.00328	0.00109	6.21×10^{-7}
1 centimeter	10	1	0.01	0.0001	0.3937	0.0328	0.0109	6.21×10^{-6}
1 meter	1000	100	1	0.001	39.37	3.281	1.094	6.21×10^{-4}
1 kilometer	10^{6}	10^{5}	1000	1	39,370	3281	1093.6	0.621
1 inch	25.4	2.54	0.0254	2.54×10^{-5}	1	0.0833	0.0278	1.58×10^{-5}
1 foot	304.8	30.48	0.3048	3.05×10^{-4}	12	1	0.333	1.89×10^{-4}
1 yard	914.4	91.44	0.9144	9.14×10^{-4}	36	3	1	5.68×10^{-4}
1 mile	1.61×10^{6}	1.01×10^{5}	1.61×10^{3}	1.6093	63,360	5280	1760	1

APPENDIX 8
Table for area conversion

Unit	cm^2	m^2	km^2	ha	in^2	ft^2	yd^2	mi^2	ac
1 sq. centimeter	1	0.0001	10^{-10}	10^{-8}	0.155	1.08×10^{-3}	1.2×10^{-4}	3.86×10^{-11}	2.47×10^{-8}
1 sq. meter	10^{4}	1	10^{-6}	10^{-4}	1550	10.76	1.196	3.86×10^{-7}	2.47×10^{-4}
1 sq. kilometer	10^{10}	10^{6}	1	100	1.55×10^{9}	1.076×10^{7}	1.196×10^{6}	0.3861	247.1
1 hectare	10^{8}	10^{4}	0.01	1	1.55×10^{7}	1.076×10^{5}	1.196×10^{4}	3.861×10^{-3}	2.471
1 sq. inch	6.452	6.45×10^{-4}	6.45×10^{-10}	6.45×10^{-8}	1	6.94×10^{-3}	7.7×10^{-4}	2.49×10^{-10}	1.574×10^{7}
1 sq. foot	929	0.0929	9.29×10^{-8}	9.29×10^{-6}	144	1	0.111	3.587×10^{-8}	2.3×10^{-5}
1 sq. yard	8361	0.8361	8.36×10^{-7}	8.36×10^{-5}	1296	9	1	3.23×10^{-7}	2.07×10^{-4}
1 sq. mile	2.59×10^{10}	2.59×10^{6}	2.59	259	4.01×10^{9}	2.79×10^{7}	3.098×10^{6}	1	640
1 acre	4.04×10^{7}	4047	4.047×10^{-3}	0.4047	6.27×10^{6}	43,560	4840	1.562×10^{-3}	1

APPENDIX 9
Table for volume conversion

Unit	mL	liters	m³	in³	ft³	gal	ac-ft	million gal
1 milliliter	1	0.001	10^{-6}	0.06102	3.53×10^{-5}	2.64×10^{-4}	8.1×10^{-10}	2.64×10^{-10}
1 liter	10^{3}	1	0.001	61.02	0.0353	0.264	8.1×10^{-7}	2.64×10^{-7}
1 cu. meter	10^{6}	1000	1	61,023	35.31	264.17	8.1×10^{-4}	2.64×10^{-4}
1 cu. inch	16.39	1.64×10^{-2}	1.64×10^{-5}	1	5.79×10^{-4}	4.33×10^{-3}	1.218×10^{-8}	4.329×10^{-9}
1 cu. foot	28.317	28.317	0.02832	1728	1	7.48	2.296×10^{-5}	7.48×10^{-6}
1 U.S. gallon	3785.4	3.785	3.78×10^{-3}	231	0.134	1	3.069×10^{-6}	10^{-6}
1 acre-foot	1.233×10^{9}	1.233×10^{6}	1233.5	75.27×10^{6}	43,560	3.26×10^{5}	1	0.3260
1 million gallons	3.785×10^{9}	3.785×10^{6}	3785	2.31×10^{8}	1.338×10^{5}	10^{6}	3.0684	1

APPENDIX 10
Table for time conversion

Unit	sec	min	hours	days	years
1 second	1	1.67×10^{-2}	2.77×10^{-4}	1.157×10^{-5}	3.17×10^{-8}
1 minute	60	1	1.67×10^{-2}	6.94×10^{-4}	1.90×10^{-6}
1 hour	3600	60	1	4.17×10^{-2}	1.14×10^{-4}
1 day	8.64×10^{4}	1440	24	1	2.74×10^{-3}
1 year	3.15×10^{7}	5.256×10^{5}	8760	365	1

APPENDIX 11
Solubility products for selected minerals and compounds

Compound	Solubility product	Mineral name
Chlorides		
$CuCl$	$10^{-6.7}$	
$PbCl_2$	$10^{-4.8}$	
Hg_2Cl_2	$10^{-17.9}$	
$AgCl$	$10^{-9.7}$	
Fluorides		
BaF_2	$10^{-5.8}$	
CaF_2	$10^{-10.4}$	Fluorite
MgF_2	$10^{-8.2}$	Sellaite
PbF_2	$10^{-7.5}$	
SrF_2	$10^{-8.5}$	
Sulfates		
$BaSO_4$	$10^{-10.0}$	Barite
$CaSO_4$	$10^{-4.5}$	Anhydrite
$CaSO_4 \cdot 2H_2O$	$10^{-4.6}$	Gypsum
$PbSO_4$	$10^{-7.8}$	Anglesite
Ag_2SO_4	$10^{-4.8}$	
$SrSO_4$	$10^{-6.5}$	Celestite
Sulfides		
Cu_2S	$10^{-48.5}$	
CuS	$10^{-36.1}$	
FeS	$10^{-18.1}$	
PbS	$10^{-27.5}$	Galena
HgS	$10^{-53.3}$	Cinnebar
ZnS	$10^{-22.5}$	Wurtzite
ZnS	$10^{-24.7}$	Sphalerite
Carbonates		
$BaCO_3$	$10^{-8.3}$	Witherite
$CdCO_3$	$10^{-13.7}$	
$CaCO_3$	$10^{-8.35}$	Calcite
$CaCO_3$	$10^{-8.22}$	Aragonite
$CoCO_3$	$10^{-10.0}$	
$FeCO_3$	$10^{-10.7}$	Siderite
$PbCO_3$	$10^{-13.1}$	
$MgCO_3$	$10^{-7.5}$	Magnesite
$MnCO_3$	$10^{-9.3}$	Rhodochrosite
Phosphates		
$AlPO_4 \cdot 2H_2O$	$10^{-22.1}$	Variscite
$CaHPO_4 \cdot 2H_2O$	$10^{-6.6}$	
$Ca_3(PO_4)_2$	$10^{-28.7}$	
$Cu_3(PO_4)_2$	$10^{-36.9}$	
$FePO_4$	$10^{-21.6}$	
$FePO_4 \cdot 2H_2O$	$10^{-26.4}$	

Source: K. B. Krauskopf, *Introduction to Geochemistry*, 2nd ed. New York: McGraw-Hill, 1979.

APPENDIX 12
Atomic weights and numbers of naturally occurring elements

Element	Symbol	Atomic number	Atomic weight
Actinium	Ac	89	227.03
Aluminum	Al	13	26.98
Antimony	Sb	51	121.75
Argon	Ar	18	39.95
Arsenic	As	33	74.92
Barium	Ba	56	137.33
Beryllium	Be	4	9.01
Bismuth	Bi	83	208.98
Boron	B	5	10.81
Bromine	Br	35	79.90
Cadmium	Cd	48	112.41
Calcium	Ca	20	40.08
Carbon	C	6	12.01
Cerium	Ce	58	140.12
Cesium	Cs	55	132.91
Chlorine	Cl	17	35.45
Chromium	Cr	24	52.00
Cobalt	Co	27	58.93
Copper	Cu	29	63.55
Dysprosium	Dy	66	162.50
Erbium	Er	68	167.26
Europium	Eu	63	151.96
Fluorine	F	9	19.00
Gadolinium	Gd	64	157.25
Gallium	Ga	31	69.72
Germanium	Ge	32	72.59
Gold	Au	79	196.97
Hafnium	Hf	72	178.49
Helium	He	2	4.003
Holmium	Ho	67	164.93
Hydrogen	H	1	1.008
Indium	In	49	114.82
Iodine	I	53	126.90
Iridium	Ir	77	192.22
Iron	Fe	26	55.85
Krypton	Kr	36	83.80
Lanthanum	La	57	138.91
Lead	Pb	82	207.19
Lithium	Li	3	6.94
Lutetium	Lu	71	174.97
Magnesium	Mg	12	24.31
Manganese	Mn	25	54.94
Mercury	Hg	80	200.59
Molybdenum	Mo	42	95.94

APPENDIX 12

continued

Element	Symbol	Atomic number	Atomic weight
Neodymium	Nd	60	144.24
Neon	Ne	10	20.18
Nickel	Ni	28	58.70
Niobium	Nb	41	92.91
Nitrogen	N	7	14.01
Osmium	Os	76	190.2
Oxygen	O	8	16.00
Palladium	Pd	46	106.4
Phosphorus	P	15	30.97
Platinum	Pt	78	195.09
Polonium	Po	84	209
Potassium	K	19	39.10
Praseodymium	Pr	59	140.91
Protactinium	Pa	91	231.04
Radium	Ra	88	226.03
Radon	Rn	86	222
Rhenium	Re	75	186.21
Rhodium	Rh	45	102.91
Rubidium	Rb	37	85.47
Ruthenium	Ru	44	101.07
Samarium	Sm	62	150.35
Scandium	Sc	21	44.96
Selenium	Se	34	78.96
Silicon	Si	14	28.09
Silver	Ag	47	107.87
Sodium	Na	11	22.99
Strontium	Sr	38	87.62
Sulfur	S	16	32.06
Tantalum	Ta	73	180.95
Tellurium	Te	52	127.60
Terbium	Tb	65	158.93
Thallium	Tl	81	204.37
Thorium	Th	90	232.04
Thulium	Tm	69	168.93
Tin	Sn	50	118.69
Titanium	Ti	22	47.90
Tungsten	W	74	183.85
Uranium	U	92	238.03
Vanadium	V	23	50.94
Xenon	Xe	54	131.30
Ytterbium	Yb	70	173.04
Yttrium	Y	39	88.91
Zinc	Zn	30	65.38
Zirconium	Zr	40	91.22

Source: K. B. Krauskopf, *Introduction to Geochemistry*, 2nd ed. New York: McGraw-Hill, 1979.

APPENDIX 13

Values of the error function of x [erf (x)] and the complementary error function of x [erfc (x)]. Note that erfc (x) = 1 − erf (x) and erfc (−x) = 1 + erf (x).

x	erf (x)	erfc (x)
0	0	1.0
0.05	0.056372	0.943628
0.1	0.112463	0.887537
0.15	0.167996	0.832004
0.2	0.222703	0.777297
0.25	0.276326	0.723674
0.3	0.328627	0.671373
0.35	0.379382	0.620618
0.4	0.428392	0.571608
0.45	0.475482	0.524518
0.5	0.520500	0.479500
0.55	0.563323	0.436677
0.6	0.603856	0.396144
0.65	0.642029	0.357971
0.7	0.677801	0.322199
0.75	0.711156	0.288844
0.8	0.742101	0.257899
0.85	0.770668	0.229332
0.9	0.796908	0.203092
0.95	0.820891	0.179109
1.0	0.842701	0.157299
1.1	0.880205	0.119795
1.2	0.910314	0.089686
1.3	0.934008	0.065992
1.4	0.952285	0.047715
1.5	0.966105	0.033895
1.6	0.976348	0.023652
1.7	0.983790	0.016210
1.8	0.989091	0.010909
1.9	0.992790	0.007210
2.0	0.995322	0.004678
2.1	0.997021	0.002979
2.2	0.998137	0.001863
2.3	0.998857	0.001143
2.4	0.999311	0.000689
2.5	0.999593	0.000407
2.6	0.999764	0.000236
2.7	0.999866	0.000134
2.8	0.999925	0.000075
2.9	0.999959	0.000041
3.0	0.999978	0.000022
∞	1.00000	0.00000

An approximate solution, correct to within 0.7%, of the error function can be determined analytically from the following equation:

$$\text{erf}(x) = \sqrt{1 - \exp\left(\frac{-4x^2}{\pi}\right)}$$

APPENDIX 14

Absolute density and absolute viscosity of water

Temperature (°C)	Density (kg/m^3)	Density (g/cm^3)	Viscosity (g/s·cm)
0	999.841	0.999841	0.017921
1	999.900	0.999900	0.017313
2	999.941	0.999941	0.016728
3	999.965	0.999965	0.016191
4	999.973	0.999973	0.015674
5	999.965	0.999965	0.015188
6	999.941	0.999941	0.014728
7	999.902	0.999902	0.014284
8	999.849	0.999849	0.013860
9	999.781	0.999781	0.013462
10	999.700	0.999700	0.013077
11	999.605	0.999605	0.012713
12	999.498	0.999498	0.012363
13	999.377	0.999377	0.012028
14	999.244	0.999244	0.011709
15	999.099	0.999099	0.011404
16	998.943	0.998943	0.011111
17	998.774	0.998774	0.010828
18	998.595	0.998595	0.010559
19	998.405	0.998405	0.010299
20	998.203	0.998203	0.010050
21	997.992	0.997992	0.009810
22	997.770	0.997770	0.009579
23	997.538	0.997538	0.009358
24	997.296	0.997296	0.009142
25	997.044	0.997044	0.008937
26	996.783	0.996783	0.008737
27	996.512	0.996512	0.008545
28	996.232	0.996232	0.008360
29	995.944	0.995944	0.008180
30	995.646	0.995646	0.008007
35	994.029	0.994029	0.007225
40	992.214	0.992214	0.006560
45	990.212	0.990212	0.005988
50	988.047	0.988047	0.005494

Source: *Handbook of Chemistry and Physics* (Cleveland, Ohio: CRC Publishing Company, 1986).

APPENDIX 15
Loading and Running Computer Programs

In order to run the free student versions of the computer programs contained in the accompanying diskette, you will need access to a computer with the following minimum configuration: IBM PC or compatible computer with DOS version 3.0 or higher, 640 Kbytes of RAM, EGA or VGA graphics card, color graphics monitor, and a 3½-in.-high density (1.4-Mbyte) disk drive. The programs will run much more quickly if your computer has a math coprocessor. Of course, the faster the machine, the more quickly the programs will run. It is not necessary to have a hard disk to run these programs, although it is useful if you wish to save a lot of data files. A mouse can be used with QUICKFLOW or AQTESOLV, but it is not necessary. The programs should automatically detect the type of graphics card in your computer.

The first thing that you should do is make a backup copy of the *Applied Hydrogeology* diskette and put the original in a safe place.

In order to run the programs, start the machine *without* the *Applied Hydrogeology* diskette in a drive. After the computer has been booted and is up and running with a DOS prompt showing, put the *Applied Hydrogeology* diskette into your 3½-in. drive. Now get your computer to display the DOS prompt for your 3½-in. disk drive. For example, if your DOS prompt is C:>, type A: and press the Enter key if your 3½-in. drive is drive A.

There are executable versions of the three programs on the diskette. To run FLOWNET, type FLOWNETD and press the Enter key. The student version of FLOWNET should boot. To run QUICKFLOW, type QF and press the Enter key. To run AQTESOLV, type AQTESOLV and press the Enter key. Naturally, you must exit one program before you run another program.

It may be possible for you to print out the screen display, depending upon the software and hardware that you have. If you have a dot-matrix graphic printer, there is a DOS program called GRAPHICS. This should have come with your DOS. If you run GRAPHICS before you run any of the preceding programs, you may be able to print the screen display using the *Print Screen* key. If you have a Hewlett Packard LaserJet™ printer, you can contact them and obtain a utility program called HPSCREEN, which can be used in place of GRAPHICS. The fully functional versions of the various programs can be configured to communicate with various printers.

There are brief manuals for each of the programs on the *Applied Hydrogeology* diskette. They are listed as FLOWPATH.DOC, QUICKFLO.DOC, and AQTESOLV.DOC. You can make a printed version of the manual with the DOS PRINT command. At the DOS prompt, type PRINT QUICKFLO.DOC and press the Enter key. Some writing will appear on the screen. Press the Enter key again and the QUICKFLOW manual should begin to print. Refer to your DOS manual for further instructions. Please be advised that the manuals may refer to the original program, not the student version. For example, the English-language version of FLOWNETD is already installed on the disk. All the files for the three programs are in one directory on the diskette. If you want to install the programs on a hard drive, they may be copied onto one directory. If you wish to copy each program onto a separate directory on your hard drive, the following files must be copied for each program:

FLOWNET

FLOWNETD.EXE

FLOWNET.TXT

ATT.BGI

CGA.BGI

EGAVGA.BGI

HERC.BGI

FLOWNET.DOC

QUICKFLOW

QF.EXE

ROMANSIN.FNT

SYSTEM72.FNT

SYSTEM16.FNT

SYSTEM24.FNT

DEMO.QFL

QUICKFLO.DOC

AQTESOLV

AQTESOLV.EXE

KDRP53.DAT

BRP427.DAT

P1UR.VEC

C2UR.VEC

S2UR.VEC

D2UR.VEC

UPPER128.GEN

LOGO.BIN

AQTESOLV.DOC

These free programs are provided without any technical support. Neither the author nor Macmillan Publishing Company makes any claim that they will work or—even if they work—that the results will be correct. No warranty is given, either expressed or implied. Please do not call the author if you have problems with these programs; he is sorry but he doesn't have the time to help you individually.

Glossary

Adiabatic expansion The process that occurs when an air mass rises and expands without exchanging heat with its surroundings.

Adsorption The attraction and adhesion of a layer of ions from an aqueous solution to the solid mineral surfaces with which it is in contact.

Advection The process by which solutes are transported by the motion of flowing ground water.

Alluvium Sediments deposited by flowing rivers. Depending upon the location in the floodplain of the river, different-sized sediments are deposited.

American Rule A ground-water doctrine that holds that an overlying property owner has the right to use only a reasonable amount of ground water.

Anisotropy The condition under which one or more of the hydraulic properties of an aquifer vary according to the direction of flow.

Aquiclude A low-permeability unit that forms either the upper or lower boundary of a ground-water flow system.

Aquifer Rock or sediment in a formation, group of formations, or part of a formation that is saturated and sufficiently permeable to transmit economic quantities of water to wells and springs.

Aquifer, confined An aquifer that is overlain by a confining bed. The confining bed has a significantly lower hydraulic conductivity than the aquifer.

Aquifer, perched A region in the unsaturated zone where the soil may be locally saturated because it overlies a low-permeability unit.

Aquifer, semiconfined An aquifer confined by a low-permeability layer that permits water to slowly flow through it. During pumping of the aquifer, recharge to the aquifer can occur across the confining layer. Also known as a leaky artesian or leaky confined aquifer.

Aquifer test *See* pumping test.

Aquifer, unconfined An aquifer in which there are no confining beds between the zone of saturation and the surface. There will be a water table in an unconfined aquifer. Water-table aquifer is a synonym.

Aquifuge An absolutely impermeable unit that will neither store nor transmit water.

Aquitard A low-permeability unit that can store ground water and also transmit it slowly from one aquifer to another.

Artificial recharge   The process by which water can be injected or added to an aquifer. Dug basins, drilled wells, or simply the spread of water across the land surface are all means of artificial recharge.

Average linear velocity   *See* seepage velocity.

Bail-down test   A type of slug test performed by using a bailer to remove a volume of water from a small-diameter well.

Bailer   A device used to withdraw a water sample from a small-diameter well or piezometer. A bailer typically is a piece of pipe attached to a wire and having a check valve in the bottom.

Barrier boundary   An aquifer-system boundary represented by a rock mass that is not a source of water.

Baseflow   That part of stream discharge from ground water seeping into the stream.

Baseflow recession   The declining rate of discharge of a stream fed only by baseflow for an extended period. Typically, a baseflow recession will be exponential.

Baseflow recession hydrograph   A hydrograph that shows a baseflow-recession curve.

Bladder pump   A positive-displacement pumping device that uses pulses of gas to push a water-quality sample toward the surface.

Borehole geochemical probe   A water-quality monitoring device that is lowered into a well on a cable and that can make a direct reading of such parameters as pH, Eh, temperature, and specific conductivity.

Borehole geophysics   The general field of geophysics developed around the lowering of various probes into a well.

Boring   A hole advanced into the ground by means of a drilling rig.

Boussinesq equation   The general equation for two-dimensional, unconfined transient flow.

Caliper log   A borehole log of the diameter of an uncased well.

Capillary forces   The forces acting on soil moisture in the unsaturated zone, attributable to molecular attraction between soil particles and water.

Capillary fringe   The zone immediately above the water table, where water is drawn upward by capillary attraction.

Casing   *See* well casing.

Cation-exchange capacity   The ability of a particular rock or soil to absorb cations.

Cementation   The process by which some of the voids in a sediment are filled with precipitated materials, such as silica, calcite, and iron oxide, and that is a part of diagenesis.

Chemical activity   The molal concentration of an ion multiplied by a factor known as the activity coefficient.

Clastic dike   Intrusion of sediment forced into fractures in rock or sediments.

Cleat   The vertical planes of fracture that are found in coal.

Collection lysimeter   A device installed in the unsaturated zone to collect a water-quality sample by having the water drain downward by gravity into a collection pit.

Combining weight *See* equivalent weight.

Common-ion effect The decrease in the solubility of a salt dissolved in water already containing some of the ions of the salt.

Condensation The process that occurs when an air mass is saturated and water droplets form around nuclei or on surfaces.

Confining layer A body of material of low hydraulic conductivity that is stratigraphically adjacent to one or more aquifers. It may lie above or below the aquifer.

Connate water Interstitial water that was not buried with a rock but that has been out of contact with the atmosphere for an appreciable part of a geologic period.

Contact spring A spring that forms at a lithologic contact where a more permeable unit overlies a less permeable unit.

Contaminant *See* pollutant.

Current meter A device that is lowered into a stream in order to record the rate at which the current is moving.

Darcian velocity *See* specific discharge.

Darcy's law An equation that can be used to compute the quantity of water flowing through an aquifer.

Debye-Hückel equation A means of computing the activity coefficient for an ionic species.

Density The mass or quantity of a substance per unit volume. Units are kilograms per cubic meter or grams per cubic centimeter.

Depression spring A spring formed when the water table reaches a land surface because of a change in topography.

Depression storage Water from precipitation that collects in puddles at the land surface.

Dew point The temperature of a given air mass at which condensation will begin.

Diagenesis The chemical and physical changes occurring in sediments before consolidation or while in the environment of deposition.

Diffusion The process by which both ionic and molecular species dissolved in water move from areas of higher concentration to areas of lower concentration.

Dipole array A particular arrangement of electrodes used to measure surface electrical resistivity.

Direct precipitation Water that falls directly into a lake or stream without passing through any land phase of the runoff cycle.

Dirichlet condition A boundary condition for a ground-water computer model where the head is known at the boundary of the flow field.

Discharge The volume of water flowing in a stream or through an aquifer past a specific point in a given period of time.

Discharge area An area in which there are upward components of hydraulic head in the aquifer. Ground water is flowing toward the surface in a discharge area and may escape as a spring, seep, or baseflow or by evaporation and transpiration.

Discharge velocity *See* specific discharge.

Dispersion The phenomenon by which a solute in flowing ground water is mixed with uncontaminated water and becomes reduced in concentration. Dispersion is caused by both differences in the velocity that the water travels at the pore level and differences in the rate at which water travels through different strata in the flow path.

Distribution coefficient   The slope of a linear Freundlich isotherm.

Drainage basin   The land area from which surface runoff drains into a stream system.

Drainage divide   *See* topographic divide.

Drawdown   A lowering of the water table of an unconfined aquifer or the potentiometric surface of a confined aquifer caused by pumping of ground water from wells.

Dupuit assumptions   Assumptions for flow in an unconfined aquifer that (1) the hydraulic gradient is equal to the slope of the water table, (2) the streamlines are horizontal, and (3) the equipotential lines are vertical.

Dupuit equation   An equation for the volume of water flowing in an unconfined aquifer; based upon the Dupuit assumptions.

Duration curve   A graph showing the percentage of time that the given flows of a stream will be equaled or exceeded. It is based upon a statistical study of historic streamflow records.

Dynamic equilibrium   A condition in which the amount of recharge to an aquifer equals the amount of natural discharge.

Effective grain size   The grain size corresponding to the one that is 10% finer by weight line on the grain-size distribution curve.

Effective porosity   *See* porosity, effective.

Effective uniform depth (EUD) of precipitation   The result that would occur if the actual precipitation over a drainage basin, which is variable from place to place, were spread out over the entire basin to an average depth.

Electrical resistance model   An analog model of ground-water flow based upon the flow of electricity through a circuit containing resistors and capacitors.

Electrical sounding   An earth-resistivity survey made at the same location by putting the electrodes progressively farther apart. It shows the change of apparent resistivity with depth.

Electromagnetic conductivity   A method of measuring the induced electrical field in the earth to determine the ability of the earth to conduct electricity. Electromagnetic conductivity is the inverse of electrical resistivity. Also known as electric conductivity and terrain conductivity.

English Rule   A ground-water doctrine that holds that property owners have the right of absolute ownership of the ground water beneath their land.

Equilibrium constant   The number defining the conditions of equilibrium for a particular reversible chemical reaction.

Equipotential line   A line in a two-dimensional ground-water flow field such that the total hydraulic head is the same for all points along the line.

Equipotential surface   A surface in a three-dimensional ground-water flow field such that the total hydraulic head is the same everywhere on the surface.

Equivalent weight   The formula weight of a dissolved ionic species divided by the electrical charge. Also known as combining weight.

Eutrophication   The process of accelerated aging of a surface-water body; caused by excess nutrients and sediments being brought into the lake.

Evaporation   The process by which water passes from the liquid to the vapor state.

Evapotranspiration   The sum of evaporation plus transpiration.

Evapotranspiration, actual   The evapotranspiration that actually occurs under given climatic and soil-moisture conditions.

Evapotranspiration, potential   The evapotranspiration that would occur under given climatic conditions if there were unlimited soil moisture.

Fault spring   A spring created by the movement of two rock units on a fault.

Field blank   A water-quality sample where highly purified water is run through the field-sampling procedure and sent to the laboratory to detect if any contamination of the samples is occurring during the sampling process.

Field capacity   The maximum amount of water that the unsaturated zone of a soil can hold against the pull of gravity. The field capacity is dependent on the length of time the soil has been undergoing gravity drainage.

Finite-difference model   A particular kind of a digital computer model based upon a rectangular grid that sets the boundaries of the model and the nodes where the model will be solved.

Finite-element model   A digital ground-water–flow model where the aquifer is divided into a mesh formed of a number of polygonal cells.

Flow net   The set of intersecting equipotential lines and flowlines representing two-dimensional steady flow through porous media.

Flow, steady   The flow that occurs when, at any point in the flow field, the magnitude and direction of the specific discharge are constant in time.

Flow, unsteady   The flow that occurs when, at any point in the flow field, the magnitude or direction of the specific discharge changes with time. Also called transient flow or nonsteady flow.

Force potential   The sum of the kinetic energy, elevation energy, and pressure at a point in an aquifer. It is equal to the hydraulic head times the acceleration of gravity.

Fossil water   Interstitial water that was buried at the same time as the original sediment.

Fracture spring   A spring created by fracturing or jointing of the rock.

Fracture trace   The surface representation of a fracture zone. It may be a characteristic line of vegetation or linear soil-moisture pattern or a topographic sag.

Free energy   A measure of the thermodynamic driving energy of a chemical reaction. Also known as Gibbs free energy or Gibbs function.

Freundlich isotherm   An empirical equation that describes the amount of solute adsorbed onto a soil surface.

Gaining stream   *See* stream, gaining.

Gamma-gamma radiation log   A borehole log in which a source of gamma radiation as well as a detector are lowered into the borehole. This log measures bulk density of the formation and fluids.

Gamma log   *See* natural gamma radiation log.

Gauss-Seidel   A particular type of method for solving for the head in a finite-difference ground-water model.

Geohydrology   A synonym for hydrogeology; also the flow of water through the earth without considering the effects of the geology.

Ghyben-Herzberg principle   An equation that relates the depth of a salt-water interface in a coastal aquifer to the height of the fresh-water table above sea level.

Glacial-lacustrine sediments   Silt and clay deposits formed in the quiet waters of lakes that received meltwater from glaciers.

Glacial outwash   Well-sorted sand, or sand and gravel, deposited by the meltwater from a glacier.

Glacial till   A glacial deposit composed of mostly unsorted sand, silt, clay, and boulders and laid down directly by the melting ice.

Gouge   Soft, ground-up rock formed between the moving surfaces of a geological fault.

Gravity drainage   The downward movement of water in the vadose zone due to gravity.

Gravity potential   A potential due to the position of ground water or soil moisture above a datum.

Ground-penetrating radar   A surface geophysical technique based upon the transmission of repetitive pulses of electromagnetic waves into the ground. Some of the radiated energy is reflected back to the surface and the reflected signal is captured and processed.

Ground water   The water contained in interconnected pores located below the water table in an unconfined aquifer or located in a confined aquifer.

Ground-water basin   A rather vague designation pertaining to a ground-water reservoir that is more or less separate from neighboring ground-water reservoirs. A ground-water basin could be separated from adjacent basins by geologic boundaries or by hydrologic boundaries.

Ground water, confined   The water contained in a confined aquifer. Pore-water pressure is greater than atmospheric at the top of the confined aquifer.

Ground-water divide   The boundary between two adjacent ground-water basins. The divide is represented by a high in the water table.

Ground-water flow   The movement of water through openings in sediment and rock; occurs in the zone of saturation.

Ground water, perched   The water in an isolated, saturated zone located in the zone of aeration. It is the result of the presence of a layer of material of low hydraulic conductivity, called a perching bed. Perched ground water will have a perched water table.

Ground water, unconfined   The water in an aquifer where there is a water table.

Grout curtain   An underground wall designed to stop ground-water flow; can be created by injecting grout into the ground, which subsequently hardens to become impermeable.

Hantush-Jacob formula   An equation to describe the change in hydraulic head with time during pumping of a leaky confined aquifer.

Hazen method   An empirical equation that can be used to approximate the hydraulic conductivity of a sediment on the basis of the effective grain size.

Head, total hydraulic   The sum of the elevation head, the pressure head, and the velocity head at a given point in an aquifer.

Hele-Shaw model   An analog model of ground-water flow based upon the movement of a viscous fluid between two closely spaced, parallel plates.

Heterogeneous   Pertaining to a substance having different characteristics in different locations. A synonym is nonuniform.

Hollow-stem auger   A particular kind of a drilling device whereby a hole is rapidly advanced into sediments. Sampling and installation of the equipment can take place through the hollow center of the auger.

Homogeneous   Pertaining to a substance having identical characteristics everywhere. A synonym is uniform.

Horizontal profiling   A method of making an earth-resistivity survey by measuring the apparent resistivity using the same electrode spacings at different grid points around an area.

Humidity, absolute   The amount of moisture in the air as expressed by the number of grams of water per cubic meter of air.

Humidity, relative   Percent ratio of the absolute humidity to the saturation humidity for an air mass.

Humidity, saturation   The maximum amount of moisture that can be contained by an air mass at a given temperature.

Hvorslev method   A procedure for performing a slug test in a piezometer that partially penetrates a water-table aquifer.

Hydraulic conductivity   A coefficient of proportionality describing the rate at which water can move through a permeable medium. The density and kinematic viscosity of the water must be considered in determining hydraulic conductivity.

Hydraulic diffusivity   A property of an aquifer or confining bed defined as the ratio of the transmissivity to the storativity.

Hydraulic gradient   The change in total head with a change in distance in a given direction. The direction is that which yields a maximum rate of decrease in head.

Hydraulic head   *See* head, total.

Hydrochemical facies   Bodies of water with separate but distinct chemical compositions contained in an aquifer.

Hydrogeology   The study of the interrelationships of geologic materials and processes with water, especially ground water.

Hydrograph   A graph that shows some property of ground water or surface water as a function of time.

Hydrologic cycle   The circulation of water from the oceans through the atmosphere to the land and ultimately back to the ocean.

Hydrologic equation   An expression of the law of mass conservation for purposes of water budgets. It may be stated as inflow equals outflow plus or minus changes in storage.

Hydrology   The study of the occurrence, distribution, and chemistry of all waters of the earth.

Hydrophyte   A type of plant that grows with the root system submerged in standing water.

Image well   An imaginary well that can be used to simulate the effect of a hydrologic barrier, such as a recharge boundary or a barrier boundary, on the hydraulics of a pumping or recharge well.

Infiltration   The flow of water downward from the land surface into and through the upper soil layers.

Infiltration capacity   The maximum rate at which infiltration can occur under specific conditions of soil moisture. For a given soil, the infiltration capacity is a function of the water content.

Interception   The process by which precipitation is captured on the surfaces of vegetation before it reaches the land surface.

Interception loss   Rainfall that evaporates from standing vegetation.

Interflow   The lateral movement of water in the unsaturated zone during and immediately after a precipitation event. The water moving as interflow discharges directly into a stream or lake.

Intermediate zone   That part of the unsaturated zone below the root zone and above the capillary fringe.

Intrinsic permeability   Pertaining to the relative ease with which a porous medium can transmit a liquid under a hydraulic or potential gradient. It is a property of the porous medium and is independent of the nature of the liquid or the potential field.

Ion exchange   A process by which an ion in a mineral lattice is replaced by another ion that was present in an aqueous solution.

Isohyetal line   A line drawn on a map, all points along which receive equal amounts of precipitation.

Isotropy   The condition in which hydraulic properties of the aquifer are equal in all directions.

Jacob straight-line method   A graphical method using semilogarithmic paper and the Theis equation for evaluating the results of a pumping test.

Joint spring   *See* fracture spring.

Juvenile water   Water entering the hydrologic cycle for the first time.

Karst   The type of geologic terrane underlain by carbonate rocks where significant solution of the rock has occurred due to flowing ground water.

Kemmerer sampler   A sampling device that can be lowered either into a deep well or into a lake in order to retrieve a water sample from a particular depth in the well or the lake.

Laminar flow   That type of flow in which the fluid particles follow paths that are smooth, straight, and parallel to the channel walls. In laminar flow, the viscosity of the fluid damps out turbulent motion. *Compare with* turbulent flow.

Land pan   A device used to measure free-water evaporation.

Langmuir adsorption isotherm   An empirical equation that describes the amount of solute adsorbed onto a soil surface.

Laplace equation   The partial differential equation governing steady-state flow of ground water.

Lapse rate, dry adiabatic   The rate that the atmospheric temperature decreases with altitude for a dry air mass.

Lapse rate, wet adiabatic   The rate that the atmospheric temperature decreases with altitude for a moist air mass. The exact rate is a function of the moisture content.

Law of mass action   The law stating that for a reversible chemical reaction the rate of reaction is proportional to the concentrations of the reactants.

Leachate Water that contains a high amount of dissolved solids and is created by liquid seeping from a landfill.

Leachate-collection system A system installed in conjunction with a liner to capture the leachate that may be generated from a landfill so that it may be taken away and treated.

Leaky confining layer A low-permeability layer that can transmit water at sufficient rates to furnish some recharge to a well pumping from an underlying aquifer. Also called aquitard.

Lineament A natural linear surface longer than a mile (1500 m).

Lithologic log A record of the lithology of the rock and soil encountered in a borehole from the surface to the bottom. Also known as a well log.

Losing stream *See* stream, losing.

Lysimeter A field device containing a soil column and vegetation; used for measuring actual evapotranspiration.

Magmatic water Water associated with a magma.

Magnetometer A geophysical device that can be used to locate items that disrupt the earth's localized magnetic field; can be used for finding buried steel.

Manning equation An equation that can be used to compute the average velocity of flow in an open channel.

Maximum contaminant level (MCL) The highest concentration of a solute permissible in a public water supply as specified in the National Interim Primary Drinking Water Standards for the United States.

Maximum contaminant level goal (MCLG) A nonenforceable health goal for solutes in drinking water; set at a level to prevent known or anticipated adverse effects with an adequate margin of safety.

Micrograms per liter A measure of the amount of dissolved solids in a solution in terms of micrograms of solute per liter of solution.

Milliequivalents per liter A measure of the concentration of a solute in solution; obtained by dividing the concentration in milligrams per liter by equivalent weight of the ion.

Milligrams per liter A measure of the amount of dissolved solids in a solute in terms of milligrams of solute per liter of solution.

Model calibration The process by which the independent variables of a digital computer model are varied in order to calibrate a dependent variable such as a head against a known value such as a water-table map.

Model field verification The process by which a digital computer model that has been calibrated and verified is tested to see if it can predict the field response of an aquifer to some transient condition.

Model verification The process by which a digital computer model that has been calibrated against a steady-state condition is tested to see if it can generate a transient response, such as the decline in the water table with pumping, that matches the known history of the aquifer.

Moisture potential The tension on the pore water in the unsaturated zone due to the attraction of the soil-water interface.

Molality A measure of chemical concentration. A 1-molal solution has 1 mol of solute dissolved in 1000 g of water. One mole of a compound is its formula weight in grams.

Molarity A measure of chemical concentration. A 1-molar solution has 1 mol of solute dissolved in one liter of solution.

Mutual-prescription doctrine A ground-water doctrine stating that in the event of an overdraft of a ground-water basin, the available ground water will be apportioned among all the users in amounts proportional to their individual pumping rates.

Natural gamma radiation log A borehole log that measures the natural gamma radiation emitted by the formation rocks. It can be used to delineate subsurface rock types.

Neumann condition The boundary condition for a ground-water–flow model where a flux across the boundary of the flow region is known.

Neutron log A borehole log obtained by lowering a radioactive element, which is a source of neutrons, and a neutron detector into the well. The neutron log measures the amount of water present; hence, the porosity of the formation.

Nonequilibrium equation *See* Theis equation.

Nonequilibrium type curve A plot on logarithmic paper of the well function $W(u)$ as a function of u.

Numerical model A model of ground-water flow in which the aquifer is described by numerical equations, with specified values for boundary conditions, that are solved on a digital computer.

Observation well A nonpumping well used to observe the elevation of the water table or the potentiometric surface. An observation well is generally of larger diameter than a piezometer and typically is screened or slotted throughout the thickness of the aquifer.

Overland flow The flow of water over a land surface due to direct precipitation. Overland flow generally occurs when the precipitation rate exceeds the infiltration capacity of the soil and depression storage is full. Also called Horton overland flow.

Packer test An aquifer test performed in an open borehole; the segment of the borehole to be tested is sealed off from the rest of the borehole by inflating seals, called packers, both above and below the segment.

Pendular water Water that clings to the surfaces of mineral particles in the zone of aeration.

Permafrost Perenially frozen ground, occurring wherever the temperature remains at or below 0°C for two or more years in a row.

Permeameter A laboratory device used to measure the intrinsic permeability and hydraulic conductivity of a soil or rock sample.

Phreatic cave A cave that forms below the water table.

Phreatophyte A type of plant that typically has a high rate of transpiration by virtue of a taproot extending to the water table.

Piezometer A nonpumping well, generally of small diameter, that is used to measure the elevation of the water table or potentiometric surface. A piezometer generally has a short well screen through which water can enter.

Piezometer nest A set of two or more piezometers set close to each other but screened to different depths.

Polar coordinates The means by which the position of a point in a two-dimensional plane is described; based upon the radial distance from the origin to the given point and the angle between a horizontal line passing through the origin and a line extending from the origin to the given point.

Pollutant   Any solute or cause of change in physical properties that renders water unfit for a given use.

Pore space   The volume between mineral grains in a porous medium.

Porosity   The ratio of the volume of void spaces in a rock or sediment to the total volume of the rock or sediment.

Porosity, effective   The volume of the void spaces through which water or other fluids can travel in a rock or sediment divided by the total volume of the rock or sediment.

Porosity, primary   The porosity that represents the original pore openings when a rock or sediment formed.

Porosity, secondary   The porosity that has been caused by fractures or weathering in a rock or sediment after it has been formed.

Potentiometric surface   A surface that represents the level to which water will rise in tightly cased wells. If the head varies significantly with depth in the aquifer, then there may be more than one potentiometric surface. The water table is a particular potentiometric surface for an unconfined aquifer.

Potentiometric surface map   A contour map of the potentiometric surface of a particular hydrogeologic unit.

Prior-appropriation doctrine   A doctrine stating that the right to use water is separate from other property rights and that the first person to withdraw and use the water holds the senior right. The doctrine has been applied to both ground and surface water.

Public trust doctrine   A legal theory holding that certain lands and waters in the public domain are held in trust for use by the entire populace. It is especially applicable to navigable waters.

Pumping cone   The area around a discharging well where the hydraulic head in the aquifer has been lowered by pumping. Also called cone of depression.

Pumping test   A test made by pumping a well for a period of time and observing the change in hydraulic head in the aquifer. A pumping test may be used to determine the capacity of the well and the hydraulic characteristics of the aquifer. Also called aquifer test.

Quantification limit   The lower limit to the range in which the concentration of a solute can be determined by a particular analytical instrument.

Radial flow   The flow of water in an aquifer toward a vertically oriented well.

Rating curve   A graph of the discharge of a river at a particular point as a function of the elevation of the water surface.

Recharge area   An area in which there are downward components of hydraulic head in the aquifer. Infiltration moves downward into the deeper parts of an aquifer in a recharge area.

Recharge basin   A basin or pit excavated to provide a means of allowing water to soak into the ground at rates exceeding those that would occur naturally.

Recharge boundary   An aquifer system boundary that adds water to the aquifer. Streams and lakes are typically recharge boundaries.

Recharge well   A well specifically designed so that water can be pumped into an aquifer in order to recharge the ground-water reservoir.

Recovery   The rate at which the water level in a well rises after the pump has been shut off. It is the inverse of drawdown.

Regolith   The upper part of the earth's surface that has been altered by weathering processes. It includes both soil and weathered bedrock.

Resistivity log   A borehole log made by lowering two current electrodes into the borehole and measuring the resistivity between two additional electrodes. It measures the electrical resistivity of the formation and contained fluids near the probe.

Retardation   A general term for the many processes that act to remove the solutes in ground water; for many solutes the solute front will travel more slowly than the rate of the advecting ground water.

Return flow   A type of overland flow that occurs when throughflow reaches the land surface and drains across the land surface before reaching a stream.

Reverse type curve   A plot on logarithmic paper of the well function $W(u)$ as a function of $1/(u)$.

Reynolds number   A number, defined by an equation, that can be used to determine whether flow will be laminar or turbulent.

Riparian doctrine   A doctrine that holds that the property owner adjacent to a surface-water body has first right to withdraw and use the water.

Rock, igneous   A rock formed by the cooling and crystallization of a molten rock mass called magma.

Rock, metamorphic   A rock formed by the application of heat and pressure to preexisting rocks.

Rock, plutonic   An igneous rock formed when magma cools and crystallizes within the earth.

Rock, sedimentary   A rock formed from sediments through a process known as diagenesis or formed by chemical precipitation in water.

Rock, volcanic   An igneous rock formed when molten rock called lava cools on the earth's surface.

Runoff   The total amount of water flowing in a stream. It includes overland flow, return flow, interflow, and baseflow.

Safe yield   The amount of naturally occurring ground-water that can be economically and legally withdrawn from an aquifer on a sustained basis without impairing the native ground-water quality or creating an undesirable effect such as environmental damage. It cannot exceed the increase in recharge or leakage from adjacent strata plus the reduction in discharge, which is due to the decline in head caused by pumping.

Saline-water encroachment   The movement, as a result of human activity, of saline ground water into an aquifer formerly occupied by fresh water. Passive saline-water encroachment occurs at a slow rate owing to a general lowering of the fresh-water potentiometric surface. Active saline-water encroachment proceeds at a more rapid rate owing to the lowering of the fresh-water potentiometric surface below sea level.

Salt-water encroachment   *See* saline-water encroachment.

Saprolite   A soft, earthy, decomposed rock, typically clay-rich, formed in place by chemical weathering of igneous and metamorphic rocks.

Saturated zone The zone in which the voids in the rock or soil are filled with water at a pressure greater than atmospheric. The water table is the top of the saturated zone in an unconfined aquifer.

Saturation ratio The ratio of the volume of contained water in a soil to the volume of the voids of the soil.

Schlumberger array A particular arrangement of electrodes used to measure surface electrical resistivity.

Screen *See* well screen.

Sediment An assemblage of individual mineral grains that were deposited by some geologic agent such as water, wind, ice, or gravity.

Seepage velocity The rate of movement of fluid particles through porous media along a line from one point to another.

Seismic refraction A method of determining subsurface geophysical properties by measuring the length of time it takes for artificially generated seismic waves to pass through the ground.

Shelby tube A sampling device that is pushed into an unconsolidated aquifer ahead of the drill bit. Typically, the Shelby tube is pushed by hydraulic means.

Single-point resistance log A borehole log made by lowering a single electrode into the well with the other electrode at the ground surface. It measures the overall electrical resistivity of the formation and drilling fluid between the surface and the probe.

Sinkhole spring A spring created by ground water flowing from a sinkhole in karst terrane.

Slug test An aquifer test made either by pouring a small instantaneous charge of water into a well or by withdrawing a slug of water from the well. A synonym for this test, when a slug of water is removed from the well, is a bail-down test.

Slurry wall An underground wall designed to stop ground-water flow; constructed by digging a trench and backfilling it with a slurry rich in bentonite clay.

Soil liquefaction A process that occurs when saturated sediments are shaken by an earthquake. The soil can lose its strength and cause the collapse of structures with foundations in the sediment.

Soil moisture The water contained in the unsaturated zone.

Solubility product The equilibrium constant that describes a solution of a slightly soluble salt in water.

Specific capacity An expression of the productivity of a well, obtained by dividing the rate of discharge of water from the well by the drawdown of the water level in the well. Specific capacity should be described on the basis of the number of hours of pumping prior to the time the drawdown measurement is made. It will generally decrease with time as the drawdown increases.

Specific discharge An apparent velocity calculated from Darcy's law; represents the flow rate at which water would flow in an aquifer if the aquifer were an open conduit.

Specific retention The ratio of the volume of water the rock or sediment will retain against the pull of gravity to the total volume of the rock or sediment.

Specific weight The weight of a substance per unit volume. The units are newtons per cubic meter.

Specific yield The ratio of the volume of water a rock or soil will yield by gravity drainage to the volume of the rock or soil. Gravity drainage may take many months to occur.

Spiked sample   A water sample to which a known quantity of a solute has been added so that the accuracy of the laboratory in analyzing the sample can be determined.

Split-spoon sample   A sample of unconsolidated material taken by driving a sampling device ahead of the drill bit in a boring. The split-spoon sampler is typically advanced by the repetitive dropping of a weight.

Spontaneous potential log   A borehole log made by measuring the natural electrical potential that develops between the formation and the borehole fluids.

Stagnation point   A place in a ground-water flow field at which the ground water is not moving. The magnitude of vectors of hydraulic head at the point are equal but opposite in direction.

Stem flow   The process by which rainwater drips and flows down the stems and branches of plants.

Stiff pattern   A graphical means of presenting the chemical analysis of the major cations and anions of a water sample.

Storage, specific   The amount of water released from or taken into storage per unit volume of a porous medium per unit change in head.

Storativity   The volume of water an aquifer releases from or takes into storage per unit surface area of the aquifer per unit change in head. It is equal to the product of specific storage and aquifer thickness. In an unconfined aquifer, the storativity is equivalent to the specific yield. Also called storage coefficient.

Storm hydrograph   A graph of the discharge of a stream over the time period when, in addition to direct precipitation, overland flow, interflow, and return flow are adding to the flow of the stream. The storm hydrograph will peak owing to the addition of these flow elements.

Stream, gaining   A stream or reach of a stream, the flow of which is being increased by inflow of ground water. Also known as an effluent stream.

Stream, losing   A stream or reach of a stream that is losing water by seepage into the ground. Also known as an influent stream.

Successive overrelaxation method   A particular type of method for solving for the head in a finite-difference ground-water model.

Suction lysimeter   A device for withdrawing pore water samples from the unsaturated zone by applying tension to a porous ceramic cup.

Surface water   Water found in ponds, lakes, inland seas, streams, and rivers.

Swallow hole   A vertical shaft in a karst terrane leading from a surface stream into an underground cavern.

Temperature inversion   A situation when a layer of warm air overlies a layer of cold air.

Tensiometer   A device used to measure the soil-moisture tension in the unsaturated zone.

Tension   The condition under which pore water exists at a pressure less than atmospheric.

Theis equation   An equation for the flow of ground water in a fully confined aquifer.

Theis type curve   *See* reverse type curve.

Thiessen method   A process used to determine the effective uniform depth of precipitation over a drainage basin with a nonuniform distribution of rain gages.

Throughflow   The lateral movement of water in an unsaturated zone during and immediately after a precipitation event. The water from throughflow seeps out at the base of slopes and then flows across the ground surface as return flow, ultimately reaching a stream or lake.

Time of concentration   The time it takes for water to flow from the most distant part of the drainage basin to the measuring point. .

Topographic divide   The boundary between adjacent surface water boundaries. It is represented by a topographically high area.

Tortuosity   The actual length of a ground-water–flow path, which is sinuous in form, divided by the straight-line distance between the ends of the flow path.

Transmissivity   The rate at which water of a prevailing density and viscosity is transmitted through a unit width of an aquifer or confining bed under a unit hydraulic gradient. It is a function of properties of the liquid, the porous media, and the thickness of the porous media.

Transpiration   The process by which plants give off water vapor through their leaves.

Turbulent flow   That type of flow in which the fluid particles move along very irregular paths. Momentum can be exchanged between one portion of the fluid and another. *Compare with* Laminar flow.

Uniformity coefficient   The ratio of the grain size that is 60% finer by weight to the grain size that is 10% finer by weight on the grain-size distribution curve. It is a measure of how well or poorly sorted sediment is.

Unsaturated zone   The zone between the land surface and the water table. It includes the root zone, intermediate zone, and capillary fringe. The pore spaces contain water at less than atmospheric pressure, as well as air and other gases. Saturated bodies, such as perched ground water, may exist in the unsaturated zone. Also called zone of aeration and vadose zone.

Vadose cave   A cave that occurs above the water table.

Vadose water   Water in the zone of aeration.

Vadose zone   *See* unsaturated zone.

Viscosity   The property of a fluid describing its resistance to flow. Units of viscosity are newton-seconds per meter squared or pascal-seconds. Viscosity is also known as dynamic viscosity.

Water budget   An evaluation of all the sources of supply and the corresponding discharges with respect to an aquifer or a drainage basin.

Water content   The weight of contained water in a soil divided by the total weight of the soil mass.

Water equivalent   The depth of water obtained by melting a given thickness of snow.

Water quality criteria   Values for dissolved substances in water based upon their toxicological and ecological impacts.

Water table   The surface in an unconfined aquifer or confining bed at which the pore water pressure is atmospheric. It can be measured by installing shallow wells extending a few feet into the zone of saturation and then measuring the water level in those wells.

Water-table cave   A cave that forms at the approximate position of the water table.

Water-table map   A specific type of potentiometric-surface map for an unconfined aquifer; shows lines of equal elevation of the water table.

Weir   A device placed across a stream and used to measure the discharge by having the water flow over a specifically designed spillway.

Well casing   A solid piece of pipe, typically steel or PVC plastic, used to keep a well open in either unconsolidated materials or unstable rock.

Well development   The process whereby a well is pumped or surged to remove any fine material that may be blocking the well screen or the aquifer outside the well screen.

Well, fully penetrating   A well drilled to the bottom of an aquifer, constructed in such a way that it withdraws water from the entire thickness of the aquifer.

Well function   An infinite-series term that appears in the Theis equation of ground-water flow.

Well log   *See* lithologic log.

Well, partially penetrating   A well constructed in such a way that it draws water directly from a fractional part of the total thickness of the aquifer. The fractional part may be located at the top or the bottom or anywhere in between in the aquifer.

Well screen   A tubular device with either slots, holes, gauze, or continuous-wire wrap; used at the end of a well casing to complete a well. The water enters the well through the well screen.

Wenner array   A particular arrangement of electrodes used to measure surface electrical resistivity.

Wilting point   The soil-moisture content below which plants are unable to withdraw soil moisture.

Winters Doctrine   A United States doctrine holding that when Indian reservations were established, the federal government also reserved the water rights necessary to make the land productive.

Xerophyte   A desert plant capable of existing by virtue of a shallow and extensive root system in an area of minimal water.

Zone of aeration   *See* unsaturated zone.

Zone of saturation   *See* saturated zone.

Answers to Selected Problems

Chapter 1 Answers

1. 12 wk
3. 34,000 ft^3
5. (a) 0.75 in^3
   (b) 2.85 × 10^6 ft^3
   (c) 8.07 × 10^5 m^3
7. 3400 g
9. 5.8°C
11. 19.6°C

Chapter 2 Answers

1. 170 m^3
3. 1.38 in.
5. 0.13 in.
7. (a) 7.5°C
   (b) 46%

Chapter 3 Answers

1. (a) 1955 c.f.s., 1280 c.f.s.
   (b) 720 c.f.s.
3. 2.95 days
5. 3.32 ft^3/s
7. (a) 135 ft^3/s
   (b) This problem can't be solved directly as the unknown, the depth, occurs in two different places in the equation. An iterative method of solution must be used. One must make an initial guess of the depth and compute the resulting discharge. The answer is compared with 135 ft^3/s. If it is less, the depth is adjusted and if it is more, the depth is reduced. The flow is then recalculated on a trial and error basis until a flow of 135 c.f.s. is obtained. Start with an initial guess somewhere between 0.5 and 5 ft.
9. 3700 ft^3/s

Chapter 4 Answers

1. weight = 314.6 N
3. (a) 168 kg/m^3
   (b) 1640 N/m^3
   (c) less dense
5. K_i = 1.24 × 10^{10} cm^2
7. (a) 6.0 × 10^{-3} cm/s or 17 ft/day
   (b) 7.0 × 10^{-8} cm^2
9. 2.5 × 10^9 ft^3
11. 6 × 10^{-10} m^2/N
13. K_n = 26 ft^2/day
    K_v = 12 ft^2/day
15. d_{10} = 0.16 mm
    C_u = 2.0

Chapter 5 Answers

1. (a) 42 m^2/s
   (b) 4.2 m
3. (a) 296.8 m
   (b) 2.29 × 10^5 N/m^2

5. 15.70 m

7. Yes. The computed Reynolds number is 0.3, which is less than 10.

9. (a) 1.4 ft³/day
   (b) 0.084 ft/day

11. (a) 151 ft³/day
    (b) 22.4 ft

13. 80 ft³/day

Chapter 6 Answers

1. (a) 19%
   (b) 0.10
   (c) 0.64
   (d) 29%
   (e) 22%
   (f) 6.6%

(g) 1.88 g/cc

(h) 28.95%

Chapter 7 Answers

1.

Distance (ft)	Drawdown (ft)
50	54.9
150	45.5
250	41.1
500	35.2
1,000	29.2
3,000	19.8
6,000	13.9
10,000	9.7

3. (a) Figure A
   (b) 72%

5.

50 ft	19.2 ft	1000 ft	0.60 ft
150 ft	10.2 ft	3000 ft	0.003 ft
250 ft	6.6 ft	6000 ft	nil
500 ft	3.2 ft	10000 ft	nil

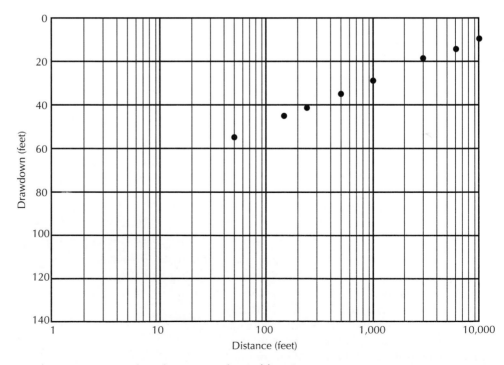

FIGURE A   Distance-drawdown curve for problem 3.

7.

Time	Drawdown	Time	Drawdown	Time	Drawdown
1 min	0.34 ft	30 min	10.1 ft	1 day	26.5 ft
2 min	1.3 ft	1 hr	13.0 ft	5 days	33.4 ft
5 min	3.5 ft	2 hr	15.9 ft	10 days	36.4 ft
10 min	5.9 ft	5 hr	19.8 ft	20 days	39.3 ft
15 min	7.4 ft	12 hr	23.5 ft	30 days	41.1 ft

9. Figure B

11. (a) $T = 5700$ ft^2/day
$S = 5 \times 10^{-5}$
(b) $T = 5400$ ft^2/day
$S = 5 \times 10^{-5}$

13. $T = 1.6 \times 10^{-1}$ cm^2/s
$S = 10^{-2}$
$K = 7 \times 10^{-4}$ cm/s

15. $T = 409$ ft^2/day

Chapter 9 Answers

1. 290 ft

3. (a) 4.91 m
(b) 0.115 m
(c) 0.115 m
(d) 0.0575 m

5. 105 ft

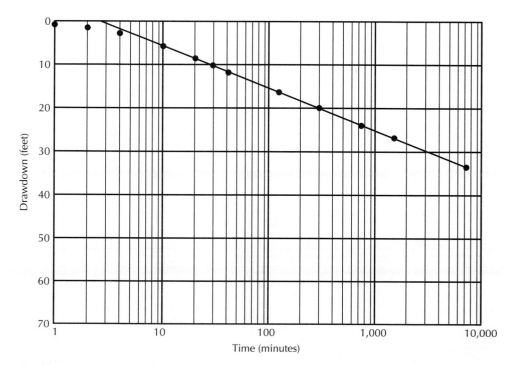

FIGURE B

TABLE A

Ion	Molality	Activity Coefficient	Activity	meq/L
Ca^{2+}	3.57×10^{-3}	0.606	2.16×10^{-3}	7.14
Mg^{2+}	1.44×10^{-3}	0.626	9.01×10^{-4}	2.88
Na^+	6.09×10^{-4}	0.874	5.32×10^{-4}	0.61
HCO_3^-	5.20×10^{-3}	0.874	4.54×10^{-3}	5.20
SO_4^{2-}	2.64×10^{-3}	0.584	1.54×10^{-3}	5.28
Cl^-	1.13×10^{-4}	0.870	9.80×10^{-5}	0.11

Chapter 10 Answers

1. 7.4557 g

3. 1.122×10^{-3} mol

5. (a) 1.992×10^{-4} mol
   (b) 3.16×10^{-5} mol

7. (b) Ionic strength = 0.0183
   (a)/(c)/(d)/(e) See Table A
   (f) cations = 10.63 anions = 10.89
   (g) K_{iap} of anhydrite = $10^{-5.475}$
   (h) $10^{-5.478}/10^{-4.5}$ = .105; 10.5% saturated with anhydrite.
   (i) K_{iap} of calcite = $10^{-7.63}$
   (j) $10^{-7.63}/10^{-8.35}$ = 5.23; 523% saturated with calcite.

9. $[H^+] = 10^{-8.7}$ $[OH^-] = 10^{-5.3}$

11. pH = 4.48

13. pH = 2.57

Chapter 11 Answers

1. 70 mg/L

3. 690 mg/L

5. 0.0041 v_x

7. (a) ± 120 ft
   (b) -37 ft

Chapter 13 Answers

1. (a) 5000 ft/s
   (b) 18,000 ft/s
   (c) 16°
   (d) 56 ft

References

ACTON V. BLUNDELL. 1843. 12 M&W. 324, 354.

AHMAD, M. U. 1974. Coal mining and its effect on water quality. *American Water Resources Proceedings,* no. 18; 138–48.

AMBROGGI, R. P. 1977. Underground reservoirs to control the water cycle. *Scientific American* 236, no. 5:21–27.

AMERICAN SOCIETY OF CIVIL ENGINEERS. 1961. *Groundwater basin management.* Manual of Engineering Practices 40.

AMERICAN WATER WORKS ASSOCIATION. 1986. Millions to plaintiffs in groundwater contamination suits. *Mainstream* 30, no. 11 (November):12.

ANDERSEN, R. L., & N. I. WENGERT. 1977. Developing competition for water in the urbanizing areas of Colorado. *Water Resources Bulletin* 13:769–73.

ANDERSON, M. P. 1976. Unsteady groundwater flow beneath strip oceanic islands. *Water Resources Research* 12:640–44.

———. 1979. "Using models to simulate the movement of contaminants through groundwater flow systems. *Critical Reviews in Environmental Control* 9, no. 2:97–156.

———. 1983. Ground-water modeling—the emperor has no clothes. *Ground Water* 21, no. 6:666–69.

———. 1984. Movement of contaminants in groundwater: Groundwater transport—Advection and dispersion. In *Groundwater contamination,* 37–45. Washington, DC: National Academy Press.

———. 1986. Field verification of ground water models. In *Evaluation of pesticides in ground water,* ed. W. Y. Garner et al., 396–412. American Chemical Society Symposium Series 315.

ANDERSON, M. P., & C. A. BERKEBILE. 1977 Hydrogeology of the South Fork of Long Island, New York: Discussion. *Bulletin, Geological Society of America* 88:895.

ANDERSON, M. P., & W. W. WOESSNER. 1992. *Applied groundwater modeling—Simulation of flow and advective transport.* San Diego: Academic Press, Inc.

ANDERSON, T. W. 1972. *Electrical-analog analysis of the hydrologic system, Tucson basin, southeastern Arizona.* U.S. Geological Survey Water-Supply Paper 1939-C.

ANDREWS, C. B. 1978. The impact of the use of heat pumps on ground-water temperatures. *Ground Water* 16:437–43.

ARCHIE, G. E. 1950. Classification of carbonate reservoir rocks and petro-physical considerations. *American Association of Petroleum Geologists Bulletin* 36:943–61.

ATWOOD, SAM. 1990. Region hunts for water to quench its thirst. *San Bernardino* (California) *Sun,* April 29.

AYRES, J. E., D. C. LAGER, & M. J. BARVENIK. 1983. The first EPA superfund cutoff wall: Design and specifications. *Proceedings, Third National Symposium on Aquifer Restoration and Ground Water Monitoring.* National Water Well Association, pp. 13–22.

BACK, W. 1960. Origin of hydrochemical facies in groundwater in the Atlantic coastal plain. *Proceedings, International Geological Congress* (Copenhagen), I:87–95.

BACK, W. 1966. *Hydrochemical facies and groundwater flow patterns in northern part of Atlantic coastal plain.* U.S. Geological Survey Professional Paper 498-A.

BACK, W., & I. BARNES. 1965. *Relation of electrochemical potentials and iron content to groundwater flow patterns.* U.S. Geological Survey Professional Paper 498-C.

BACK, W., R. N. CHERRY, & B. B. HANSHAW. 1966. Chemical equilibrium between water and minerals of a carbonate aquifer. *National Speleological Society Bulletin* 28:119–26.

BACK, W., & B. B. HANSHAW. 1965. Chemical geo-hydrology. In *Advances in hydroscience,* vol. 2, ed. V. T. Chow, 49–109. New York: Academic Press.

———. 1970. Comparison of the chemical hydrogeology of the carbonate peninsulas of Florida and Yucatan. *Journal of Hydrology* 10:330–68.

BAIR, E. S., & S. E. NORRIS. 1990. Field reports— Expert testimony for the plaintiffs in the case that brought Ohio ground-water law into the 20th century. *Ground Water* 28, no. 5:767–774.

BAKER, C. H., JR., & D. G. FOULK. 1980. *WATSTORE user's guide,* vol. 2, Ground-Water File, U.S. Geological Survey Open-File Report 75–589.

BALDWIN, A. D., JR., & J. MILLER. 1979. Use of a gamma logger to delineate glacial and bedrock stratigraphy in southwestern Ohio. *Ground Water* 17, no. 4:385–90.

BALLARD, R., & J. G. A. FISKELL. 1974. Phosphorus retention in coastal plain soils: I. Relation to soil properties. *Soil Science Society of America, Proceedings* 38:250–55.

BANKS, DAVID. 1992. Optimal orientation of water supply boreholes in fractured rock. *Ground Water* 30, no. 6:895–900.

BANKS, H. O. 1953. Utilization of underground storage reservoirs. *Transactions, American Society of Civil Engineers* 118:220–34.

BARCELONA, M. J., ET AL. 1984. A laboratory evaluation of ground water sampling mechanisms. *Ground Water Monitoring Review* 4, no. 2:32–41.

———. 1985. *Practical guide for ground-water monitoring.* Illinois State Water Survey Contract Report 374.

BARCELONA, M. J., J. P. GIBB, & R. A. MILLER, 1983. *A guide to the selection of materials for monitoring well construction and ground water sampling.* Illinois State Water Survey Contract Report 327, EPA-600/52-84-024.

BARKER, J. F., G. C. PATRICK, & D. MAJOR. 1987. Natural attenuation of aromatic hydrocarbons in a shallow sand aquifer. *Ground Water Monitoring Review* 7, no. 1:64–71.

BARNES, J. C. & C. J. BOWLEY. 1968. Snow cover distribution as mapped from satellite photography. *Water Resources Research* 4:257–72.

BARNETT, P. M. 1984. Mixing water and the commerce clause: The problems of practice, precedent and policy in *Sporhase v. Nebraska. Natural Resources Journal* 24:161–94.

BARNS, H. H. 1967. *Roughness characteristics of natural channels.* U.S. Geological Survey Water-Supply Paper 1849.

BARRIO-LAGE, G., F. Z. PARSONS, & R. S. NASSAR. 1987. Kinetics of the depletion of trichloroethene. *Environmental Science and Technology* 21, no. 4:366–370.

BARRIO-LAGE, G. F., F. Z. PARSONS, R. S. NASSAR, & P. A. LORENZO. 1986. Sequential dehalogenation of chlorinated ethenes. *Environmental Science and Technology* 20, no. 1:96–99.

———. 1987. Biotransformation of trichloroethene in a variety of subsurface materials. *Environmental Toxicology and Chemistry* 6:571–578.

BARROW, N. J., & T. C. SHAW. 1975a. The slow reactions between soil and anions: 2. Effect of time and temperature on the decrease in phosphate concentration in the soil solution. *Soil Science* 119:167–77.

———. 1975b. The slow reactions between soil and anions: 3. The effects of time and temperature on the decrease in isotopically exchangeable phosphate. *Soil Science* 119:190–97.

BARTHOLIC, J. F., J. R. RUNKELS, & E. B. STENMARK. 1967. Effects of a monolayer on reservoir temperature and evaporation. *Water Resources Research* 3:173–80.

BASS BECKING, L. G. M., I. R. KAPLAN, & D. MOORE. 1960. Limits of the natural environment in terms of pH and oxidation-reduction potential. *Journal of Geology* 68:243–84.

BASSET V. SALISBURY MANUFACTURING CO., INC. 1862. 43 N.H. 569, 82 Am. Dec. 179.

BAYDON-GHYBEN, W. 1888–1889. *Nota in verband met de voorgenomen putboring nabij Amsterdam.* Koninklyk Instituut Ingenieurs Tijdschrift (The Hague), pp. 8–22.

BAYS, C. A. 1950. Prospecting for ground water—Geophysical methods. *Journal of American Water Works Association* 42:947–56.

BEATTY, M. T., & J. BOUMA. 1973. Application of soil surveys to selection of sites for on-site disposal of liquid household wastes. *Geoderma* 10:113–22.

BEETON, A. M. 1965. Eutrophication of the St. Lawrence Great Lakes. *Limnology and Oceanography* 10:240–54.

BEETON, A. M. 1966. *Indices of Great Lakes eutrophication.* Great Lakes Research Division Publication No. 15, University of Michigan.

BEHNKE, J. J. 1969. Clogging in surface spreading operations for artificial ground-water recharge. *Water Resources Research* 5:870–76.

BENTLEY, H. W., & S. N. DAVIS. 1980. Feasibility of ^{36}Cl dating of very old ground water, *EOS. American Geophysical Union Transactions* 61:230.

BERES, MILAN, JR., & F. P. HAENI. 1991. Applications of ground-penetrating-radar methods in hydrogeologic studies. *Ground Water* 29, no. 3:375–386.

BERNER, R. A. 1971. *Principles of chemical sedimentology*. New York: McGraw-Hill.

BERRY, M. P. 1974. The importance of perceptions in the determination of Indian water rights. *Water Resources Bulletin* 10:137–43.

BERUCH, J. C., & R. A. STREET, 1967. Two-dimensional dispersion. *Journal, Sanitary Engineering Division, American Society of Civil Engineers* 93, SA6:17–39.

BIANCHI, W. C., & HASKELL, E. E., JR. 1968. Field observations compared with Dupuit-Forcheimer theory for mound heights under a recharge basin. *Water Resources Research* 4:1049–57.

BLAIN, L. M., & H. S. EVANS. 1990. In-stream flows for fish protection: supremacy of federal authority pursuant to the Federal Power Act. *Water Resources Bulletin* 26, no. 6:1023–4.

BLANEY, H. F. 1956. Discussion of paper by H. L. Penman, ''Estimating evaporation.'' *Transactions, American Geophysical Union* 37:46–48.

———. 1959. Monthly consumptive use requirements of irrigated crops. *Proceedings, American Society of Civil Engineers, Irrigation and Drainage Division* 85(IR3): 1–12.

BOATWRIGHT, B. A., & D. W. AILMAN. 1975. The occurrence and development of Guest sink, Hernando County, Florida. *Ground Water* 13, no. 4:372–75.

BORN, S. M., S. A. SMITH, & D. A. STEPHENSON. 1979. The hydrogeologic regime of glacial-terrain lakes, with management and planning applications. *Journal of Hydrology* 43 no. 1/4:7–44 (Special issue: *Contemporary Hydrogeology—The George Maxey Memorial Volume*, ed. William Back & D. A. Stephenson.)

BOULTON, N. S. 1954. The drawdown of the water table under non-steady conditions near a pumped well in an unconfined formation. *Proceedings of the Institute of Civil Engineers* (London) 3, no. 3:564–79.

———. 1955. *Unsteady radial flow to a pumped well allowing for delayed yield from storage*. International Association of Scientific Hydrology Publication 37, pp. 472–77.

———. 1963. Analysis of data from non-equilibrium pumping test allowing for delayed yield from storage. *Proceedings of the Institute of Civil Engineers* (London) 26:269–82.

———. 1973. The influence of the delayed drainage on data from pumping tests in unconfined aquifers. *Journal of Hydrology* 19:157–69.

BOULTON, N. S., & J. M. A. PONTIN. 1971. An extended theory of delayed yield from storage applied to pumping tests in unconfined anisotropic aquifers. *Journal of Hydrology* 14:53–65.

BOULTON, N. S., & T. D. STRELTSOVA. 1975. New equations for determining the formation constants of an aquifer from pumping test data. *Water Resources Research* 11:148–53.

BOUSSINESQ, J. 1904. Recherches théoriques sur l'écoulement des nappes d'eau infiltrées dan le sol et sur le débit des sources. *Journal de Mathématiques Pures et Appliquées* 10:5–78.

BOUWER, H. 1977. Land subsidence and cracking due to ground-water depletion. *Ground Water* 15:358–64.

———. 1989. The Bouwer and Rice slug test—an update. *Ground Water* 27, no. 3:304–309.

BOWER, C. A., G. OGATA, & J. M. TUCKER. 1968. Sodium hazard of irrigation waters as influenced by leaching fraction and by precipitation on solution of calcium carbonate. *Soil Science* 106:29–34.

BOUWER, H., & R. C. RICE. 1976. A slug test for determining hydraulic conductivity of unconfined aquifers with completely or partially penetrating wells. *Water Resources Research* 12:423–28.

BOWMAN, J. A., & G. R. CLARK. 1989. Transitions in midwestern ground water. *Water Resources Bulletin* 25, no. 2:413–20.

BRACE, W. 1980. Permeability of crystalline and argillaceous rocks. *International Journal of Rock Mechanics and Geomechanical Abstracts* 17:241–51.

BRACE, W. F., B. W. PAULDING, JR., & C. SCHOLZ. 1966. Dilatancy in the fracture of crystalline rock. *Journal of Geophysical Research* 71:3939–53.

BRADBURY, K. R., & E. R. ROTHSCHILD. 1985. A computerized technique for estimating hydraulic conductivity of aquifers from specific capacity data. *Ground Water* 23, no. 2:240–45.

BRADBURY, K. R., & R. W. TAYLOR. 1984. Determination of the hydrogeologic properties of lakebeds using offshore geophysical surveys. *Ground Water* 22:690–95.

BRANDON, L. V. 1965. Evidences of ground-water flow in permafrost regions. *Proceedings of the 1963 International Conference on Permafrost*, Lafayette, Indiana, pp. 176–77.

BREDEHOEFT, J. D., & G. F. PINDER. 1973. Mass transport in flowing groundwater. *Water Resources Research* 9:194–210.

BREDEHOEFT, J. D., & R. A. YOUNG. 1970. The temporal allocation of ground water—A simulation approach. *Water Resources Research* 6:3–21.

BRENOEL, M., & R. A. BROWN. 1985. Remediation of a leaking underground storage tank and enhanced bioreclamation. *Proceedings, Fifth National Symposium on Aquifer Restoration and Ground Water Monitoring.* National Water Well Association, pp. 527–37.

BROOK, G. A. 1977. Surface and groundwater hydrogeology of a highly karsted sub-Arctic carbonate terrain in northern Canada. In *Karst hydrogeology,* ed. J. S. Tolson and F. L. Doyle, 99–108. Huntsville, Ala.: UAH Press.

BROOKSHIRE, D. S., G. L. WATTS, & J. L. MERRILL. 1985. Current issues in the quantification of federal reserved water rights. *Water Resources Research* 21, no. 11:1777–84.

BROWN, D. L. 1971. Techniques for quality-of-water interpretations from calibrated geophysical logs, Atlantic coastal areas. *Ground Water* 9, no. 4:25–38.

BROWN, J. C. 1979. Definition of the basalt groundwater flow system, Horse Heaven Plateau, Washington. Geological Society of America, *Abstracts with Programs* 11, no. 7:395.

BROWN, J. H., & A. C. BARKER. 1970. An analysis of throughflow and stemflow in mixed oak stands. *Water Resources Research* 6:316–23.

BROWN, R. F. 1966. *Hydrology of the cavernous limestones of the Mammouth Cave area, Kentucky.* U.S. Geological Survey Water-Supply Paper 1837. 64.

BRUCKER, R. W., J. W. HESS, & W. B. WHITE. 1972. Role of vertical shafts in the movement of groundwater in carbonate aquifers. *Ground Water* 10, no. 6:5–13.

BRUINGTON, A. E. 1972. Saltwater intrusion into aquifers. *Water Resources Bulletin* 8:150–60.

BRUINGTON, A. E., & F. D. SEARES. 1965. Operating a sea water barrier project. *Journal, Irrigation and Drainage Division, American Society of Civil Engineers* 91:117–40.

BRUNSING, T. P., & J. CLEARY. 1983. Isolation of contaminated ground water by slurry-induced ground displacement. *Proceedings, Third National Symposium on Aquifer Restoration and Ground Water Monitoring.* National Water Well Association, pp. 28–36.

BRUTSAERT, W. F., & T. G. GEBHARD, JR. 1975. Conjunctive availability of surface and groundwater in the Albuquerque area, New Mexico: A modeling approach. *Ground Water* 13:345–53.

BRUTSAERT, W., G. W. GROSS, & R. M. MCGEHEE. 1975. C. E. Jacob's study on the prospective and hypothetical future of the mining of the ground water deposited under the southern high plains of Texas and New Mexico. *Ground Water* 13:492–505.

BRYAN, K. 1919. Classification of springs. *Journal of Geology* 27:522–61.

BUSH, P. W., & R. H. JOHNSON. 1986. Floridan regional aquifer—study system. In *Regional Aquifer-System Analysis Program of the United States Geological Survey, Summary of Projects, 1978–84,* 17–29. U.S. Geological Survey Circular 1002.

BUSH, P. W., J. A. MILLER, & M. L. MASLIA. 1986. Floridan regional aquifer system, phase II study. In *Regional Aquifer-System Analysis Program of the U.S. Geological Survey, Summary of Projects, 1978–84,* 248–54. U.S. Geological Survey Circular 1002.

BUTLER, J. J., JR. 1990. The role of pumping test in site characterization: some theoretical considerations. *Ground Water* 28 no. 3:394–402.

CALIFORNIA V. FEDERAL ENERGY REGULATORY COMMISSION. 1990. 110 S.Ct. 2024.

CALIFORNIA WATER RESOURCES CONTROL BOARD. 1984. Alleged Waste and Unreasonable Use of Water by Imperial Irrigation District, Decision 1600.

CAMPBELL, P., R. C. BOST, & R. W. JACOBSEN. 1984. Subsurface organic recovery and contaminant migration simulation. *Proceedings, Fourth National Symposium on Aquifer Rstoration and Ground Water Monitoring.* National Water Well Association, pp. 82–91.

CANTER, L. W. 1982. Overview of aquifer restoration. *Proceedings, Second National Symposium on Aquifer Restoration and Ground Water Monitoring.* National Water Well Association, pp. vii–x.

CANTER, L. W., & R. C. KNOX. 1985. *Ground water pollution control.* Chelsea, Mich.: Lewis Publishers, Inc.

CAPPAERT V. UNITED STATES. 1976. 426 US 128.

CARNEY, MICHAEL. 1991. European drinking water standards. *Journal, American Water Works Association* 83, no. 6:48–55.

CARPENTER, P. J., R. S. KAUFMANN, & BETH PRICE. 1990. Use of resistivity soundings to determine landfill structure. *Ground Water* 28, no. 4:569–575.

CARR, J. R. 1976. Tacoma's well field might be world's most productive. *Johnson Drillers Journal* (September and October):1–4.

CASAGRANDE, A. 1940. Seepage through dams. *Journal of Boston Society of Civil Engineers*, pp. 295–337.

CEDERGREN, H. R. 1989. *Seepage, drainage and flow nets*. New York: Wiley-Interscience.

CEDERSTROM, D. J. 1963. *Ground-water resources of the Fairbanks area, Alaska*. U.S. Geological Survey Water-Supply Paper 1590.

———. 1971. Ground water in the Aden sector of Southern Arabia. *Ground Water* 9, no. 2:29–34.

CEDERSTROM, D. J., P. M. JOHNSON, & S. SUBITZKY. 1953. *Occurrence and development of ground water in permafrost regions*. U.S. Geological Survey Circular 275.

CH2M-HILL, INC. 1985. *Remedial investigation, Seymour Recycling Corporation*. Report to U.S. Environmental Protection Agency.

CH2M-HILL, INC. 1986. *Feasibility study report, Seymour Recycling Corporation hazardous waste site*. Report to U.S. Environmental Protection Agency.

CHANGNON, D., T. B. MCKEE, & N. J. DOESKEN. 1991. Hydroclimate variability in the Rocky Mountains. *Water Resources Bulletin* 27:733–743.

CHERRY, J. A., R. W. GILLHAM, & J. F. BARKER. 1984. Contaminants in groundwater: Chemical processes. In *Groundwater contamination*, 46–63. Washington, D.C.: National Academy Press.

CHERRY, J. A., & P. E. JOHNSON. 1982. A multilevel device for monitoring in fractured rock. *Ground Water Monitoring Review* 2, no. 3:41–44.

CHILDS, E. C. 1967. Soil moisture theory. In *Advances in hydroscience*, vol. 4, ed. V. T. Chow, 73–117. New York: Academic Press.

CHORLEY, R. J. 1978. The hillslope hydrologic cycle. In *Hillslope hydrology*, ed. M. J. Kirkby, 1–42. Chichester, Sussex, England: John Wiley.

CITY OF LOS ANGELES V. CITY OF SAN FERNANDO. 1975. 14 Calif. 3d 199.

CITY OF LOS ANGELES V. GLENDALE. 1943. 23 C2d 68.

CLINE V. AMERICAN AGGREGATES. 1984. 15 Ohio St. 3rd, 474, N.E.2d.

CLINE, P. V., & D. R. VISTE. 1984. Migration and degradation patterns of volatile organic compounds. *Proceedings, Seventh Annual Madison Waste Conference*, University of Wisconsin, Madison, Wisconsin, pp. 14–29.

———. 1985. Migration and degradation patterns of volatile organic compounds. *Waste Management & Research* 3:351–360.

COATES, M. S., C. B. HAIMSON, W. J. HINZE, & W. R. VAN SCHMUSS. 1983. Introduction to the Illinois deep hole project. *Journal of Geophysical Research* 88, B9.

COHEN, P. 1965. *Water resources of the Humboldt River Valley near Winnemucca, Nevada*. U.S. Geological Survey Water-Supply Paper 1975.

COLLINS, M. A. 1972. Ground-surface water interactions in the Long Island aquifer system. *Water Resources Bulletin* 6:1253–58.

———. 1978. Discussion. *Water Resources Bulletin* 14:484–85.

COLLINS, M. A., & L. W. GELHAR. 1971. Seawater intrusion in layered aquifers. *Water Resources Research* 7:971–79.

COLLINS, M. A., L. W. GELHAR, & J. L. WILSON, III. 1972. Hele-Shaw model of Long Island aquifer system. *Journal of Hydraulics Division, Proceedings of the American Society of Civil Engineers* 98, HY9:1701–14.

COLLINS, R. B. 1986. Indian reservation water rights. *Journal, American Water Works Association* 78, no. 10:48–54.

CONKLING, H. 1946. Utilization of ground-water storage in stream system development. *Transactions, American Society of Civil Engineers* 3:275–305.

CONRAD, S. H., J. L. WILSON, W. R. MASON, & W. J. PEPLINSKI. 1992. Visualization of residual organic liquid trapped in aquifers. *Water Resources Research* 28, no. 2:467–78.

COOPER, H. H., JR. 1959. A hypothesis concerning the dynamic balance of fresh water and salt water in a coastal aquifer. *Journal of Geophysical Research* 64:461–67.

———. 1966. The equation of groundwater flow in fixed and deforming coordinates. *Journal of Geophysical Research* 71:4785–90.

COOPER, H. H., JR., J. D. BREDEHOEFT, & I. S. PAPADOPULOS. 1967. Response to a finite diameter well to an instantaneous charge of water. *Water Resources Research* 3:263–69.

COOPER, H. H., JR., & C. E. JACOB. 1946. A generalized graphical method for evaluating formation constants and summarizing well-field history. *Transactions, American Geophysical Union* 27:526–34.

CORBETT, D. M., ET AL. 1945. *Stream-gauging procedure*. U.S. Geological Survey Water-Supply Paper 888, pp. 43–51.

COUNTS, H. B., & E. DONSKY. 1963. *Salt-water encroachment, geology and ground water resources of* the Savannah area, Georgia and South Carolina. U.S. Geological Survey Water-Supply Paper 1611.

COURT, A. 1960. Reliability of hourly precipitation data. *Journal of Geophysical Research* 65:4017–24.

COUTRE, R. A., M. G. STEITZ, & M. J. STEINDLER. 1983. Sampling of brine in cores of Precambrian granite from northern Illinois. *Journal of Geophysical Research* 88, B9:7331–34.

COX, D. C. 1954. Water development for Hawaiian sugar cane irrigation. *Hawaiian Planters Record* 54: 175–97.

CRAIG, H. 1961. Isotopic variations in meteoric water. *Science* 133:1702–3.

CRAUN, G. F. 1979. Waterborne-disease status report emphasizing outbreaks in ground-water systems. *Ground Water* 17:183–91.

———. 1981. Outbreaks of waterborne disease in the United States: 1971–78. *Journal, American Water Works Association* 73:360–69.

———. 1984. Health aspects of ground water pollution. In *Groundwater Pollution Microbiology,* ed. G. Britton and C. P. Gerba. New York: John Wiley.

CRIDDLE, W. D. 1958. Methods of computing consumptive use of water. *Proceedings, American Society of Civil Engineers, Irrigation and Drainage Division* 84(IR1):1–27.

CROSBY, J. W., III, & J. V. ANDERSON. 1971. Some applications of geophysical well logging to basalt hydrogeology. *Ground Water* 9, no. 5:12–20.

CSALLANY, S., & W. C. WALTON. 1963. *Yields of shallow dolomite wells in northern Illinois.* Illinois State Water Survey Report of Investigation 46.

CUSHMAN, R. V. 1967. *Recent developments in hydrogeologic investigation in the karst area of central Kentucky.* International Association of Hydrogeologists Memoirs, Congress of Istanbul, pp. 236–48.

DAGAN, G. 1967. A method of determining the permeability and effective porosity of unconfined anisotropic aquifers. *Water Resources Research* 3:1059–71.

———. 1986. Statistical theory of groundwater flow and transport: Pore to laboratory, laboratory to formation and formation to regional scale. *Water Resources Research* 22, no. 9 (Supplement):120S–34S.

DAGAN, G., V. CVETKOVIC, & A. M. SHAPIRO. 1992. A solute flux approach to transport in heterogeneous formations, 1, the general framework. *Water Resources Research* 28, no. 5:1369–76.

D'APPOLONIA CONSULTING ENGINEERS, INC. 1980. *User's manual.* GEOFLOW Ground Water Model.

DANIELS, J. J., G. R. OLHOEFT, & J. H. SCOTT. 1983. Interpretation of core and well log physical property data from drill hole UPH-3, Stephenson County, Illinois. *Journal of Geophysical Research* 88, B9:7346–54.

DANIELSON, J. A., & A. R. QAZI. 1972. Stream depletion by wells in the South Platte Basin—Colorado. *Water Resources Bulletin* 8:359–66.

DARCY, H. 1856. *Les fontaines publiques de la ville de Dijon.* Paris: Victor Dalmont.

DAVIS, S. N. 1969. Porosity and permeability in natural materials. In *Flow through Porous Media,* ed. R. J. M. DeWiest. 53–89. New York: Academic Press.

DAVIS, S. N., & H. W. BENTLEY. 1982. Dating groundwater, a short review. In *Nuclear and Chemical Dating Techniques: Interpreting the Environmental Record,* ed. Lloyd A. Curie, 187–222. American Chemical Society Symposium Series No. 176.

DAVIS, S. N., G. M. THOMPSON, H. W. BENTLEY, & G. STILES. 1980. Ground-water tracers—A short review. *Ground Water* 18:14–23.

DAVIS, S. N., & L. J. TURK. 1964. Optimum depth of wells in crystalline rocks. *Ground Water* 2, no. 2: 6–11.

DAWSON, K. J. & J. D. ISTOK. 1991. *Aquifer testing—Design and Analysis of pumping and slug tests.* Chelsea, Mich.: Lewis Publishers.

DELORIA, SAM. 1985. A Native American view of Western water development. *Water Resources Research* 21, no. 11:1785–86.

DENNE, J. E., H. L. YARGER, P. A. MACFARLANE, P. W. KNAPP, M. A. SOPHOCLEOUS, J. R. LUCAS, & D. W. STEEPLES. 1984. Remote sensing and geophysical investigations of glacial buried valleys in northeastern Kansas. *Ground Water* 22: 56–65.

DEVRIES, H. 1959. Measurement and use of natural radiocarbon. In *Researches in Geochemistry,* vol. 1, ed. P. H. Abelson, 169–89. New York: John Wiley.

DOBRIN, M. B. 1976. *Introduction to geophysical prospecting,* 3d ed. New York: McGraw-Hill.

DOMENICO, P. A. 1972. *Concepts and models in groundwater hydrology.* New York: McGraw-Hill.

DOONAN, C. J., G. E. HENDRICKSON, & J. R. BYERLAY. 1970. *Ground water and geology of Keweenaw Peninsula, Michigan.* Michigan Department of Natural Resources, Geological Survey Division, Water Investigation 10.

DOTY, C. B., & C. C. DAVIS. 1991. *The effectiveness of ground-water pumping as a restoration technology.*

Oak Ridge National Laboratory Report ORNL/TM-11866.

DREVER, J. I. 1988. *The geochemistry of natural waters.* Englewood Cliffs, N.J.: Prentice Hall.

DRUBACK, G. W., & S. V. ARLOTTA. 1985. Subsurface pollution containment using a composite system vertical cut-off barrier. *Proceedings, Fifth National Symposium on Aquifer Restoration and Ground Water Monitoring.* National Water Well Association, pp. 400–11.

DUCOMMUN, J. 1828. On the cause of fresh water springs, fountains, and c. *American Journal of Science,* 1st ser., vol. 14:174–76. As cited in S. N. Davis, Flotation of fresh water on sea water, a historical note. *Ground Water* 16, no. 6 (1978): 444–45.

DUDLEY, J. G., & D. A. STEPHENSON. 1973. *Nutrient enrichment of ground-water from septic tank disposal systems.* Inland Lake Renewal and Shoreland Management Demonstration Project Report, University of Wisconsin—Madison.

DUNLAP, W. J., ET AL. 1977. *Sampling of organic chemicals and microorganisms in the subsurface.* U.S. Environmental Protection Agency, SW-611.

DUNNE, T. 1978. Field studies of hillslope flow processes. In *Hillslope hydrology,* ed. M. J. Kirkby, 227–94. Chichester, Sussex, England: John Wiley.

DUNNE, T., & R. D. BLACK. 1970. An experimental investigation of runoff production in permeable soils. *Water Resources Research* 6:478–90.

DUNNE, T., & L. B. LEOPOLD. 1978. *Water in environmental planning.* San Francisco: W. H. Freeman, p. 88.

DUPUIT, J. 1863. *Études théoriques et pratiques sur le mouvement des eaux dans les canaux découverts et à travers les terrains perméables,* 2d ed. Paris: Dunod.

DUTCHER, L. C., & A. A. GARRETT. 1963. *Geological and hydrologic features of the San Bernardino area, California.* U.S. Geological Survey Water-Supply Paper 1419.

EAKIN, T. E. 1966. A regional interbasin groundwater system in the White River area, southeastern Nevada. *Water Resources Research* 2, no. 2:251–71.

EAKIN, T. E., & D. O. MOORE. 1964. *Uniformity of discharge of muddy river springs.* U.S. Geological Survey Professional Paper 501-D, pp. 171–76.

EAKIN, T. E., D. PRICE, & J. R. HARRILL. 1976. *Summary appraisals of the nation's ground water resources—Great Basin region.* U.S. Geological Survey Professional Paper 813-G.

EL PASO V. REYNOLDS. 1983. Civ. No. 80-730 HB.

ERSKINE, A. D. 1991. The effect of tidal fluctuation on a coastal aquifer in the UK. *Ground Water* 29, no. 4:556–562.

EVANS, R. B., R. C. BENSON, & J. RIZZO. 1982. Systematic hazardous waste site assessments. *National Conference on Management of Uncontrolled Hazardous Waste Sites.* Silver Spring, Md.: Hazardous Materials Control Research Institute, pp. 17–22.

EVERETT, L. G. 1981. Monitoring in the vadose zone. *Ground Water Monitoring Review* 1, no. 2:44–51.

EVERETT, L. G., E. W. HOYLMAN, GRAHAM WILSON, & L. G. MCMILLION. 1984. Constraints and categories of vadose zone monitoring devices. *Ground Water Monitoring Review* 4, no. 1:26–32.

EWERS, R. O., ET AL. 1978. The origin of distributary and tributary flow within karst aquifers. Geological Society of America, *Abstracts with Programs* 10, no. 7:398–99.

EXXON MINERALS COMPANY. 1982. *Environmental impact report,* Crandon Project, Volume X, Appendia 4.1A.

FARVOLDEN, R. N., & J. A. CHERRY. 1991. Are geology departments prepared for the 21st century? *Geology* 19, no. 5:419.

FEDERAL COUNCIL FOR SCIENCE AND TECHNOLOGY. 1962. *Scientific hydrology.* Washington, D.C.: U.S. Government Printing Office.

FENN, D., ET AL. 1977. *Procedures manual for ground water monitoring at solid waste disposal facilities.* U.S. Environmental Protection Agency, SW-611.

FERGUSON, B. K., & P. W. SUCKLING. 1990. Changing rainfall-runoff relationships in the urbanizing Peachtree Creek watershed, Atlanta, Georgia. *Water Resources Bulletin* 26:313–322.

FERRIS, J. G., ET AL. 1962. Theory of aquifer tests. U.S. Geological Survey Water-Supply Paper 1536-E.

FERRIS, K. 1980. Arizona's groundwater code: Strength in compromise. *Journal, American Water Works Association* 78, no. 10:79–84.

FETH, J. F. 1973. *Water facts and figures for planners and managers.* U.S. Geological Survey Circular 601-I.

FETTER, C. W., JR. 1972a. The concept of safe groundwater yield in coastal aquifers. *Water Resources Bulletin* 8:1173–76.

———. 1972b. Position of the saline water interface beneath oceanic islands. *Water Resources Research* 8:1307–14.

———. 1973. Water resources management in coastal plain aquifers. *Proceedings of the International Wa-*

ter Resources Association, First World Congress on Water Resources, pp. 322–31.

———. 1975. Use of test wells as water-quality predictors. Journal, American Water Works Association 67:516–18.

———. 1976. Hydrogeology of the South Fork of Long Island, New York. Bulletin, Geological Society of America 87:401–6.

———. 1977a. Attenuation of wastewater elutriated throuh glacial outwash. Ground Water 15:365–71.

———. 1977b. Hydrogeology of the South Fork of Long Island, New York: Reply. Bulletin, Geological Society of America 88:896.

———. 1981a. Interstate conflict over ground water: Wisconsin-Illinois. Ground Water 19:201–13.

———. 1981b. Determination of the direction of ground water flow. Ground Water Monitoring Review 1, 3:28–31.

———. 1983. Potential sources of contamination in ground water monitoring. Ground Water Monitoring Review 2:60–64.

———. 1985. Final hydrogeologic report, Seymour Recycling Corporation hazardous waste site, Seymour, Indiana. Report to United States Environmental Protection Agency.

———. 1989. Transport and fate of organic compounds in ground water. In Recent advances in ground-water hydrology. eds. J. E. Moore, A. A. Zaporozec, and S. C. Csallany, 174–84. Minneapolis, Minn.: American Institute of Hydrology.

———. 1992. Remedial investigation for ground water contamination at the Seymour superfund site. Hazardous Materials Control 5, no. 1:51–55.

———. 1993. Contaminant hydrogeology. New York: Macmillan.

FETTER, C. W., JR., & R. G. HOLZMACHER. 1974. Ground water recharge with treated waste water. Journal of Water Pollution Control Federation 46:260–70.

FETTER, C. W., JR., & H. YOUNG. 1978. Unpublished results of a computer model study of the Deep Sandstone Regional Aquifer.

FITTER, A. H., & C. D. SUTTON. 1975. The use of the Freudlich isotherm for soil phosphate sorption data. Journal of Soil Science 26:241–46.

FITZWATER, P. L., C. L. BRASSOW, & C. W. FETTER, JR. 1983. Assessment of ground-water contamination and remedial action for a hazardous waste facility in the Gulf Coast. Proceedings, Third National Symposium on Aquifer Restoration and Ground Water Monitoring. National Water Well Association, pp. 135–41.

FLATHMAN, P. W., ET AL. 1985. In-situ physical/biological treatment of methylene chloride (dichloromethane) contaminated ground water. Proceedings, Fifth National Symposium on Aquifer Restoration and Ground Water Monitoring. National Water Well Association, pp. 571–97.

FLATHMAN, P. W., J. R. QUINCE, & L. S. BOTTOMLEY. 1984. Biological treatment of ethylene glycol-contaminated ground water at the Naval Air Engineering Center, Lakehurst, New Jersey. Proceedings, Fourth National Symposium on Aquifer Restoration and Ground Water Monitoring. National Water Well Association, pp. 111–19.

FLIPSE, W. J., JR., ET AL. 1984. Sources of nitrate in ground water in a sewered housing development, central Long Island, New York. Ground Water 22:418–26.

FLORES, E. Z., A. L. GUTJAHR, & L. W. GELHAR. 1978. A stochastic model of the operation of a stream-aquifer system. Water Resources Research 14:30–38.

FONTES, J. CH. 1980. Environmental isotopes in groundwater hydrology. In Handbook of environmental isotope hydrology, volume 1, The terrestrial environment, ed. P. Fritz and J. Ch. Fontes, 25–140. Amsterdam: Elsevier Scientific Publishers.

FORCHHEIMER, P. 1914. Hydraulik. Leipzig: B. G. Teubner.

FORD, D. C., & R. O. EWERS. 1978. The development of limestone cave systems in the dimensions of length and depth. Canadian Journal of Earth Science 15:1783–98.

FORD, P. A., P. J. TURINA, & D. E. SEELY. 1983. Characterization of hazardous waste sites—A methods manual, volume II, Available sampling methods. U.S. Environmental Protection Agency, EPA-600/4-83-040.

———. Lessons learned in a hydrogeological case at Sheffield, Illinois. Proceedings, symposium on low-level waste disposal, site characterization and monitoring. Oak Ridge National Laboratory, NUREG/CP-0028, CONF-820674, vol. 2, pp. 237–44.

FOSTER, K. E. 1978. The Winters doctrine: Historical perspective and future applications of reserved water rights in Arizona. Ground Water 16:186–91.

FOX, I. A., & K. R. RUSHTON. 1976. Rapid recharge in a limestone aquifer. Ground Water 14, no. 1:21–27.

FOXWORTHY, B. L. 1983. Pacific Northwest region. In Ground water resources of the United States. ed.

David K. Todd, 590–629. Berkeley, Calif.: Premier Press.

FRAZIER V. BROWN. 1861. 12 Ohio St. 294.

FREEZE, R. A. 1971. Three-dimensional, transient, saturated-unsaturated flow in a groundwater basin. *Water Resources Research* 7:347–66.

FREEZE, R. A., & J. A. CHERRY. 1979. *Groundwater.* Englewood Cliffs, N.J.: Prentice Hall.

———. 1989. What has gone wrong. *Ground Water* 27, no. 4:458–64.

FREEZE, R. A., & P. A. WITHERSPOON. 1966. Theoretical analysis of regional groundwater flow: 1. Analytical and numerical solutions to the mathematical model. *Water Resources Research* 2, no. 4:641–56.

———. 1967. Theoretical analysis of regional groundwater flow: 2. Effect on water-table configuration and subsurface permeability variation. *Water Resources Research* 3, no. 2:623–34.

FRETWELL, J. D., & M. T. STEWART. 1981. Resistivity study of a coastal karst terrain, Florida. *Ground Water* 19:156–62.

FUSILLO, T. V., J. J. HOCHREITER, JR., & D. G. LORD. 1985. Distribution of volatile organic compounds in a New Jersey coastal plain aquifer system. *Ground Water* 23:354–60.

GALE, J., E. B. ROBERTS, & R. M. HAGEN. 1967. High alcohols and antitranspirants. *Water Resources Research* 3:437–41.

GAMBOLATI, G. 1976. Transient free surface flow to a well: An analysis of theoretical solutions. *Water Resources Research* 12:27–39.

GAMBOLATI, G. 1977. Deviations from the Theis solution in aquifers undergoing three-dimensional consolidation. *Water Resources Research* 13:62–68.

GARRELS, R. M., & C. L. CHRIST. 1965. *Solutions, minerals and equilibria.* New York: Harper & Row.

GASS, T. E., & J. LEHR. 1977. Ground-water energy and the ground-water heat pump. *Water Well Journal* 31: no. 4:42–47.

GELHAR, L. W. 1986. Stochastic subsurface hydrology from theory to applications. *Water Resources Research* 22, no. 9 (Supplement):135S.

GERAGHTY AND MILLER SOFTWARE NEWSLETTER. 1992. G & M Survey Results. Volume 4 (summer): 1–2.

GIBB, J. P., R. M. SCHULLER, & R. A. GRIFFIN. 1981. *Procedures for the collection of representative water quality data from monitoring wells.* Illinois State Geological Survey and Illinois State Water Survey Cooperative Ground Water Report 7.

GILKESON, R. H., & K. CARTWRIGHT. 1983. The application of surface electrical and shallow geothermic methods in monitoring network design. *Ground Water Monitoring Review* 3, no. 3:30–42.

GILKESON, R. H., P. C. HEIGOLD, & D. E. LAYMON. 1986. Practical application of theoretical models to magnetometer surveys of hazardous wsste disposal sites—A case history. *Ground Water Monitoring Review* 6, no. 1:54–61.

GLEASON, V. E. 1978. The legalization of ground water storage. *Water Resources Bulletin* 14:532–41.

GLOVER, R. E. 1964. The pattern of fresh-water flow in a coastal aquifer. In *Sea water in coastal aquifers.* U.S. Geological Survey Water-Supply Paper 1613., pp. 32–35.

GOLDFARB, WILLIAM. 1984. Mono Lake and the public trust doctrine. *Bulletin, American Water Resources Association* 20:292–93.

———. 1988. *Water law,* 2d ed. Chelsea, Mich.: Lewis Publishers.

GORDON, M. E., & P. M. HUEBNER. 1983. An evaluation of the performance of zone of saturation landfills in Wisconsin. In *Proceedings, sixth annual Madison waste conference—Municipal and industrial waste.* University of Wisconsin, pp. 23–53.

GOYAL, V. C., SRI NIWAS, & P. K. GUPTA. 1991. Theoretical evaluation of modified Wenner array for shallow resistivity exploration. *Ground Water* 29, no. 4:582–586.

GRBIC-GALIC, D., & T. M. VOGEL. 1987. Transformation of toluene and benzene by mixed methanogenic cultures. *Applied and Environmental Microbiology* 53, no. 2:254–260.

GREENHOUSE, J. P., & D. J. SLAINE. 1983. The use of reconnaissance electromagnetic methods to map contaminant migration. *Ground Water Monitoring Review* 3, no. 2:47–59.

GREGORY, H. E. 1916. *The Navajo country.* U. S. Geological Survey Water-Supply Paper 380.

GREYDANUS, H. W. 1978. Management aspects of cyclic storage of water in aquifer systems. *Water Resources Bulletin* 14:477–83.

GRIFFIN, R. A. 1985. Illinois State Geological Survey, personal communication.

GRUBB, H. F. 1986. Gulf coastal plain regional aquifer-system study. In *Regional aquifer-system analysis program of the United States Geological Survey.* U.S. Geological Survey Circular 1002: 152–61.

GRUBB, STUART. 1993. Analytical model for estimation of steady-state capture zones of pumping wells in confined and unconfined aquifers. *Ground Water* 31, no. 1:27–32.

GUNDERSON, L. H. 1989. Accounting for discrepancies in pan evaporation calculations. *Water Resources Bulletin* 25:573–579.

GUTENTAG, E. D., F. J. HEIMES, N. C. KROTHE, R. R. LUCKEY, & J. B. WEEKS. 1984. *Geohydrology of the high plains aquifer in parts of Colorado, Kansas, Nebraska, New Mexico, Oklahoma, South Dakota, Texas and Wyoming.* U.S. Geological Survey Professional Paper 1400-B.

HACKETT, GLEN. 1987. Drilling and constructing monitoring wells with hollow stem augers—Part I: Drilling considerations. *Ground Water Monitoring Review* 7, no. 4:51–62.

———. 1988. Drilling and constructing monitoring wells with hollow stem augers—Part II: Monitoring well installation. *Ground Water Monitoring Review* 8, no. 1:60–68.

HAIMES, Y. Y., & Y. C. DREIZIN. 1977. Management of ground-water and surface-water via decomposition. *Water Resources Research* 13:69–77.

HAIMSON, B. C., & T. W. DOE. 1983. State of stress, permeability, and fractures in the Precambrian granite of Northern Illinois. *Journal of Geophysical Research* 88, B9:7355–71.

HANTUSH, M. S. 1956. Analysis of data from pumping tests in leaky aquifers. *Transactions, American Geophysical Union* 37:702–14.

———. 1960a. Analysis of data from pumping tests in anisotropic aquifers. *Journal of Geophysical Research* 71:421–26.

———. 1960b. Modification of the theory of leaky aquifers. *Journal of Geophysical Research* 65:3713–25.

———. 1961. Aquifer test on partially penetrating wells. *Proceedings of the American Society of Civil Engineers,* 87:171–95.

———. 1964. Hydraulics of wells. In *Advances in hydroscience,* vol. 1, ed. V. T. Chow, 281–432. New York: Academic Press.

———. 1966. Wells in homogeneous anisotropic aquifers. *Water Resources Research* 2:273–79.

———. 1967. Growth and decay of ground-water mounds in response to uniform percolation. *Water Resources Research,* 3:227–34.

HANTUSH, M. S., & C. E. JACOB. 1954. Plane potential flow of ground-water with linear leakage. *Transactions, American Geophysical Union* 35:917–36.

HARBECK, G. E. 1962. *A practical field technique for measuring reservoir evaporation utilizing mass-transfer theory.* U.S. Geological Survey Professional Paper 272-E: 101–5.

HARBECK, G. E., & F. W. KENNON. 1954. The water budget control. In *Water-loss investigations: Lake Hefner studies technical report.* U.S. Geological Survey Professional Paper 269.

HARDT, W. F., & C. B. HUTCHINSON. 1978. Model aids planners in predicting rising ground-water levels in San Bernardino, California. *Ground Water* 16:424–31.

HARMESON, R. H., R. L. THOMAS, & R. L. EVANS. 1968. Coarse media filtration for artificial recharge. *Journal, American Water Works Association* 60:1396–1403.

HARR, M. E. 1962. *Groundwater and seepage.* New York: McGraw-Hill.

HARSHBARGER, J. W. 1968. Ground-water development in desert areas. *Ground Water* 6, no. 5:2–4.

HAUPTMAN, M. G., J. RUMBAUGH, & N. VALENBURGH. 1990. Use of groundwater modeling during Superfund cleanup. *Proceedings, 11th National Superfund Conference.* Hazardous Materials Control Research Institute, Silver Spring, MD, pp. 110–116.

HAZEN, A. 1911. Discussion: Dams on sand foundations. *Transactions, American Society of Civil Engineers,* 73:199.

HEATH, R. C. 1983. *Basic ground-water hydrology.* U.S. Geological Survey Water-Supply Paper 2220.

HEATH, R. C. 1984. *Ground water regions of the United States.* U.S. Geological Survey Water-Supply Paper 2242.

HEATH, R. C. 1988. Hydrogeologic setting of regions. In *Hydrogeology,* ed. W. Back, J. S. Rosenshein and P. R. Seber, *The Geology of North America* O-2:15–23. Boulder, Colo.: Geological Society of America.

HEIGOLD, P. C., ET AL. 1979. Aquifer transmissivity from surficial electrical methods. *Ground Water* 17, no. 4:338–45.

HELVEY, J. D. 1967. Interception by eastern White Pine. *Water Resources Research,* 3:723–30.

HELWIG, O. J. 1978. "Regional ground-water management." *Ground Water* 16:318–21.

HEM, J. D. 1960. *Restraints in dissolved ferrous iron imposed by bicarbonate redox potentials, and pH.* U.S. Geological Survey Water-Supply Paper 1459-B.

HEM, J. D. 1985. *Study and interpretation of the chemical characteristics of natural water,* 3d ed. U.S. Geological Survey Water-Supply Paper 2254.

HEM, J. D., & W. H. CROPPER. 1959. *Survey of the ferrous-ferric chemical equilibria and redox potential.* U.S. Geological Survey Water-Supply Paper 1459-A.

HENDRICKS, D. W., & V. E. HANSEN. 1962. Mechanics of transpiration. *American Society of Civil Engineers, Journal of Irrigation and Drainage Division* 88 (June): 67–82.

HENDRY, M. J. 1982. Hydraulic conductivity of a glacial till in Alberta. *Ground Water* 20:162–69.

HERWALDT, B. L., ET AL. 1992. Outbreaks of waterborne disease in the United States: 1989–90. *Journal, American Water Works Association* 84, no. 4:129–135.

HERZBERG, A. 1901. Die Wasserversorgung einiger Nordseebader. *Journal Gasbeleuchtung und Wasserversorgung* (Munich), 44:815–19, 842–44.

HERZOG, B. L., ET AL. 1982. A study of trench covers to minimize infiltration at waste disposal sites. Illinois State Geological Survey Contract Report No. 1981-5, Nuclear Regulatory Commission, NUREG/CR-2478.

HERZOG, B. L., & W. J. MORSE. 1984. A comparison of laboratory and field determined values of hydraulic conductivity at a hazardous waste disposal site. In *Proceedings, seventh annual Madison waste conference—municipal and industrial waste.* University of Wisconsin, pp. 30–52.

HIBBERT, A. R. 1967. Forest treatment effects on water yield. In *Forest hydrology,* ed. W. E. Sopper and H. W. Lull, 527–43. Oxford, England: Pergamon Press.

———. 1971. Increases in streamflow after converting chaparral to grass. *Water Resources Research* 7: 71–80.

HICKEY, J. J. 1989. An approach to the field study of hydraulic gradients in variable-salinity ground water. *Ground Water* 27, 4:531–539.

HILLEL, D. 1971. *Soil and water.* New York: Academic Press.

HIRSHLEIFER, J., J. C. DEHAVEN, & J. W. MILLIMAN. 1960. *Water supply.* Chicago: University of Chicago Press.

HITCHCOCK, A. S., & H. D. HARMON, JR. 1983. Applications of geophysical techniques as a site screening procedure at hazardous waste site. *Proceedings of the Third National Symposium on Aquifer Restoration and Ground-Water Monitoring.* Dublin, Ohio: National Water Well Association, pp. 307–12.

HIX, GARY. 1992. Squeaky clean drill rigs. *Ground Water Monitoring Review* 13, no. 3:94–96.

HOFFMAN, J. 1983. Flooded San Bernardino fears even a minor quake. *Los Angeles* (California) *Herald Examiner,* April 17.

HOLLARD, R. E., & A. M. BEETON. 1972. Significance to eutrophication of spatial differences on nutrients and diatoms in Lake Michigan. *Limnology and Oceanography* 17:88–96.

HOLZSCHUH, J. C., III. 1976. A simple computer program for the determination of aquifer characteristics from pump test data. *Ground Water* 14:283–85.

HOOPES, J. A., & D. R. F. HARLEMAN. 1967. Wastewater recharge and dispersion in porous media. *Journal, Hydraulics Division, American Society of Civil Engineers* 93, HY5:51–71.

HORDON, R. M. 1977. Water supply as a limiting factor in developing communities; Endogenous vs. exogenous sources. *Water Resources Bulletin,* 13:933–39.

HORTON, K. A., ET AL. 1981. The complementary nature of geophysical techniques for mapping chemical waste disposal sites: Impulse radar and resistivity. *National Conference on Management of Uncontrolled Hazardous Waste Sites.* Silver Spring, Md.: Hazardous Materials Control Research Institute, pp. 158–64.

HORTON, R. E. 1933. The role of infiltration in the hydrologic cycle. *Transactions, American Geophysical Union* 14:446–60.

———. 1940. An approach toward a physical interpretation of infiltration capacity. *Soil Science Society of America, Proceedings* 4:399–417.

HUBBERT, M. K. 1940. The theory of ground-water motion. *Journal of Geology* 48, no. 8:785–944.

HUBBERT, M. K. 1956. Darcy's law and the field equations of flow of underground fluids. *Transactions, American Institute of Mining and Metallurgical Engineers* 207:222–39.

HUBER V. MERKEL. 1903. 117 Wis. 355, 94 N.W. 354.

HUFF, F. A. 1955. Comparison between standard and small orifice rain gauges. *Transactions, American Geophysical Union* 30:689–94.

HUFF, F. A., & R. G. SEMONIN. 1975. Potential of precipitation modification in severe droughts. *Journal of Applied Meteorology* 14:974–79.

HUNT, B. W. 1971. Vertical recharge of unconfined aquifer. *Journal of Hydraulics Division, American Society of Civil Engineers,* 97:1017–30.

HUNTOON, P. W. 1977. Cambrian stratigraphic nomenclature and ground-water prospecting failures in the Hualapai Plateau, Arizona. *Ground Water* 15: 426–33.

————. 1981. Fault controlled ground-water circulation under the Colorado River, Marble Canyon, Arizona. *Ground Water* 19:20–27.

————. 1985. Fault severed aquifers along the perimeters of Wyoming artesian basin. *Ground Water* 23: 176–81.

————. 1986. Incredible tale of Texasgulf Well 7 and fracture permeability, Paradox Basin, Utah. *Ground Water* 24:643–53.

HVORSLEV, M. J. 1951. *Time lag and soil permeability in ground water observations*. U.S. Army Corps of Engineers Waterway Experimentation Station, Bulletin 36.

INTERNATIONAL GREAT LAKES LEVELS BOARD. 1973. Regulation of Great Lakes water levels. Appendix A, *Hydrology and hydraulics*. Report to the International Joint Commission, December 7.

IRWIN V. PHILLIPS. 1855. 5 Cal. 140.

JACOB, C. E. 1940. On the flow of water in an elastic artesian aquifer. *Transactions, American Geophysical Union* 21:574–86.

————. 1950. Flow of ground-water. In *Engineering hydraulics*, ed. H. Rouse, 321–86. New York: John Wiley.

JACOBY, G. C., JR., G. D. WEATHERFORD, & J. W. WEGNER. 1976. Law, hydrology and surface water supply in the upper Colorado River basin. *Water Resources Bulletin* 12:973–84.

JAMES, R. V., & J. RUBIN. 1972. Accounting for apparatus-induced dispersion in analysis of miscible displacement experiments. *Water Resources Research* 8:717–21.

JAVANDEL, I., C. DOUGHTY, & C.-F. TSANG. 1984. *Ground water transport: Handbook of mathematical models*. American Geophysical Union, Water Resources Monograph 10.

JOHNSON, A. I. 1967. *Specific yield—Compilation of specific yields for various materials*. U.S. Geological Survey Water-Supply Paper 1662-D.

JOHNSON, A. I., R. C. PRILL, & D. A. MORRIS. 1963. *Specific yield—Column drainage and centrifuge moisture content*. U.S. Geological Survey Water-Supply Paper 1662-A.

JOHNSON DIVISION, UNIVERSAL OIL PRODUCTS COMPANY. 1966. *Ground water and wells*. St. Paul: Minn.

JOHNSON, T. J., ET AL. 1983. Hydrologic investigations of failure mechanisms and migration of organic chemicals at Wilsonville, Illinois. *Proceedings, Third National Symposium on Aquifer Restoration and Ground Water Monitoring*. National Water Well Association, pp. 413–20.

JOHNSON, T. M., K. CARTWRIGHT, & R. M. SCHULLER. 1981. Monitoring of leachate migration in the unsaturated zone in the vicinity of sanitary landfills. *Ground Water Monitoring Review* 1, no. 3:55–63.

JONES, LADON, TRACY LEMAR, & CHIN-TA TSAI. 1992. Results of two pumping tests in Wisconsin-age weathered till in Iowa. *Ground Water* 30, no. 4: 529–538.

JRB ASSOCIATES, INC. 1982. *Handbook, remedial actions at waste disposal sites*. U.S. Environmental Protection Agency, EPA-625/6-82-006.

JURY, W. A., W. R. GARDNER, & W. H. GARDNER. 1991. *Soil physics*. New York: John Wiley.

KARDOS, L. T. 1967. Waste-water renovation by the land—A living filter. In *Agriculture and the quality of our environment*, ed. N. C. Brady, 241–50. American Association for the Advancement of Science Publication 85.

KAUFMAN, R. F., G. G. EADIE, & C. R. RUSSELL. 1976. Effects of uranium mining and milling on ground water in the Grants uranium belt, New Mexico. *Ground Water* 14:296–308.

KAZMANN, R. G. 1956. "Safe Yield" in ground-water development, reality or illusion?" *Proceedings of the American Society of Civil Engineers* 82, IR3:1103:1-1103–12.

————. 1970. *The present and future ground water supply of the Baton Rouge area*. Louisiana Water Resources Research Institute Bulletin 5.

KEEFER, W. R., & R. F. HADLEY. 1976. *Land and natural resource information and some potential environmental effects of surface mining of coal in the Gillette Area, Wyoming*. U.S. Geological Survey Circular 743.

KEELY, J. F. 1984. Optimizing pumping strategies for contaminant studies and remedial actions. *Proceedings, Fourth National Symposium on Aquifer Restoration and Ground Water Monitoring*. National Water Well Association, pp. 33–42.

KEELY, J. F., & KWASI BOATENG. 1987. Monitoring well installation, purging, and sampling techniques—Part I: Conceptualizations. *Ground Water* 5, no. 3:300–313.

KEITH, L. H., ET AL. 1983a. Principles of environmental analysis. *Analytical Chemistry* 55:2210–18.

————. 1983b. Dealing with the problem of obtaining accurate ground-water quality analytical results. In

Proceedings of the third national symposium on aquifer restoration and ground-water monitoring. Worthington, Ohio: National Water Well Association, pp. 272–82.

KELLER, G. V. 1960. *Physical properties of tuffs in the Oak Spring formation, Nevada.* U.S. Geological Survey Professional paper 400-B.

KENNEDY, V. C. 1965. *Mineralogy and cation exchange capacity of sediments from selected streams.* U.S. Geological Survey Professional Paper 433-D.

KEYS, W. S. 1967. *Borehole geophysics as applied to groundwater.* Canadian Geological Survey Economic Geology Report 26, pp. 598–614.

———. 1986. Analysis of geophysical logs of water wells with a microcomputer. *Ground Water* 24:750–60.

KEYS, W. S., & R. F. BROWN. 1978. The use of temperature logs to trace the movement of injected water. *Ground Water* 16:32–48.

KEYS, W. S., & L. M. MACCARY. 1971. Application of borehole geophysics to water-resources investigations. In *Techniques of water-resources investigations.* U.S. Geological Survey, Book 2, Chap. E1.

KIMBLER, O. K. 1970. Fluid model studies of the storage of freshwater in saline aquifers. *Water Resources Research* 6:1522–27.

KIRCHMER, C. J. 1983. Quality control in water analyses. *Environmental Science and Technology* 17, no. 4:178A–181A.

KIRKBY, M. J., & R. J. CHORLEY. 1967. Throughflow, overland flow, and erosion. *Bulletin, International Association Scientific Hydrology,* 12:5–21.

KLUSMAN, R. W., & K. W. EDWARDS. 1977. Toxic metals in ground water of the front range, Colorado. *Ground Water* 15:160–69.

KMET, P., K. J. QUINN, & C. SLAVIK. 1981. Analysis of design parameters affecting the collection efficiency of clay lined landfills. *Proceedings, Fourth Annual Madison Waste Conference.* Madison, Wis.: University of Wisconsin.

KOERNER, R. M., A. E. LORD, JR., & J. J. BOWDERS. 1981. Utilization and assessment of a pulsed RF system to monitor subsurface liquids. *National Conference on Management of Uncontrolled Hazardous Waste Sites.* Silver Spring, Md.: Hazardous Materials Control Research Institute, pp. 165–70.

KOERNER, R. M., ET AL. 1982. Use of NDT methods to detect buried containers in saturated silty clay soil. *National Conference on Management of Uncontrolled Hazardous Waste Sites.* Silver Spring, Md.: Hazardous Materials Control Research Institute, pp. 12–16.

KOHLER, M. A., T. J. NORDENSON, & W. E. FOX. 1955. *Evaporation from ponds and lakes.* U.S. Weather Bureau Research Paper 38.

KOHLER, M. A., & L. H. PARMELE. 1967. Generalized estimates of free-water evaporation. *Water Resources Research,* 3:997–1005.

KOHOUT, F. A., ET AL. 1977. Fresh groundwater stored in aquifers under the continental shelf: Implications from a deep test well, Nantucket Island, Massachusetts. *Water Resources Bulletin,* 13:373–86.

KONIKOW, L. F., & J. D. BREDEHOEFT. 1978. *Computer model of two-dimensional solute transport and dispersion in ground water.* U.S. Geological Survey, Techniques of Water-Resources Investigations, Book 7, Chap. C2.

KRAMER, V. H., N. VALKENBURG, & M. HAUPTMAN. 1990. Aquifer testing is essential during remedial investigations. *Proceedings, 11th National Superfund Conference.* Hazardous Materials Control Research Institute, Silver Spring, MD, pp. 580–584.

KRAMER, W. H. 1982. Ground water pollution from gasoline. *Ground Water Monitoring Review* 2, no. 2:18–22.

KRAUSKOPF, K. B. 1967. *Introduction to geochemistry.* New York: McGraw-Hill.

KREITLER, C. W., & D. C. JONES. 1975. Natural soil nitrate: the cause of the nitrate contamination of ground water in Runnels County, Texas. *Ground Water* 13, no. 1:53–61.

KREITLER, C. W., S. E. RAGONE, & B. G. KATZ. 1978. Nitrogen isotope ratios of groundwater nitrate, Long Island, New York. *Ground Water* 16:404–9.

KRILL, R. M., & W. C. SONZOGNI. 1986. Chemical monitoring of Wisconsin's ground water. *Journal, American Water Works Association* 78, no. 9:70–75.

KRUSEMAN, G. P., & N. A. DERIDDER, 1991. *Analysis and evaluation of pumping test data,* 2d ed. International Association for Land Reclamation and Improvement, Publication 47, Wageningen, The Netherlands.

KRYNINE, D. P., & W. R. JUDD. 1957. *Principles of engineering geology and geotechnics.* New York: McGraw-Hill.

KWADER, T. 1986. The use of geophysical logs for determining formation water quality. *Ground Water* 24:11–15.

LACHENBRUCH, A. H., M. C. BREWER, G. W. GREENE, & B. V. MARSHALL. 1962. Temperature in permafrost. In *Temperature and its measurement and control in science and industry,* vol. 3.

American Institute of Physics. New York: Reinhold Publishing.

LAMOUREAUX, P. E., & W. J. POWELL. 1960. *Stratigraphic and strucural guides to the development of water wells and well fields in limestone terrane.* International Association of Scientific Hydrology, Pub. 52, pp. 363–75.

LATTMAN, L. H. 1958. Technique of mapping geologic fracture traces and lineaments on aerial photographs. *Photogrammetric Engineering* 24:568–76.

LATTMAN, L. H., & R. H. MATZKE. 1961. Geological significance of fracture traces. *Photogrammetric Engineering* 27:435–38.

LATTMAN, L. H., & R. R. PARIZEK. 1964. Relationship between fracture traces and the occurrence of groundwater in carbonate rocks. *Journal of Hydrology* 2:73–91.

LEE, C. H. 1915. The determination of safe yield of underground reservoirs of the closed basin type. *Transactions, American Society of Civil Engineers* 78:148–51.

LEE, D. R. 1976. A device for measuring seepage flux in lakes and estuaries. *Limnology and Oceanography* 22:140–47.

LEGRAND, H. E. 1954. *Geology and groundwater in the Statesville area, North Carolina.* North Carolina Department of Conservation and Development, Division of Mineral Resources Bulletin 68.

———. 1962. Perspective on problems in hydrogeology. *Bulletin, Geological Society of America,* 73:1147–52.

LEGRAND, H. E., & W. A. PETTYJOHN. 1981. Regional hydrogeological concepts of homoclinal flanks. *Ground Water* 19:303–10.

LEGRAND, H. E., & V. T. STRINGF'ELD. 1971. Water levels in carbonate rock terranes. *Ground Water* 9, no. 3:4–10.

LEHR, J. H. 1978. Ground water: Nature's investment banking system. *Ground Water* 16:143–53.

———. 1981. A problem yes—A disaster no. *Ground Water* 19:2–3.

LENZO, F. C. 1984. Air-stripping for VOCs in water: Pilot, design, construction. *Proceedings, Fourth National Symposium on Aquifer Restoration and Ground Water Monitoring.* National Water Well Association, pp. 100–10.

LIAKOPOULOS, A. C. 1965. Variation of the permeability tensor ellipsoid in homogeneous, anisotropic soils. *Water Resources Research* 1, no. 1:135–42.

LINDQUIST, E. 1933. *On the flow of water through porous soil.* Premier Congres des grands barrages (Stockholm), pp. 81–101.

LINSLEY, R. K., JR., M. A. KOHLER, & J. L. H. PAULHUS. 1975. *Hydrology for engineers.* New York: McGraw-Hill.

LIPPY, E. C., & S. C. WALTRIP. 1984. Waterborne disease outbreaks—1946–1980: A thirty-five year perspective. *Journal, American Water Works Association* 76:60–67.

LLOYD, J. W., & M. H. FARAG. 1978. Fossil groundwater-gradients in arid regional sedimentary basins. *Ground Water* 16:388–93.

LUSCZYNSKI, N. J. 1961. Head and flow in ground water of variable density. *Journal of Geophysical Research* 66, 12:4247–56.

LUZIER, J. E., & J. A. SKRVIAN. 1975. *Digital simulation and projection of water level declines in basalt aquifers of the Odessa-Lind area, East Central Washington.* U.S. Geological Survey Water-Supply Paper 2036.

LYNCH, E. R., ET AL. 1984. Design and evaluation of in-place structures utilizing ground water cutoff walls. *Proceedings, Fourth National Symposium on Aquifer Restoration and Ground Water Monitoring.* National Water Well Association, pp. 1–7.

MACCARY, L. M. 1983. "Geophysical Loggings in Carbonate Aquifers." *Ground Water,* 21:334–42.

MACCARY, L. M., & T. W. LAMBERT. 1962. *Reconnaissance of ground-water resources of the Jackson purchase region, Kentucky.* U.S. Geological Survey Hydrologic Atlas HA-13.

MACKAY, D. M., & J. A. CHERRY. 1989. Groundwater contamination: pump and treat remediation. *Environmental Science and Technology* 23, no. 6: 630–636.

MACKICHAN, K. A., & J. C. KAMMERER. 1961. *Estimated use of water in the United States, 1960.* U.S. Geological Survey Circular 456.

MADDOCK, T., III. 1974. The operation of a stream-aquifer system under stochastic demands. *Water Resources Research* 10:1–10.

MANGER, G. E. 1963. *Porosity and bulk density of sedimentary rocks.* U.S. Geological Survey Bulletin 1144-E.

MARCHER, M. V., R. H. BINGHAM, & R. E. LOUNSBURY. 1964. *Ground water geology of the Dickson, Lawrenceburg and Waverly areas of the Western Highland Rim, Tennessee.* U.S. Geological Survey Professional Paper 1764.

MASCH, F. E., & K. J. DENNY. 1966. Grain-size distribution and its effect on the permeability of unconsolidated sands. *Water Resources Research* 2:665–77.

MATTHEWS, C. A. 1991. Using ground water basins as storage facilities in southern California. *Water Resources Bulletin* 27, no. 5:841–847.

MATTINGLY, G. E. G. 1975. Labile phosphate in soils. *Soil Science* 119:369–75.

MAXEY, G. B. 1968. Hydrogeology of desert basins. *Ground Water* 6, no. 5:10–22.

MAYO, A. L., & R. H. KLAUK. 1991. Contributions to the solute and isotopic groundwater geochemistry, Antelope Island, Great Salt Lake, Utah. *Journal of Hydrology* 127:307–335.

MAYO, A. L., A. B. MULLER, & D. R. RALSTON. 1985. Hydrogeology of the Meade Thrust Allochthon, Southeastern Idaho, U.S.A., and its relevance to stratigraphic and structural groundwater flow control. *Journal of Hydrology* 76:27–61.

MAYO, A. L., P. S. NIELSEN, M. LOUCKS, & W. H. BRIMHALL. 1992. The use of solutes and isotope chemistry to identify flow patterns and factors which limit acid mine drainage in the Wasatch Range, Utah. *Ground Water* 30, no. 2:243–249.

MCDONALD, H. R., & D. WANTLAND. 1961. Geophysical procedures in ground water study. *Transactions, American Society of Civil Engineers* 126:122–35.

MCDONALD, M. G., & A. W. HARBAUGH. 1984. *A modular three-dimensional finite-difference groundwater flow model*. U.S. Geological Survey.

MCGREEVY, PATRICIA. 1987. *San Bernardino* (California) *Sun,* July 17.

MCGUINNESS, C. L. 1963. *The role of ground water in the national water situation*. U.S. Geological Survey Water-Supply Paper 1800.

MCHARG, I. L. 1969. *Design with nature*. Garden City, N.Y.: The Natural History Press.

MCLEOD, R. S. 1984. Evaluation of "Superfund" sites for control of leachate and contaminant migration. *Proceedings, The 5th National Conference on Management of Uncontrolled Hazardous Waste Sites,* Hazardous Materials Control Research Institute, pp. 114–21.

MCWHORTER, D. B., R. K. SKOGERBOE, & G. V. SKOGERBOE. 1974. Potential of mine and mill spoils for water quality degradation. *American Water Resources Association Proceedings,* no. 18: 123–37.

MEDEIROS, W. E., & O. A. L. DE LIMA. 1990. A geoelectrical investigation for ground water in crystalline terrains of Central Bahia, Brazil. *Ground Water* 28, no. 4:518–523.

MEINZER, O. E. 1923a. *The occurrence of groundwater in the United States, with a discussion of principles*. U.S. Geological Survey Water-Supply Paper 489.

———. 1923b. *Outline of groundwater hydrology, with definitions*. U.S. Geological Survey Water-Supply Paper 494.

———. 1927. *Large springs of the United States*. U.S. Geological Survey Water-Supply Paper 557.

MEISLER, H., ET AL. 1986. Northern Atlantic coastal plain regional aquifer-system study. In *Regional aquifer-system analysis program of the United States Geological Survey*. U.S. Geological Survey Circular 1002, pp. 168–94.

MENDOZA, C. A., & E. O. FRIEND. 1990a. Advective-dispersive transport of dense organic vapors in the unsaturated zone 1: Model development. *Water Resources Research* 26, no. 3:379–387.

———. 1990b. Advective-dispersive transport of dense organic vapors in the unsaturated zone 2: Sensitivity analysis. *Water Resources Research* 26, no. 3: 388–398.

MENDOZA, C. A., & T. A. MCALARY. 1990. Modeling of ground water contamination caused by organic solvent vapors. *Ground Water* 28, no. 2:199–206.

MERCER, J. W., & C. R. FAUST. 1981. *Ground-water modeling*. Dublin, Ohio: National Water Well Association.

MERCER, M. W., & C. O. MORGAN. 1982. *Storage and retrieval of ground-water data at the U.S. Geological Survey*. U.S. Geological Survey Circular 856:9.

MEYBOOM, P. 1961. Estimating groundwater recharge from stream hydrographs. *Journal of Geophysical Research* 66:1203–14.

———. 1967. Mass-transfer studies to determine the ground-water regime of permanent lakes in Hummocky Moraine of western Canada. *Journal of Hydrology* 5:117–42.

MEYER, W. R. 1963. *Use of a neutron moisture probe to determine the storage coefficient of an unconfined aquifer*. U.S. Geological Survey Professional Paper 450-E, pp. 174–76.

MIFFLIN, M. D. 1968. *Delineation of ground-water flow systems in Nevada* (Technical Report Ser. H-W, Pub. No. 4). Reno: University of Nevada, Desert Research Institute.

MILLER, D. W., & N. VALKENBURG. 1992. Seymour: A Superfund success story. *Hazardous Materials Control* 5, no. 1:42–45.

MILLER, J. A. 1986. *Hydrogeological framework of the Floridan aquifer system in Florida and in parts of Georgia, Alabama and South Carolina.* U.S. Geological Survey Professional Paper 1043-B.

MITCHELL, J. 1932. The origin, nature and importance of soil organic constituents having base exchange properties. *Journal of American Society of Agronomy* 24:256–75.

MITCHELL, W. D. 1954. *Stage-fall-discharge relations for steady flow in prismatic channels.* U.S. Geological Survey Water-Supply Paper 1164.

MOLZ, F. J., I. REMSON, A. A. FUNGAROLI, & R. L. DRAKE. 1968. Soil moisture availability for transpiration. *Water Resources Research* 4:1161–70.

MOONEY, H. M., & W. W. WETZEL. 1956. *The potentials about a point electrode and apparent resistivity for a two-, three-, and four-layer earth.* Minneapolis: University of Minnesota Press.

MOORE, J. E. 1991. *A guide for preparing hydrologic and geologic projects and reports.* Minneapolis: American Institue of Hydrology.

MORAVCOUA, V., L. MASINOVA, & V. BERNATOVA. 1968. Biological and bacteriological evaluation of pilot plant artificial recharge experiments. *Water Research* 2:265–76.

MOREL-SEYTOUX, H. J. 1975. A simple case of conjunctive surface–ground-water management. *Ground Water* 13:506–15.

MOTA, L. 1954. Determination of dips and depths of geological layers by the seismic refraction method. *Geophysics:* 19:242–54.

MULLER, A. B., & A. L. MAYO. 1983. Ground-water circulation in the Meade Thrust Allochthon evaluated by radiocarbon techniques. *Radiocarbon,* 25:357–72.

———. 1986. ^{13}C variation in limestone on an aquifer-wide scale and its effects on groundwater ^{14}C dating models. *Radiocarbon* 28, no. 3:1041–1054.

MURRAY, C. R. 1960. Origin of porosity in carbonate rocks. *Journal of Sedimentary Petrology,* 30:59–84.

———. 1968. *Estimated use of water in the United States, 1965.* U.S. Geological Survey Circular 556.

MURRAY, C. R., & E. B. REEVES. 1972. *Estimated use of water in the United States, 1970.* U.S. Geological Survey Circular.

———. 1977. *Estimated use of water in the United States, 1975.* U.S. Geological Survey Circular.

MYLNE, M. F. 1989. Combination of radar and gauge data in a rainfall archive system. *Water Resources Bulletin* 25:535–539.

NATIONAL AUDUBON SOCIETY V. SUPERIOR COURT OF ALPINE COUNTY. 1983. 33 Cal.3d 419, 658 P.2d 709, 189 Cal. Rptr. 346: Cert. denied, 104 S.Ct. 413.

NATIONAL GROUND WATER ASSOCIATION. 1992. Personal communication.

NATIONAL WATER WELL ASSOCIATION. 1986. *Newsletter of the Association of Ground Water Scientists and Engineers* 2 (February): 1.

NELSON, M. J. K., ET AL. 1986. Aerobic metabolism of trichloroethylene by a bacterial isolate. *Applied and Environmental Microbiology* 52, no. 2:383–384.

NELSON, M. J. K., ET AL. 1987. Biodegradation of trichloroethylene and involvement of an aromatic biodegradative pathway. *Applied and Environmental Microbiology* 52, no. 5:949–954.

NEUMAN, S. P. 1972. Theory of flow in unconfined aquifers considering delayed response to the water table. *Water Resources Research* 8:1031–45.

———. 1974. Effect of partial penetration on flow in unconfined aquifers considering delayed gravity response. *Water Resources Research* 10:303–12.

———. 1975. Analysis of pumping test data from anisotropic unconfined aquifers considering delayed gravity response. *Water Resources Research* 11:329–42.

———. 1987. On methods of determining specific yield. *Ground Water* 25, no. 6:679–684.

———. 1990. Universal scaling of hydraulic conductivities and dispersivities in geologic media. *Water Resources Research* 26, no. 8:1749–1758.

NEUMAN, S. P., & P. A. WITHERSPOON. 1969. Applicability of current theories of flow in leaky aquifers. *Water Resources Research* 5:817–29.

NEWCOMB, R. C. 1972. *Quality of the ground water in basalt of the Columbia River group, Washington, Oregon and Idaho.* U.S. Geological Survey Water-Supply Paper 1999-N.

NGUYEN, V., & G. F. PINDER. 1984. Direct calculation of aquifer parameters in slug test analysis. In *Groundwater hydraulics,* J. Rosenshein and G. D. Bennett, ed. 222–39. American Geophysical Union Water Resources Monograph 9.

NIELSEN, D. M., & G. L. YEATES. 1985. A comparison of sampling mechanisms available for small-diameter ground water monitoring wells. *Ground Water Monitoring Review* 5, no. 2:83–99.

NIGHTINGALE, H. I., & W. C. BIANCHI. 1973. Ground-water recharge for urban use: Leaky Acres project. *Ground Water* 11, no. 6:36–43.

NILES SAND AND GRAVEL CO., INC. V. ALAMEDA COUNTY WATER DISTRICT. 1974. 37 Calif. App. 3d 924: Cert. denied 419 US 869.

NORBECK. P. N., L. L. MINK, & R. E. WILLIAMS. 1974. Ground water leaching of jig tailing deposits in the Coeur d'Alene district of northern Idaho. *American Water Resources Association Proceedings,* no. 18:149–57.

NORRIS, S. E. 1963. *Permeability of glacial till.* U.S. Geological Survey Professional Paper 450-E, pp. 150–51.

NORRIS, S. E. 1972. The use of gamma logs in determining the character of unconsolidated sediments and well construction features. *Ground Water* 10, 6:14–21.

NORRIS, S. E., & R. E. FIDLER. 1965. *Relation of permeability to grain size in a glacial-outwash aquifer at Piketown, Ohio.* U.S. Geological Survey Professional Paper 525-D, pp. 203–6.

NORRIS, S. E., & A. M. SPIEKER. 1966. *Ground-water resources of the Dayton area, Ohio.* U.S. Geological Survey Water-Supply Paper 1808.

NOSS, R. R., & E. T. JOHNSON. 1984. Field monitoring of the Adams, Massachusetts, landfill leachate plume. *Proceedings, Fourth National Symposium and Exposition on Aquifer Restoration and Ground Water Monitoring.* National Water Well Association, pp. 356–62.

NYER, E. K., V. KRAMER, & N. VALKENBURG. 1991. Biochemical effects on contaminate fate and transport. *Ground Water Monitoring Review* 11, no. 2:80–82.

OBERLANDER, P. A. 1989. Fluid density and gravitational variations in deep boreholes and their effect on fluid potential. *Ground Water* 27, 3:341–350.

OGATA, A. 1970. *Theory of dispersion in a granular medium.* U.S. Geological Survey Professional Paper 411-I.

OLIVEIRA, D. P., & N. SITAR. 1985. Ground water contamination from underground solvent storage tanks, Santa Clara, California. *Proceedings, Fifth National Symposium and Exposition on Aquifer Restoration and Ground Water Monitoring.* National Water Well Association, pp. 691–708.

OLSEN, S. R., & F. S. WATANABE. 1957. A method to determine a phosphorus adsorption maximum of soils as measured by the Langmuir isotherm. *Soil Science Society of America, Proceedings* 21:144–49.

PALMER, A. N. 1984. Recent trends on karst geomorphology. *Journal of Geological Education* 32:247–53.

PAPADOPULOS, I. S., J. D. BREDEHOEFT, & H. H. COOPER, JR. 1973. On the analysis of "slug test" data. *Water Resources Research* 9:1087–89.

PARIZEK, R. R. 1976. On the nature and significance of fracture traces and lineaments in carbonate and other terranes. In *Karst hydrology and water resources,* ed. V. Yevjevich, 47–108. Fort Collins, Colo.: Water Resources Publications.

PARIZEK, R. R., & E. A. MEYERS. 1967. Recharge of groundwater from renovated sewage effluent by spray irrigation. *American Water Resources Association, Proceedings of the Fourth Conference,* pp. 426–43.

PARK, S. K., D. W. LAMBERT, & TIEN-CHANG LEE. 1990. Investigation by DC resistivity methods of a ground-water barrier beneath the San Bernardino Valley, southern California. *Ground Water* 28, no. 3:344–349.

PARKER, G. G. 1975. Water and water problems in the southwest Florida water management district and some possible solutions. *Water Resources Bulletin* 11:1–20.

PARKER, G. G., R. H. BROWN, D. G. BOGART, & S. K. LOVE. 1955. Salt water encroachment. In *Water resources of southeastern Florida,* 571–711. U.S. Geological Survey Water-Supply Paper 1255.

PARMELEE, M. A. 1993. Milwaukee takes steps to ensure water quality. *AWWA Mainstream, American Water Works Association* 37:1, 8.

PARSONS, F., P. R. WOOD, & J. DEMARCO. 1984. Transformations of tetrachloroethane and trichloroethane in microorganisms and ground water. *Journal, American Water Works Association* 76, no. 2:56–59.

PASADENA V. ALHAMBRA. 1949. 33 Calif. (2d) 908, 207 Pac. (2d) 17.

PEAK, W. 1977. Institutionalized inefficiency: The unfortunate structure of Colorado's water resources management system. *Water Resources Bulletin* 13:551–62.

PENMAN, H. L. 1948. Natural evaporation from open water, bare soil, and grass. *Proceedings of the Royal Society* (London), ser. A, 193:120–45.

———. 1956. Estimating evaporation. *Transactions, American Geophysical Union* 37, no. 1:43–50.

PETER, Y. 1970. Model tests for a horizontal well. *Ground Water* 8, no. 5:30–34.

PETERS, H. J. 1972. Ground-water management. *Water Resources Bulletin* 8:188–97.

PETERSON, F. L. 1972. Water development on tropic volcanic islands—Type example: Hawaii. *Ground Water* 10, no. 5:18–23.

PETERSON, F. L., J. A. WILLIAMS, & S. W. WHEATCRAFT. 1978. Waste injection into a two-phase flow field: Sand box and Hele-Shaw model study. *Ground Water* 16:410–16.

PEYTON, G. R., ET AL. 1986. Effective porosity of geologic materials. *Proceedings of the Twelfth Annual Research Symposium,* U.S. Environmental Protection Agency. EPA/600/9-86:21–8.

PHILIP, J. R. 1969. Theory of infiltration. In *Advances in hydroscience,* vol. 5, ed. V. T. Chow, 215–96. New York: Academic Press.

PICKENS, J. F., ET AL. 1981. A multilevel device for ground water sampling. *Ground Water Monitoring Review* 1, no. 1:48–51.

PIGNATELLO, J. J. 1986. Ethylene dibromide mineralization in soils under aerobic conditions. *Applied and Environmental Microbiology* 51, no. 3:588–592.

PINDER, G. F., & H. H. COOPER, JR. 1970. A numerical technique for calculating the transient position of the saltwater front. *Water Resources Research* 6:875–82.

PINDER, G. F., & W. G. GRAY. 1977. *Finite element simulation in surface and subsurface hydrology.* New York: Academic Press.

PIONKE, H. B., & J. B. URBAN. 1985. Effect of agricultural land use on ground water quality in a small Pennsylvania watershed. *Ground Water* 23:68–80.

PIPER, A. M. 1944. A graphic procedure in the geochemical interpretation of water analyses. *Transactions, American Geophysical Union* 25:914–23.

P. L. 97-293, TITLE III, 96 STATE. 1274 (1982), San Xavier Papago Reservation of Arizona.

PLINES, P., D. LANGMUIR, & R. S. HARMON. 1974. Stable carbon isotope ratios and the existence of a gas phase in the evolution of carbonate ground waters. *Geochemica et Cosmochimica Acta* 38: 1147–64.

PLOTKIN, S. E., H. GOLD, & I. L. WHITE. 1979. Water and energy in the western coal lands. *Water Resources Bulletin* 15:94–107.

PLUHOWSKI, E. J., & I. H. KANTROWITZ. 1964. *Hydrology of the Babylon-Islip area, Suffolk County, Long Island, New York.* U.S. Geological Survey Water-Supply Paper 1768.

PLUMMER, L. N. 1977. Defining reactions and mass transfer in part of Floridan aquifer. *Water Resources Research* 13:801–12.

PLUMMER, L. N., & E. BUSENBERG. 1982. The solubilities of calcite, aragonite and vaterite in CO_2-H_2O solutions between 0 and 90°C, and an evaluation of the aqueous model of the system $CaCO_3$-CO_2-H_2O. *Geochemica et Cosmochemica Acta* 46: 1011–40.

PLUMMER, L. N., T. M. L. WIGLEY, & D. L. PARKHURST. 1978. The kinetics of calcite dissolution in CO_2-water systems at 5° to 60°C and 0.00 to 1.0 atm CO_2. *American Journal of Science* 278: 176–216.

POLUBARINOVA-KOCHINA, P. Y. 1962. *Theory of ground water movement,* trans. R. J. M. DeWiest. Princeton, N. J.: Princeton University Press.

PONTIUS, F. W. 1992a. A current look at the federal drinking water regulations. *Journal, American Water Works Association* 84, no. 3:36–50.

———. 1992b. New regs set for phase V contaminants. *Journal, American Water Works Association* 84, no. 8:30–32, 108–109.

———. 1992c. Proposed sulfate MCL reconsidered. *Journal, American Water Works Association* 84, no. 12:20, 128–129.

POULOS, S. J., & A. C. LAWS. 1985. Gradient control for containment of pollutants. *Proceedings, Fifth National Symposium on Aquifer Restoration and Ground Water Monitoring.* National Water Well Association, pp. 390–99.

PRATT, P. F., & F. L. BLAIR. 1969. Sodium hazard of bicarbonate irrigation waters. *Soil Science Society of America, Proceedings* 33:880–83.

PRICE, A. G., & T. DUNN. 1976. Energy balance computations of snowmelt in a subarctic area. *Water Resources Research* 12:686–94.

PRICKETT, T. A. 1965. Type curve solution to aquifer tests under water table conditions. *Ground Water* 3, no. 3:5–14.

———. 1975. Modeling techniques for ground water evaluation. In *Advances in hydroscience,* vol. 10, ed. V. T. Chow, 1–143. New York: Academic Press.

———. 1979. Ground water computer models—State of the art. *Ground Water* 17:167–73.

PRICKETT, T. A., & C. G. LONQUIST. 1971. *Selected digital computer techniques for groundwater resources evaluation.* Illinois State Water Survey Bulletin 55.

PRICKETT, T. A., T. G. NAYMIK, & C. G. LONNQUIST. 1981. *A "random-walk" solute transport model for selected ground water quality evaluations.* Illinois State Geological Survey, Bulletin 65.

PRILL, R. C., A. I. JOHNSON, & D. A. MORRIS. 1965. *Specific yield—Laboratory experiments showing the effect of time on column drainage.* U. S. Geological Survey Water-Supply Paper 1662-B.

PRUDIC, D. E. 1982. Hydraulic conductivity of a fine-grained till, Cattaraugus County, New York. *Ground Water* 20:194–204.

RADOSEVICH, G. E., & M. B. SABEY. 1977. Water rights, eminent domain, and the public trust. *Water Resources Bulletin* 13:747–57.

RADSTAKE, FRANK, ET AL. 1991. Applications of forward modeling resistivity profiles. *Ground Water* 29, no. 1:13–17.

RAHN, P. H., & H. A. PAUL. 1975. Hydrogeology of a portion of the sand hills and Ogallala aquifer, South Dakota and Nebraska. *Ground Water* 13:428–37.

RALSTON, D. R. 1973. Administration of ground-water as both a renewable and non-renewable resource. *Water Resources Bulletin,* 9:908–17.

RALSTON, D. R., & A. G. MORILLA. 1974. Ground water movement through an abandoned tailings pile. *American Water Resources Association Proceedings,* no. 18:174–83.

RAZACK, M., & D. HUNTLEY. 1991. Assessing transmissivity from specific capacity data in a large and hetrogeneous alluvial aquifer. *Ground Water* 29, no. 6:856–861.

REED, P. C., K. CARTWRIGHT, & D. OSBY. 1981. Electrical earth resistivity surveys near brine holding ponds in Illinois. Illinois State Geological Survey Environmental Geology Notes 95.

REHM, B. W., G. H. GROENEWOLD, & S. R. MORAN. 1978. The hydraulic conductivity of lignite and associated geological materials and strip mine spoils, western North Dakota. Geological Society of America, *Abstracts with Programs* 10, no. 7:477.

RICHARDS, L. A., ED. 1954. *Diagnosis and improvement of saline and alkali soil.* U.S. Department of Agriculture Agricultural Handbook 60.

RIPPLE, C. D., J. RUBIN, & T. E. A. VAN HYLCKAMA. 1972. *Estimating steady-state evaporation rates from bare soils under conditions of high water table.* U.S. Geological Survey Water-Supply Paper 2019-A.

ROBBINS, G. A., B. G. DEYO, M. R. TEMPLE, J. D. STUART, & M. J. LACY. 1990a. Soil-gas surveying for subsurface gasoline contamination using total organic vapor detection instruments—Part 1: The-

ory and laboratory experimentation. *Ground Water Monitoring Review* 10, no. 3:122–31.

———. 1990b. Soil-gas surveying for subsurface gasoline contamination using total organic vapor detection instruments—Part 2: Field experimentation. *Ground Water Monitoring Review* 10, no. 4:110–17.

ROBBINS, G. A., & M. M. GEMMELL. 1985. Factors requiring resolution in installing vadose zone monitoring systems. *Ground Water Monitoring Review* 5, no. 3:75–80.

ROBERTS, P. V., J. SCHREINER, & G. D. HOPKINS. 1982. Field study of organic water quality changes during groundwater recharge in the Palo Alto baylands. *Water Research* 16:1025–1035.

ROBERTS, W. J., & J. B. STALL. 1967. *Lake evaporation in Illinois.* Illinois State Water Survey Report of Investigation 57.

ROGERSON, T. L., & W. R. BYRNES. 1968. Net rainfall under hardwoods and red pine in central Pennsylvania. *Water Resources Research,* 4:55–58.

ROLLO, J. R. 1960. *Ground water in Louisiana.* Louisiana Geological Survey and Louisiana Department of Public Works Water Resources Bulletin 1.

ROSE, H. E. 1945a. An investigation into the laws of flow of fluids through beds of granular materials. *Proceedings of the Institute of Mechanical Engineers* 153:141–48.

———. 1945b. On the resistance coefficient—Reynolds number relationship for fluid flow through a bed of granular materials. *Proceedings of the Institute of Mechanical Engineers* 153:154–68.

ROTHSCHILD, E. R., R. J. MANSER, & M. P. ANDERSON. 1982. Investigation of aldicarb in ground water in selected areas of the central sand plain of Wisconsin. *Ground Water* 20:437–45.

ROUX, P. H., & W. F. ALTHOFF. 1980. Investigation of organic contamination of ground water in South Brunswick Township, New Jersey. *Ground Water* 18:464–71.

RUBIN, Y., & G. DAGAN. 1992. A note on head and velocity covariances in three-dimensional flow through heterogeneous anisotropic porous media. *Water Resources Research* 28, no. 5:1463–70.

RUHE, R. V. 1977. "Summary of geohydrologic relationships in the Lost River watershed, Indiana, applied to water use and environment." In *Hydrologic problems in karst regions,* ed. R. R. Dilamarter and S. Csallany, 64–78. Bowling Green, Ky: Western Kentucky University.

RUMER, R. R., JR. 1969. Resistance to flow through porous media. In *Flow through porous media,* ed. R. J. DeWiest, 91–108. New York: Academic Press.

RUTLAND, W. W., J. A. CHERRY, & STAN FEENSTRA. 1991. The depth of fractures and active ground water flow in a clayey till plain in southwestern Ontario. *Ground Water* 29, no. 3:405–417.

SALO, J. E., D. HARRISON, & E. M. ARCHIBALD. 1986. Removing contaminants by ground water recharge basins. *Journal, American Water Works Association* 78, no. 9:76–81.

SARTZ, R. S., W. R. CURTIS, & D. N. TOLSTAD. 1977. Hydrology of small watersheds in Wisconsin's driftless area. *Water Resources Research* 13, no. 3:524–30.

SASMAN, R. T., ET AL. 1977. *Water level decline and pumpage in deep wells in the Chicago region, 1971–75.* Illinois State Water Survey Circular 125.

SCALF, M. R., ET AL. 1981. *Manual of ground water quality sampling procedures.* Worthington, Ohio: National Water Well Association.

SCHAFER, D. C. 1978. Casing storage can affect pumping test data. *Johnson Drillers Journal* (January-February): 1–5, 10–11.

SCHAFER, J. M. 1984. Determining optimum pumping rates for creation of hydraulic barriers to ground water pollutant migration. *Proceedings, Fourth National Symposium on Aquifer Restoration and Ground Water Monitoring.* National Water Well Association, pp. 50–63.

SCHICHT, R. J., & J. R. ADAMS. 1977. *Effects of proposed 1980 and 1985 lake water allocations in the deep sandstone aquifer in northeastern Illinois.* Illinois State Water Contract Report for Illinois Division of Water Resources.

SCHICHT, R. J., J. R. ADAMS, & J. B. STALL. 1976. *Water resources availability, quality and cost in northeastern Illinois.* Illinois State Water Survey Report of Investigation 83.

SCHNEEBELI, G. 1955. Experiences sur la limite de validité de la loi de Darcy et l'apparition de la turbulence dans un écoulement de filtration. *La Houille Blanche* 10, no. 2:141–49.

SCHOELLER, H. 1955. Geochemie des eaux souterraines. *Revue de L'Institute Francais du Petrole* 10:230–44.

———. 1962. *Les eaux souterraines.* Paris: Mason et Cie.

SCHWILLE, FREDRICH. 1988. *Dense chlorinated solvents in porous and fractured media, Model Experiments.* Translated from the German by J. F. Pankow. Chelsea, Michigan: Lewis Publishers, 146 pp.

SEABURN, G. E. 1970a. *Preliminary results of hydrologic studies at two recharge basins on Long Island, New York.* U.S Geological Survey Professional Paper 627-C, pp. 1–17.

———. 1970b. *Preliminary analysis of rate of movement of storm runoff through the zone of aeration beneath a recharge basin on Long Island, New York.* U.S. Geological Survey Paper 700-B, pp. 196–98.

SEARS, JAN-CHRISTRIAN. 1987. *San Bernardino* (California) *Sun,* August 1987.

SEGOL, G., G. F. PINDER, & W. G. GRAY. 1975. A Galerkin–finite element technique for calculating the transient position of the saltwater front. *Water Resources Research* 11:343–47.

SELLERS, J. H. 1973. Tax implications of ground-water depletion. *Ground Water* 11, no. 4:27–35.

SETZER, J. 1966. Hydrologic significance of tectonic fractures detectable on air photos. *Ground Water* 4, no. 4:23–29.

SETZER, L. H., & R. H. MATZKE. 1961. "Hydrologic significance of tectonic fractures detectable on air photos." *Ground Water* 4, no. 4:23–29.

SGHIA-HUGHES, KENNETH, ART TADDEO, & SAMUEL FOGEL. 1991. Installation of a vadose zone bioremediation system at the Seymour superfund site. *Proceedings, 12th National Superfund Conference,* Hazardous Materials Control Research Institute, Silver Springs, Md., pp. 224–229.

SHARP, J. M., JR. 1984. "Hydrogeologic characteristics of shallow glacial drift aquifers in dissected till plains (north-central Missouri). *Ground Water* 22:683–89.

SHAVER, ROBERT B., & STEVE W. PUSC. 1992. Hydraulic barriers in Pleistocene buried valley aquifers. *Ground Water* 30 no. 1:21–28.

SHEPHERD, R. G. 1989. Correlations of permeability and grain size. *Ground Water* 27, 5:633–638.

SHERRILL, M. G. 1978. *Geology and ground water in Door County, Wisconsin, with emphasis on contamination potential in the silurian dolomite.* U.S. Geological Survey Water-Supply Paper 2047.

SHOWN, L. M., G. C. LUSBY, & F. A. BRANSON. 1972. Soil moisture effects of conversion of sagebrush cover to bunchgrass cover. *Water Resources Bulletin* 8:1265–72.

SHUSTER, E. T., & W. B. WHITE. 1971. Seasonal fluctuations in the chemistry of limestone springs: A possible means for characterizing carbonate aquifers. *Journal of Hydrology* 14:93–128.

SIDDIQUI, S. H., & R. R. PARIZEK. 1971. Hydrogeologic factors influencing well yields in folded and faulted carbonate rocks in central Pennsylvania. *Water Resources Research* 7:1295–1312.

SIEGEL, D. I. 1988a. The recharge-discharge functions of wetlands near Juneau, Alaska: Part I, hydrogeological investigations. *Ground Water* 26, no. 4: 427–34.

———. 1988b. The recharge-discharge functions of wetlands near Juneau, Alaska: Part II, geochemical investigations. *Ground Water* 26, no. 5:580–586.

SINGH, R. 1976. Prediction of mound geometry under recharge basins. *Water Resources Research* 12: 775–80.

SLAWSON, G. C., JR., K. E. KELLY, & L. G. EVERETT. 1982. Evaluation of ground-water pumping and bailing methods—Application in the oil shale industry. *Ground Water Monitoring Review* 2, no. 3:27–31.

SMITH, H. F. 1967. *Artificial recharge and its potential in Illinois.* International Association of Scientific Hydrology Publication No. 72 (Haifa), pp. 136–42.

SMITH, W. O. 1967. Infiltration in sands and its relation to groundwater recharge. *Water Resources Research* 3:539–55.

SNIEGOCKI, R. T., & R. F. BROWN. 1970. Clogging in recharge wells, causes and cures. *Proceedings of the Artificial Ground-water Recharge Conference, The Water Resources Association* (England), pp. 337–57.

SNIPES, D. S., ET AL. 1986. Ground-water problems in the Mesozoic Pax Mountain fault zone. *Ground Water* 24:375–81.

SOCHA, B. J. 1983. *Fracture trace analysis for water-well site locations in Precambrian igneous and metamorphic rock in central Wisconsin.* Wisconsin Geological and Natural History Survey Misc. Paper 83-5.

SOLLEY, W. B., E. B. CHASE, & W. B. MANN IV. 1983. *Estimated use of water in the United States in 1980.* U.S. Geological Survey Circular 765.

SOLLEY, W. B., C. F. MERK, & R. R. PIERCE. 1988. *Estimated use of water in the United States in 1985.* U.S. Geological Survey Circular 1004.

SOLLEY, W. B., & R. R. PIERCE. 1992. Preliminary estimates of water use in the United States, 1990. U.S. Geological Survey Open File Report 92-63.

SOMMERS, D. A. 1970. Put hydrogeology into planning. *Ground Water* 8, no. 6:2–7.

SPIEKER, A. M. 1968. *Effect of increased pumping of ground water in the Fairfield–New Baltimore area, Ohio—A prediction by analog-model study.* U.S. Geological Survey Professional Paper 605-C.

SPORHASE V. NEBRASKA. 1982. 102 S.Ct. 3456.

STATE OF MONTANA/ASSINIBOINE AND SIOUX TRIBES OF FORT PECK INDIAN RESERVA-TION COMPACT, 1985. Ratified in S.B. 467, 49th Leg., 1985 Montana Laws.

STEPHENSON, D. A. 1974. Hydrogeology is more than a classical science. *Ground Water* 12, no. 3:148–51.

STEPHENSON, D. A., B. L. CUTRIGHT, & W. W. WOESSNER. 1991. Hydrogeology: It is. *GSA Today* 1 no. 5:93–95, 99.

STEPHENSON, D. A., A. H. FLEMING, & D. M. MICKELSON. 1988. Glacial deposits. In *Hydrogeology,* ed. Back, W., J. S. Rosenshein and P. R. Seber. *The Geology of North America* O-2:301–314. Boulder Colo.: Geological Society of America.

STEWART, J. W. 1962. *Relation of permeability and jointing in crystalline metamorphic rocks near Jonesboro, Georgia.* U.S. Geological Survey Professional Paper 450-D, pp. 168–70.

———. 1964. *Infiltration and permeability of weathered crystalline rocks, Georgia Nuclear Laboratory, Dawson County, Georgia.* U.S. Geological Survey Bulletin 1133-D.

STEWART, J. W., L. R. MILLS, D. D. KNOCHENMUS, & G. L. FAULKNER. 1971. *Potentiometric surface of Floridan aquifer, southwest Florida water management district.* U.S. Geological Survey Hydrologic Investigations Atlas HA-440.

STEWART, M. T. 1982. Evaluation of electromagnetic methods for rapid mapping of salt-water interfaces in coastal aquifers. *Ground Water* 20:538–45.

STEWART, M. T., & M. C. GAY. 1986. Evaluation of transient electromagnetic soundings for deep detection of conductive fluids. *Ground Water* 24:351–56.

STEWART, M. T., M. LAYTON, & T. LIZANEC. 1983. Application of surface resistivity surveys to regional hydrogeologic reconnaissance. *Ground Water* 21:42–48.

STIFF, H. A., JR. 1951. The interpretation of chemical water analysis by means of patterns. *Journal of petroleum technology,* 3:15–17.

STONE, R., & D. F. SNOEBERGER. 1977. Cleat orientation and areal hydraulic anisotropy of a Wyoming coal aquifer. *Ground Water* 15:434–38.

STONER, J. B. 1981. Horizontal anisotropy determined by pumping in two river basin coal aquifers, Montana. *Ground Water* 19:34–40.

STOUT, G. E., & E. A. MUELLER. 1968. Survey of relationships between rainfall rate and radar reflectivity in the measurement of precipitation. *Journal of Applied Meteorology* 7:465–74.

STREETER, V. L. 1962. *Fluid mechanics.* New York: McGraw-Hill.

STRELTSOVA, T. D. 1972. Unsteady radial flow in an unconfined aquifer. *Water Resources Research* 8:1059–66.

———. 1973. Unsteady radial flow in an unconfined aquifer. *Water Resources Research* 9:236–42.

———. 1976a. Comments on "Analysis of pumping test data from anisotropic unconfined aquifers considering delayed gravity response," by Shlomo P. Neuman. *Water Resources Research* 12:113–14.

———. 1976b. Analysis of aquifer-aquitard flow. *Water Resources Research* 12:415–22.

STRINGFIELD, V. T. 1966. *Artesian water in tertiary limestone in the southeastern states.* U.S. Geological Survey Professional Paper 517.

STRINGFIELD, V. T., & H. E. LEGRAND. 1960. *Hydrology of limestone terranes in the coastal plain of the southeastern United States.* Geological Society of America Special Paper 93.

SUTER, M., ET AL. 1959. *Preliminary report of groundwater resources of the Chicago region, Illinois.* Cooperative Ground-water Report 1, Illinois State Water Survey and Illinois State Geological Survey.

SVERDRUP, K. A. 1986. Shallow seismic refraction survey of near-surface ground water flow. *Ground Water Monitoring Review* 6, no. 1:80–83.

SWARTZENDRUBER, D. 1969. The flow of water in unsaturated soils. In *Flow through porous media,* ed. R. J. M. DeWiest, 215–92. New York: Academic Press.

SWEENEY, J. J. 1984. Comparison of electrical resistivity methods for investigation of ground water conditions at a landfill site. *Ground Water Monitoring Review* 4, no. 1:52–59.

SWENSON, F. A. 1968. New theory of recharge to the artesian basin of the Dakotas. *Bulletin, Geological Society of America* 79:163–82.

SYERS, J. K., ET AL. 1973. Phosphate sorption by soils evaluated by the Langmuir adsorption equation. *Soil Science Society of America, Proceedings* 37:358–63.

TABIDIAN, M. A., DARRYLL PEDERSON, & P. A. TABIDIAN. 1992. A paleovalley aquifer system and its interaction with the Big Blue River of Nebraska during a major flood. In *Proceedings of the National Symposium on The Future Availability of Ground Water Resources,* ed. R. C. Borden & W. L. Lyke, 165–72. Bethesda, Md: American Water Resources Association.

TAKASAKI, K. J. 1978. *Summary appraisals of the nation's ground water resources—Hawaii region.* U.S. Geological Survey Professional Paper 813-M.

TANK, R. W. 1983. *Legal aspects of geology.* New York: Plenum Press.

TARLOCK, A. D. 1985. An overview of the law of groundwater management. *Water Resources Research* 21, no. 11:1751–1766.

TERZAGKI, K. 1950. Permafrost. *Boston Society of Civil Engineers Journal* 39:1–50.

THEIS, C. V. 1935. The lowering of the piezometer surface and the rate and discharge of a well using ground-water storage. *Transactions, American Geophysical Union* 16:519–24.

———. 1938. The significance and nature of the cone of depression in ground-water bodies. *Economic Geology* 38:889–902.

———. 1963. Estimating the transmissivity of a water table aquifer from the specific capacity of a well. U.S. Geological Survey Water Supply Paper 1536-I:332–336.

THIEM, G. 1906. *Hydrologische methoden.* Leipzig: Gebhardt, p. 56.

THOMAS, H. E. 1951. *Conservation of ground water.* New York: McGraw-Hill.

———. 1952. Ground water regions of the United States—Their storage facilities. U.S. 83rd Congress, House Interior and Insular Affairs Committee, *The Physical and Economic Foundation of Natural Resources,* vol. 3.

———. 1972. Water-management problems related to ground-water rights in the Southwest. *Water Resources Bulletin* 8:110–17.

———. 1978. Cyclic storage, where are you now? *Ground Water* 16:12–17.

THOMAS, J. L., & D. KLARICH. 1979. Montana's experience in reserving Yellowstone River water for instream beneficial uses—Legal framework. *Water Resources Bulletin* 15:60–74.

THORNTHWAITE, C. W. 1944. Report of the committee on transpiration and evaporation, 1943–44. *Transactions, American Geophysical Union* 25:687.

THORNTHWAITE, C. W., & J. R. MATHER. 1955. *The water balance,* Publication 8, 1–86. Centeron, N.J.: Laboratory of Climatology.

———. 1957. *Instructions and tables for computing potential evapotranspiration and the water balance,* Publication 10, 185–311. Centerton, N.J.: Laboratory of Climatology.

THORSON, J. E. 1986. Public rights at the headwaters. *Journal, American Water Works Association* 78, no. 10:72–79.

THORUD, D. B. 1967. The effect of applied interception on transpiration rates of potted ponderosa pine. *Water Resources Research* 3:443–50.

TODD, D. K. 1953. Sea water intrusion in coastal aquifers. *Transactions, American Geophysical Union* 34:749–54.

———. 1980. *Groundwater hydrology,* 2d ed. New York: John Wiley.

TOMSHO, ROBERT. 1992. States, Indians seek settlement of water issues. *Wall Street Journal,* November 25.

TÓTH, J. A. 1962. A theory of ground-water motion in small drainage basins in central Alberta, Canada. *Journal of Geophysical Research* 67, no. 11: 4375–87.

———. 1963. A theoretical analysis of ground-water flow in small drainage basins. *Journal of Geophysical Research* 68, no. 16:4795–4811.

TRELEASE, F. J. 1970. New water laws for old and new countries. In *Contemporary developments in water law,* ed. C. W. Johnson and S. H. Lewis, 40–54. University of Texas Center for Research in Water Resources, Symposium No. 4.

TRESCOTT, P. C., G. F. PINDER, & S. P. LARSON. 1976. *Finite difference model for aquifer simulation in two dimensions with results of numerical experiments.* U.S. Geological Survey Techniques of Water-Resources Investigations, Book 7, Chap. C1.

TRIMBLE, G. R., JR., & S. WEITZMAN. 1954. Effect of a hardwood forest canopy on rainfall intensities. *Transactions, American Geophysical Union* 35:226–34.

TSANG, CHIN-FU. 1991. The modeling process and model verification. *Ground Water* 29, no. 6:825–831.

TWENTER, F. R. 1962. *Geology and promising areas for ground-water development in the Hualapai Indian Reservation, Arizona.* U.S. Geological Survey Water-Supply Paper 1576-A.

U.S. ENVIRONMENTAL PROTECTION AGENCY. 1976. *Quality criteria for water.* Washington, D.C.

———. 1977. *The report to Congress—Waste disposal practices and their effects on ground water.* Washington, D.C.

———. 1984. *A ground water protection strategy for the Environmental Protection Agency.* Washington, D.C.

U.S. WATER NEWS. 1992a. Congress makes progress on tribal settlements, but many remain (December).

———. 1992b. Western supply picture redrawn by federal bill (December).

URIE, D. H. 1967. Influences of forest cover on ground-water recharge timing and use. In *International symposium on forest hydrology,* ed. W. E. Sopper & H. W. Hull, 313–24. Oxford, England: Pergamon Press.

URISH, D. W. 1983. The practical application of surface electrical resistivity to detection of ground-water pollution. *Ground Water* 21:144–52.

VACHER, H. L. 1987. Personal communication.

———. 1988. Dupuit-Ghyben-Herzberg analysis of strip-island lenses. *Geological Society of America Bulletin* 100:580–591.

———. 1989. The three point problem in the context of elementary vector analysis. *Journal of Geological Education* 37:280–87.

VACHER, H. L., & T. N. WALLIS. 1992. Comparative hydrogeology of fresh-water lenses of Bermuda and Great Exuma Island, Bahamas. *Ground Water* 30, no. 1:15–20.

VAN DER HEIJDE, P., YEHUDA BACHMAT, JOHN BREDEHOEFT, BARBARA ANDREWS, DAVID HOLTZ, & SCOTT SEBASTIAN. 1985. *Ground water management: The use of numerical models,* 2d ed. American Geophysical Union, Water Resources Monograph 5.

VEIHMEYER, F. J., & A. H. HENDIRCKSON. 1955. Does transpiration decrease as soil moisture decreases? *Transactions, American Geophysical Union* 36:425–48.

VIJAYACHANDRAN, P. K., & R. D. HARTER. 1975. Evaluation of phosphorus adsorption by a cross section of soil types. *Soil Science* 119:119–26.

VISHER, F. N., & J. F. MINK. 1964. *Ground water resources, southern Oahu, Hawaii.* U.S. Geological Survey Water-Supply Paper 1778.

VISOCKY, A. P., M. G. SHERRILL, & K. CARTWRIGHT. 1985. *Geology, hydrology and water quality of the Cambrian and Ordovician systems in northern Illinois.* Cooperative Ground Water Report 10, Illinois State Geological Survey and Illinois State Water Survey.

VOGEL, T. M., C. S. CRIDDLE, & P. L. MCCARTY. 1987. Transformations of halogenated aliphatic compounds. *Environmental Science and Technology* 21, no. 8:722–736.

VOGEL, T. M., & P. L. MCCARTY. 1985. Biotransformation of tetrachloroethylene to trichloroethylene, dichloroethylene, vinyl chloride and carbon dioxide under methanogenic conditions. *Applied and Environmental Microbiology* 49, no. 5:1080–1083.

———. 1987a. Abiotic and biotic transformations of 1,1,1,-trichloroethane under methanogenic conditions. *Environmental Science and Technology* 21, no. 12:1208–1213.

———. 1987b. Rate of abiotic formation of 1,1-dichloroethylene from 1,1,1-trichloroethane in groundwater. *Journal of Contaminant Hydrology* 1:299–308.

VOGEL, T. M., & M. REINHARD. 1986. Reaction products and rates of disappearance of simple bromoalkanes, 1,2-dibromopropane and 1,2-dibromoethane in water. *Environmental Science and Technology* 20, no. 10:992–997.

VORSTER, PETER. 1985. A water balance forecast model for Mono Lake, California. Master's Thesis in geography, California State University, Hayward.

VOSS, C. I. 1984. A finite-element simulation model for saturated-unsaturated, fluid-density-dependent ground water flow with energy transport or chemically reactive single-species solute transport. U.S. Geological Survey Water-Resources Investigations Report 84-4369.

WAIT, R. L., ET AL. 1986. Southeastern coastal plain regional aquifer-system study. In *Regional aquifer-system analysis program of the United States Geological Survey,* 205–22. U.S. Geological Survey Circular 1002.

WALLIS, T. N., H. L. VACHER, & M. T. STEWART, 1991. Hydrogeology of freshwater lens beneath a Halocene strandplain, Great Exuma, Bahamas. *Journal of Hydrology* 125:93–109.

WALTER, G. R., ET AL. 1990. Gas in a hat. *Proceedings, 11th National Superfund Conference,* 557–564. Silver Spring, Md.: Hazardous Materials Control Research Institute.

WALTON, W. C. 1960. *Leaky artesian aquifer conditions in Illinois.* Illinois State Water Survey Report of Investigation 39.

———. 1962. *Selected analytical methods for well and aquifer evaluation.* Illinois State Water Survey Bulletin 49, p. 81.

———. 1965. *Groundwater recharge and runoff in Illinois.* Illinois State Water Survey Report of Investigation 48.

———. 1978. Comprehensive analysis of water-table aquifer test data. *Ground Water* 16:311–17.

———. 1979. Progress in analytical groundwater modeling. *Journal of Hydrology* 43:149–59.

———. 1984. Analytical ground water modeling with programmable calculators and hand-held computers. In *Ground water hydraulics,* ed. J. Rosenshein & G. D. Bennett, 298–312. American Geophysical Union Monograph 9.

WALTON, W. C., & S. CSALLANY. 1962. *Yields of deep sandstone wells in northern Illinois.* Illinois State Water Survey Report of Investigation 43.

WANG, H. F., & M. P. ANDERSON. 1982. *Introduction to groundwater modeling—Finite difference and finite element methods.* San Francisco: W. H. Freeman.

WATSON, K. K. 1967. A recording field tensiometer with rapid response characteristics. *Journal of Hydrology,* 5:33–39.

WAYMAN, C. H. 1967. Adsorption on clay mineral surfaces. In *Principles and applications of water chemistry,* ed. S. D. Faust and J. V. Hunter, 127–67. New York: John Wiley.

WEATHERFORD, G. D., & S. J. SCHUPE. 1986. Reallocating water in the west. *Journal, American Water Works Association* 78, no. 10:63–71.

WEEKS, E. P. 1969. Determining the ratio of horizontal to vertical permeability by aquifer-test analyses. *Water Resources Research* 5, no. 1:196–214.

WEEKS, J. B. 1986. High plains regional aquifer study. In *Regional aquifer-system analysis program of the U.S. Geological Survey, summary of projects, 1978–84,* 30–49. U.S. Geological Survey Circular 1002.

WEIDMAN, S., & A. R. SCHULTZ. 1915. *The underground and surface water supplies of Wisconsin.* Wisconsin Geological and Natural History Survey Bulletin 35.

WELBY, C. W. 1984. Ground-water yields and inventory for land-use planning in crystalline rock areas of Wake County, North Carolina. *Water Resources Bulletin* 20:875–82.

WELTY, CLAIRE, & L. W. GELHAR. 1992. Simulation of large-scale transport of variable density and viscosity fluids using a stochastic mean model. *Water Resources Research* 28 no. 2:815–28.

WENTINK, G. R., & J. E. ETZEL. 1972. Removal of metal ions by soil. *Journal of the Water Pollution Control Federation* 44:1561–74.

WENZEL, L. K. 1942. *Methods for determining permeability of water-bearing materials with special reference to discharging well methods.* U.S. Geological Survey Water-Supply Paper 887.

WHITE, A. F., & N. J. CHUMA. 1987. Carbon and isotope mass balance studies of Oasis Valley 40-Mile Canyon ground-water basin, southern Nevada. *Water Resources Research* 23, no. 4:571–82.

WHITE, D. E. 1957a. Magmatic, connate and metamorphic water. *Bulletin, Geological Society of America* 68, no. 12:1659–82.

———. 1957b. Thermal waters of volcanic origin. *Bulletin, Geological Society of America* 68, no. 12: 1637–58.

WHITE, R. B., & R. B. GAINER, 1985. Control of ground water contamination at an active uranium mine. *Ground Water Monitoring Review* 5, no. 2: 75–81.

WHITE, W. B. 1969. Conceptual models for carbonate aquifers. *Ground Water* 7, no. 3:15–22.

———. 1970. The central Kentucky karst. *Geographical Review* 60:88–115.

WIGLEY, T. M. L. 1975. Carbon 14 dating of groundwater from close and open systems. *Water Resources Research* 11:324–28.

WILKINSON, C. F. 1986. Western water law in transition. *Journal, American Water Works Association* 78, 10:34–47.

WILLIAMS, J. R. 1970. *Ground water in the permafrost regions of Alaska.* U.S. Geological Survey Professional Paper 696.

WILSON, J. T., ET AL. 1986. In-situ biorestoration as a ground water remediation technique. *Ground Water Monitoring Review* 6, no. 4:56–64.

WILSON, J. W. 1970. Integration of radar and rain gauge data for improved rainfall measurements. *Journal of Applied Meteorology* 9:489–97.

WILSON, L. G. 1983. Monitoring in the vadose zone: Part II. *Ground Water Monitoring Review* 3, no. 1:155–66.

WILSON, M. P., D. N. PETERSON, & T. F. OSTRYE. 1983. Gravity exploration of a buried valley in the Appalachian plateau. *Ground Water* 21:589–96.

WINOGRAD, I. J., & W. THORDARSON. *Hydrogeologic and hydrochemical framework, south central great basin, with special reference to the Nevada test site.* U.S. Geological Professional Paper 712-C.

WINSAUER, W. O., ET AL. 1952. Resistivity of brine-saturated sands in relation to pore geometry. *American Association of Petroleum Geologists Bulletin* 36:253–77.

WINTER, T. C. 1973. *Hydrogeology of glacial drift, Mesabi iron range, northeastern Minnesota.* U.S. Geological Survey Water-Supply Paper 2029-A.

———. 1976. *Numerical simulation analysis of the interaction of lakes and groundwaters.* U.S. Geological Survey Professional Paper 1001.

———. 1977. Classification of the hydrogeologic settings of lakes in the north central United States. *Water Resources Research* 13:753–67.

———. 1978. Numerical simulation of steady-state three-dimensional groundwater flow near lakes. *Water Resources Research* 14:245–54.

———. 1981. Effects of water-table configuration on seepage through lakebeds. *Limnology and Oceanography* 26:925–34.

———. 1983. The interaction of lakes with variably saturated porous media. *Water Resources Research,* 19:1203–18.

WINTERS V. UNITED STATES. 1908. 207 US 564.

WISCONSIN DEPARTMENT OF NATURAL RESOURCES. 1986. Final environmental impact statement, Exxon Coals and Minerals Co. Zinc-Copper Mine, Crandon, Wisconsin.

WISCONSIN V. MICHAELS PIPELINE CONSTRUCTORS, INC. 1974. 63 Wis.2d 278.

WISCONSIN ET AL. V. ILLINOIS ET AL. 1967. 388 US 426.

WOBBER, F. J. 1967. Fracture traces in Illinois. *Photogrammetric Engineering* 33:499–506.

WYLLIE, M. R. J., & M. B. SPANGLER. 1952. Application of electrical resistivity measurements to problems of fluid flow in porous media. *American Association of Petroleum Geologists Bulletin* 36:359–403.

YANIGA, P. M., C. MATSON, & D. J. DEMKO. 1985. Restoration of water quality in a multiaquifer system via in-situ biodegradation of the organic contaminants. *Proceedings, Fifth National Symposium on Aquifer Restoration and Ground Water Monitoring.* 510–26. National Water Well Association.

YATES, M. V. 1985. Septic tank density and groundwater contamination. *Ground Water* 23:586–91.

YAZICIGIL, H., & L. V. A. SENDLEIN. 1982. Surface geophysical techniques in ground-water monitoring. *Ground Water Monitoring Review* 2, no. 1:56–62.

YIN, ZHI-YONG, & G. A. BROOK. 1992a. The topographic approach to locating high-yield wells in crystalline rocks: Does it work? *Ground Water* 30, no. 1:96–102.

———. 1992b. Reply to the preceding discussions by Harry E. LeGrand and Ralph C. Heath of "The topographic approach to locating high-yield wells in crystalline rocks: Does it work? *Ground Water* 30, no. 4:619–621.

YOUNG, H. L. 1976. *Digital computer model of the sandstone aquifer in southeastern Wisconsin.* Southeastern Wisconsin Regional Planning Commission, Technical Report 16.

YOUNG, H. L., ET AL. 1986. Northern midwest regional aquifer system study. In *Regional aquifer-system analysis program of the U.S. Geological Survey,* 72–87. U.S. Geological Survey Circular 1002.

YOUNG, R. A., & J. D. BREDEHOEFT. 1972. Digital computer simulation for solving management prob-

eastern Wisconsin Regional Planning Commission, Technical Report 16.

YOUNG, H. L., ET AL. 1986. Northern midwest regional aquifer system study. In *Regional aquifer-system analysis program of the U.S. Geological Survey,* 72–87. U.S. Geological Survey Circular 1002.

YOUNG, R. A., & J. D. BREDEHOEFT. 1972. Digital computer simulation for solving management problems of conjunctive ground-water and surface-water systems. *Water Resources Research* 8:533–56.

ZIMMERMAN, U., ET AL. 1960. Tracers determine movement of soil moisture and evapotranspiration. *Science* 152:346–47.

ZOHDY, A. A. R. 1965. Geoelectric and seismic refraction investigations near San José, California. *Ground Water* 3, no. 3:41–48.

ZOHDY, A. A. R., G. P. EATON, & D. R. MABEY. 1974. Application of surface geophysics to ground-water investigations. In *Techniques of water-resources investigations.* U.S. Geological Survey, Book 2, Chap. D1.

ZURAWSKI, A. 1978. *Summary appraisals of the nation's groundwater resources—Tennessee region.* U.S. Geological Survey Professional Paper 813-L.

Index